2ND EDITION

The Flowering of Australia's Rainforests

Pollination Ecology and Plant Evolution

Geoff Williams

CSIRO PUBLISHING

'The diversity of life in a place is not simply a random assortment of things; it is a fairly predictable set of organisms connected by certain ecological processes.'

From *The Forgotten Pollinators* (Stephen L. Buchmann and Gary Paul Nabhan)

A catalogue record for this book is available from the National Library of Australia.

ISBN: 9781486314270 (hbk)
ISBN: 9781486314287 (epdf)
ISBN: 9781486314294 (epub)

How to cite:

Williams G (2021) *The Flowering of Australia's Rainforests: Pollination Ecology and Plant Evolution*. 2nd edn. CSIRO Publishing, Melbourne.

Published by:

CSIRO Publishing
36 Gardiner Road, Clayton VIC 3168
Private Bag 10, Clayton South VIC 3169
Australia

Telephone: +61 3 9545 8400
Email: publishing.sales@csiro.au
Website: www.publish.csiro.au

Front cover: (left) Maadja Walk, Cape Tribulation, North Queensland (photo by Hugh Nicholson); (top right) *Eupomatia laurina* (Eupomatiaceae) and *Elleschodes weevils* (photo by Geoff Williams); (bottom right) *Stenocarpus sinuatus* (Proteaceae) (photo by Geoff Williams).
Back cover: (top) *Graphium agamemnon* (photo by Kevin Mitchell); (bottom) *Graphium macleayanum* (photo by Kevin Mitchell).

Plate photographs by Geoff Williams; with additional plate contributions by Hugh Nicholson and Jack Hasenpusch (as credited in Appendix 4).

Cover design by Cath Pirret
Typeset by Envisage Information Technology
Printed by Ingram Lightning Source

Acknowledgement

CSIRO acknowledges the Traditional Owners of the lands that we live and work on across Australia and pays its respect to Elders past and present. CSIRO recognises that Aboriginal and Torres Strait Islander peoples have made and will continue to make extraordinary contributions to all aspects of Australian life including culture, economy and science. CSIRO is committed to reconciliation and demonstrating respect for Indigenous knowledge and science. The use of Western science in this publication should not be interpreted as diminishing the knowledge of plants, animals and environment from Indigenous ecological knowledge systems.

Jan21_RP_ILS

Contents

Acknowledgments

I express my special thanks to Tracey Kudis, Eloise Moir-Ford, Mark Hamilton and Lauren Webb (CSIRO Publishing). Their meticulous attention to detail and design, and not least their forbearance with the task I presented them with, is greatly appreciated. I am also grateful to the following for assistance of diverse kinds in this and/or the first edition (2010): Dr Michael Batley, Dr Daniel Bickel, Dr David McAlpine, Derek Smith, Russell Cox, Dr Shane McEvey, Alexandra Hegedus, Dr David Britton, Jacquie Recqui (Australian Museum, Sydney), Dr Ken Walker (Museum Victoria, Melbourne), Dr Laurence Mound, Mr Tom Weir (CSIRO, Canberra), Dr Terry Houston (Western Australian Museum, Perth), Dr Helene Martin (University of New South Wales, Kensington), Dr Hans Bänziger (Chiang Mai University, Thailand), John Hunter (New South Wales Department of Environment and Climate Change, Coffs Harbour), Bronwyn Ellis (Forests NSW), Tony Mayes (Lansdowne), Nan and Hugh Nicholson (The Channon), Terry Evans (New South Wales Department of Environment and Climate Change, Gloucester), Dr Beth Gott (Monash University), Dr Shoko Sakai (Smithsonian Tropical Research Institute, Panama), Dr Irene Terry (University of Utah, Salt Lake City), and not least my partner Thusnelda Williams (Lorien Wildlife Refuge and Conservation Area, Lansdowne, New South Wales). The staff of the Australian Museum Research Library gave patient assistance in accessing articles.

To Kevin Mitchell (Wherrol Flat), Jack Hasenpusch (Innisfail), and Hugh Nicholson (The Channon), I express my appreciation for their image contributions to this edition (cited with respective photo captions). Their photographic contributions greatly enhanced the visual engagement of the project.

I am especially indebted to my colleague Paul Adam, University of New South Wales (Kensington), for his encouragement with this second edition, and here and there in the following text echoes of his insight and earlier involvement with the project can be discerned. For this my understanding of Australian rainforests has been greatly increased.

Finally, without the long hours of field study undertaken by the many field researchers whose papers are cited in the Bibliography, knowledge of the biota and pollination ecology of the world's threatened rainforests would be much the poorer.

Preface

This second edition of *The Flowering of Australia's Rainforests* incorporates a body of newly published research on the pollination and evolution of vascular plants, in particular that of the Angiosperms. In this it builds on the foundation established by Paul Adam and myself in the first edition (Williams and Adam 2010) and our earlier review of rainforest pollination and plant–pollinator interactions (Williams and Adam 1994).

The original edition dealt primarily with the pollination ecology of Australia's rainforests, especially that of the subtropical formations. I continue that theme, but have modified the structure, most notably embedding the reference citations within the text. Individual chapters have been expanded. The photo Plate content has also been increased, but nevertheless, these cover only a small portion of the diversity of form and colour found in Australia's flowering rainforests plants. There are glaring omissions. However, photographs in Nicholson and Nicholson (1985–1994) and Harden *et al.* (2014), as well as that of numerous websites, provide widespread and accessible complementary coverage of Australia's rainforest flora.

As with the original edition, there is coverage of plants that do not belong solely or even tenuously to the rainforest environment, and there is the revisiting of issues that are worth narrating from different perspectives. Thus there is no continuous single plot to the storyline. Like the very ecosystem being addressed, the story is a mosaic in the telling.

The above opening aside, the essential purpose of the work is to project some of the specialist knowledge that is available on the pollination ecology of Australian rainforest ecosystems into a more popular and accessible arena. I have continued to place together snapshots of what is known of the pollination of Australia's, and the world's, rainforest plants, and the evolutionary pathways by which they have come into being. In addition a little of the pollination ecology of non-rainforest plants has also been considered, for this places that of rainforest species in a broader ecological landscape context. However, I have ommitted consideration of boreal rainforest and most temperate rainforests (see DellaSala 2011; Reid and Hill 2005); the cool temperate rainforests and mixed wet forests of Australia's south-east being the notable exception (Plates 122–129). In instances where non-Australian plants are mentioned, or discussed at length, I have cited their country of origin.

In addition, the agents and mechanisms that facilitate pollination have been reviewed as these have an inherent interest, more than just that of flower visitor and pollinator. Interactions between plant and pollinator are a kind of biodiversity in themselves, for these are the manifest ecological relationships that link species to species. The intimate relationship, though accidental it may be, of pollinators with plants is poorly documented and the inter-related ecological partnerships in which they participate are increasingly confronted with the prospect of decline and extinction (Braby 2019; Montero-Castaño and Vilà 2012; Sánchez-Bayo and Wyckhuysb 2019), equal to the individual loss of the participating plants and animals. Such threats of extinction are highlighted by the catastrophic fires that devastated the Gondwana Rainforest World Heritage Area and other south-eastern Australian forests in 2019–20, the introduction of the fungal pathogen 'Myrtle rust', *Austropuccinia psidii*, in 2010 (see Makinson 2018; Fernandez Winzer *et al.* 2020) (Plate 149), the susceptability of rare and threatened plants to the root-rot pathogen *Phytophthora cinnamoni* (Wan *et al.* 2019), and the vulnerability of *Nothofagus cunninghamii* in Tasmanian and Victorian cool temperate rainforests to 'Myrtle wilt' disease, this caused by the fungal pathogen *Chalara australis* (Packham *et al.* 2008). From silent and preserved museum specimens we may be able to tabulate the extinction of plant and animal species, but what of the partners each partnered? What shared ecologies and interdependencies of plant and animal might fade without observation and record? Paul Adam and I posed these questions in the first edition of *The Flowering of Australia's Rainforests*. More than a decade on we confront the prospect of species extinction and cataclysmic ecosystem cascades with increased threat and urgency.

Although the spatial and temporal distribution of resources in rainforest is discussed, it is stressed that there are still relatively few studies and data on the pollination of the Australian rainforest flora, nor on the impact that invasive plants may pose to the reproductive ecology of native rainforest species. Consequently, to continue to overview rainforest pollination processes,

Rainforest as a concept

The concept of rainforest encompasses several 'ecosystems' simplistically considered to have a 'closed' canopy; this with a projective foliage cover of 70–100% in the upper stratum. In Australia the term *rainforest* (see Schimper 1903) is employed to characterise a variety of floristic formations of considerable structural and physiognomic diversity; these ranging from the complex monsoon systems of tropical North Queensland (Plates 135–138), subtropical rainforests of southern Queensland and New South Wales (Plates 117, 119, 121), the floristically and structurally simple vine scrubs and thickets of inland New South Wales and Queensland, the Northern Territory and Western Australia (Plates 139, 140), the temperate mesic forests of Victoria and Tasmania (Plates 125–129) and the 'closed' mangrove systems that extend in decreasing floristic diversity from northern Australia to southern temperate regions (Plate 132) (Adam 1992; Duke 2006). Over this range rainforest communities differ dramatically in terms of floristic composition, evolutionary and recruitment history, and responses to the influence of latitude, altitude, topography, rainfall patterns, soil type and relictual status owing to climate events, and the human impacts of clearing and the use of fire. Their definition and classification is discussed in detail in Adam (1992), Floyd (1990) and Webb *et al.* (1984).

Rainforests, worldwide, vary greatly in structure and composition, and are popularly thought of as lush evergreen tropical jungles, such as those of Amazonia, West Africa and South-east Asia–Western Pacific, these teeming with a great diversity and abundance of spectacular and often bizarre life forms. Subtropical rainforests occupy a lesser place in such a consideration, with vine thickets and mangrove forests usually placed outside of our perception of Australian rainforest altogether.

But the recognition of 'rainforest' casts a wider net. In a world context, boreal and temperate rainforests (synonyms variously include Celtic rainforests, Atlantic woodlands, high-latitude rainforests) have received little recognition, their existance and ecological importance appreciated by a relative few (DellaSala 2011). They are to be found in North America (e.g. western Canada, Alaska, northern California, south-west Oregon), South America (Chile, Argentina), Japan, Taiwan, Korea, Russia's Far East, Tasmania and Victoria, New Zealand, South Africa (so-called Knysna-Tsitsikamma forests), Caucasus (Georgia, Azerbaijan, Iran), Balkans, and western Europe (e.g. Scotland, Ireland, west Wales, Norway, Alps). Nearly all the rainforests have been eliminated from Europe, and just half survives in North America and Chile. Though definitions differ, and transitional formations can blur floristic boundaries, boreal and temperate rainforests are essentially forests that receive high levels of precipitation, but unlike tropical rainforests (and to a degree subtropical rainforest also), these forests are in high latitudes where temperature remains cool.

breeding systems and adaptive strategies I have had to partly rely (as in Williams and Adam 2010) on studies undertaken overseas: for the mechanisms of plant evolution and interdependencies are not very different regardless of where they occur in the global rainforest environment – only individual species, and some higher taxa (such as genera and subfamilies), are conspicuous by their presence or absence in the interplay of plant and pollinator that unfolds elsewhere in the world.

As previously mentioned special attention is given to Australia's subtropical rainforests in New South Wales and southern Queensland, for historically these have been somewhat overshadowed by the Wet Tropics rainforests of North Queensland (Plates 135–138) (Adam 1992; Stork *et al.* 2008; Webb and Kikkawa 1990). But it is increasingly clear for all rainforests that generalist rather than specialist plant–pollinator relationships are more common, and this may confer resilience to the ecosystem in the face of environmental perturbations. However, individual plant species, particularly those with highly co-adapted plant–pollinator relationships, and plant species in which breeding success is strongly dependent on the density and dispersal of the population, may be at risk if rainforests are fragmented – for under different environmental conditions ecosystem responses can involve sharp changes (e.g. biodiversity collapse) when given thresholds are reached (Solé *et al.* 2004). As rainforest becomes increasingly fragmented, will disruption of plant–pollinator relationships result in a domino effect with further decline of ecosystem biodiversity? – though generalist behaviours of pollinators may mitigate the risk of pollination failure for reproductively specialised plants (e.g. Braun and Gottsberger 2012). Some of the major disruptions to ecosystems are predicted to be consequences of the increased atmospheric concentration of carbon dioxide and other greenhouse gases. At various times in the geological past carbon dioxide concentrations were higher than they are today (e.g. Bond and Scott 2010; Lamont and He 2012) but the rate of change, in conjunction with other human-induced pressures on ecosystems, raises serious questions about a species' ability to survive through evolutionary change or migration.

Rainforests are widely recognised as great storehouses of terrestrial biodiversity and the continuing loss and degrading of rainforests is a matter of international concern; as continues in the Amazon and South-east Asia, and worldwide for oil palm production (with consequent devastation of the original rainforest biota [e.g. Alonso-Rodríguez *et al.* 2017]). Australian rainforests are now very limited in extent but contain a large proportion of national biodiversity. During the 1980s substantial advances were made towards the conservation of Australian rainforests. These advances were made ahead of knowledge of the ecology of rainforest; the scientific basis for the long-term management of often-fragmented stands is still to be developed. An important topic of concern is regeneration, and increasingly with regards to impacts from fire (e.g. Miller and Murphy 2017; Zimmer *et al.* 2015), but details of particular aspects of the regeneration process are poorly known (Floyd 1990; Harrington *et al.* 1997; Summerbell 1991). Individual plant species may be vulnerable to disturbances, often subtle ones, which can result in a reduction in breeding success and local extinction. For example, Fernandez Winzer *et al.* (2020) note that the impact of Myrtle rust on fire-damaged plants may be significant at the species level (Plate 149), which may have flow-on effects at the community level, especially after repeated infections, and that these impacts may be exacerbated by climate change as the predicted increase in fire intensity and frequency will result in more frequent new leaf growth, providing more opportunities for Myrtle rust infection. Over large areas of rainforest there may be temporal changes in species distribution at the local scale in which overall diversity is maintained, but as habitat becomes fragmented patches might diverge markedly and species and their associated fauna may face extinction.

The combinations of patterns of disturbances and availability of propagules are so complex that detailed prediction of the course of regeneration in individual patches is impossible at present. National parks and nature reserves established for the preservation of a particular combination of biota may fulfil their desired function for only a limited period. To assess the viability of reserves and manage them to meet particular conservation objectives, we need a much greater understanding of regeneration processes and plant breeding systems. We also need to be mindful that disturbance to and localised extinction of plant species, especially if these are the larval and adult foodplants of pollinators (e.g. moths and butterflies), will sequentially impact on the availability of adults to undertake pollination. This will be particularly critical where taxonomically circumscribed plant–pollinator mutualisms are involved.

Introduction

Pollination is an essential process in the life cycle of most flowering plants, thus maintenance of this process is necessary for the long-term retention of rainforest ecosystems and their associated biota. Although pollination is distinct from fertilisation, it is in most cases the necessary precursor to the fertilisation of plant ovules. In tropical humid and warm environments ~95% of flowering plant species are pollinated by invertebrate and vertebrate animals. Globally, ~88% of wild plant species are thought to be pollinated by animals (Wardhaugh 2015). This represents a considerable level (and value) of ecosystem services. Lautenbach *et al.* (2012) value these at $US230–$US410 billion annually. However, many recent research papers have suggested that pollinator abundance and diversity is declining in the face of several environmental pressures, in particular clearing of vegetation habitat, widespread use of chemicals, diseases, the introduction of invasive species and climate change.

The flowering plants – the angiosperms – now dominate the Earth's terrestrial flora and in the angiosperms pollination is the transfer of pollen from anther to stigma. This dispersing of pollen, the pollination process by a different name, results in the retarding of the subdivision of plant populations caused by selection and genetic drift (see discussions in Duncan *et al.* 2016; Kramer *et al.* 2008; Shapcott *et al.* 2016; Van der Niet *et al.* 2014). The agents that effect this transfer can be wind, water or even simple gravity; these passive modes of pollination are referred to as 'abiotic'. Pollination can also be by active 'biotic' vectors such as insects, birds and mammals and potentially reptiles – before the dominance of mammals and the evolution of birds, reptiles and their kin may have played a broader role as the pollinators of the first true flowers, as they still do in the dispersal of some seeds. Pollen transport by reptiles is no thing of the imagination or past ages, for examples still survive. Zoophily is considered ancestral in early fossil angiosperms (Hu *et al.* 2008).

Pollination precedes sexual reproduction in most vascular plants, a notable exception being the ferns. Though commonly confused or used synonymously with fertilisation, pollination means nothing more than the transport of pollen grains (the haploid gametophyte stage of the life cycle) to the surface of an appropriate stigma. Once there, the function of the pollen grain is to germinate and, by the pollen tube so produced, facilitate the movement of the sperm nucleus to the ovule complex of the flower or (in the case of gymnosperms) the cone structure (strobilus). Fertilisation and the subsequent formation of seeds and fruiting bodies are usually dependent upon the pollination process. From this chain of events, pollen can lay claim to a precursory and fundamental participation in the later dispersal of seeds and the founding of new plant populations and points of distribution; in effect, to the movement of the species. For it is through the dispersal of seeds (and to a lesser degree by the transport of vegetative propagules, such as broken tree branches, uprooted stolon sections and leaf clusters) that plants are able to move and radiate out from established zones of habitation, as well as initial points of colonisation.

However, pollen plays further roles, not constrained by those of part-agent of procreation and later migration of resultant seed. Pollen grains provide nutrition for animals, being a primary food source for numerous vertebrates and invertebrates that visit the anthers in which angiosperm pollen is produced (the pollen 'dust' of gymnosperms equally represents a source of food to pollen feeders). Even some bee species, the popular icons of pollination, consume rather than convey the pollen they seek out. Fossil evidence appears to indicate that, by the Late Carboniferous-Mesozoic, insects had become well adapted to feeding on plant reproductive structures; they evolved specialised feeding strategies by the time that angiosperms appear in the fossil record.

Pollinators do not attend to flowers, like supplicant handmaidens. They are there almost exclusively to collect food. Sometimes they directly feed on pollen or nectar (and occasionally resins and other floral resources), in which case pollen is deposited upon them incidentally, from their perspective, to the consumption of flower products. However, sometimes pollinators are lured to the flower by the deceitful ploy of an attracting fragrance or an enticing blossom shape and colour, none of which offer reward (e.g. Bänziger and Hansen 2000; Dafni and Calder 1987). In some instances, even those in which the ecological relationship between plants and their pollinators is highly specialised, pollination may be at the price of damage to, or the eating of, structures such as stamens, petals (collectively called the corolla) and ovules (as by many beetles, fig wasps and gracillariid moths), or the consumption of a large proportion of available pollen itself, as by thrips (Thysanoptera). Such is the 'price paid for babies' as the American forest ecologist Dan Janzen so aptly said, when considering the pollination ecology of fig species and the destruction of a percentage of fig ovules during the life cycle of their minute wasp pollinators (Janzen 1979).

There is not, at least as far as one can divine, prescience as to the outcome of the pollinator's visit to flowers. It is a function seemingly undertaken without understanding of their reproductive role and the co-evolutionary processes that they might be driving (e.g. Zhang and Wang 2017). Thus pollinators are inadvertent agents of pollen transfer, their visits to flowers driven not by altruism but by self-interest. Upon their backs, or ensconced among the many hairs and pollen-trapping structures and sculpturing that ornament their bodies, the contingent function of pollination, and subsequent fertilisation and production of plant progeny, is but an ecological 'freeloader'. Biotic pollinators are no more than the unwitting parties in the sex life of plants.

One more role presents itself, for the external casing of pollen grains – the exine – generally fossilises well (Faegri *et al.* 1992; Friis *et al.* 2011) (an exception being the plant family Lauraceae, owing to a thin exine wall); but without the glamour of momentous exposure from some eroding fossil bed. Without the aid of a microscope the minute grains of pollen are hidden from us, and the information they might convey about the structure of Earth's past vegetation rests silent. The wicked claws of terrible dinosaurian beasts, the rare find of a hominid tooth or ancient leaf frond, speak of an individual species, and at best assemblages of fossilised bone strata and trace fossils hint at food webs, animal populations and communities. In comparison to such isolated discoveries, leaf beds clearly indicate past plant communities, though caution is needed in interpreting their nature in any detail. The definition of fossil pollen aggregations, though some types are difficult to specifically identify beyond family, allows us to reconstruct not only a semblance of the regional flora that existed in earlier environments, but much more, for whole landscapes can be constructed – primordial ones that existed before the arrival of axe and fire tool, and the onset of aridity and climatic oscillations (Benson 1991; Martin 1998). Such that we can bring forth a grand apparition, even if part-made and skewed, of what once was. But caution is needed in the extrapolation of fossil pollen finds as pollen grains characteristically disperse and accumulate from a wide area, especially when driven by coalescing tongues of water or wind to a collective grave. Though this allows us to make a general reconstruction, it does no more than infer the nature of more localised vegetation communities.

1 Categorising rainforest plants

THE DAWNING OF VASCULAR PLANTS, AND THOSE THAT ARE DEAD

> *'The colonisation and radiation of multicellular plants on land that started over 470 million years ago was one of the defining events in the history of this planet.'*
>
> *(Pires and Dolan 2012)*

Originating from aquatic forebears, vascular plants (plants with woody conducting tissue) came to occupy primordial land habitats and are now the dominant plant forms on the Earth's landmasses (Rensing 2018). Vascular plants probably originated from multicellular green algae (e.g. Charophyaceae [see Nishiyama 2007]) during the Silurian period ~440 million years ago (mya) (Taylor *et al.* 2009), with perhaps the most rapid evolution unfolding during the broader Silurian-Devonian time interval (Niklas and Crepet 2020). Developments that defined the successful invasion of land by vascular plants included the possession of spores that were protected from drying out, the protection of the plant body from the extremes of evaporation, and the evolution of the phloem and xylem which allowed the efficient conduction of water and nutrients (Brodribb *et al.* 2010; Raven *et al.* 1992); with laminate (leaf-like) sporophylls and bracts occurring independently in several distantly related Early Devonian plant lineages (Hao and Xue 2013). Ultimately several lineages developed seeds, an outstanding feature which provided a major element in the dominance of terrestrial vascular plants. The seed protects and nourishes (via the seed-coat or testa, and the nutritive endosperm and perisperm) the embryonic plant in uncertain and sometimes harsh environments, allowing a period of dormancy during conditions not ideal for germination and subsequent growth.

Today more than a quarter of a million species of vascular plants are recognised, divided into four extinct and about eight living groups (authors differ on the number and placement of taxa). The extinct divisions are represented by the Rhyniophyta (e.g. *Rhynia* [Taylor *et al.* 2009]), which are known first from Silurian fossils, and the Zosterophyllophyta (*Zosterophyllum* [Taylor *et al.* 2009]), which date at least from the Late Silurian and are considered by some palaeobotanists to be the ancestors of the living club mosses (Lycophyta [= Lycopsida]) (e.g. see Crane 1989). Zosterophyllophyta were diverse by the Late Silurian, and by the Early Devonian are known from Gondwana (Taylor *et al.* 2009). A third early group of now extinct vascular plants is the Trimerophytophyta (*Psilophyton* [Taylor *et al.* 2009]), first recorded from the Lower-Middle Devonian ~360 mya.

Rhyniophyta have traditionally been considered the most ancient and simplest group of vascular plants but more recent information indicates that some taxa are not true Rhyniophyta at all but share characteristics of bryophytes (plants lacking vascular tissue; i.e. liverworts, hornworts, mosses [see Bell *et al.* 2020]) and vascular plants (Taylor *et al.* 2009). The Rhyniophyta were dichotomously (in successive dual divisions) branched plants, believed to have inhabited mudflats and marshes. Rhyniophytans were seedless and their bodies did not possess distinguishable roots or leaves. Known Zosterophyllophyta species were also dichotomously branched, and may have been aquatic with the lower branches possibly anchored in mud. This conjecture is based on the restriction of stomata (structures in the leaf surface allowing gaseous exchange during respiration and photosynthesis, and the transpiration of water) to the uppermost branches. The lateral and downward growth of some branches may have provided support within the substrate, consequently allowing further outward growth of the parent. Trimerophytophyta may have been derived from the Rhyniophyta (Taylor *et al.* 2009) and are possibly the progenitors of the ferns, progymnosperms and horsetails (Sphenopsida). However, the Trimerophytophyta possess a more complex branching pattern and a more massive vascular strand than the Rhyniophyta, which likely permitted the development of a relatively large-sized plant. These three groups of early vascular plants produced only a single kind of spore in their sporangia and so their reproductive systems were like those of nearly all the present-day ferns, in addition to the less-known living groups Psilopsida, Sphenopsida and several of the Lycophyta.

The fourth group of extinct ancient vascular plants, the progymnosperms, stands aside from the Rhyniophyta, Trimerophytophyta and Zosterophyllophyta. The progymnosperms (e.g. *Archaeopteris* and *Triloboxylon* [Taylor *et al.* 2009]) are considered precursors to true gymnosperms, being intermediate in aspects of their development between gymnosperms and the Trimerophytophyta – they reproduced by spores but possessed a more complex branching and vascular system. Known from the Palaeozoic era, some, such as *Archaeopteris*, resembled tall leafy branched conifers. *Archaeopteris* is widespread in the fossil record, particularly the Northern Hemisphere, and is also known from Australia. During the Carboniferous period, ~300 mya, the 'seed ferns' (e.g. Medullosales, Glossopteridales, Caytoniales, Corystospermales), appeared in the Australian fossil record. Related to the progymnosperms, seed ferns (which are neither a natural nor monophyletic group [Taylor *et al.* 2009]), became extinct by the end of the Cretaceous. However, amongst their various members is the genus *Caytonia*, which was originally considered as representing a new group of angiosperms, possessing features showing possible affinity with an angiosperm carpel.

LIVING VASCULAR PLANTS

All living vascular land plants are now grouped within the Equisitopsida (see Angiosperm Phylogeny Group III 2009). Species that occur today within the divisions of the plant world can be grouped into those that are seedless and those that produce seeds of various kinds (Kramer *et al.* 2013; Kubitzki 2013). Among the former the most numerous are the Filicopsida (true ferns) and the more poorly represented are Psilopsida (fork ferns), Lycophyta (club mosses, quillworts) and Sphenopsida (horsetails) (Niklas and Crepet 2020; Taylor *et al.* 2009). Only the Sphenopsida do not naturally occur in Australia (although they were present millions of years ago). However, one sphenopsid species (*Equisetum arvense*) is locally naturalised (Harden 1990) and is regarded as a potential major weed. All four classes lack flowers and reproduce by the production of spores.

In Australia the Psilopsida consist of two genera, *Psilotum* and *Tmesipteris*, encountered in rainforest as epiphytes, particularly on the trunks of tree ferns (Harden 1990). *Psilotum* is sometimes referred to as a 'living fossil' as it resembles some of the first terrestrial plants, such as the extinct *Cooksonia* (however, *Cooksonia* may represent a separate lineage because the connective tissues are different from vascular plants [Niklas and Crepet 2020; Taylor *et al.* 2009]). Some modern Psilopsida are also terrestrial in habit, though all are small in size. *Psilotum nudum* can be found growing on shaded rock overhangs in rainforest, and is even recorded from rock outcrops adjacent to Sydney Harbour. Psilopsida do not possess true roots, rather being differentiated into a creeping rhizome and stems with reduced leaves that resemble scales (as in *Psilotum*) or more conspicuous leaves (*Tmesipteris*). But even in *Tmesipteris* the 'leaves' are not true leaves, but outgrowths of the stem epidermis. In the Lycophyta the herbaceous plants comprise true roots, with leaves arranged spirally about the aerial stems or arising in a grass-like manner. The class includes terrestrial and epiphytic species found in rainforest as well as other types of plant communities. Lycophyta fossils are known from the Late Silurian (Taylor *et al.* 2009), and during the Carboniferous the Lycophyta included tree-like species (lepidodendrids) that reached 25 m in height and dominated much of the part of the Carboniferous that is popularly called the 'Age of Coal' (predominantly Northern Hemisphere, as the Australian coal basins are younger). Among the fossil forms recorded from Australia is the enigmatic Upper Silurian-Lower Devonian genus *Baragwanathia* (see discussion in Taylor *et al.* 2009 regarding its implication for the early origin of vascular plants). Sphenopsida were widespread during the Carboniferous and Devonian periods, and though still widely distributed now consist of a single living family (Equisetaceae) and a solitary though cosmopolitan genus, *Equisetum*. Sphenopsidans possess true roots and a rhizome and leafy stems; the extinct Australian flora included *Schizoneura*, a characteristic element of the Permian flora of Gondwana (Taylor *et al.* 2009), whose thin stems were branched, with sporangia formed as long cones at the ends of small side branches. The leaves of sphenopsidans are reduced and scale-like, in appearance reminiscent of those of the angiosperm she-oaks (Casuarinaceae). In *Gondwanophyton*, a taxon known from the Early Permian of India and Australia (McLoughlin 1992), the relatively large kidney-shaped leaves are arranged in paired whorls around a narrow stem. However, the reproductive structures of *Gondwanophyton* are unknown. Though now greatly impoverished in diversity and of diminished form, during the Carboniferous and Devonian the Sphenopsida once included towering trees (*Calamites*) over 15 m in height (Taylor *et al.* 2009). Their greatest development, like that of the tree-like lepidodendrids, took place on the landmasses of the Northern Hemisphere where tropical conditions prevailed during the Carboniferous. The climate of Australia (within the great southern supercontinent of Gondwana, comprising Africa, South America, Antarctica, Australia, India, New Zealand and New Caledonia) was then cold and dry, allowing little opportunity for the development of wet forests.

The Devonian saw the emergence and radiation of the true ferns (Hao and Xue 2013; Pires and Dolan 2012; Rensing 2018). Present-day Filicopsida comprise rainforest-inhabiting species of diverse form. These include the tall trunked tree ferns (Dicksoniaceae and Cyatheaceae) typical of many Australian temperate rainforests at high and intermediate altitudes, in addition to the multitude of herbaceous terrestrial and epiphytic species that festoon tree trunks, boulders, stream lines and moist forest floors. In families such as the Polypodiaceae and Dicksoniaceae the fronds of several species characteristic of rainforests are both conspicuous and robust. Yet in some, such as the Hymenophyllaceae and Adiantaceae, there are ferns of a much more delicate construction (Harden 1990). All Filicopsida possess true roots, stems and leaves.

There are five extant seed-bearing divisions of the plant world; Cycadophyta (= Cycadopsida), Pinophyta (= Coniferopsida), Ginkgophyta, Gnetophyta (= Gnetales) and Magnoliophyta (Magnoliopsida) (see Taylor *et al.* 2009). The Magnoliophyta comprise the flowering plants (angiosperms) with the remaining four groups characterised as the 'gymnosperms' (Kramer *et al.* 2013). Gymnosperms arose in the Devonian (c. 400–360 mya), much earlier than the flowering seed plants (the angiosperms) (Enright *et al.* 1995). Several authors have proposed that the Gnetophyta and angiosperms, plus the extinct orders Bennettitales, Pentoxylales and Caytoniales, constitute a closely related group, this hypothesis based on the idea that the angiosperm flower is homologous with the reproductive organ of a gymnosperm (i.e. the 'pseudanthial', or 'anthophyte' theory [Donoghue and Doyle 2000; Doyle and Donoghue 1992; Friis *et al.* 2011]). The Cycadophyta (cycads) and Pinophyta (conifers) are seed-bearing, though non-flowering, classes commonly referred to as gymnosperms. Both have true roots, stems and differentiated leaves, reproduce sexually and, as in all gymnosperms, possess seeds that do not develop within an ovary (thus being considered 'naked'). If we include fossil evidence, living gymnosperms represent only a fraction of the known diversity, though ~750 species of gymnosperms (~550 if cycads are excluded) survive today (Dodd *et al.* 1999, Enright *et al.* 1995). The shrubby, palm-like Cycadophyta and the 'reptilian' Dinosauria collectively lent their names to the Mesozoic, popularly defining the period as the 'Age of Dinosaurs and Cycads'. This was a time when both were conspicuous in the landscape.

Cycads, from the Palaeozoic to the present, have never been taxonomically diverse, and though morphologically conservative (Gorelick and Olson 2011), extant cycads are not relictual 'living fossils'. Cycads are often cited as reaching their greatest diversity during the Jurassic-Cretaceous (~199–65 mya). Extinctions occurred towards the end of the Mesozoic, however, fossil-calibrated molecular phylogenies have shown that cycads underwent an almost synchronous global rediversification beginning about the late Miocene (Nagalingum *et al.* 2011) followed by a slowdown in diversification towards the Holocene (Recent). The later decline of cycads has been attributed to competition with flowering plants (Norstog and Nicholls 1997) and the extinction of their putative dinosaur seed dispersal agents; but numerical analyses testing co-radiation between dinosaurs and cycads are inconclusive (Nagalingum *et al.* 2011, see also Butler *et al.* 2010). The initiation of cycad rediversification has apparently occurred over a short (~5 my) timeframe for all the large extant genera (i.e. *Ceratozamia, Cycas, Encepalartos, Macrozamia, Zamia*), but there has been almost no diversification in the last 2 million years (Nagalingum *et al.* 2011). Whilst the cycad lineage remains unquestionably ancient the living taxa are likely not much older than 12 million years, with gymnosperm crown groups of Cenozoic age (extending from ~70 million years ago) being significantly younger than their angiosperm counterparts (median age: 32 mya versus 50 mya [Crisp and Cook 2011]), and undergoing major extinctions in the Cenozoic, with diversification of surviving gymnosperms being slower than angiosperms during the Neogene period (23.03–2.58 mya). In addition, the recent radiation of cycads suggests that the coevolution of living species and their insect pollinators should be examined over a much smaller timeframe and may explain low levels of genetic diversity recorded within cycad species.

About 11 genera and between 100–300 species (Gorelick and Olson 2011) of Cycadophyta are still found in tropical, subtropical and warm temperate regions, with the greatest diversity occurring in South and Central America, Africa and Australia. Other separate groups of living gymnosperms include the lone surviving maidenhair tree *Ginkgo biloba* (which is sometimes placed in a separate family, Ginkgoaceae/Ginkgophyta, but previously within the Pinophyta), and the Gnetophyta (= Gnetales), which contain a disparate grouping of the genera *Gnetum*, *Ephedra* and *Welwitschia* (Ickert-Bond and Renner 2016). Insect pollination is probably ancestral in Gnetophyta (Jörgensen and Rydin 2015), and historically, Gnetophyta were considered a sister group to true angiosperms. However, molecular sequence data demonstrates no close relationship between them (see Stuessy 2004); though this differs from morphological data (Burleigh and Mathews 2004; Doyle and Donoghue 1992). Equally, there has been controversy over an alternative interpretation as to whether gnetophytes and pines are sister-groups (Hajibabaei *et al.* 2006; Rydin *et al.* 2002). The seed plant analysis of Nixon *et al.* (1994) considers 'the Gnetophyta as paraphyletic with the angiosperms nested within them…' (see Taylor *et al.* 2009) but Doyle (2008) aligns Gnetophyta with conifers; though with Bennettitales, and *Pentoxylon* and *Caytonia* (plus glossopteroids) with angiosperms. Within the Gnetophyta, *Gnetum*, *Ephedra* and *Welwitschia* differ significantly from one another and so are placed in separate orders, the last two being taxa adapted to desert conditions while *Gnetum* is restricted to tropical rainforests. Species of *Ephedra* are pollinated by wind and/or insects. *Ephedra foeminea* is pollinated by Diptera and Lepidoptera (mostly nocturnal species), in which pollination coincides with the full moon, its reflection in the pollination drops being the only apparent means of nocturnal attraction (Rydin and Bolinder 2015). Some botanists have suggested *Welwitschia* to be wind-pollinated, but insects are the principal agents of pollination for *Welwitschia* (being pollinated by flies – e.g. Calliphoridae, Muscidae, Sarcophagidae, Ulidiidae, Syrphidae, and possibly bees and sphecid wasps [Wetschnig and Depisch 1999]), since its pollen is sticky (but lacks pollenkitt, see Hesse 1984) and unlikely to be mobilised and dispersed by wind. Pollenkitt, an exuded substance composed primarily of lipids, hydrocarbons, terpenoids and carotenoid pigments (Willmer 2011) that facilitates adherence of pollen to the bodies of biotic pollen vectors, is lacking in *Welwitschia* (as well as *Ephedra campylopoda* and *Ephedra americana*); therefore the 'stickiness' of its pollen does not depend on pollenkitt (Hesse 1984). Pollenkitt is all-present in the angiosperms (at least in those that are not wind pollinated) but is missing in all classes of gymnosperms, whether they be wind-pollinated or dependent upon insects for pollination (Hesse 1984).

Pollination in *Gnetum* is principally by insects, but some members are also thought to be adapted to being pollinated by wind (anemophilous). Two species of *Gnetum* (*Gnetum cuspidatum* and *Gnetum gnemon*), from Sarawak, grow in humid rainforest understorey conditions that are unlikely to provide an environment conducive to wind-pollination (Kato 1996). Both produce pollination droplets and/or nectar, and emit strong odours that attract nocturnal insects. In the case of *Gnetum gnemon* the insects attracted are moths (families Geometridae and Pyralidae). *Gnetum cuspidatum* attracts several groups of insects but especially small flies (family Lauxaniidae), possibly due to the fungus-like odours produced by the plant's strobili (cone-like reproductive structures). The absence of petals in *Gnetum* imposes a restriction on its members' capacity to store nectar, unlike angiosperms, in a structure that shields the resource from evaporation. Only the high humidity and still air of the tropical rainforest understorey buffer the exuded pollination droplets and nectar from the effects of evaporation. Nocturnal presentation of such rewards, at a time of heightened daily humidity levels, to pollinators acts as a strategy to maximise reward availability. Such strategies of reward production and insect pollinator recruitment and reliance suggest that Gnetophyta and angiosperms have drawn upon an ancestral mode of insect-dominated pollination that existed before their emergence or evolutionary divergence.

However, neither the surviving species of *Ginkgo* (*Ginkgo biloba*) nor the Gnetophyta are now native to Australia (though once ginkgoaleans were [e.g. Drinnan and Chambers 1985]). Although the Gnetophyta are not found among the modern Australian flora, their present mode of pollination is pertinent when we consider the ecological ancestry of individual Australian rainforest pollination strategies. The emission of particular odours that appear to target certain insect groups, the production of nectar and nectar-like substances ('pollination drops'; analogues of angiosperm nectars) that serve as pollinator rewards, and the occupation of the dim world of the rainforest understorey (with its peculiar ecological constraints and circumstances [e.g. Stocker 1988; Turton 1988]) give clues and signposts, and hint at answers to past plant–pollinator relationships that have likely long slipped from the diversity of our surviving communities.

POLLINATION OF CYCADS AND THE DICHOTOMY OF CONTENTION

Of the gymnosperms, the Cycadophyta and *Ginkgo*, the latter a wind-pollinated taxon (Del Tredici 2007), share the feature of possessing motile sperm cells (spermatozoids). These are spherical in shape and bear cilia along the length of two or more spiral bands. It is by the pulsating movements of the cilia that the spermatozoid is able to swim along the pollen tube to the ovule and so effect fertilisation. Many cycads develop a distinct trunk, particularly *Lepidozamia* (endemic to eastern mainland Australia) (Plate 3) and species of *Cycas* (distributed widely from Africa, Asia, Australia and the South Pacific). And indeed the endemic *Cycas angulata* is often the tallest plant growing in some tropical Australian open woodland communities. However, *Lepidozamia hopei*, a rainforest species restricted to north-east Queensland, can lay claim to possibly the tallest cycad (~20 m) in the world. Cycads are usually dioecious (have distinct male or female cone-bearing plants); however, there are instances of sex change, 'sequential hermaphroditism', in several species (Gorelick and Olson 2011; Norstog and Nicholls 1997; Osborne and Gorelick 2007). Asexual reproduction by the production of bulbils arising as offshoots of the stem, has also been recorded (Raju and Rao 2011), including in *Cycas* as a means of vegetative reproduction following severing of the trunk (G. Williams pers. obs.). Bulbils germinate either on the same plant or fall off to germinate and produce new plants. In cycads pollen needs to be transported from male to female cones on separate plants for pollination, and subsequent fertilisation and seed production, to occur. They are relatively slow-growing and possess palm-like or fern-like leaves. The number of recognised Australian cycad species has substantially increased in recent years (e.g. Hill *et al.* 2004; see also Fragnière *et al.* 2015 for a discussion of latitudinal diversity gradients, distribution and conservation status) from previous estimates. However, many of the known species have restricted distributions, and occur as small populations that are vulnerable to habitat destruction and collection, with possibly little exchange of genetic material (Forster *et al.* 1994). Terry *et al.* (2008) found that although plants of the threatened central Queensland species *Macrozamia platyrhachis* are locally abundant the species is highly restricted in distribution with almost no seed dispersal. In addition fires have destroyed most seeds and seedlings, and significantly impacted upon the pollinators (*Cycadothrips chadwicki*) such that these occur in low numbers.

Three familes occur in Australia (Cycadaceae, Zamiaceae and Stangeriaceae) and though the cycad flora is widely distributed in warmer regions, including Central Australia, it is not an element characteristic of rainforests. In Australia the largest number of species are placed in the family Zamiaceae, one of which (*Lepidozamia peroffskyana*) can be a conspicuous tall-trunked species in the understorey of subtropical rainforest-sclerophyll forest ecotones (G. Williams 1993). Extant *Lepidozamia* comprises two species, *L. peroffskyana* and *L. hopei*, however, fossil *Lepidozamia* have also been recorded from Australia (Hill 1980). Several species of *Macrozamia* (e.g. *M. douglasii, M. lucida, M. macleayi* – Zamiaceae), as well as *Cycas megacarpa* (Cycadaceae) and *Bowenia serrulata* (Stangeriaceae), are encountered in rainforest at subtropical latitudes (see Curran *et al.* 2008; Hall and Walter 2018; Harden *et al.* 2006) (Plate 2). In addition, and apart from the already mentioned *Lepidozamia hopei,* species of *Cycas* (*C. silvestris, C. media*) and *Bowenia* (e.g. *Bowenia spectabilis*) grow in or near rainforest (including littoral rainforest, vine thickets) of tropical north-eastern Queensland (Cooper and Cooper 2004). The endemic genus *Bowenia* is notable because it is the only cycad taxon to have bipinnate (twice-divided pinnate) leaves. Interestingly, molecular studies by Bogler and

Francisco-Ortega (2004) found that *Lepidozamia* is more closely related to African *Encephalartos* than Australian *Macrozamia*, and that *Cycas*, a widespread genus, was somewhat divergent amongst cycads suggesting long isolation.

The Jurassic and Early Cretaceous were periods in which true cycads (Cycadophyta) flourished though a related evolutionary line, the Bennettitales or Cycadeoideopsida, died out by the end of the Cretaceous (Taylor *et al.* 2009). This origin for Bennettitales significantly pre-dates angiosperms in the fossil record. The gymnospermous Bennettitales are of particular interest because their reproductive organs bore a striking resemblance to the flowers of angiosperms, resulting in speculation by some palaeobotanists as to the possible existence of an evolutionary relationship between the two groups. Indeed, the flower-like nature of their reproductive organs is structured sufficient to include them within a definition of 'flowering plants' (see Bateman *et al.* 2006). The unisexual (single-sex) or bisexual bennettitalean cones are evenly distributed over the stem and the short receptacles are surrounded by bracts. The male sporophylls (sporangia-bearing units of the cone) are positioned around the receptacle base and the female sporophylls sit closer to the centre of the 'flower'; the sporangium is a structure or cell within which reproductive spores are produced. The Permian and Jurassic fossil genus *Crossozamia* has cones similar in structure to those of the extant genus *Cycas*, held to be the most primitive of living cycads. A true fossil cycad characteristic of the Jurassic and Early Cretaceous was *Pentoxylon*, which possessed seeds held within a fleshy fruit, unlike modern cycads which generally produce seed structures that more closely resemble cones (Taylor *et al.* 2009). The 'cones', however, are modified, and more or less reduced, leaves with attached sporangia. Though the present world cycad flora is relatively small, with a high proportion of species being endangered, the surviving flora is taxonomically diverse and widespread geographically as well as occupying a range of habitats that include islands, open forest, rainforest, grasslands and mangrove communities.

As with *Welwitschia* (Gnetophyta), the history of cycad pollination studies has involved divergent views on the role of wind and insects, contrary views that remained entrenched well into the 20th century (Hall *et al.* 2004; Hall and Walter 2011; 2018; Vorster 1995). The fossil cycad *Lagenostoma lomaxi* (see Taylor *et al.* 2009) was suggested to be insect-pollinated, and this condition of entomophily was first contended to apply to cycads more generally. However, cycad taxa such as *Stangeria* (from South Africa), *Zamia* (Central and South America) and *Macrozamia* and *Bowenia* (Australia) were later considered anemophilous, or if insects had been postulated as pollinators their role was dismissed as sporadic or simply fortuitous. *Encephalartos villosus* (South Africa) was believed to be pollinated by insects, but the related *Encephalartos altensteinii* (also from South Africa) was held to be both insect- and wind-pollinated (Marloth 1914). But the suggestion that cycads, irrespective of the taxon, were primarily if not exclusively anemophilous gained increased acceptance though no rigorous studies had been undertaken. This school of thought was probably strongly influenced by the obvious role of wind pollination in the second major gymnosperm group, the conifers, and the similarly massive amounts of pollen released by male cycad cones. The cone scales of conifers are widely separated when mature, so that the ovules are exposed to air currents in which their pollen is dispersed. The female cones of cycads (except for *Cycas*) enclose their ovules and when mature all, or at least some, of the cone scales are closed. Beetles (Coleoptera) were commonly associated with cycad cones but were largely assumed to have no true role as pollinators, instead being considered vagrants or consumers of pollen. In most instances the beetles were weevils (Curculionidae), with species of *Amophocerus, Porthetes, Antiliarhinus* and *Derelomus* present in *Encephalartos* cones (e.g. Donaldson 1997; Vorster 1995).

Wind tunnel experiments indicated that variation in the cone morphology of different cycad genera has a significant ability to influence patterns of pollen deposition. In *Lepidozamia peroffskyana*, for example, the micropyles are not exposed to open air but are sheltered behind a barrier of interlocking sporophylls (Hall *et al.* 2004) such that pollen receipt is obstructed. In addition pollen is not light and dry, but rather individual grains tend to clump which acts against transport in air currents. The differential aerodynamics of pollen movement also influences pollen deposition and the nature of the pollination syndrome (i.e. insect- versus wind-pollination). For Central American *Dioon edule* and the Mexican *Zamia furfuracea* most pollen lands some centimetres from the protected ovules – too large a distance to result in fertilisation (e.g. Niklas and Norstog 1984). In *Cycas circinalis* pollination is by wind (Raju and Rao 2011). Air currents have the ability to transfer pollen in such a way that secondary movement of the pollen grains by insects or water could direct pollen to the micropyle (the small pore in an ovule, allowing entry of the pollen tube for penetration of the nucleus). But wind currents alone may not necessarily account for the deposition of pollen within the cones of cycads generally, to effect pollination. The issue was far from resolved. The periodic presence of great numbers of beetles within and on the male and female cones of some cycads led to increased conjecture on the pollination ecology of modern cycads and the function of the beetles, irrespective of whether they were pollinators (e.g. Marloth 1914); and indeed, the discovery of a new fossil boganiid beetle, *Cretoparacucujus cycadophilus,* (specialised for pollen feeding) from the Mid Cretaceous amber of Myanmar indicates a probable ancient origin of beetle pollination of cycads at least in the Early Jurassic, long before the radiation of angiosperm pollinators in the Cretaceous (Cai *et al.* 2018; see also Liu *et al.* 2018 for a discussion of Middle Jurassic *Palaeoboganium* [Boganiidae]).

Horticultural studies in America, of wind pollination of closely spaced African *Encephalartos* and Australian *Macrozamia*, did not result in the production of fertile seed unless artificial pollination was used, thus appearing to rule out the agency of wind. In species of *Zamia*, experimentally dismissed as being anemophilous, *Rhopalotria* weevils and the languriid beetle *Pharaxonotha zamiae* were shown to be pollinators (Norstog and Fawcett 1989; Tang 1987), while insects (especially the weevil genus *Tranes*) were found to be pollinators of the endemic genera *Macrozamia* and *Lepidozamia* (Chadwick 1993; Hall *et al.* 2004; Terry 2001; Terry *et al.* 2005, 2007) (Plates 3, 4). Both species of *Bowenia* (*Bowenia spectabilis*, *Bowenia serrulata*), following artificial exclusion of wind- and water-borne pollen, have been shown to be pollinated by insects, specifically small weevils of the genus *Milotranes* (G. Wilson 2002). *Encephalartos* was also shown to be pollinated by languriid and boganiid (*Metacucujus*) beetles (Donaldson 1997). Languriidae are recorded in association with cycads from every continent in which cycads occur (Hall *et al.* 2004) and Boganiidae are associated with cycads in South Africa and Western Australia (Crowson 1981; Donaldson 1997). Hall *et al.* (2004) consider that because languriids have a global distribution it is probable that languriid-cycad pollination relationships developed before Late Mesozoic continental drift, and that the Jurassic biogeography of Languriidae and Boganiidae cycad-associated beetles (Cai *et al.* 2018; Liu *et al.* 2018) provides good evidence that cycad-insect pollination systems had developed before the Cretaceous evolution of angiosperms.

In *Macrozamia*, which is occasionally an ecotonal associate of subtropical Australian littoral (rainforest encountered in proximity to the seashore) and dry rainforest (Plate 2), pollen is shed as early as late winter and the cones thereafter remain open. The megasporophylls (cone units, cones scales that bear the ovules) of mature female plants separate vertically (though see Miao *et al.* 2017 for a reassessment of 'megasporophyll' structure). This leaves an opening that remains accessible for several days; during this short period the plant is considered to be receptive, allowing pollination to take place. The weevil *Tranes lyterioides* inhabits both male and female cones of *Macrozamia communis* (Chadwick 1993), occurring in large numbers on the female cones during the night. Although *Tranes lyterioides* consumes pollen (apparently later excreted as a poorly digested pellet of compacted grains), much pollen gathers about the body surface. This is available for pollination as the weevils have no difficulty entering the spaces formed when female cones are in the receptive stage. Other species of *Tranes* are also associated with *Macrozamia*, and *Tranes lyterioides* occurs on the cones of *Lepidozamia peroffskyana* (Ornduff 1989) (though the obligate *Tranes* pollinator of *Lepidozamia* may prove distinctive [Hall *et al.* 2004; R. Oberprieler pers. comm.]), the pollen of both *Macrozamia* and *Lepidozamia* readily adhering onto the weevils' bodies. *Tranes lyterioides* mates and completes its lifecycle in the male cone of *Macrozamia communis*, but the recruitment attractant to female cones is not fully understood. Regardless, the intimate interrelationship between *Tranes lyterioides* and *Macrozamia communis* (and possibly other Australian cycads) is likely to represent an ancient one, as weevils (or at least 'Curculionoidea' [see discussion in Grimaldi and Engel 2005]) trace an ancestry to at least the Late Jurassic, and possibly the Triassic (c. 225–190 mya); contemporaneous with the dawning of the cycads. An additional *Macrozamia*–beetle association lending support to ancient relationships between cycads and Coleoptera is that of the basal buprestid beetle genus *Xyroscelis*. *Xyroscelis* comprises two species (*Xyroscelis crocata*, Western Australia, *Xyroscelis bumanna*, New South Wales) restricted to *Macrozamia* (Williams and Watkins 1986).

Thrips are recorded as pollinators of cycads, including the Central Australian *Macrozamia macdonnellii* and

Macrozamia lucida (a Queensland species) (Mound and Terry 2001; Terry *et al.* 2005, 2007), and claim an ancient lineage among terrestrial invertebrates. Peñalver *et al.* (2012) record fossil female thrips from the Early Cretaceous amber of Spain covered in *Cycadopites* pollen. The females bore unique ring setae interpreted as specialised structures for pollen grain collection; the association providing evidence of specialised collection and transportation of pollen, and likely gymnosperm pollination, by 110–105 million years ago, or earlier (Peñalver *et al.* 2012).

Indeed, in a few instances, both beetles (*Tranes*) and thrips (*Cycadothrips*) participate jointly as pollinators, the greater pollen-carrying ability of each beetle being equalled by the greater abundance of thrips. Terry *et al.* (2005) considered that in *Macrozamia lucida* and *Macrozamia machinii* the pollination potential, due to respective pollen loads, of 25 *Cycadothrips* approximated that of one *Tranes* weevil. Beetles and thrips are believed to be among the pollinators of the earliest and most primitive flowering plants (and both frequent the flowers of the relictual North Queensland angiosperm *Idiospermum australiense* [Calycanthaceae], [Worboys and Jackes 2005]). The pollination association of weevils and thrips with *Macrozamia* demonstrates that cycad relationships with pollinating insects have evolved more than once in the history of cycads. Notwithstanding the possibility that pollination by *Tranes* represents a host shift onto cycads from original angiosperm hosts (because weevils did not evolve until the Early Cretaceous [Hall *et al.* 2004; Oberprieler 1995]), nevertheless it is tempting to see the relationship between insects and cycads as providing an established pollinator pathway already in existence at the dawning of the angiosperms, a scenario strengthened by the cycad–languriid and boganiid association.

However, the role of insects in the pollination of cycads is not fully settled. The view that the Cycadaceae are largely wind-pollinated, and that in the Zamiaceae wind primarily functions to transfer pollen to female cones, from where insects then function to move pollen to the ovules, is not so clearly demonstrated. In southern Japanese island populations of *Cycas revoluta* (a native of Japan and Ryuku Islands) pollination is achieved by nitidulid beetles and by wind, however, anemophily appears to be resticted to female plants spaced no further than a 2 m radius from males (Kono and Tobe 2007). The nitidulids are not host specific, feed on plant tissues, but serve as pollinators when pollen is released. In *Cycas panzhihuaensis* (native to China) and *Cycas seemannii* (widespread in south-west Pacific) pollination also is by both wind and beetles (Keppel 2001); a strategy also demonstrated by Hall and Walter (2018) for the Central Australian *Cycas ophiolitica*. However, although Hall and Walter note the primary mode of pollen transfer is by an undescribed weevil, *Hapalips* sp. (Erotylidae) and *Ulomoides* sp. (Tenebrionidae), wind pollination in *Cycas ophiolitica* may occur sporadically (but only at high spatial densities), yet this limited potential for ambophily could be significant given the history of dispersal and pollinator shifts among particular Cycadaceae as well as other cycads. In the Australian cycad *Macrozamia riedlei*, controlled investigations that account for the transfer of pollen, either by wind or insects, also have shown that biotic and abiotic agencies can potentially facilitate pollination individually or as a dual strategy (see Mound and Terry 2001; Mound *et al.* 1998). It is even possible that pollen is transported first by wind to receptive female cones, from where massed populations of beetles snare some of the pollen upon their bodies, carrying grains deep within the megastrobilus to the ovules. Such a situation may represent an intermediate transitory state in the evolution of cycad pollination ecology. But in what direction is the transition occurring?

HEAT PRODUCTION AND ODOUR EMISSION IN CYCADS

The intricacy of cycad reproductive ecology does not end here. Male and female cycad cones are able to emit resinous and fruity-smelling odours, and numerous cycads studied to date are able to generate heat (thermogenesis), ranging from 1°C to 17°C above surrounding air temperature (e.g. Seymour *et al.* 2004; Skubatz *et al.* 1993). Heat is produced even if the cone is removed from the plant. Thermogenesis is a widely reported phenomenon in cycads, its generation generally following a circadian rhythm and its strength and duration being related to cone size (Tang 1987). Thermogenesis in male cycad cones (e.g. Skubatz *et al.* 1993), and an associated release of volatile chemicals, takes place during the stage when pollen is shed, the nature of the complex chemical compounds produced having some relationship to those found in angiosperm fragrances. Heat production in females is generally considered to relate to maturation and pollination receptivity.

The strong odours emitted by both male and female *Cycas revoluta* cones when plants are receptive for pollination contain an estragole (4-allylanisole). This is the main volatile compound produced, and acts as both an attractant and a deterrent to insects (Kono and Tobe 2007). Heat and odour production in *Cycas sphaerica*, a threatened species restricted to the Eastern Ghats of India, serves to attract weevils (*Derelomus*). These breed in male cones and in looking for these weevils enter female cones, and achieve pollination (Raju and Jonathan 2010). Raju and Rao (2011) found that *Cycas circinalis* (India) cones of both sexes showed only weak thermogenesis (c. 1.6°C above ambient temperature) during maturation and neither endogenous heat production, nor the weak odour produced at the same time, played any role in pollination.

Thermogenic sexual dimorphism is also reported (e.g. in Australian *Macrozamia machinii*, *Macrozamia lucida* [Roemer *et al.* 2005], southern African *Encephalartos natalensis* [Suinyuy *et al.* 2010]). Male cones of *Encephalartos natalensis* are thermogenic during pollen shedding but females are neither thermogenic before or during the period of receptivity. In *Macrozamia machinii* temperatures in male cones rise 6°C above ambient air temperature, but thermogenesis in female cones is insignificant. Heat production in male *Macrozamia machinii* cones is suggested as a means to enhance the production of odour compounds leading to increased pollinator (weevil, *Tranes*) activity, mating and egg-laying whilst female cones may benefit from reduced levels of visitation and freedom from damage by weevil larvae (Seymour *et al.* 2004). This is counterintuitive as females need to recruit pollinating weevils, and these in moving to female cones are unlikely to discriminate against these in favour of male cones for egg laying and larval development. Several species of weevils (e.g. *Enteles vigorsii*, *Demyrsus meleiodes*), in addition to *Tranes* pollinators, have been reared from female *Lepidozamia peroffskyana* cones indicating that cycad cones are more broadly utilised as brood sites by Coleoptera (G. Williams 2020).

Australian cycads exhibit great variety in the daily cycles of heat production. *Bowenia spectabilis* has two peaks, one in the early morning and a lesser second peak in late afternoon. The Queensland woodland species *Cycas media* has no peaks in production, rather, heat levels are maintained continuously during the event. *Lepidozamia* and species of *Macrozamia* have a single daily peak, the situation recorded from most cycads. Thermogenesis in *Macrozamia platyrhachis* peaks in the early evening, that of *Macrozamia machinii* near sunset. However, in *Macrozamia lucida* and *Macrozamia macleayi* peak temperature occurs near midday (Roemer *et al.* 2008; Seymour *et al.* 2004; Terry *et al.* 2008). In species with large cones, such as *Lepidozamia peroffskyana* (Plate 3), the peak in heating is broad, but in species with small cones the production of heat is pronounced and brief.

During the production of heat large quantities of plant starch are metabolised and lipids (e.g. fats, sterols, terpene) are consumed; male cones contain a high concentration of starch and this is likely related to the generation of heat (Tang 1987). The role of heat production has drawn considerable interest. Hypotheses on cycad heat production have included the volatilisation of odours to attract pollinators, facilitation of pollinator movement between cones, promotion of the dehiscence of pollen sacs, and the engendering of the shedding of pollen (e.g. Raju and Jonathan 2010). Or heat could be a by-product of an alternative metabolic process, an activity which allows cones to burn off excess cell sugars that might otherwise disrupt osmotic balances (as in osmosis, and the dynamics of pressure in the movement of water within the plant). More than one function could be involved.

The role of heat in the release of odours and its potential relationship to pollination has been subject to increasing investigation, with fragrance release believed to attract pollinators. Indeed, this is the case in *Macrozamia lucida*, where daily pulse cycles of heat production and volatile chemical emissions serve to sequentially attract and repel pollinating *Cycadothrips chadwicki* thrips. During midday, *Macrozamia lucida* is able to generate heat levels of up to ~12°C above ambient air temperatures; in combination with a massive rise (of nearly a million-fold) in the concentration of volatile emissions (of β-myrcene and ocimene) this obliges pollen-carrying thrips to vacate male cones. Because high levels of β-myrcene emissions are toxic to the thrips, this is considered the primary inducement for thrips to leave, for most of the adults that remain, die. Thrips larvae generally avoid that fate by moving to the exterior of the male cones. As cone temperature and consequent β-myrcene emissions decline, thrips are attracted back to cones: some of these are females (which produce only ~20% of the emission levels achieved by males). They are able to deposit pollen in concentrations close to the micropyle of each ovule (these produce an amino acid and carbohydrate-containing droplet that may be a food reward for visiting insects), thus achieving pollination

(Terry *et al.* 2007). This daily push–pull pattern of high β-myrcene emission levels which repel thrips, and low β-myrcene emission levels (in combination with ocimene components) which thrips favour, may give important clues to the attractant relationships between cycads and other pollinating insects.

In addition to change in the concentration of volatile emissions produced during thermogenesis, there is considerable variation between species in the kinds of chemical constituents and their complexity. Some of the scent components have yet to be determined but blends of monoterpenoids, benzenoids, methyl salicate, sesquiterpenes, aliphate hydrocarbons, alcohols, ketones and esters have been identified (Pellmyr *et al.* 1991). Those from the male cones of *Zamia* include compounds recorded from certain species of orchid flowers. Cycads also exhibit differences in fragrance components released by male and female cones, with some compounds absent from one sex, or significant differences in the proportion of particular compounds produced. Suinyuy *et al.* (2010) isolated 32 compounds from *Encephalartos natalensis*, 17 of which were common to both sexes, 12 found only in male cones and 2 only in females. This variation between the compounds (not just their concentration) isolated from male and female cones could also serve as strong recruitment signals to insects, for the sensory abilities of many groups, especially beetles, are highly developed. Compound classes found in cycads can also function as powerful deterrents against herbivore attack, and no difference has been found in this regard whether cycads are insect or wind pollinated (Pellmyr *et al.* 1991). Pellmyr *et al.* (1991) consider that the compound classes occurring in cycads suggest convergent evolution in cycads and magnoliid angiosperms of the scent cues that attract pollinating insects and that early chemical attractants for their respective pollinators evolved from those serving as herbivore deterrents.

However, cone production in cycads is frequently episodic, and sometimes correlated with fires which may stimulate cone production in both sexes. Coning is not a dependable annual occurrence, and only a proportion of individuals within cycad populations may produce cones during any one coning event (Ornduff 1989, 1990). In populations of *Lepidozamia peroffskyana* in Lansdowne State Forest on the north coast of New South Wales mass coning events are frequently punctuated by many years in which few if any mature plants cone, and coning in this population is generally unrelated to fires (G. Williams pers. obs.). In the absence of alternative hosts, where do the pollinators go? What environmental signals trigger the emergence of pollinators to coincide with the presence of cones?

AUSTRALIAN CONIFERS AND THEIR PROBLEM OF POLLINATION

The world history of the conifers can be traced back at least to the Late Carboniferous–Permian (e.g. Permian Gondwanan family Buriadiaceae), at least 270 mya, with the genus *Voltziopsis* (Voltziaceae) being known only from Permian–Triassic Southern Hemisphere landscapes (Taylor *et al.* 2009). Conifers can be characterised by great longevity and a low potential rate of population increase compared to that of angiosperms, however, they are among the largest and most visually striking trees in the Southern Hemisphere; yet their loss of worldwide dominance (despite the pre-eminence of conifers in some Northern Hemisphere landscapes) and the parallel ascendancy of the angiosperms since the Early Cretaceous is one of the most significant biotic changes in the Earth's history (Enright *et al.* 1995). Conifers are now virtually eliminated from the tropics, the exception being the Podocarpaceae with their greatest species diversity occurring in wet equatorial forests. Biffin *et al.* (2011) hypothesise that efficient light harvesting by highly flattened leaves of shade tolerant podocarp genera facilitated persistance with canopy-forming angiosperms, and that leaf evolution in Southern Hemisphere conifers tracked the radiation of angiosperms from the Cretaceous through the Palaeogene.

But even with their loss of pre-eminence the conifers can claim the tallest living vascular plant, the redwood (*Sequoia sempervirens*) growing in Oregon and California. All conifers are pollinated by wind, and in general are sensitive to fire (see Zimmer *et al.* [2015] for a discussion of fuel loads produced by the endemic 'Wollemi pine' *Wollemia nobilis*). Many characteristically develop cones (these being structured to facilitate receipt and germination of pollen, though some such as the Northern Hemisphere yews (Taxaceae) have the ovules borne in fleshy cup-like structures (Raven *et al.* 1992). In *Podocarpus* (Podocarpaceae), at maturity the hard seeds are carried exposed atop a broad fleshy-axis (Harden 1990). A feature of the southern coniferous flora is their poor ability, relative to that of their Northern Hemisphere cousins, to tolerate extremes of cold. There are, for example, no southern equivalents of the great northern boreal or mountainous forests (DellaSala 2011). Unlike the needle-leaved northern Pinaceae, many of the Southern Hemisphere conifers have broad leaves (e.g. *Agathis, Wollemia, Araucaria bidwillii* in addition to *Podocarpus*) (see Biffin *et al.* 2011). There is reliable fossil pollen evidence for Southern Hemisphere conifers which demonstrates that conifer diversity has been reducing for more than 30 million years (R. Hill 1995). The living Australian Pinophyta include trees and shrubs (including dwarf forms), some of the former (e.g. *Araucaria, Agathis* and *Wollemia*) being distinctive emergents in tall rainforests or, as with *Araucaria,* extending above low-crowned vine thickets (an inland form of rainforest possessing a characteristic low canopy) (Plate 115). Both *Agathis robusta* and *Araucaria cunninghamii* (Plate 1) occur in well developed forests growing on sandmasses in south-east Queensland. Native conifers are frequently dioecious and so must effect pollen transport from male to female cone-bearing trees. Several families occur in Australia, but one of these (Pinaceae) is naturalised. The Cupressaceae are monoecious (individual plants with separate male and female cones) or dioecious, and include species that can form extensive forests in sclerophyll environments, yet several species (e.g. *Callitris macleayana*) are mostly restricted to or occasionally encountered (e.g. *Callitris baileyi, C. rhomboidea*) in some rainforests (Harden *et al.* 2006). The Cupressaceae are the most widely distributed of the families naturally occurring in Australia, and this dominance within the Australian coniferous flora holds true for the family throughout the Southern Hemisphere. Ten of the 19 living Cupressaceae genera are entirely Gondwanan in their current distribution, and in Australia species range from rainforests to open forest and arid zone vegetation (Bowman and Harris 1995; R. Hill 1995). Although the fossil history of the Cupressaceae is uncertain, several taxa (e.g. *Elatides, Athrotaxites*) are reliably dated from Mid- to Late Mesozoic deposits (Taylor *et al.* 2009).

The Taxodiaceae, in the Southern Hemisphere, comprise a solitary living genus, *Athrotaxis* (although some authors place this genus in the Cupressaceae). *Athrotaxis* is restricted to Tasmania, where three species are found (Plate 128). The family once possessed a more extensive distribution, with taxa being recorded more widely through both hemispheres. The genus includes *Athrotaxis selaginoides*, an emergent endemic tree in cool temperate rainforest of western and southern Tasmania. This species is found growing in association with evergreen trees and the deciduous tree *Nothofagus gunnii* in several rainforest communities. It relies on the formation of gaps or major disturbance for regeneration opportunities in the former forest formation but in forests with a high proportion of *Nothofagus gunnii Athrotaxis selaginoides* is able to regenerate continuously; this likely because of the higher light levels below *N. gunnii*-dominated forest canopies (Cullen 1987).

The Podocarpaceae were in existence by the beginning of the Mesozoic (c. 250–65 mya) and arose to prominence in the Cretaceous (R. Hill 1995). Podocarpaceae are common in Australian Tertiary deposits but with the exception of Tasmania the family is not conspicuous in the modern vegetation of Australia, the family having higher diversity in New Guinea, New Caledonia and New Zealand (Greenwood 1987). In number of species *Podocarpus* dominates the worldwide Podocarpaceae, with taxa known throughout the Southern Hemisphere. Several Podocarpaceae genera are endemic to Australia (e.g. *Microstrobus* and *Microcachrys*) while *Lagarostrobos* consists of two species, one each endemic to Australia and New Zealand. The Tasmanian endemic pine *Phyllocladus aspleniifolius* (Podocarpaceae/Phyllocladaceae) is mainly restricted to western and south-western Tasmania and is a common component of cool temperate rainforest (Plate 128), though also occurring frequently in wet sclerophyll forests (synonym 'tall open forest'; this is essentially a rainforest transitional stage community, see box in Chapter 6). Though occurring from sea level to ~900 m its abundance within individual rainforest stands is strongly influenced by soil type, being a dominant species on poorer soils (Adam 1992; Boland *et al.* 1984).

Although more widely distributed in the Mesozoic, the Araucariaceae are now largely restricted to the Southern Hemisphere. Three genera – *Araucaria, Agathis* (which may have been derived from *Araucaria* in the Late Jurassic or Early Cretaceous) and *Wollemia* – are represented in Australia. *Araucarioides*, a fossil foliage type, has been described from the Eocene of Tasmania and on the basis of epidermal characters is considered intermediate between *Agathis* and *Araucaria* (Taylor *et al.* 2009). *Araucaria* and *Agathis* reach their greatest diversity of species in New Caledonia. Both *Araucaria* and *Agathis* are absent from more complex north-eastern Australian rainforests owing to their inability to regenerate below the dense, shade-inducing, floristically complex angiosperm canopies (Kershaw and McGlone 1995). The remaining genus, *Wollemia,* is represented solely by the relict species *Wollemia nobilis* (Jones *et al.* 1995) and has a highly localised distribution in warm temperate rainforest north-west of Sydney where it is found as a canopy co-dominant associated with *Ceratopetalum apetalum, Doryphora sassafras* and *Syzygium smithii* (Benson and Allen 2007). *Wollemia* has exceptionally low genetic variation (Peakall *et al.* 2003) and possesses pollen indistinguishable from fossil pollen of the extinct araucarian genus *Dilwynites*, a genus that in Australia dates to the Late Cretaceous (Dettmann and Jarzen 2000). However, *Dilwynites* is a 'form taxon' – while the pollen is manifest, what the tree was, is less clear (Williams and Adam 2010).

In north-east Australia the Podocarpaceae and Araucariaceae are confined to rainforest. The Cupressaceae have some rainforest representation but also inhabit open forest (Kershaw and McGlone 1995). Although the current occurrence of Araucariaceae in north-east Australia is now relictual and depauperate, Tertiary fossil deposits indicate a previously much greater dominance of the family until ~140 000 years ago (Kershaw and McGlone 1995). A similar pre-Tertiary representation in the fossil record exists for the north-east representation of the Podocarpaceae, but with a reduction in taxonomic diversity during the Late Cenozoic.

By ~26 000 years ago conifer representation in north-east Australia had markedly declined, with only relictual representation of taxa (Kershaw and McGlone 1995). This decline is partly ascribed to climatic changes widespread over the Australian landscape; the increased occurrence of fire is implicated as a possible factor in limiting later regeneration and wider distribution of conifers following renewed onset of more favourable rainfall patterns ~8000 years ago. The forests that abutted the new rainforests were primarily dominated by eucalypts (Plate 117), fire-adapted and fire-promoting in nature (see Campbell and Clarke 2006; Keith 2017), within which only conifers of the Cupressaceae claimed a significant degree of co-habitation.

In south-east Australia there is a dichotomy between the conifers of Tasmania and those of the mainland. Tasmania has a relatively rich coniferous flora (Plate 128). Its cool moist rainforests include endemic and distinctive conifers (e.g. *Athrotaxis selaginoides* and *Lagarostrobos franklinii*), while several species (e.g. *Microcachrys tetragona*) are characteristic of subalpine communities. Several previously more dominant genera (e.g. *Phyllocladus*) were possibly subsumed by the expansion of *Nothofagus* (see Hill *et al.* 2015) in cool temperate rainforest from ~13 000 years ago (Kershaw and McGlone 1995); later Mid Holocene contraction of *Nothofagus* rainforest might have favoured localised regeneration of *Phyllocadus*-dominated rainforest. However, in low, open-canopied implicate rainforest (in which there is a dense understorey of tangled stems) *Lagarostrobos, Phyllocladus* and *Athrotaxis* occur as co-dominants with various angiosperms (Jarman *et al.* 1987).

The wet forests of the immediately adjacent Australian mainland support a single native conifer species, *Podocarpus lawrenceii*, which is restricted to eastern Victoria (Harden *et al.* 2006). The species is unusual in that besides being a component of a more diverse canopy flora, individuals also occur as emergents. In New South Wales and southern Queensland, several more widely distributed (e.g. *Podocarpus elatus, Araucaria cunninghamii, Callitris macleayana*) or localised (*Wollemia nobilis, Araucaria bidwillii* and *Agathis robusta*) rainforest conifer taxa are to be found (Plate 1) (Harden *et al.* 2006). Historically, the south-east coniferous flora was more diverse during the beginning of the Quaternary but, along with regional rainforest more broadly, the conifers progressively declined in extent during the Late Tertiary as a response to reduced rainfall and a general increase in the fluctuation of climatic conditions. Fossil evidence indicates that several conifer genera (*Dacrydium* and *Dacrycarpus*), now extinct in south-east mainland Australia, survived at least into the Early Pleistocene, but beyond this rainforest conifers were largely represented by *Podocarpus*. The reduction in rainforest conifers during the Late Miocene to Pliocene corresponded with the marked onset of landscape aridity (Martin 1998). Already faced with climatic variability in which survival was uncertain, the necessity of conifers to be pollinated by wind could have further doomed overly fragmented and relictual populations to localised if not widespread extinction. Interestingly, Leonardi *et al.* (2012), in a study of genetic diversity and gene flow amongst fragmented *Fagus sylvaticus* (a non-Australian wind-pollinated angiosperm) populations concluded that gene flow via pollen is effective, even at relatively long distances, in balancing the effect of fragmentation if population size is not small.

Pollination by wind is termed 'anemophily' and is a pollination strategy shared by flowering and non-flowering plants. Conifers are wholly dependent upon wind and air currents for pollination; to better facilitate dispersal, the individual pollen grains of several coniferous taxa (e.g. Podocarpaceae) possess bladder-like structures to promote buoyancy. In the families Araucariaceae, Cupressaceae and Taxodiaceae these bladders are absent, the dispersal of Araucarian pollen being further affected by the relatively large size of grains. Nevertheless, studies on the dispersal of conifer pollen in the Northern Hemisphere have shown pollen grains can be transported over distances greater than 2000 km, frequently as a part of airborne microbial and dust clouds (Kellogg and Griffin 2006), though their viability, and thus ability to effect pollination and fertilisation (not the simple dispersal of dead grains), has been questioned. Some researchers have postulated that the female cones of conifers create vortices that facilitate the precipitation of pollen of their own kind from the air stream, the pollen of other species being left air-borne (e.g. Niklas 1985). Though the concept sounds remarkable, the possibility of particular species extracting preferred pollen deserves examination.

Conifer pollen is dry, without a sticky pollenkitt on its surface, however, the physical characters of conifer pollen can strongly limit dispersal (Sousa and Hattemer 2003). Wind is an unreliable agent at the best of times, for though its occurrence may have an amount of seasonal predictability the direction to which adequate amounts of pollen can be carried may defy account. Widely separated individual plants may be subject to inadequate levels of pollen transfer or be beyond the threshold of pollen receipt, a problem faced by all dioecious and self-incompatible plants, be they gymnosperm or angiosperm. Therefore, widely separated individuals within any local population might contribute nothing to the genetic interplay of the population – a situation not confronted when native conifers are grown in plantations (e.g. *Araucaria cunninghamii*, in northern New South Wales). In effect, a core group of more closely spaced individuals may come to represent the truly interactive, and thus reproductively sustainable, representation of a species in any given area, the outliers being artefacts of seed dispersal that will eventually succumb without producing progeny. However, a distinctive feature of the introduced Pinaceae is that cones are monoecious, such that the uncertain transfer of pollen between otherwise strongly dioecious individuals (as faced by individual sexes of often widely separated rainforest taxa) is potentially mediated by a higher representation of receptive female reproductive bodies within any given population. The naturally endangered araucarian *Wollemia nobilis* is also monoecious, with male and female cones generally located at the end of branches and thus offering less obstruction for the release and receipt of wind-dispersed pollen. *Wollemia* also reproduces vegetatively from basal adventitious stems and epicormic buds, allowing greater reproductive flexibility for a plant that is highly restricted in range and reproductively constrained by small size of its natural population. Male and female strobili of the Tasmanian endemic *Phyllocladus aspleniifolius* (Podocarpaceae/Phyllocladaceae) are produced separately, sometimes on separate trees (Boland *et al.* 1984). The period over which male strobili release pollen has been found to differ in *Phyllocladus aspleniifolius* populations with respect to altitude; P. Barker (1995) records variation between 11 days at sites of 600 m to 24 days at those of 840 m, with release across all his study sites spanning 5–6 weeks. However, the number of male strobili may differ from year to year and between sites, being rare at some sites in some years. The development of female cones generally begins one or two weeks before pollen release and was found to be staggered in a similar manner to male strobili. But female cones were at times rare or absent.

In north-east Australia the majority of the Podocarpaceae occur as trees of the canopy, or as emergents. However, *Podocarpus dispermus* is an understorey tree in complex forests (Kershaw and McGlone 1995), the occupation of such a zone presenting difficulties for pollen transfer between plants. In floristically complex rainforest, where adjacent trees have a high probability of not being conspecific (of the same species), pollen flow to other individuals of the same species is not predictable. Along with an array of wrong species to which pollen might easily be directed, the problem of pollen transfer is further exacerbated when species are both dioecious and wind-pollinated, for no biotic pollinator is available to increase the prospect of directed pollen carriage between receptive conspecific plants.

In south-east Australia the related *Podocarpus elatus* (a long-lived, but slow growing, species that likely survived historic population contraction in dispersed refugia [Mellick *et al.* 2011]) can be found growing as a canopy and understorey tree in rainforests on coastal headlands and on geologically recent Holocene sands immediately behind sand dunes. The male cones are clustered, producing large quantities of pollen, a portion of which is carried to the solitary female cones. In producing large amounts of pollen, a greater chance of achieving pollination is assured, though at the cost of vast quantities of pollen failing to reach receptive female cones. Basically it is messy sex, but this large investment in pollen, much of which is wasted in terms of reproductive outcomes, is the price paid for survival of the species and the population. Even if costly and inefficient in terms of 'pollen produced' versus 'quantity of pollen that reaches its target', the longevity of plants (which can span several human lifetimes) means that even low levels of pollination can effect species reproduction and survival.

POLLEN FEEDERS OF ARAUCARIACEAE

Evidence of pollen-feeding invertebrates dates to the Permian (Grimaldi and Engel 2005; Krassilov *et al.* 2007) and the record of pollen-feeding insects is well represented by fossils dated to the Late Jurassic and Early Cretaceous (Friis *et al.* 2011). Knowledge of the impact of pollen-feeding insects and their broader ecological interaction with their pollen hosts may be important for the management of relictual and geographically isolated araucarian populations, for example, that of the endangered *Wollemia nobilis.* Several Lepidoptera (Geometridae, Tortricidae), Hemiptera (Aphididae, Diaspididae, Pseudococcidae) and Thysanoptera (Thripidae) have been recorded attacking seedlings and cuttings under recovery plan greenhouse cultivation (Trueman *et al.* 2007); however, insect–*Wollemia* associations in natural habitat situations, including that of pollen feeders, are generally unknown.

Limited records exist for other araucarian genera. One thrips (*Pseudanaphothrips* sp.) is known to feed in male cones of Araucariaceae (Mound and Palmer 1990) and species of the primitive weevil family Nemonychidae are recorded from the male cones of Australian *Araucaria* and *Agathis* (Lawrence and Ślipiński 2013). Adult and larval nemonychids eat the pollen of Araucariaceae but are not known to be pollinators. The weevils are moderately flattened and are clothed in curved hairs (setae), usually with some erect hairs scattered less densely over the body. The Australian species are placed in the subfamily Rhinorhynchinae, which is also recorded from New Caledonia, New Zealand, Papua New Guinea and South America. Its distribution parallels that of the Araucariaceae. The Australian chrysomelid beetles *Palophagus bunyae* and *Palophagus australiensis* (Chrysomelidae-Megalopodinae) are associated with the cones of *Araucaria* (Kuschel and May 1990) the adults and larvae of which apparently feed exclusively on pollen. This association is at variance with the suggestion that early chrysomeloids preferred 'Cycadales' rather than 'Coniferales' (Willemstein 1987).

2 Rise of the angiosperms, and archaic vascular plants in Australia's rainforests

The angiosperms (variously called the Magnoliophyta or Anthophyta) – the 'flowering plants' – take their name from the Greek words *angeion* (vessel; the carpel), and *sperma* (seed), for as a group their seeds are borne in a mature ovary (fruit). In angiosperms the classical defining reproductive structure is the flower, characterised by the presence of carpels, parts of the flower in which the seeds will develop following pollination and fertilisation (Bateman *et al.* 2006; Friis *et al.* 2011). There is a huge difference in the number of living angiosperms (c. 369 000 species) and that of the gymnosperms (c. 750 species) (Christenhusz and Byrng 2016) and this is attributed to the evolution of the flower (Friis *et al.* 2011).

In recent years an increasing number of molecular studies (which can only sample living taxa) have influenced plant classification, resulting in major revisions of plant phylogeny and a reconsideration of plant evolution (e.g. Doyle 2012; Herendeen *et al.* 2017; Sauquet *et al.* 2017; 2018; Sokoloff *et al.* 2018. These have conflicted with earlier classifications (e.g. Cronquist 1988; Dahlgren 1980; Takhtajan 1997). Though more recent classifications do not necessarily agree in detail (e.g. Judd *et al.* 2007; Xi *et al.* 2014; Zeng *et al.* 2014), they have gained an increasing codified hegemony (e.g. Angiosperm Phylogeny Group III 2009; Angiosperm Phylogeny Group IV 2016). Nevertheless, there remain several unresolved salient issues (e.g. defining what a flower actually is, and how this influences our understanding of divergence from seed-bearing gymnospermous progenitors) as to the assessment and construction of modern molecular-based plant phylogenies (see discussions in Bateman *et al.* 2006; Sauquet and Magallón 2018; Sauquet *et al.* 2017, 2018; Smyth 2018), in which monophyletic ancestry has been generally assumed; with the omnipresent restriction of pollenkitt to angiosperms and its absence in gymnosperms giving support to the hypothesis that angiosperms are a single phylogenetic group (Hesse 1984). There is still debate as to the affinities of several taxa (e.g. inclusions within Acanthaceae, Aristolochiaceae, Boraginaceae, Cactaceae, Icacinaceae, Lamiaceae, Xanthorrhoeaceae/Asphodelaceae and higher ranks [see Angiosperm Phylogeny Group IV 2016]); such that a prior more liberal definition of some taxa for inclusion within various groups and clades could be perceived as the convenient pursuit of congruence. In addition, the revised phylogenetic classification of angiosperms, and the broader circumscription of families (Angiosperm Phylogeny Group III 2009, largely followed in Angiosperm Phylogeny Group IV 2016), has given rise to issues as to how these might be systematically listed (Haston *et al.* 2009; Hawthorne and Hughes 2008) – a problem for herbarium curators.

Today ~64 orders and 416 families of angiosperms are recognised (Angiosperm Phylogeny Group IV 2016), with more than 13 000 genera comprising nearly 370 000 species (Christenhusz and Byrng 2016). Historically, these have been divided into two subclasses commonly referred to as monocotyledons (Liliidae/Liliopsida) and dicotyledons (Magnoliidae/Magnoliopsida) (e.g. Harden 1990), the two being simplistically distinguished by the presence of a single seed leaf (the cotyledon) in 'monocots' and two seed leaves in 'dicots'. However, there are several angiosperms with the 'wrong' or an enigmatic number of cotyledons. For example, the Australian *Idiospermum australiense*, sometimes termed the 'green dinosaur', can sprout two to six leaf shoots per seed (Edwards *et al.* 2001). Monocots also lack a vascular cambium, and this in combination (though not uniform) with a single seed leaf, are characters shared by several early-divergent angiosperms and eudicots (Bateman *et al.* 2006). Additional supposed defining features of monocots (e.g. trimerous flowers and sulcate pollen) are also common among early-divergent angiosperms (Friis *et al.* 2011).

Adding further variety to the Magnoliophyta is the non-Australian family Trochodendraceae, consisting of just two species (*Trochodendron aralioides*, *Tetracentron sinense*) restricted to Asia, though fossil evidence indicates the family was more widely distributed in the Cretaceous and Tertiary (Taylor *et al.* 2005). Both share the unusual characteristic of a secondary xylem without vessels, a feature rare in angiosperms and considered to be indicative of *primitiveness*. However, more recent molecular sequencing-based classifications place angiosperms into three or four basal groups (Austrobaileyales, Nymphaeaceae, Amborellaceae, and sometimes including the morphologically distinct Hydatellaceae [Friis *et al.* 2011) (the so-termed earlier ANITA ***A****mborella*, **N**ymphaeaceae, ***I****llicium* [Illiciaceae/Schisandraceae], **T**rimeniaceae, **A**ustrobaileyaceae grade and subsequent subsuming ANA Amborellales, Nymphaeales, Austrobaileyales grade) that are generally considered to represent taxa (of different rank) that diverged very early in the lineage of extant angiosperms, and several additional clades of varying later ancestry represented by magnoliids (Canellales, Laurales, Magnoliales, Piperales), Chloranthaceae, monocots (within which the Hydatellaceae were first included), Ceratophyllaceae and eudicots (e.g. Angiosperm Phylogeny Group IV 2016; Qiu *et al.* 2005; Saarela *et al.* 2007; Zanis *et al.* 2002). Chloranthaceae-like pollen represents some of the earliest, well recognised pollen of flowering plants (Endress 2001). Magnoliids constitute ~3% of extant angiosperm species, monocots 22% (Herendeen *et al.* 1995) and eudicots 75% (Magallón *et al.* 1999), but cladistic studies do not support the magnoliids (and therefore also the dicotyledons) as a monophyletic (having a shared ancestor) group (Endress 2001). Using different analytical approaches several studies have considered the magnoliids to be paraphyletic (excluding some recently evolved taxa that lie within the group) (e.g. Renner 1999), with molecular investigation nesting Magnoliales (long considered the most archaic order of flowering plants) among the eumagnoliids (Sauquet *et al.* 2003).

The ANA/ANITA grade taxa have low species diversity and most exhibit geographically scattered distributions restricted to tropical regions. Of the basal groups, Amborellaceae, the sister group (+ or – Nymphaeales [families Cabombaceae, Nymphaeaceae]) to the remaining angiosperms, are known only by a single species restricted to New Caledonia (Gandolfo *et al.* 2004; Hilu *et al.* 2003). Hydatellaceae (*Trithura, Hydatella*) are dwarf aquatic plants that have been considered a sister group to Nymphaeales (Rudall *et al.* 2007; Saarela *et al.* 2007) (and are thought restricted to India and Australasia) and the Austrobaileyales comprise three families – Austrobaileyaceae, Trimeniaceae (Plate 73) and Illiciaceae/Schisandraceae; all but the last occurring in Australia (Angiosperm Phylogeny Group III 2009; Angiosperm Phylogeny Group IV 2016).

Approximately 175 000 angiosperm species are 'dicotyledons'. The actual number of taxa (families, genera, species etc.) is subject to constant review, reflecting the discovery of new species and continuing critical study of the flora, which sometimes results in the reassessment of relationships and histories. In Australia there are ~230 families, some 2600 genera and over 18 000 species recognised (Harden 1990). This flora includes families that are considered evolutionarily conservative (primitive or basal taxa, retaining traits thought to be ancestral [e.g. Lauraceae], though supposed 'primitive' traits may represent the evolution of secondary simplification of floral characters), those with present-day (and sometimes known fossil records) that indicate restricted southern or Gondwanan geological histories (e.g. Menispermaceae, Ebenaceae, Proteaceae and Eucryphiaceae [the later now placed within the Cunoniaceae] and Nothofagaceae) (Forster and Hyland 1997; Rose *et al.* 2018; Sauquet *et al.* 2009; Taylor and Hill 1996), endemics (e.g. Akaniaceae – *Akania bidwillii*), and numerous cosmopolitan families distributed over various continents in both hemispheres. Before the break-up of the Gondwanan supercontinent, initiated within the Jurassic, elements of the angiosperm flora that are now restricted in distribution were more widespread. There are now major floristic and faunistic variations between Australian tropical, subtropical and cool temperate rainforests, with the significance of wind, invertebrate and vertebrate pollination strategies being sifted by latitude and altitude (e.g. Williams *et al.* 2010).

Separation of taxa within and above family level and membership within orders (e.g. the Rosales, to include Rhamnaceae, Urticaceae, Moraceae and others) has been contentious and not more fully resolved until recently (Zhang *et al.* 2011, and see evolving discussion in Angiosperm Phylogeny Group III 2009, IV 2016). Molecular evidence indicates that historic membership of some families (e.g. Euphorbiaceae) was artificial, the members not necessarily being closely related. At a higher order of plant classification, the ranking of monocots and eudicots, under earlier classifications, remained unclear; their recognition within a consensus system of formalised classification is likely to undergo further resolution of detail (Angiosperm Phylogeny Group IV 2016).

Charles Darwin considered the origins of the angiosperms a great and perplexing mystery as their appearance (including the almost simultaneous differentiation of the ANA/ANITA grade angiosperms and Chloranthaceae, as well as other magnoliids, early monocots and early eudicots), within the geological scheme of things, was relatively sudden, occurring at least in the first half of the Cretaceous. Particular groups achieving great diversity in the early development of flowering plants included the asterids (Bremer *et al.* 2004) and the rosids. The rosids, which contain a quarter of all angiosperm species and includes most lineages of living temperate and tropical forest trees, rapidly evolved 108–91 mya (Wang *et al.* 2009); this evolution corresponding with several dramatic phases of angiosperm radiation more generally (Taylor *et al.* 2009). Thus by the early Late Cretaceous, many of the lineages of living angiosperms (including the core eudicots) had already differentiated from ancestral stock. In addition, a significant diversity of flower types and flower structures developed within that time period (Friis *et al.* 2011).

Historically, the earliest angiosperms recognised confidently from the fossil record dated to some 127 mya (*Archaefructus* spp.) (Gandolfo *et al.* 2004), with several recent candidates (e.g. *Yuhania, Nanjinganthus*) for a placement in the Early-Middle Jurassic >170 mya (Fu *et al.* 2018; Liu and Wang 2017). Stuessy (2004) suggests angiosperms evolved slowly from seed ferns sometime in the Jurassic, beginning first with the carpel, followed later by double fertilisation and finally by the appearance of flowers, and that these transformations may have taken more than 100 million years to complete; but with the diversification, rather than being a constant steady increase, advancing episodically with rapid periods of diversification (as in rosids; Casuarinaceae, Proteaceae) interspersed by periods of apparent inertia (R. Hill 2004; Crepet and Niklas 2009). However, the recent evaluation of palaeobotanical evidence by Ramírez-Barahona *et al.* (2020) indicates that the transition to widespread angiosperm-dominated terrestrial landscapes was delayed until the Palaeocene (66–56 mya), and that the diversification of living flowering plants was geographically heterogeneous and occurred long after their phylogenetic diversity originated in the Cretaceous.

However, the fossil record gives us relatively little insight to the nature of the true forebears of the angiosperms and Friis *et al.* (2011) considered that the current molecular data appeared insufficient for the unambiguous identification of relationships among living seed plants. In the Palaeozoic and Early Mesozoic, periods of geological time that preceded the Cretaceous, there existed several gymnosperms that possessed angiosperm-like features (or as in the putative angiosperm *Nanjinganthus dendrostyla* with an enclosed seed that distinguishes it from angiosperms [Fu *et al.* 2018]). It is from some form of ancestral gymnosperm stock that the earliest of the angiosperms are now generally agreed to have developed. Most molecular-based studies indicate that all living gymnosperms appear to form a natural group, this suggesting an early divergence of the line that led to true angiosperms, but morphology-based phylogenies indicate there was a succession of gymnosperms (now largely extinct) that preceded the origin of angiosperms.

The habitat and geological location for the origin of angiosperms remains highly speculative, for before the Cretaceous and even well into it, the vegetated landscape was one of gymnosperms, cycad-like plants, seed ferns and their precursors. Within the pre-Cretaceous landscape may have existed, uncommonly, the very first of the flowering plants, probably herbaceous and shrub-like plants of the understorey rather than trees or, like the Hydatellaceae, aquatics. By ~80 mya, flowering plants already included a large number of the families that exist now. From their initial appearance they had come to dominate the vegetation, be it evergreen forest, open woodland, heath, saltmarsh or tundra. They had evolved aquatic forms adapted to live and reproduce in freshwater and saltwater habitats, and had evolved rapid reproductive cycles relative to other seed plants (J. Williams 2009, 2012). To put this time period into geological context, it was ~80 mya that the landmass now called New Zealand split off from Antarctica (Adam 1992), but the separation of Australia from Antarctica lay a further 50 million years or so in the future. In the process of progressive dominance, angiosperms produced species adapted to the extremes of cold and heat, aridity and inundation, though fossil evidence seems to suggest that they may have been at first most successful at lower elevations (Grimaldi and Engel 2005). During the movements of continents and the fluctuation of the Earth's climate over long periods of geologic time (see Appendix 5), some species failed to adapt and so became extinct or lingered as relics. Others found their place of opportunity, diversified and radiated as the moment allowed. Only in some temperate regions (altitudinally as well as geographically) do angiosperms now generally play a subsidiary role to their coniferous precursors, yet even then their diversity can outstrip all other plants present, with only the overarching canopy being the last preserve of the gymnosperms. The forests of Antarctica are dead (e.g. Stilwell and Long 2011), having been covered by ice since the Late Miocene (c. 10 mya), and remain silent within fossil strata. Families such as the Cunoniaceae/Eucryphiaceae, Nothofagaceae, Loranthaceae and Proteaceae give an echo of the Antarctic forests, for members of these survived on other Gondwanan continents after the grim blanketing of the Antarctic landmass. Yet it is via Antarctica that angiosperms are posited to have invaded Australia (Dettmann 1990), though this conflicts with the 'southern' origins of the most basal angiosperm lineages.

Both conjecture and new evidence surround our understanding of the appearance and ongoing diversification of the flowering plants (e.g. Ramírez-Barahona *et al.* 2020; Sauquet *et al.* 2017, 2018; Soltis *et al.* 2011; Sokoloff *et al.* 2018; Wang 2018; Xiang *et al.* 2014; Zeng *et al.* 2014), for some botanists would argue for a possible Triassic or Permian origin (Li *et al.* 2019; McLoughlin 2001; Salomo *et al.* 2017), this >100 million years earlier than the Cretaceous; with angiosperm forebears postulated as possibly first appearing as early as 350 mya (Cromie 1999). Such a pre-Cretaceous lineage, would not necessarily have included the angiosperm 'crown group', the clade containing all living taxa and with all the characteristic features that define the extant group (Friis *et al.* 2011). Most molecular-based analyses of divergence time for angiosperms indicate that their age is considerably older than that inferred directly from the fossil record (Li *et al.* 2019; Sanderson and Doyle 2001), as Cromie (1999) posited. But in cases where the basal branches of a lineage have been 'pruned by extinction' throughout their history, as Grimaldi and Engel (2005) suggest is the situation for the base of the angiosperm 'tree', then the use of of living sister groups for estimating divergence, and rates of divergence, will be misleading.

Based on phylogenetic studies angiosperms have been seen as arising from multiple ancestral groups (Hunter 2003), perhaps as many as six (i.e. Amborellales, Nymphaeales, Austrobaileyales, magnoliids, early monocots, early eudicots), though not from a single gymnosperm-derived stock: if Gnetophyta are placed next to or within the living conifers (Burleigh and Mathews 2004; Frohlich and Chase 2007), which is popularly (if not universally) accepted, then at least two independent origins of the flower amongst living angiosperms are required; more, if we consider extinct taxa (Bateman *et al.* 2006; Friis *et al.* 2011).

The origin of angiosperms, the evolution of floral structure and symmetry (e.g. Endress 2010; Endress and Doyle 2007; Friis *et* al. 2011; Ronse De Craene *et al.* 2003) and chemistry (e.g. as circumscribed within the 'eumagnoliids', i.e. Canellales, Piperales, Magnoliales and Laurales [Friis *et al.* 2011]), and the putative driving relationships between flowers and their pollinators (and how evolutionary changes in pollination mode might account for present day variation in the species diversity of angiosperm familes), remain central matters of investigation and debate (e.g. Dodd *et al.* 1999; Gervasi and Schiestl 2017; Gottsberger 2015; Hu *et al.* 2008; Spigler and Kalisz 2017). For example, over time biotic pollination, and seed dispersal, may promote the reproductive isolation necessary for allopatric (originating from or occupying different areas) speciation to occur. This is because pollinators can be species-specific in the way they visit flowers and subsequently transfer pollen, and because seed dispersers are able to deposit seeds at places far removed from the parent population; thus the early evolution of biotic pollination, and changes in the mode of pollination (biotic/abiotic) may have contributed to the triggering of diversification of angiosperms (Dodd *et al.* 1999). Most floral traits have potential to evolve rapidly in response to novel selection pressures (thus providing resilience of plant reproductive systems in the event of changing pollinator communities) (Opedal 2019), but nevertheless floral form and function may be conservative over large evolutionary time scales, even in environments full of opportunities for ecological interactions and niche specialisation (Vasconcelos *et al.* 2019).

Despite considerable advances in our knowledge of Cretaceous angiosperms we still confront a phylogenetic enigma, for their lineage confounds and confuses (Frohlich and Chase 2007), with pre-Cretaceous taxa either representing other plant groups or lacking features that might confidently assign them to the angiosperms (Coiro *et al.* 2019). Angiosperm lineage is certainly problematic, at least with how the fossil record chooses to divulge it (Antonelli and Sanmartín 2011), for we were not spectators at their 'dawning', and '[g]iven that we can never gain access to the 'one true tree of life', by definition we cannot assess its accuracy, which is an absolute rather than a relative property' (Bateman *et al.* 2006). Definitions of 'classic' angiospermy and gymnospermy compound issues of classification, for a definition of what constitutes 'a flower' crosses the boundaries ascribed to the reproductive structures of both groups, and the angiosperm flower in a strict sense remains ill-defined (Bateman *et al.* 2006; Friis *et al.* 2011; but see Sauquet *et al.* 2017 for a modelled construct). One aspect of angiosperm history remains evident and unassailable – that much of the original diversity of flowering plants is now extinct (Friis *et al.* 2011).

The greater proportion of extant angiosperm diversity is contributed by families, with great numbers of species (though these differ markedly from one another in species richness [Dodd *et al.* 1999]), that appear relatively late in the fossil record (e.g. Asteraceae with over 20 000 species, known from early Cenozoic, ~70 mya). This presents us with problems on how we might interpret plant evolution and lineage diversification (Antonelli and Sanmartín 2011), as well as the evolution of character traits that reflect and influence plant–pollinator relationships (Thien *et al.* 2000). If, as might well be the case, living flowering plant taxa are simply the surviving relics of formerly diverse groups, our conclusions may prove illusory, formed from a misreading of the fossil record.

It deserves emphasis that many early angiosperm species cannot be placed with certainty in extant angiosperm clades ('clades' being branches of a devised phylogenetic 'tree', their hypothetical membership formed by sequential loss or acquisition of particular morphological characters or genetic attributes) (Taylor *et al.* 2009). Rather, they seem to belong to extinct branches. The earliest episodes of angiosperm diversification may have occurred at a point that pre-dates that of the common ancestor of living forms. If so, there is a potentially insurmountable challenge to our efforts to reassemble an accurate understanding of angiosperm evolution and plant–pollinator ecology.

ARCHAIC AUSTRALIAN RAINFOREST ANGIOSPERMS

Australia may have been an evolutionary locus in the history of ancient flowering plants, particularly, but certainly not exhaustively, for the Monimiaceae/Atherospermataceae and Winteraceae (Barlow 1981; Hunter 2003).

The massive extinction event that occurred at the end of the Cretaceous Period (the so-called K/T or K-Pg boundary extinction; c. 65.5 mya) marked one of the three largest mass extinctions in the past 500 million years and resulted in the loss of ~80% of known living species. Angiosperms did not escape unscathed. Of the dominant animal groups living at the time, only insects seem to have endured without major levels of elimination. The cataclysm, considered to be caused by the episodic impact of the Chicxulub asteroid near Mexico's Yucatan Peninsula (Schulte *et al.* 2010) had its most dramatic impact on the landmasses of the Northern Hemisphere. In combination with pressure from a concomitant period of climate change and volcanism (e.g. Keller *et al.* 2018; Punekar *et al.* 2016; Schoene *et al.* 2019), it not only seems, arguably, to have brought about the end of the long reign of the dinosaurs (though non-avian dinosaur diversity had faded before the impact), pterosaurs, large marine reptiles, marine ammonites, reef-building rudists, numerous early bird lineages and others (e.g. Hart 1996), but resulted in the extinction of a significant proportion of the first plant taxa. As a consequence of this extinction a second radiation of plants (e.g. Orchidaceae) began in the following Tertiary Period. There was a parallel second radiation among insects (e.g. Lepidoptera [Espeland *et al.* 2018; Wahlberg *et al.* 2013]), of which some important modern forms (e.g. Schizophora-Diptera) largely did not appear until the Tertiary (Grimaldi and Engel 2005; Kukalová-Peck 1991). Thus the Cretaceous–Tertiary 'impact boundary' is one that separates a now generally extinct Cretaceous-evolved flora, from one whose Tertiary progenitors have largely survived into our own time. The Late Cretaceous flora of the Southern Hemisphere is thought to have been less affected by the extraterrestrial impact (and coinciding climate and volcanism pressures) than that in the devastated Northern Hemisphere, and it is conjectured that key surviving southern groups were able to evolve and diversify (e.g. within the Myrtaceae,

Menispermaceae and Proteaceae [Christophel and Lys 1986]); though not without latter extinction events, such as *Metrosideros* (Myrtaceae) (Plate 56) from the Australian mainland (Tarran *et al.* 2016).

There is evidence to indicate that rainforest in Australia may have been less extensive than the dominating 'extensive rainforest' paradigm holds, and that Australia's open sclerophyll habitat possesses a greater antiquity than previously considered (Carpenter *et al.* 2015). Nevertheless, during the course of the Tertiary, dramatic increases in the extent of aridity across the Australian landscape caused rainforest to retreat (implying movement through geographical variable landscapes in the face of increasing aridity) to refugial areas (Crisp and Cook 2013; VanDerWal *et al.* 2009), or to the survival of representative examples in smaller wet areas, along the eastern seaboard (Hilbert *et al.* 2007). This brought the renewed extinction and fragmentation of angiosperm populations. Regional changes to dominant seasonal rainfall cycles, and alternate drier and wetter phases of climatic conditions over the last 2 million years, led to further reduction in the extent of Australian rainforest and a continuing modification and filtering of the composition of the surviving plant communities. Opportunity for the dispersal of rainforest species between Australia and South America via Antarctica was possible during the Early Tertiary (Adam 1992) but some elements of the flora are likely to have evolved within Australia, to which several (e.g. Akaniaceae, Austrobaileyaceae and Surianaceae) are now restricted or largely so, or in Australia and what are now islands of the south-west Pacific (e.g. Trimeniaceae). Families such as the Winteraceae may have spread from Australia via Antarctica to South America (though see discussion in Grímsson *et al.* [2018] for implications of macro- and microfossil pollen finds in Western Greenland that extend Winteraceae back to the Upper Cretaceous). The Monimiaceae as traditionally circumscribed have an apparent Australasian origin but underwent a subsequent radiation into South-east Asia (see Lorence 1985). The South American *Mollinedia* appears to have dispersed via a trans-Pacific route from Australasia (Renner *et al.* 2010); Renner *et al.* (2010) note that the break-up of East Gondwana has left a discernable signature consisting of sister clades in Chile and Australia (with parallels exhibited in pelecorhynchid and nemistrinid Diptera, and stigmoderine Coleoptera) but found no support for an earlier hypothesis (e.g. Raven and Axelrod 1974) that the break-up of West Gondwana (Africa + South America) explains the extant disjunct distribution of Monimiaceae. Surianaceae are known from the fossil record of North America, but the family is now extinct there. Living species comprise three endemic Australian genera (*Cadellia, Guilfoylia, Stylobasium*), a single Mexican genus *Recchia*, and the pantropical *Suriana* (Bello *et al.* 2009; Telford 1993).

The earliest angiosperm evidence in Australia comes from a single fossil found in the Koonwarra Fossil Bed of eastern Victoria, however, the plant macrofossils at this site are otherwise characterised by many forms with *Ginkgo* and *Araucaria* associations, in addition to other kinds of seed plants and ferns (Drinnan and Chambers 1985; Friis *et al.* 2011). Diverse angiosperm leaves have been found in the Winton Formation, this sequence distributed through south-western Queensland, north-western New South Wales and north-eastern South Australia. Plant fossils isolated from the Winton Formation include the angiosperm floral structure *Lovellea wintonensis,* collected from western Queensland (Dettmann *et al.* 2009).

The wealth of flowering plants, the assorted and complex plant communities that now surround us and the current geographic placement of species and higher taxa reflect a long process of selective pressure and dispersal history, as well as the impacts of sea level rise and retreat and the oscillation of latitudinal temperature gradients. They constitute species and communities that are likely to be profoundly different from those of their distant ancestors.

Primitive and possibly ancestral angiosperms (e.g. Austrobaileyaceae, Himantandraceae and Calycanthaceae-'Idiospermaceae') do survive within Australian rainforests. Many are found either as rare and highly localised species, that by their very rarity suggest they are hovering on the edge of extinction, or are common (e.g. many Lauraceae and Monimiaceae), their 'primitive' botanical assignation no apparent indicator of failure within the floristically complex world of modern rainforests. Indeed the reverse, for several families (Nothofagaceae and Cunoniaceae) presently define and dominate certain rainforest types. The highest concentration of the most primitive angiosperm families (and endemic plant genera) in Australia is to be found in the rainforests of north-east Queensland; a secondary refugial area is the rainforests of south-east Queensland and northern New South Wales (Adam 1992; Hilbert *et al.* 2007). The distribution of primitive plant taxa is correlated with seasonally predictable climates and the retention of environmental conditions that have survived from the Early Tertiary.

Leaf size applied to rainforest classification

Leaf size provides an important character for defining and recognising structural and physiognomic features of rainforest, and for categorising rainforest stands. Leaves are classified into five size classes, the size determined by the length of the leaf blade. The classes are nanophyll (<2.5 cm), microphyll (2.5–<7.5 cm), notophyll (>7.5–<12.5 cm), mesophyll (>12.5–<25 cm) and macrophyll (leaves >25 cm). Altitude, latitude, moisture and soil nutrient status influence leaf size of trees in the canopy stratum. Dominant leaf size in the rainforest canopy has been used as a diagnostic feature in defining rainforest formations but it is not always possible to assign most species into a single category, thus categories frequently reflect combinations of leaf size classes (e.g. 'noto-micro'). A discussion of the categorisation of rainforests using leaf classes is given in Adam (1992).

Though such habitats are represented by various rainforest types (e.g. notophyll vine forest), these are likely to be threatened by global warming, with an attendant loss of species and mutualistic dependencies such as those that exist between primitive plants and their pollinators.

Living plants that retain flower morphologies thought to characterise primitive traits are widespread in temperate, subtropical and tropical regions, but are especially well represented in the South Pacific and in the northern and subtropical regions of Australia. Many such taxa are characteristic of Australian rainforests, which are putative centres of ancestral diversity and evolution in these groups. Families with Early Cretaceous lineages include Annonaceae, Austrobaileyaceae, Eupomatiaceae, Lauraceae, Monimiaceae (including the often separated family Atherospermataceae, to include such genera as *Atherosperma, Daphnandra, Doryphora*), Trimeniaceae and Winteraceae (e.g. Friis *et al.* 2011; Taylor *et al.* 2009). The Trimeniaceae and Winteraceae are the first living members of the 'Lower' Magnoliidae to occur in the Australian fossil pollen record. Austrobaileyaceae are confined to the Wet Tropics of North Queensland and the Trimeniaceae are restricted to Australia and the south-west Pacific (e.g. New Caledonia and Fiji). The other families are present in a wider floristic and geographical range of rainforests. The Annonaceae exhibit a relict distribution in Australia, and though species diversity is low there are numerous genera (Hnatiuk 1990). Floral evidence for the Annonaceae is known from the Late Cretaceous but the origin of the family is uncertain (Friis *et al.* 2011). Several Australian Annonaceae are likely post-Miocene immigrants from the north (the Sundaland biogeographic region, or a Melanesian island arc) but others may represent descendents of an old east Gondwana flora (Jessup 1988). Annonaceae possess actinomorphic (radially symmetrical; see Neal *et al.* [1998] for a discussion of symmetry in flowers), usually bisexual (hermaphroditic) flowers, and include several Australasian rainforest genera (e.g. *Meiogyne* and *Fitzalania*). In *Meiogyne stenopetala*, an endemic rainforest shrub restricted to north-eastern New South Wales and south-eastern Queensland, the sepals are short but the petals are distinctively long and spreading (Harden 1990). The Eupomatiaceae are confined to Australia and New Guinea (though fossil *Detrusandra* from the Late Cretaceous of North America is close to Eupomatiaceae [Friis *et al.* 2011], suggesting a possible once wider distribution for the family). The three known species (*Eupomatia laurina, Eupomatia barbata* and *Eupomatia bennettii* [Cooper and Cooper 2004; Endress 2003; Jessup 2002]) also possess actinomorphic and bisexual flowers (but are functionally protogynous). The Winteraceae (e.g. *Tasmannia, Drimys* and *Bubbia*) are principally distributed in the south-west Pacific, South America and Madagascar, and in Australia range from Queensland to Tasmania. Flowers of Winteraceae are mainly actinomorphic, bisexual or dioecious, but those of the Trimeniaceae are actinomorphic, bisexual or male. The numerous genera of Australian Monimiaceae (or Atherospermataceae) (e.g. *Atherosperma, Austromatthea, Daphnandra, Doryphora, Hedycarya, Palmeria, Steganthera, Tetrasynandra* and *Wilkiea*) are usually actinomorphic, bisexual or unisexual (plants are either monoecious or dioecious). Several genera (e.g. *Atherosperma, Doryphora* and *Daphnandra*) possess staminodes (modified floral structures that resemble stamens or petals). In *Wilkiea* the perianth is fused (and later shed as a single unit), with access to the stamens or carpels of the pale-coloured male or female flowers via a small hole at the apex of the cone-like structure (Williams *et al.* 2001). Fossil Monimiaceae/Atherospermataceae pollen is recorded from the Cretaceous and early fossil pollen types resemble that of extant *Hedycarya* and the endemic

Daphnandra. The Lauraceae have an extensive world fossil record (including flowers and seeds), are known from the Mid Cretaceous, and were widely distributed by the Early Tertiary. There is a rich Lauraceae fossil flower record from the Cenozoic (e.g. *Androglandula tennessensis* [USA], *Litseopsis rottensis, Lindera rottensis* [Germany]). Flowers of *Cinnamomum prototypum, C. felixi* and *Trianthera eusideroxyloides* are preserved as inclusions in Eocene-Oligocene Baltic amber (Friis *et al.* 2011). Phylogenetic studies undertaken by Rohwer (2000) indicated an early division into a Gondwanan group and a Laurasian group. *Potomacanthus lobatus* (with bisexual flowers) from the Early Cretaceous (USA) provides an early unequivocal record of Lauraceae (Friis *et al.* 2011). The flowers of Lauraceae continue the predominantly actinomorphic planal formation of the Lower Magnoliidae, are either bisexual or unisexual, often inconspicuous in size and colour, and usually possess staminodes. The family is a large one, containing ~32 genera and 2500 species (Harden 1990), mainly distributed in the tropical and subtropical regions of South-east Asia and South America. Several species are of economic importance. For example, *Cinnamomum zeylanicum* is the source of cinnamon, and the avocado is harvested from *Persea americana*. North Queensland, and the rainforests of north-east New South Wales and south-east Queensland, are centres of endemism for the Lauraceae (Hunter 2003). Genera occurring in Australian rainforests include *Beilschmiedia, Cryptocarya, Endiandra, Litsea* and *Neolitsea*.

Of interesting geographical and phylogenetic placement is *Idiospermum australiense*. Its small and isolated populations are restricted to the rainforests of the Daintree region in North Queensland, in which this so-called 'primitive' plant has lived for perhaps 120 million years (Worboys and Jackes 2005). It is a tree of the rainforest canopy, discovered in the late 19th century near Cairns then lost to science until 1971, owing to the clearing of the site on which it was found. The flowers of *Idiospermum australiense* are creamy white (later changing from pinkish to reddish-purple) and strongly fragrant. The inner perianth whorl provides a tightly enclosed refuge for its insect pollinators. The seeds are large, toxic to most animals and apparently dispersed by gravity (which might explain why the species has such a limited distribution). First placed as the sole known living member of the family Idiospermaceae, *Idiospermum australiense* has been moved to the family Calycanthaceae and is that family's only known Southern Hemisphere member (Taylor *et al.* 2009).

Australian rainforests include a second group of families with lineages that diverged early after the evolution of the Magnoliidae (see Massoni *et al.* 2015). These include the Akaniaceae (*Akania*), Cunoniaceae (e.g. *Ackama, Ceratopetalum, Karrabina, Schizomeria* and *Vesselowskya* [Plate 26]), Davidsoniaceae/Cunoniaceae (*Davidsonia*), Escalloniaceae/Grossulariaceae (e.g. *Anopterus*), Escalloniaceae/Rousseaceae (e.g. *Abrophyllum, Cuttsia* and *Quintinia*), Cunoniaceae/Eucryphiaceae (*Eucryphia*), Myrtaceae (e.g. *Allosyncarpia, Backhousia, Gossia, Eucalyptus* and *Syzygium*), Nothofagaceae (*Nothofagus*), Pittosporaceae (e.g. *Hymenosporum* and *Pittosporum*), Proteaceae (e.g. *Alloxylon, Eidothea, Floydia, Grevillea, Hicksbeachia, Macadamia, Orites* and *Stenocarpus*), Sapindaceae (e.g. *Arytera, Mischocarpus, Sarcopteryx* and *Toechima*) and Surianaceae (*Guilfoylia*) (Hunter 2003). These families predominantly possess shallow, brush-like or bell-shaped actinomorphic flowers, but many of those of the Proteaceae, for example *Stenocarpus sinuatus*, are zygomorphic (flowers that are divisible into two halves along a single plane only). In proteaceous genera such as the relictual *Eidothea*, known only from northern New South Wales and North Queensland (Rossetto and Kooyman 2005), flowers are actinomorphic. Even among the earliest evolved floras there is the dichotomous expression of floral syndromes (see Fenster *et al.* 2004; Friis *et al.* 2011) that reflect adaptation to attracting unspecialised flower visitors, and those that are ecological specialists (Proteaceae) seeking to attract a narrower guild of pollinators.

Within species possessing actinomorphic flowers there is great diversity in gross flower formation. In *Eucryphia lucida* (Cunoniaceae/Eucryphiaceae – see Taylor and Hill [1996]), for example, the final 'flower' is formed solely by petals, the fused sepal cap being shed as the petals develop. The reverse is found in *Karrabina benthamiana* (= *Geissois benthamii*) (Cunoniaceae), with petals being absent and the flower formed only by sepals (Harden 1990). In *Syzygium smithii* (Myrtaceae) the petals and sepals are deciduous, its flowers usually left to a display facilitated only by the retention of numerous whorls of stamens (G. Williams 1995). Other variations include flowers that have no depth effect or are tubular in formation, variable numbers of sepal and petal whorls, ovules that are exposed or hidden, stamens that are sessile or free, and variation in stamen and style length.

Species within the above families sometimes constitute the dominant element of individual Australian forest types. For example, the myrtles *Syzygium floribundum,*

Lophostemon confertus, Backhousia myrtifolia (Plates 55, 60) and *Tristaniopsis laurina* can comprise narrow bands of *gallery rainforest* on stream banks within otherwise dry eucalypt-dominated forest (Plate 131) (Floyd 1990; G. Williams 1993), *Allosyncarpia ternata* forms patches of closed forest in northern Australia (Fordyce *et al.* 2000) and various species of Sapindaceae (e.g. *Cupaniopsis anacardioides* and *Arytera divaricata*) are prominent in the canopies of rainforests growing on Quaternary sands adjacent to the seashore (Adam 1992; Floyd 1990). *Ceratopetalum apetalum* (Cunoniaceae) is a rainfall-dependent wind-dispersed species (Heslewood *et al.* 2014), often a co-dominant canopy tree in montane warm temperate Australian rainforests but at lower elevations is frequently restricted to stream lines and associated slopes. The genus is known only from Australia and New Guinea (Harden 1990), with fossil flowers assigned to *Ceratopetalum* recorded from the Cenozoic of Australia (Friis *et al.* 2011). *Eucryphia* and *Nothofagus* are often dominants in cool temperate rainforests in southern latitudes. This present floristic prevalence reminds us that taxa which are considered primitive and even possibly ancestral, in that they reflect long fossil lineages and morphologies thought similar to the first flowering plants, are not necessarily competitively disadvantaged in modern rainforests or predominantly non-rainforest landscapes.

Families that exemplify the post-Cretaceous diversification of the Australian rainforest flora include the Elaeocarpaceae, Euphorbiaceae, Menispermaceae, Myrtaceae, Nothofagaceae, Proteaceae and Sterculiaceae. The Rutaceae are an additional family that has survived, then diversified, in Australian rainforests following the worldwide Cretaceous-Tertiary extinction of angiosperm taxa (Scriber *et al.* 2007). Fossil flowers and fruits of Myrtaceae are known from the Middle Eocene of South Australia (Basinger *et al.* 2007). Named *Tristaniandra alleyi*, the capsular-fruit fossils indicate some affinity with the living genus *Tristaniopsis*. Capsular-fruited Myrtaceae are generally restricted in distribution to Australasia, and appear to have a Gondwanic origin sometime in the later stages of the Cretaceous, or the Palaeogene (~65–23 mya). Nothofagaceae were well established and diverse in the Late Cretaceous (Hill and Dettmann 1996). Now comprising 35 species in the genus *Nothofagus*, members of the Nothofagaceae are usually monoecious trees with extant species found in Australia, New Guinea, New Caledonia, New Zealand and South America (Harden 1990). Their flowers are small and unisexual. Molecular modelling suggests that the extant species date from a radiation that most likely commenced between 55–40 million years ago, with dates for divergences between Australasian and South American taxa being consistent with the rifting of Australia plus South America from Antarctica (Cook and Crisp 2005). The Elaeocarpaceae (e.g. *Elaeocarpus, Peripentadenia* and *Sloanea*) appear to have arisen in the Cretaceous, the family being widespread and diverse in the early Tertiary (Christophel and Greenwood 1987). The fossil Elaeocarpaceae flora of the Southern Hemisphere is better recorded than that of the Northern Hemisphere (Friis *et al.* 2011). Though earlier placed within the Malvales (where it was considered the most ancient member [see Cronquist 1988]), the family is now placed within the Oxalidales (Angiosperm Phylogeny Group IV 2016). Australia has the greatest diversity of living Elaeocarpaceae with most species being found in moist environments. The flowers of Elaeocarpaceae are actinomorphic and usually bisexual, but those of some genera (e.g. *Aristotelia* and *Elaeocarpus*) are also unisexual. *Elaeocarpus* (e.g. Plate 29) is an important element in northern megathermal and southern mesothermal rainforests, with the genus being a recognisable feature of ~30% of rainforest in the Wet Tropics Bioregion (Baba 2013). The genus was present in the eastern Australian flora as early as the Early Oligocene, with modification to the distribution range and diversity levels occurring since the Neogene (Dettmann and Clifford 2001). *Sloanea* is widely distributed from Asia to America, and the five species that occur in Australia are the most primitive members of the genus. *Peripentadenia* is restricted to north-east Queensland. The Euphorbiaceae (e.g. *Fontainea, Macaranga, Mallotus*; and *Austrobuxus, Bridelia, Cleistanthus* as earlier placed) are a very large and diverse group of flowering plants, and first appear in the fossil record in the Early Tertiary (Palaeocene ~65–55 mya [Friis *et al.* 2011]). Their flowers are actinomorphic and unisexual but the petals are sometimes absent, or the sepals and petals are not clearly distinguished within the formation of the perianth. Members of the Euphorbiaceae are the source of cassava, rubber and castor oil; several are highly poisonous. A number of Australian species (e.g. *Baloghia inophylla*, Plate 33) possess the faculty of ballistically expelling their seeds when these are mature. The Menispermaceae (e.g. *Carronia, Stephania, Legnephora, Sarcopetalum* and *Hypserpa*) are mainly climbing plants, the family including few trees or shrubs within its ranks. Most species are dioecious and flowers are actinomorphic. Menispermaceae have a pantropical distribution and are an important family in the basal eudicots. Fossils of *Menispermapyllum* from southern New South Wales are most closely allied to the extant genus *Legnephora* and give evidence for a presence of Menispermaceae in Gondwana during the Tertiary (R. Hill 1989), however, the family is poorly known from the fossil record. The estimated origin of the family is arguably dated to the Mid Jurassic (Jacques *et al.* 2011). The Proteaceae are restricted to the Southern Hemisphere, where they are one of the most diverse families, and are believed to have originated in Australia as trees in closed forests during the Mid Cretaceous (Sauquet et al. 2007). Proteaceous fossils (i.e. *Parafatsia*) are recorded from the Middle Eocene of Australia (Taylor *et al.* 2009). *Parafatsia* is thought to have been derived from evergreen woody plants that evolved in montane warm and humid environments of the Southern Hemisphere. *Musgraveinanthus* is an inflorescence axis from the Cenozoic of Australia (Taylor *et al.* 2009; Friis *et al.* 2011). Approximately 50 genera and 900 species are recognised from Australia (Harden 1991), this representing more than half of the world flora. They include numerous genera that exemplify the Australian flora (e.g. *Banksia, Telopea, Macadamia, Grevillea, Hakea*). The Sterculiaceae (sometimes subsumed within the Malvaceae) are a basal family within the Malvales and are particularly diverse within Australia. Most Australian genera are endemic or almost so (e.g. *Brachychiton,* with 31 of its 32 species being endemic) (Harden 1990). Flowers are actinomorphic and either bisexual or unisexual. The petals are usually minute or absent, the petaloid calyx being fused at the base or tubular (as in *Argyrodendron* and *Brachychiton,* the latter a genus also found in dry forest and woodlands [Guymer 1988]).

DEVELOPMENT OF THE ANCESTRAL ANGIOSPERM FLOWER

Sauquet *et al.* (2017), in modelling the nature of the ancestral angiosperm flower, hypothesise that it was 'bisexual, and radially symmetric, with more than two whorls of three separate perianth organs each (undifferentiated tepals), more than two whorls of three separate stamens each, and more than five spirally arranged separate carpels', however, before the angiosperms there existed a group of plants, the now extinct Bennettitales, whose fossilised bisexual flower-like structures strongly suggest adaptations to attract pollinators. Other extinct groups, such as Caytoniales, Czekanowskiales, Glossopteridales, Pentoxylales and Gigantopteridales, possess characters that have led paleobotanists to place them close to angiosperms; sometimes possessing traits (e.g. partially closed carpels in Czekanowskiales – a condition shared with such living angiosperms as *Drimys* and *Degeneria*) that arguably could place them within the angiosperm tree (Friis *et al.* 2011; Hilton and Bateman 2006; Taylor *et al.* 2009). However, our understanding of plant phylogenetic lineages is subject to the tyranny (or patchiness) of the fossil record and the difficulties of placing fossil taxa within modern families (see, for example, issues in Bremer *et al.* 2004; Crayn *et al.* 2006; and broader discussion given in Taylor *et al.* 2009), and as long as the relationships of extinct angiosperms and their forebears remains unclear (e.g. Friis *et al.* 2011) the hypothetical reconstruction of the earliest angiosperm floral structures needs to be cautious. Given the current understanding that flowering plants are derived from multiple ancestral, non-angiosperm, pathways, there is unlikely to be a single prototypic angiosperm flower: diversity of form ruled from the beginning but only a limited number of stylised and idealistic models can be constructed.

The earliest angiosperm flower fossils include examples from Portugal, North America, the Yixian Formation of China, and western Queensland of Australia (Dettmann *et al.* 2009; Friis *et al.* 2011). Sometimes whole flowers have been discovered, their three-dimensional forms preserved in charcoalised and lignite deposits, and occasionally in limestone (Taylor *et al.* 2009). In Australia, the fossil flowers of angiosperms now known as *Austrodiospyros* (Ebenaceae) and *Musgraveinanthus* (possibly a member of the Proteaceae) have been found in Tertiary clay lenses of western Victoria (Taylor *et* al. 2009; Friis *et al.* 2011). The flowers of *Austrodiospyros* resemble several living Ebenaceae and represent the earliest known example of the family from Australia; supporting the hypothesis of a Southern Hemisphere origin for Ebenaceae and wide diversification in the Late Cenozoic (see Taylor *et al.* 2009). More lately Rose *et al.* (2018) give Ebenaceae a combined Neotropical/Africa origin, this dated at ~57 mya, and with extant Australian *Diospyros* lineages dated no older than 20 mya.

Fossil flowers from the Cretaceous have mainly been minute in size, and fossils from the Lower and Mid Cretaceous have included tiny flowers crowded into spike-like or compound inflorescences (e.g. *Myricanthium amentaceum*) (Friis *et al.* 2011), the size of individual flowers typically in the order 1 mm in diameter (Endress 2001). Early Cretaceous flowers include bisexual and unisexual forms but at this early stage of angiosperm evolution the production of either may not have been strongly fixed. Late Cretaceous flowers are typically small and actinomorphic (the ancestral state), with simple petals, but during this period more elaborate and complex flowers appear (Friis *et al.* 2011).

Although new fossil flowers continue to be discovered, flowers are relatively rare in fossil strata as they are inherently delicate, their fragile structures not easily preserved within sediments, no matter how fine. It is upon the critical and comparative examination of modern flowers that our assessment of the evolution of the angiosperm flower is heavily reliant.

Defining a 'flower'

Altough the term 'flower' is readily employed by both botanist and lay folk alike, as if there is immediate understanding of form and affinity, there is actually little uniformity in the different definitions of a flower that have been espoused (see Bateman *et al.* 2006). As Bateman and his colleagues emphasise most definitions regard reproduction as the primary function of a flower, however they note that 'reproduction' does not solely differentiate the angiosperm flower from the reproductive structures encountered in other groups of land plants. Indeed several defining features in the characterisation of the flower (see Friis *et al.* 2011; Sauquet *et al.* 2017; Smyth 2018) may be absent in some angiosperms, especially early divergent angiosperms. These features include hermaphroditism, fully enclosed carpels and a distinctly whorled floral arrangement. Many extinct and living gymnospermous taxa, which by definition are considered to lack a differentiated perianth, unequivocally possess bract-like organs. Bracts are normally regarded as extra-floral structures, however, they can also be considered part of the flower and in some angiosperms the bract has been integrated into the flower (e.g. *Amborella, Austrobaileya, Trimenia*) such that there is a continuum between bract and perianth. In others (e.g. Araceae) the inflorescence bracts are petal-like and thus serve to advertise for pollinators, the role typically performed by petals. If a definition of a flower partly incorporates the possession of leaf-like accessory reproductive structures (not just those of sepals and or petals) the flower is no longer seen as a unique and defining attribute of the angiosperms (Bateman *et al.* 2006).

The earliest flower carpels were folded together lengthwise (as in, or similar to, present *Degeneria* [Degeneriaceae]), enveloping several ovules; the margins were not sealed (Raven *et al.* 1992). In a generalised primitive carpel there was no localised stigmatic surface to snare pollen, but from this initial state there developed sealed carpels with localised stigmatic surfaces at the crest of the carpel. In primitive angiosperms the stigmatic surface is relatively large, being positioned directly on the unsealed carpels. In the basal taxon *Amborella* the carpels are barrel-shaped with a large cap-like stigma, with carpels reminiscent of *Amborella* possessed by the Trimeniaceae (Endress 2001). More specialised types possess a stigma that is much smaller, positioned atop the carpel style (the elongate sterile column of the carpel, through which the tube of the germinating pollen grain, carrying the sperm, moves from stigma to the ovary). In the first angiosperms the ovules were probably positioned in rows near or along the edges of the carpel's inner surface. These original angiosperms possessed a gynoecium (the ovule-bearing structure of the flower, made up of one or more carpels) which contained separate carpels, however, their presumed evolution entailed a later joining or reduction in the number of carpels, or both. Primitive states of carpel structure are exhibited by the Australian rainforest genus *Tasmannia* (Winteraceae) (Plate 78), in which the carpels are longitudinally folded with the stigmatic surface extending along the carpel suture (Harden 1990). Sterile carpels are often encountered in male flowers. In the 'dinosaur' flower genus *Magnolia* (Magnoliaceae), a non-Australian genus but well known as a garden ornamental, the short bulbous carpels bear large stigmatic surfaces placed upon stout recurved styles, the anthers borne below the upright column of carpels so as not to reduce the possibility of self-pollination (with pollen from the same flower or tree).

In several Early Cretaceous fossil angiosperm flowers the gynoecium possessed only a single carpel, or several carpels that were not fused together (apocarpous, as in the magnolialean *Endressinia* [Friis *et al.* 2011] which possesses ~20 free carpels). Such flowers generally have a small poorly differentiated stigmatic surface. Their reduced spatial area available for pollen deposition, coupled with the limited number of carpels (which allow for few ovules), would result in reduced levels of fertilisation and production of seed (although even high levels of fertilisation can result in low seed production). The small stigmatic area available for pollen placement suggests these plants were pollinated by insects, for angiosperms that are wind-pollinated possess stigmas with greater surface areas that provide targets more effective at capturing pollen. However, fossil angiosperm flowers (e.g. *Sinocarpus*) discovered from the Early Cretaceous Yixian Formation in China (Friis *et al.* 2011) have individual carpels in which several ovules are present, indicating that a dramatic increase in reproductive potential occurred early in the evolution of some flowering plants. The eventual development of syncarpous (fused) carpels (e.g. see Remizowa *et al.* 2010), found more commonly in Late Cretaceous groups, resulted in increased reproductive potential – a single pollination event (if germinating pollen grains shared the same pathway) had the potential to result in greater levels of fertilisation.

Ancestral stamens may have been generally blade-like in shape, indeed *Magnolia* and the primitive northern Australian rainforest vine taxon *Austrobaileya* (Austrobaileyaceae) possess stamens that are so shaped in profile. However, to date fossil evidence is lacking for Austrobaileyaceae (Friis *et al.* 2011). Ancestral staminal structures are hypothesised to have had pollen-producing organs positioned along the centre of the blade, and the blade later evolved into a long anther-bearing filament. Alternatively, stamens may have developed from branch systems that became fused, and were possibly leaf shaped. In several instances stamens became fused in column-like structures (e.g. extant *Stephania* [Menispermaceae]), the stamens and styles became fused and columnar (as in Aristolochiaceae) or, as in some Nymphaeaceae (water lilies), became fused to the perianth itself. Occasionally, stamens have become secondarily evolved away from the function of bearing pollen-carrying anthers. They have become sterile bodies (staminodes or staminodia) and can function to secrete nectar that attracts pollinators, a role also played by other nectar-producing floral bodies (nectaries) that do not trace their ancestry to a modified stamen. These functions are discussed later. Staminodes are present in the flowers of many genera (e.g. *Legnephora* [Menispermaceae]) and, although their role has not been exhaustively investigated, in certain flowers the staminodes appear to be non-secretory. Some act as sacrificial food units eaten by flower visitors; as in Eupomatiaceae (by pollinating *Elleschodes* weevils), and Neotropical Cyclanthaceae (where staminodes are harvested by non-pollinating *Staminodeus* weevils) and may represent, morphologically, a transitional state between petal and stamen. In Calycanthaceae (e.g. *Idiospermum*), a family now restricted to North America, eastern Asia and northern Australia (Taylor *et al.* 2009), the staminodes may be distinctive organs, somewhere between stamens and carpels, functioning primarily in the closure of the floral cup (Staedler *et al.* 2007). Stamen structure in several Early Cretaceous flowers suggests adaptation to insect pollination, as living floras with similar stamen architecture are pollinated by insects. In these early angiosperms the pollen sacs are small, whereas fossil wind-pollinated flowers have pollen sacs that are significantly larger.

Early Cretaceous flowers can be differentiated into two main groups: flowers in which there is no apparent perianth (e.g. *Archaefructus*, a possibly aquatic taxon) and those in which the perianth segments show little or no differentiation (e.g. *Virginianthus* [Friis *et al.* 2011]). By the Late Cretaceous flowers with fully differentiated petals and sepals occur in the fossil record, and in some instances large bracts protect the perianth. Over time the perianth has developed into numerous highly specialised forms, often formed by the fusion of the petals and or sepals (e.g. as in extant *Davidsonia* [Davidsoniaceae/Cunoniaceae], Degeneriaceae and Myristicaceae) or by the development of an unbranched pointed spadix (e.g. Araceae), often enveloped by a coloured spathe (bract structure found in some monocotyledons). In several rainforest-inhabiting families (e.g. Peperomiaceae and Piperaceae) the perianth is completely absent, or appears so without the aid of magnification (e.g. *Gymnostachys* [Araceae]), the perianth segments of the central inflorescence axis are greatly reduced in size and scale-like (Harden 1990). In the basal Eupomatiaceae the petals and sepals are absent, each relatively large white 'flower' being formed from spirally arranged fleshy stamens and petal-like staminodes, the first concealed by a fused conical hat-like bract (calyptra) which is shed as a single unit at flower opening (anthesis) (Hamilton 1897). Floral structures comparable to the calyptrate bracts of Eupomatiaceae are found in the closely related Annonaceae, Himantandraceae and Magnoliaceae. In Annonaceae and Magnoliaceae bracts initially envelop the developing floral buds, while in the Himantandraceae (which consists of two species in the genus *Galbulimia* restricted to the rainforests of north-western Australia and the Papuan region [Friis *et al.* 2011]) the enclosing structures have been interpreted as true bracts or as fused petals and sepals. However, a great many present-day flowers possess simple open structures which provide no obstacle to animals that land upon the flower. These have unobstructed access to the receptacle (the tip of the floral stem from which arises the perianth segments) and the many exposed flower parts, such as delicate filaments and anthers. Sepals are often photosynthetic and can be featureless with respect to colour, so that in effect they resemble leaves. Sepals are thought to be directly evolved from leaves. In contrast, petals are believed to have evolved from stamens and their primarily role is to attract pollinators to the flower. But there are exceptions, for example the Nymphaeaceae in which the petals are thought to be derived from sepals. In the rainforest vine genus *Pararistolochia*, a member of the basal Aristolochiaceae, the flowers are formed by a fused tubular elongation of the calyx (collectively comprised of the sepals); the petals are absent (Plate 15). In *Ceratopetalum* (Cunoniaceae) petals have probably been lost (e.g. as in fossil *Ceratopetalum westermannii*) because of replacement of ancestral pollinators, or in response to fruit specialisation (Barnes and Hill 1999). In separating the apparently differentiating pathways of sepals and petals, it should be noted that sepals normally have three or more vascular strands of xylem and phloem tissue (as in leaves), but most petals have one.

Amongst the basal angiosperms (the ANA/ANITA grade and other basal lineages), there are several shared floral characters, though there remains diversity in both form and breeding system. Living Chloranthaceae (4 genera, c. 75 species) have tiny flowers (c. 1–6 mm) and few floral organs, and the structure of their flowers are similar in some respects to *Amborella* (Amborellaceae). The perianth is mostly lacking (except in *Hedyosmum*), with short, scale-like organs that are likely tepals (Friis *et al.* 2011). One of the unique features of the family is the 3-parted stamen found in *Chloranthus* (Taylor *et al.* 2009). Though not part of the ANA/ANITA grade the Chloranthaceae are placed immediately above, between magnoliids and monocots (Angiosperm Phylogeny Group IV 2016). Their flowers are bisexual and protogynous, with species variously recorded as self-compatible, self-incompatible and agamospermous, insect pollinated (*Sarcandra, Chloranthus*) and wind pollinated (*Hedyosmum*) (Endress 2001; Ma *et al.* 1997). *Sarcandra glabra*, a species native to Japan, is pollinated by beetles (Mordellidae, Cerambycidae, Coccinellidae, Chrysomelidae), bees and Hemiptera. However, flowers are self-compatible and may experience a level of selfing when pollen falls from apical to lower flowers (Tosaki *et al.* 2001). The flowers of Trimeniaceae (Austrobaileyales) (2 genera, 8 species) are 5–7 mm in diameter, whilst the flowers of the single surviving member of the Amborellaceae (*Amborella trichopoda*) are ~3–5 mm in diameter, are functionally unisexual and dioecious, and the floral organs are arranged on a flat, expanded base. Most female flowers of *Amborella* have 1–2 staminodes, and consequently are structurally bisexual (Endress 2001). However, the flowers of *Austrobaileya scandens*, also placed within the Austrobaileyales, are relatively large, some 5 cm in diameter. The flowers of *Austrobaileya* and *Trimenia* possess floral characters that grade between bracts (modified leaves) and perianth (Bateman *et al.* 2006). The monotypic Ceratophyllaceae (Ceratophyllales) comprise a single cosmopolitan genus, *Ceratophyllum*, of herbaceous water plants (~6 species). Fossil fruits are known from the Palaeocene (Taylor *et al.* 2009). Cronquist (1988) placed the order Ceratophyllales in the Nymphaeales (Cabombaceae, Nymphaeaceae); however, in more recent molecular analyses the family is not at the base but is placed as a sister taxon of monocots and eudicots (e.g. Friis *et al.* 2011). The Ceratophyllaceae have no close affinities with living angiosperms and the family may be a survivor of an ancient, otherwise extinct, angiosperm lineage (Taylor *et al.* 2009). Flowers are unisexual and minute (c. 0.5–1.5 mm), and pollination takes place underwater. Female flowers bear a single carpel with one ovule, the ovary is barrel-shaped and the stigma is not distinctly differentiated from the style. The floral structure is greatly reduced, and tepals are lacking and the organs surrounding the stamens or carpel may be interpreted as bracts because a lateral flower is sometimes formed between these organs and the pollination organs (Les 1993). The Nymphaceaceae, comprising six genera and ~60 species, with a worldwide distribution in tropical and subtropical regions (Schneider and Williamson 1993), exhibit flowers, with variable organ numbers, that range from smaller sizes (c. 1–2 cm, *Ondinea*) up to the very large blossoms of *Victoria* (50 cm), an iconic Amazonian taxon. Flowers of Nymphaceae are complex, usually with several zones of appendages including those that are sterile and which are variously interpreted as either staminodes or pistillodes (paracarpels) (Gandolfo *et al.* 2004). Breeding systems within the Nymphaeaceae vary from self-compatability, partial cleistogamy (the condition by which cross-pollination is suppressed by the partial closing of the flower), and with evidence of self-incompatability (Endress 2001). As has already been noted Nymphaeaceae possess flowers that are quite large. Large flowers with numerous organs were once considered as primitive, from which smaller flowers then evolved by reduction, however, neither phylogenetic studies of living plants nor the fossil record support a basal position for large-flowered angiosperms (Les *et al.* 1999; Mohr and Friis 2000). It is apparent that angiosperm flowers have evolved in several different directions, as evidenced by the small to medium-sized flowers with limited number of organs in those plants assigned to the ANA/ANITA grade, such that early evolution of small flowers in the first angiosperms is more probable (Endress 2001).

Over the last 130 million years changes in phyllotaxy (the arrangement of floral parts and leaves on their axes) are likely to have played a major role in the evolution and diversity of angiosperms (Staedler *et al.* 2007), however, phyllotaxis was unstable in early angiosperm evolution and it is uncertain if the original condition of floral phyllotaxis was spiral or whorled. Because woody eumagnoliids appear to be nested within angiosperms Endress and Doyle (2007) considered it most likely that spiral floral phyllotaxis in Magnoliales and Laurales is a derived rather than a primitive state (as in Magnoliaceae, *Degeneria, Eupomatia,* Calycanthaceae, Atherospermataceae), and that in comparison with monocots and eudicots (taxa with tricolpate and derived pollen), floral phyllotaxis is unusually diverse in basal angiosperms. The eumagnoliids, however, are an informal cluster of plant orders that comprise ~1/3 of all angiosperm species, but a formal grouping of the proposed taxa (to embrace Canellales, Magnoliales, Laurales, Piperales and monocotyledons, that include advanced taxa such as grasses, orchids and palms) is not currently possible since their relationship to the eudicots is yet to be fully resolved (Friis *et al.* 2011; Hilu *et al.* 2003). In basal angiosperms phyllotaxis is 'evolutionarily plastic'; the perianth and androecium are both 'spiral' in *Amborella, Trimenia, Austrobaileya,* and Illiciaceae/Schisandraceae, but 'whorled' in Nymphaeales. The perianth is whorled in *Hedyosmum* (Chloranthaceae; however, androecium phyllotaxis is undefined as male flowers consist of a single stamen). In *Ceratophyllum* (Ceratophyllaceae, Ceratophyllales) whorled and spiral flowers are recorded from the same species (Endress 2001; Endress and Doyle 2007). For the 'remaining 99.9%' of angiosperm species, the floral phyllotaxis ancestral condition was whorled, either as a new feature or derived from the retention of the ancestral state, the consolidation of whorled phyllotaxis in derived angiosperms allowing further adaptive radiation. Change in flower phyllotaxis is difficult to ascertain owing to their small and more tightly packed organs. Additionally, there are changes between distinctive organs during development, especially at the transition between androecium and gynoecium (Endress and Doyle 2007). In basal angiosperms a change from decussate (pairs of opposite members, these being alternatively angled) to spiral phyllotaxis may occur at the vegetative to floral transition (e.g. *Austrobaileya, Trimenia,* Calycanthaceae). Additional changes involve expressions from distichous (arranged in two ranks) to spiral phyllotaxis, as in Eupomataceae, and from distichous to a whorled state (e.g. Annonaceae). A common expression of change is seen amongst the core eudicots where the calyx is spiral in formation but the remainder of the organs are whorled, however, Endress and Doyle (2007) note that in many Magnoliaceae, the perianth is whorled but the androecium and gynoecium are spiral.

Throughout the history of angiosperms, asynchronous evolutionary pressures have been placed on reproductive strategies. Early-divergent angiosperms were and are generalists; but with early specialisation for either flies, beetles, thrips and bees (Gottsberger 2015). Beetle pollination (cantharophily) is frequently observed in the Magnoliales, for example in some South American species of the genus *Drimys* (Winteraceae), and varying suites of insects that include flies and bees, in addition to beetles, are pollinators of Nymphaeaceae (Endress 2001; Seymour and Matthews 2006). Though beetles, among the suggested earliest pollinators (with flies and thrips) of the first angiosperms, have remained a dominant and abundant invertebrate group (see Crowson 1981; Lawrence and Ślipiński 2013), the first likely dispersers of angiosperm seeds – reptiles and related vertebrates – have become extinct or lost their pre-eminent role (though some plants still depend on reptiles as primary or secondary dispersers of seed, particularly in Gondwana). The decline of reptiles was largely completed by the Late Mesozoic (65 mya) and their place, and a consequent switch in primary seed dispersal modes, was taken by birds and mammals (though ants are important secondary dispersers of small seeds [see A. Beattie 1982; Hölldobler and Wilson 1990]). Although this change in seed dispersers could have had significant co-evolutionary ramifications (see Dodd *et al.* 1999; Ricklefs and Renner 1994; Rule *et al.* 2012), continued reliance by unspecialised primitive angiosperms on very small beetles and other tiny insects (that did not harm the reproductive organs of flowers) as pollinators might have resulted in little selective pressure to modify the architecture of flowers and the placement of sporophylls; for example, to enlarge, flatten or increase the number of sporophylls and thus potentially retain the earliest floral traits. Where early pollinators were large or destructive, or occurred in large numbers, this would be expected to result in secondary (derived) development of the flower state, such as an increase in number, size and aggregation of the sporophylls, or evolution of medium-sized or large flowers, and the evolution of solitary flowers, derived from an original inflorescence condition (Friis *et al.* 2011). From the perspective of increased visual signal to pollinators, however, the trend is likely to have been from individual flowers to inflorescences. It is likely that during even the earliest development of flowering plants a variety of different pollination strategies and plant breeding systems co-existed under similar ecological conditions, and that this adaptive divergence was expressed within closely related taxa, as is found among closely related species today. Shifts in reproductive strategies, including switching of primary or preferred pollinators, are hypothesised to have had significant influences on angiosperm evolution and diversification (Willmer 2011). And whilst the rise of the angiosperms resulted in new pollinator associations, their ascendancy brought about the demise of most of the pollinator associations that had evolved in the preceding 65 million years, with the transfer of only some of the existing associations onto angiosperms, with these existing today as relics (Labandeira 2010).

Although Sauquet *et al.* (2017) proposed an ancestral angiosperm floral architecture, there is still no single established hypothesis as to the first structural and functional flower characteristics (e.g. discussions in Sauquet *et al.* 2018; Sokoloff *et al.* 2018). Traits that might have been present in the first angiosperms could have included flowers that were small or moderately sized and open in structure, located at the end of branches and possessing both functional stamens and carpels (bisexual/hermaphroditic) (Friis *et al.* 2011), but in which carpel receptivity preceded maturity of the stamens and availability of pollen (protogyny, as in Eupomatiaceae [Endress 2003]). Early in the development of angiosperms, formation of lateral inflorescences, flowers with relatively few truncate stamens and bladder-shaped carpels positioned loosely on a relatively short floral axis, may have occurred. The flowers were likely ecologically generalist, attracting a diverse assemblage of non-specialist insects (including silverfish) and dispersing seeds mainly by reptiles (saurochory) or wind (Willmer 2011). Smaller seed classes may have been dispersed by ants (myrmechory); while ants may not disperse seeds over large distances they potentially disperse seeds to microsites that are especially beneficial for germination and seedling growth (Horvitz 1991). Later developmental trends in entomophilous and still primitive angiosperms included enlargement of flowers (with retention of the bisexual/hermaphrodite and protogynous condition), with attendant increase in flower robustness, evolution of single flowers carried in lateral and terminal branch positions, production of pollinator-attracting odours, partial or full protection of ovaries, dense grouping of stamens and greater numbers of carpels. There was also a switch from saurochory to primary seed dispersal by mammals or birds (Friis *et al.* 2011). Thus it has been conjectured that the flower seen in *Magnolia* might reflect a derived morphology, rather than a truly prototypic condition found in some members of the earliest Magnoliales (Gottsberger 1974). Yet even in other basal groups of the Magnoliales there is development away from a bisexual/hermaphrodite condition to dioecy (e.g. *Tasmannia* [Winteraceae], *Palmeria, Hedycarya* and *Wilkiea* [Monimiaceae]) or an androdioecious state (e.g. *Trimenia* [Trimeniaceae]) where normally hermaphroditic populations also possess some functionally male members (Bernhardt *et al.* 2003). In the Australian rainforest vine *Trimenia moorei*, for example, hermaphroditic flowers are borne terminally on the stems, while the lower flowers are usually male (Plate 73), a condition that may favour out-crossing (by increasing flower attraction to pollinators and mobilising larger quantities of pollen) and minimise loading of self-pollen upon apically placed hermaphroditic flowers.

CHEMICAL WARFARE AND THE EVOLUTION OF FLOWERS

A further influence on flower evolution and consequent development of floral structures is the interaction between pollinators and predators in the form of herbivores – not the variously sized vertebrates and invertebrates that graze upon foliage but the creatures that attack floral parts or feed upon flower resources without necessarily contributing to pollination. For example, Mendez and Munzinger (2010) record high levels of floral bud predation by insects on *Pouteria laetevirens* (Sapotaceae), a tree endemic to the evergreen rainforests of New Caledonia. Buds also suffered from pathogen infection. Among the Australian rainforest flora, larvae of the moth *Alucita phricodes*

(Alucitidae) feed on flowers of the vines *Pandorea jasminoides* and *Pandorea pandorana* (Plate 19) (Common 1990), those of the moth *Cosmoclostis aglaodesma* (Pterophoridae) feed on flowers of the tree *Gmelina leichhardtii* (Lamiaceae/Verbenaceae), and larvae of *Oxyodes tricolor* (Noctuidae) feed on flower buds of *Arytera divaricata* (Sapindaceae) (see Plate 107). Attack can also be directed against seeds; some plants in 15 of 23 species of Leguminosae investigated by Janzen (1969) in Central America had at least 50% of their seed crop destroyed by bruchid beetles. It is uncertain as to whether continual herbivore attack on floral parts and seed crops results in a reduction of plant fitness in individual species (Hendrix 1990) though, intuitively, relentless herbivory over time (potentially in combination with disease and or fluctuating environmental conditions) should result in a reduction of seedling recruitment and population abundance.

Plant chemistry is one of the main factors mediating interactions with pollinators and herbivores (Kessler and Halitschke 2009; Peakall *et al.* 2010; Scriber *et al.* 2007), though even herbivores considered to be specialists on particular plant familes have been shown under experimental conditions to exhibit considerable food plant plasticity (and thus detoxification capabilities); as with the butterfly *Papilio aegeus* whose larvae are in nature restricted to Rutaceae (e.g. *Citrus, Flindersia, Geijera*), but under experimental conditions are able to process many existing species of basal angiosperms without having direct ancestors that historically fed on them (Scriber *et al.* 2007). Kessler and Halitschke (2009) suggested that plants might be able to avoid conflicts between attracting pollinators and the need for herbivore defence by tissue-specific regulation of pollinator reward chemistry, as well as herbivore induced changes in flower chemistry and structure. However, they emphasise that there are major gaps in our understanding of how plant defence and pollination are coordinated through secondary metabolism.

Although herbivores play a crucial role in structuring plant rainforest communities and in maintaining their diversity (Bagchi *et al.* 2014), the need for a plant to attract pollinators and at the same time defend itself against predator/herbivore attack involves conflicting selection pressures (Pellmyr *et al.* 1991; Rosenthal and Janzen 1979). If plant defensive compounds and those involved in the plant pigmentation pathway trace back to the same chemical precursors (see Herrera *et al.* 2002; Schaefer and Rolshausen 2005), as has been suggested, then defensive selection responses to herbivores might be constrained (owing to contrary selection from pollinators favouring pigmented flowers) (e.g. Armbruster *et al.* 1997). Favouring one response diminishes optimisation of the other. The presence of secondary compounds in flowers may mean that a plant seeking to maximise its defence against herbivores may differ in its attractiveness to pollinators, or subgroups within its pollinator guild. Defensive compounds isolated from floral structures include glucosinolates, alkaloids and nicotine, and their presence can influence how, or if, potential pollinators will use floral resources such as nectar and pollen. Plants investing in secondary compounds to strengthen their defences in the plant–herbivore conflict may do so at a cost of reduced attractiveness to pollinators. Pollinators may somehow need to accommodate plant toxins when their young utilise floral resources for their development (as in some *Aristolochia*). Clearly there is a risk that increased selection for defence may not permit the maintenance of appropriate levels of pollinator attraction. Flower evolutionary selection pressure is exerted not only by herbivores but by contrary selection responses (or the requirement for them) from pollinators (De Luca and Vallejo-Marín 2013; Van der Niet *et al.* 2014), which constrains defensive responses against attack.

A specific example of the chemical defences erected by flowering plants against attack is the highly specialised fig syconium, a structure that is both fruit and a flower collective (see Chapter 8). Serving as a cradle for its own young and that of its co-adapted wasp pollinators, tissue of the fig syconium contains alkaloids, bergapten, flavonol glucosides, polyphenols, sterols, tannins and triterpenoids. Immature syconia are protected against herbivore attack by derivatives of latex, which contains the protease ficin. Within this impressive defensive array of chemicals are substances that are directly toxic and those that act to reduce digestibility. The competing influences of selection pressure are deterring predators while not adversely affecting the fig's role as partial sacrifice in allowing development of its fig wasp pollinators. Ultimately, the fig's purpose is to be eaten and its seeds dispersed, and so its defences are only good to a degree – among other predators, drosophilid flies are known to oviposit and breed within the developing syconia, in so doing destroying a proportion of the ovules.

3 Being a flower

INFLUENCE OF FLOWER COLOUR, FRAGRANCE AND STRUCTURE

In the great majority of flowers the defining visual characters are provided by the sepals and petals which together or separately form the perianth (Raven *et al.* 1992; Willmer 2011). Occasionally the bracts (modified leaves at the base of the flower stalk or spike) provide the primary colour signal: in the northern Australian *Curcuma australasica* (Zingiberaceae) the bracts at first are bright pink but fade to green as the enclosed flowers develop and open. The colour and pattern of individual floral parts can vary. In *Austrobaileya scandens* (Austrobaileaceae) for example, the bract-like or sepal-like tepals are green, the inner petal-like tepals are greenish or yellowish in colour with brown-purple spots. The broad stamens are yellowish with dark inner staminodes and the gynoecium is bright yellow (Endress 2001). In Australian tropical *Hornstedtia scottiana* (Zingiberaceae) the red coloured bracts not only serve to enhance the recruitment signal conveyed to pollinators but also function to protect flower buds and developing fruit (Ippolito and Armstrong 1993). More familiar in Australian gardens are the bright colours of the bracts of *Bouganvillea* (Nyctaginaceae). The constituent parts of the angiosperm perianth are sterile and thus play no direct part in plant fertilisation, however, their colour and form function to attract pollinators, thus facilitating pollination and sexual reproduction. Occasionally, the perianth is greatly reduced or absent and it is the bright or distinct coloration of the exposed pollen or anthers (as in *Polyscias sambucifolia* [Araliaceae], *Alectryon coriaceus, A. subcinereus* [Sapindaceae] and *Gymnostachys anceps* [Araceae]) that most likely act as the principal visual means of attraction for pollinators and other flower visitors. Other flowers possess long coloured filaments (the bodies that carry the pollen-bearing anthers) which magnify the flowers' visual attraction. For example, the mauve filaments of *Crinum pedunculatum* (Amarylliidaceae/Liliaceae) (Plate 6), the reddish-pink filaments of *Lysiphyllum hookeri* (Caesalpiniaceae) and the pink-tipped filaments of *Barringtonia asiatica* (Lecythidaceae), their flowers otherwise being white in colour. In *Capparis arborea* (Capparaceae), a tree particularly of dry rainforests and some mangrove margins in subtropical New South Wales, the small anthers sit atop greatly lengthened filaments that dwarf the white petals (Plate 20). The stigma projects above the brushy stamens. In the Myrtoideae (Myrtaceae) petals and or stamens act as visual attractants but the stamens (as in *Syzygium wilsonii, S. moorei*) generally are the most conspicuous structures in the open flower (Lughadha and Proença 1996). Similar structure and colour is exhibited by *Cynometra lenticellata* (Plates 37, 38), a tropical member of the Fabaceae.

The flowers of most Australian rainforest trees are coloured white, cream or greenish-white; this relative colour homogeneity being dominant across lineages from different biogeographical origins (Delmas *et al.* 2020). This is also the dominating theme in subtropical rainforest, with the prevalence of white and greenish-white coloured flowers progressively trending to a greater incidence in rainforests at more southerly (higher) latitudes. Greater proportions of similar coloured 'plain' flowers, as opposed to colourful ones, can also be expressed in relation to increases in altitude, as Boulter *et al.* (2011) noted for the rainforest flora of Lamington National Park, south-eastern Queensland. In North Queensland rainforest most plants (72%) also possess white or green-white flowers, with 12% being yellow-orange, 8% pink-red, 6% blue-purple and 1% brown (Boulter *et al.* 2008).

Although early angiosperms may not have had prominent colour signals (e.g. Kevan and Backhaus 1998) the colour of flowers is one of the principal forms of visual attraction in the extant angiosperms; to which male and female pollinators, and different lineages of related pollinator taxa, may respond differentially (e.g. de Jager and Ellis 2012; Ödeen and Håstad 2010). Colour advertises the flower's presence to pollinators, and in some species differences in blossom colour can differentiate sterile and fertile flowers. Change of colour in any floral part can signal loss of function or the absence or prediction of floral rewards owing to the flower having been pollinated (Dyer *et al.* 2006). For example, in the large ephiphytic and lithophytic (epilithic) Australian rainforest orchid *Dendrobium speciosum* there is a more intense colouring, and closure, of the perianth following pollination (Slater 1991). Colour change in angiosperm flowers may also be age-dependent (Delph and Lively 1989), as in the tree *Guoia semiglauca* (Sapindaceae) in which mature anthers turn from pink to mauve when no longer functional, and signal senescence (G. Williams 1995). The larger the flower or the inflorescence (as in the large endemic rhizomatous herb *Neoastelia spectabilis* – Asteliaceae [Plate 17]), the greater the visual 'conspicuousness' signal that the colour will be able to project, presumably over a greater distance (see Møller and Eriksson 1995; Neal *et al.* 1998). In smaller solitary flowers or small flower clusters (e.g. *Cayratia clematidea, Cissus antarctica, C. sterculifolia* – Vitaceae, *Elatostema* spp. – Urticaceae [Plates 74, 75, 76, 77]), dependent on attracting pollinators by colour alone, the value of colour by itself must be much reduced, operating over only a short distance no matter how bright the colour of the blossoms.

In flowers, both the surface structure of the petals and the pigments they contain determine coloration (van der Kooi *et al.* 2014). However, the great diversity of colours exhibited by angiosperm flowers is produced by a relatively small number of pigments, yet, except for the Nymphaeaceae, blue pigments are generally lacking in basal angiosperms (Endress 2001; Willmer 2011). Most colour pigments in the range of pink, red, violet and blue are due to the presence of anthocyanins (see Brockington *et al.* 2011 for a discussion of betalain pigments and replacement by anthocyanins). Anthocyanins are water-soluble, and are a class of flavonoids. They occur in solution in cell sap and are present in stems, leaves and fruits as well as in flowers. The pH (the quantitive expression of acidity and alkalinity of a solution) of the cell sap influences colour, with the selective pressure emanating from different types of pollination considered to be important in the evolution of anthocyanin biosynthesis (Shoeva *et al.* 2017). Flavonoids are commonly encountered throughout the angiosperms but are less common in other vascular plants. In leaves, flavonoids block destructive far ultraviolet radiation and facilitate the process of photosynthesis by selectively admitting light in the blue-green and red wavelengths. Anthoxanthins are less common in flowers but modify colours produced by other compounds. Yellow, purple and some reds are due to carotinoids and flavones, white results from multiple reflections in the spaces between uncoloured cells. Reflections in the ultraviolet range are caused by several chemical compounds, but are mostly related to the presence of carotenoids. Consequently, the final flower colour is determined by the range of pigmentation compounds present and the contributing factors of cellular pH and the structural and reflective qualities of floral parts.

The colour of a flower is not always constant through its life (Ohashi *et al.* 2015) and, as previously stated, may change with receptivity and age. Some colours present at anthesis later fade, apparently not being light-resistant. In other species the colour of flowers becomes more intense. Those of the Australian rainforest vine *Passiflora aurantia* (Passifloraceae), for example, are at first creamy-white but later change to pink and red (Harden *et al.* 2007). In the large bisexual woody vine *Melodinus australis* (Apocynaceae), the tubular corollas can be coloured orange–yellow to white (Plate 8). Such colour variation may be due to changes in the acidity of cell sap and the biochemical conversion of floral compounds responsible for pigmentation.

In several flowers there is a marked contrast in the colours that collectively comprise each flower's visual signal (e.g. the richly contrasting purple – mauve perianth and yellow anthers of *Osbeckia australiana* [Melastomataceae] and *Solanum aviculare* [Solanaceae] (Plate 72) and the rich orange stamens and wax-white perianth of *Calophyllum inophyllum* [Clusiaceae]). This vibrant contrast, in addition to the actual colours, sends a strong recruitment message to potential pollinators. In other flowers visual contrast is achieved through varying saturation of a single colour, by marked changes or by more subtle gradients across individual flower segments (e.g. as in *Archidendron grandiflorum* [Mimosaceae], *Castanospermum australe* [Fabaceae] and *Hibiscus heterophyllus* [Malvaceae]). Further contrast may be obvious to organisms with different spectral sensitivity, as there is great variation in how different kinds of animals (including different species related in higher taxon ranks, such as orders) perceive colour within the spectrum (e.g. Dafni *et al.* 1997; Ödeen and Håstad 2010). Bird-visited flowers and insect-visited flowers, for example, have been shown to differ significantly from each other in the chromatic signals they project, and that the differences are concentrated near wavelengths of optimal colour discrimination by each class of flower visitor (Shrestha *et al.* 2013). Removing, or loss of, the 'recruitment signal' can result in greatly reduced visits by potential pollinators (Gowda *et al.* 2001).

Flower contrast, and consequent enhancement of the flower's recruitment signal, has been achieved by the colour of some nectars; in effect the 'coupling' of the signal and the reward. Nectar is usually clear, or nearly so, however, more than 60 species of plants in 14 angiosperm families are known to produce coloured nectar (Hansen *et al.* 2006). These include *Aloe* (Asphodelaceae) from South Africa, *Schiedea* (Caryophyllaceae) in Hawaii and *Jaltomata* (Solanaceae) in South America. Amongst those plants that produce coloured nectar are several endemic species from Mauritius that are pollinated by lizards (*Phelsuma* spp., Gekkonidae); *Nesocodon mauritianus* (Campanulaceae) producing a deep red nectar, *Trochetia boutonia* (Malvaceae) producing red nectar, and *Trochetia blackburnii* producing yellow-coloured nectar. Experimental evidence showed that the gecko *Phelsuma ornata* preferred coloured over clear nectar in artificial flowers (Hansen *et al.* 2006).

Contrast with foliage is also used to highlight the recruitment signal displayed by flowers (Bischoff *et al.* 2013), and so solicit pollinators (though different pollinators may not necessarily respond in like manner). Examples are the rich deep red flowers of *Stenocarpus sinuatus* (Proteaceae) (Plate 69), *Graptophyllum excelsum* (Acanthaceae) and *Metrosideros nervulosa* (Myrtaceae); the latter an endemic of Lord Howe Island – *Metrosideros* is now naturally extinct on the Australian mainland but is known from fossils (i.e. *Metrosideros leunigii*) dated to the Eocene-Oligocene (~40–30 mya) of Tasmania; this suggesting a possible Australian origin of the genus (Tarran *et al.* 2016). The large showy white flowers of *Capparis arborea* (Capparaceae) and *Cynometra lenticellata* (Fabaceae) (Plates 20, 37, 38) are conspicuous against the dark green broad-leaved foliage of their parent plant. Contrast of bright flowers against dark foliage is common among trees of the upper canopy stratum (as in the bird- and insect-attracting myrtaceous *Xanthostemon chrysanthus*, Plate 61 [Franklin and Noske 2000) but is also encountered among plants of the shaded rainforest understorey (e.g. *Tapeinocheilos ananassae* [Costaceae], *Pittosporum revolutum* [Pittosporaceae] and *Lembrassia australiana* [Gesneriaceae]). Even the massed white flowers of understorey plants, as with the vine *Ripogonum discolor* (Ripogonaceae, Plate 70), or more sparsely presented white flowers of *Drymophila moorei* (Alstroemeriaceae/Luzuriagaceae, Plate 5), can be conspicuous against a dim background of dense green leaves. In the Australian epiphytic orchid *Rhinerrhiza divitiflora* (Plate 64) the individual flowers of the long pendent inflorescence are mainly mottled (camouflage-like) greenish-red, however, the labellum is conspicuously white in colour suggesting this serves as primary visual means of attraction. In the climbing saprophytic orchid *Pseudovanilla foliata* the large labellum is a showy orange–red, the flower otherwise being pale creamish-yellow in colour; and the bract-like leaves being inconspicuous (Plate 63).

Patterning of flowers is also common (e.g. the rainforest vine *Austrobaileya scandens* whose flowers are green, flecked with purple), the patterns frequently serving as a pollinator guide to the flower's reproductive organs. Such refinements are normally present in flowers where there is some marked level of specialisation in the plant–pollinator relationship. For example, deep gullet (e.g. *Pandorea* [Bignoniaceae], Plate 19) or lobed flowers (e.g. *Alpinia* [Zingiberaceae], Plate 79, *Pseuderanthemum* [Acanthaceae] and Orchidaceae Plates 63, 64) are adapted to be pollinated by a certain taxonomic class, or a size class, of pollinators. Sometimes the presence of pollinator guides is apparent only in the ultraviolet and thus is invisible to the eyes of humans (Lunau *et al.* 2020). A unique contrast signal may be provided by floral electric fields for charge fields are frequently greater on flowers than elsewhere on the plant (Vaknin *et al.* 2000; Willmer 2011). These electrostatic fields can act as recruitment cues and exhibit variation in pattern and structure to which flower visitors may discriminate during foraging; as has been reported for bumblebees (*Bombus*) (Clarke *et al.* 2013).

Even when flowers are individually small, they are commonly borne as massed clusters (e.g. *Cuttsia viburnea* [Escalloniaceae/Rousseaceae], *Cupaniopsis anacardioides* [Sapindaceae]). The signals emitted by these aggregations of tiny flowers attract large numbers of insects and other animals. For example, the small greenish coloured flowers of the rainforest margin vine *Cayratia clematidea* are borne in large conspicuous clusters and attract a diversity, and often abundance, of insects that in subtropical rainforest communities can include thynnid (e.g. *Rhagigaster* sp., *Zeleboria* sp.), gasteruptiid (*Gasterupton* spp.), pompilid and vespid wasps (*Polistes humilis*), tachinid and syrphid flies (*Dideopsis aegrota, Mesembrius hilaris*), and pierid butterflies (*Appias paulina*) (Plates 75, 91, 109) (G. Williams pers. obs.). Movement of inflorescences and single blossoms, by the agency of wind, can enhance the contrast between flowers and foliage and thus increase their capacity to be perceived by pollinators. In some species there is a differential carriage (presentation) of unisexual flowers, and the reason for this is not readily understood. In monoecious trees of *Guoia semiglauca* (Plate 71) male flowers are commonly exposed and massed in prominent branched clusters developed within leaf-axils (axillary panicles) which are not obscured by foliage, but female flowers are partially hidden within or under foliage (G. Williams 1995). Perhaps the prominently positioned male flowers, having served to attract pollinators that then mobilise pollen resources, become expendable; the female flowers are afforded a degree of protection as fertilisation proceeds.

Though flowers can enhance their recruitment signal by adding contrasting colours and patterns to their basic palette, flower colour (including ultraviolet) and colour pattern conspicuousness and form do not necessarily provide the only recruitment signals to pollinators. Flower size and fragrance also provide advertisement signals to prospective pollinators (e.g. Dobson 2006; Willmer 2011), with variation in floral odour and flower size within discrete populations of individual plant species possibly being an adaptation to local pollinator assemblages (Sun *et al.* 2014). In addition pollinators may demonstrate a preference to symmetrical over asymmetrical shaped flowers (Møller and Eriksson 1995). The behaviour of pollinating hawkmoths is affected by differences in corolla shape, but the size of the flower may be more important than shape differences in floral discrimination by hawkmoths, and other pollinators (Kaczorowski *et al.* 2012). The size of the flower projects recruitment signals that can be perceived in either frontal or side profiles. Large flowers individually may project an enhanced signal but the strength of the signal is also related to the pollinator's angle of approach as well as projection length and contour of the flower, and the ratio between them (Dafni 1994; Dafni *et al.* 1997).

Pollinators may rely on floral scent to distinguish 'honestly signalling' flowers from deceptive ones (Peakall and Whitehead 2014), and floral scent, and the ability of pollinators to learn to associate floral signals with food rewards, may have played an important role in the early evolution of flowers (Hetherington-Rauth and Ramírez 2016; Peakall *et al.* 2010). It can also function to increase short-term fidelity which would result in an advantage to both plant and pollinator instead of sole reliance on a visual recruitment signal (Wright and Schiestl 2009). Angiosperms can exhibit sexual dimorphism with respect to scent composition and scent production (Ashman 2009), and this may serve to influence patterns of pollinator visitation and pollen movement.

The total amount of fragrance emitted by flowers can vary significantly throughout anthesis (Silva *et al.* 2019). However, daily patterns of fragrance emission may not be fully in phase with the potential suite of pollinators. For example, in the case of *Hoya australis* (Asclepiadaceae), a species of rainforest, vine thickets and open forests extending north from Queensland to Melanesia, the emission of strong floral fragrances increases at dusk and persists into the night. Though this suggests that nocturnal pollinators are involved (Forster 1992a), the recorded pollinator at cultivated plants is a day-active skipper butterfly *Ocybadistes walkeri sothis* (Hesperiidae). Many plants produce strong floral scents, ranging from highly aromatic (e.g. *Eupomatia*) or subtly fragrant (*Sarcochilus* [Orchidaceae]) to odours which are foetid (Rafflesiaceae, Aristolochiaceae and *Alphitonia excelsa* [Rhamnaceae]) (e.g. Hamilton 1897; Raguso 2004, 2020). The fragrances produced by some families may be generally similar. For example the flowers of most Australian Myrtaceae are usually considered sweet but the flowers of *Syzygium cormiflorum* (Plate 59) have a faint unpleasant smell (Crome and Irvine 1986). The emission of particular floral scents and rewards can be episodic (as in Passifloraceae) or exhibit circadian rhythmicity (Altenburger and Matile 1990; Forster 1992a); the production and release cued by events such as photoperiod, specific daily temperature thresholds, autochthonous production of heat (as in *Alocasia* [Araceae]), the receptivity of the stigma in bisexual species (as in *Crinum pedunculatum* [Amaryllidaceae/Liliaceae]), or the activity patterns of their co-adapted (if not co-evolved) pollinators. This is the case with some hawkmoths (Sphingidae) that are active only during twilight (Baker 1961; Moulds *et al.* 2020). If co-evolutionary processes are reflected in the plant–pollinator relationship (e.g. Sapir and Ghara 2017; Zhang and Wang 2017), what came first – the activity patterns of the pollinator or the anthesis pattern of the flower? Or did they develop in concert, or fortuitously meet by a chance pre-existing disposition?

The fragrances that flowers produce, including pollen odours, are believed to have preceded the evolution of colour as an attractant. Precursors to fragrances might have been secondary plant compounds that originally acted to deter herbivorous insects, or herbivores in general (Rosenthal and Janzen 1979). The orchid *Epipactis helleborine* exhibits physiological and morphological adaptations for attracting social wasp pollinators. Wounded plants infested with herbivores emit 'green leaf volatiles' (aldehydes, alcohols, acetates) that attract prey-hunting wasps (*Vespula germanica, V. vulgaris*) which then may serve as agents of pollination (Brodmann *et al.* 2008).

Flower odours and fragrances serve as secondary attractants to recruit pollinators; during the chemical process of odour production a great amount of plant starch may be utilised and there is a resource cost to the plant. Floral odours are particularly strong in plants exclusively or primarily adapted to beetle pollination (see Chapter 8), and the fragrances emitted by a single flowering *Eupomatia* or *Alocasia* can permeate the air for many metres. The flowers of Aristolochiaceae, which are pollinated by Diptera (e.g. Phoridae, Sepsidae, Calliphoridae), produce odours that mimic those of mammalian waste products and decomposition (Bänziger and Disney 2006); the release of the constituent compounds is thought to be at least partly responsible for the attraction of their pollinators. Plants that do not offer a reward may use volatile chemicals to attract pollinators; the most extreme examples being recorded from carnivorous plants (Jürgens 2009).

The strength of the fragrance that a flower emits is not necessarily related to its size – quite large flowers may produce scents poorly discernible to humans (though not necessarily to pollinators), while many small flowers can individually produce odours that are readily detected. Some Australian subtropical rainforest flowers appear not to produce a detectable fragrance at all (e.g. *Alectryon coriaceus* [Sapindaceae], Plate 94; and *Daphnandra micrantha* [Monimiaceae/Atherospermataceae], Plate 47) (G. Williams 1995). In those that do, it is not always certain if the role of the fragrance is to attract fragrance-sensitive animals in general, for fragrances may equally function to differentiate between species that co-flower so that preferential pollinator visits are made. Such consequential foraging visits result in periods of reproductive isolation for the plant species (its competitors being temporarily removed from the foraging cycle of shared pollinators), with the potential maximisation of species-specific pollen being transported between flowers and plants of its population.

In the strongly scented Amazonian understorey palm *Geonoma macrostachys* (Knudsen *et al.* 1999), individual compounds produced exhibit patterns of a steady daily decrease in production or a morning increase followed by a steady decrease through the remainder of the day. Although the overall daily amount of floral scent production decreases, the variation in production of individual compounds may have significant outcomes for pollinator activity (in this case beetles, flies and bees), serving to modulate recruitment patterns and avoidance of already pollinated flowers. But shared floral odour by species may enable hybridisation, and the ability of pollinators to perceive floral volatiles over large distances will result in long-range movement of pollinators and likely commensurate movement of pollen (Whitehead and Peakall 2009). For most angiosperms the recruitment signals that serve to alert and orient a pollinator towards a given flower are likely to be a subtle interplay of various attraction stimuli, operating over different times of day or night; as with the Australian vine *Trimenia moorei* (Trimeniaceae), in which fragrance release appears restricted to the morning (Bernhardt *et al.* 2003). Different stimuli may operate at different spatial scales, some to attract pollinators in the general area, others to bring the pollinator to the flower.

It is mainly from osmophores (specialised floral tissue bodies) that fragrances are produced. Only a small proportion of the spectrum of volatile compounds present in the fragrance emitted by any one species may suffice to permit anthophilous (flower-loving/flower-visiting) insects to quickly discriminate among a cohort of co-flowering plant species. Bees, for example, are particularly adept at learning from chemical choices (e.g. Hetherington-Rauth and Ramírez 2016).

The chemical constituents of floral fragrances may involve hundreds of compounds, though floral chemistry might have limited utility in studies of plant phylogeny (Knudsen *et al.* 2006). They include aromatic compounds, terpenoids and monoterpenes, the last including cineole, geraniol, menthol, limonene, myrcene and camphor. Among the aromatic compounds are eugenol, methyl salicylate and vanillin; up to 18 aromatic compounds have been detected from a single species of Neotropical orchid and more than 20 components have been isolated from the fragrance plume emitted by *Trimenia moorei* (Bernhardt *et al.* 2003). Jürgens (2009) considered it is likely that aromatic aldehydes, esters and alcohols (e.g. methyl benzoate), alone or in combination with some of the monoterpenoids (e.g. linalool), can serve as signals to attract nectivorous insects for these compounds are found commonly in flowers associated with nectar-feeding insects, and especially those associated with Hymenoptera and Lepidoptera. Riffell *et al.* (2009) report more than 70 compounds emitted by night-blooming *Datura* (Solanaceae), but only nine (as a mixture) were required to attract their hawkmoth pollinators. In the fig *Ficus semicordata* (distributed from India, China through South-east Asia) 4-methylanisole is the main compound in floral scent that attracts the obligate pollinator, the fig-wasp *Ceratosolen graveleyi*, at the precise stage required for pollination and oviposition (Chen *et al.* 2009). The compound is emitted by male and female *Ficus semicordata* when receptive but is absent in volatile chemicals emitted by figs four days after pollination, thus it seems to serve as a species-specific chemical cue. In the Himalayan herb *Rheum nobile* (Polygonaceae) 2-methyl butyric acid methyl ester is emitted by plants only during anthesis and serves as a specific olfactory attraction cue to its obligate pollinator, a species of the fungus gnat genus *Bradysia* (Sciaridae) (Song *et al.* 2014).

Defining the brew that lures the beast is still very much frontier science, but in 'carrion' flowers (those that emit putrid odours) the production of dimethyl disulfide is a primary component of the odour attractant strategy. Dimethyl disulfide, and related volatile compounds, are also present in several New World angiosperms visited and pollinated by small phyllostomatid bats (Bestmann *et al.* 1997).

The functional relationship between flowering plants and their pollinators has progressed beyond tricks of colourful and fragrant beguilement. Certain species, and indeed whole families (e.g. Orchidaceae), have evolved systems so that when the appropriate pollinator enters the flower it trips a mechanism, resulting in the deposition of a pollen bundle upon the pollinator, or the dusting of the pollinator's body with pollen previously held within the flower's dehiscent anthers. The placement of pollen grains or discrete pollen masses on the pollinator can be quite specific. Commonly, there is the downward movement of stamens resulting from the pollinator applying force to an enlargened petal (the labellum in Orchidaceae), as it attempts to push its way to the nectar at the base of the perianth. An interesting development of the strategy for placing pollen on a potential pollinator is found in the Australian pioneer (a species that occurs early in regeneration or colonisation) rainforest tree *Alphitonia excelsa* (Rhamnaceae) (see Appendix 1). *Alphitonia excelsa* attracts a broad taxonomic guild of insect pollinators and visitors and produces sticky grains of pollen. When the individually small flowers open the stamens are encapsulated by the petals, and when these are depressed by a foraging insect the pollen grains are exuded in an elongated mass, clear of the sheathing petals, onto the insect's body (Williams and Adam 2001).

Gross flower structure by itself has a strong determining role in differentiating the kinds of animals that are able to visit a flower, how and if they may land upon it and whether they can function as pollinators. So-called 'bird-flowers', for example, are often large and may have secondary perches so as to accommodate their avian visitors (Cronk and Ojeda 2008). In the moth-pollinated orchid *Plantanthera bifolia*, widely distributed from Europe into Asia, plants with long floral spurs have more flowers pollinated, have greater quantities of pollen removed and produce more and larger fruit; spur length, not the size of the overall perianth being critical for reproductive success (Boberg and Ågren 2009). Individually small flowers can deter large flower visitors because their size is incapable of supporting the weight even if the flowers are borne close to the leaf axils. Or, as with the laurel (Lauraceae) genera *Cryptocarya* and *Endiandra* (as in *E. muelleri*, Plate 40), the corolla can so closely confine the stamens and stigma that large flower visitors cannot easily gain access to pollen or sweet exudates (G. Williams 1995). In the woody climber *Melodinus australis* (Apocynaceae) (Plate 8) the small anthers are located near the base of the tubular corolla, the depth and narrow structure of the corolla seemingly denying access to all but long-tongued visitors. In addition, the quantity of nectar small flowers produce (if they produce any at all) (as in *Cissus* – Vitaceae; Plates 76, 77) is probably insufficient for a large animal to achieve a profitable energy budget. That is, foraging expends more energy than the flower and its adjacent cohorts are likely to return. In contrast, large flowers are generally accessible to many flower visitors, but unless these are able to effect pollination their visits have no reproductive benefit for the plant. Such animals rob nectar or destroy pollen without necessarily touching stigmas or anthers.

Actinomorphic flowers are recorded from the Early Cretaceous (Friis *et al.* 2011) and are most normally adapted for visitation by a range of pollinators, frequently possessing open floral structures with no significant depth – this inferring little specialisation towards pollinators of any kind. However, in *Davidsonia pruriens*, the open actinomorphic pale pinkish-white flowers are covered by stiff irritable hairs that are easily dislodged when touched. Such irritant hairs are also found on other organs (e.g. leaves and fruit). The hairs deter animals that might destroy the calyx segments and probably have a filtering effect on the flower visitors able to serve as pollinators. Any insect would need to be able, in flight, to avoid the outer surface of the calyx whorl as it attempted to land on the flower.

Zygomorphic flowers (e.g. those of Orchidaceae, Fabaceae and some Proteaceae) usually exhibit specialised pollination strategies involving a restricted taxonomic range of pollinators and precise mechanical methods of pollen dislodgement from the flower and placement upon the pollinator. It is interesting to note that plants with more complex flowers (e.g. Zingiberales), implying an adaptation to specialist pollinators (e.g. birds [Cronk and Ojeda 2008]), were already widely distributed by the end of the Cretaceous: flowers with pronounced zygomorphic symmetry (perianth bilateral symmetry) are known from the Early Cenozoic (i.e. Late Palaeocene – Early Eocene of North America [Crepet and Taylor 1985]). Zygomorphy has evolved repeatedly from radial symmetry (actinomorphy) throughout angiosperms. The zygomorphous floral state has been linked previously with plant-pollinator specialisation and higher species diversification, with perianth zygomorphy recognised in 32 orders and 110 families, and with a minimum of 130 origins and 69 reversals to actinomorphy (Reyes 2016).

Individual blossoms and massed inflorescences placed terminally on weak elongate branches are incapable of supporting any but the lightest pollinators, and favour pollinators capable of hovering while they feed (hawkmoths, hummingbirds etc.). Those with flowers at ground level (geoflorous) suggest their pollinators are mammals (e.g. *Balanophora fungosa* (Balanophoraceae; but see Pierce and Ogle 2017) (Plate 18). The highly modified fig 'flower', and those of the rainforest shrub *Breynia oblongifolia* (Phyllanthaceae/Euphorbiaceae) (Plate 65) and the small trees and shrubs of the genus *Wilkiea* (Monimiaceae) (Plate 48), present flowers of the most frustrating and seemingly unappealing kind – entry is by means of a tiny hole (an ostiole) and so access is denied to all but the most minute creatures.

As previously discussed, some flowers possess elongate, lipped or distinctly lobed petals (e.g. Lobeliaceae, Zingiberaceae and Orchidaceae), upon which specialised pollinators can alight. Even the diminutive herb *Pratia purpurascens* (Lobeliaceae), of Australian subtropical rainforest margins, provides a floral lobe upon which minute long-tongued 'bee' flies (e.g. *Geron* species [Plate 97]) can land to feed from the nectar held inside (G. Williams pers. observations). Such flower adaptations, that facilitate the approach of their chosen pollinators, are particularly well developed in the rainforest orchid *Pseudovanilla foliata* (Plate 63) and in species of native gingers of the genus *Alpinia* (Plate 79). The intricate variations in functional structure and pollen-depositing mechanisms are many and generally minute, so go unnoticed except by the most keen-eyed and patient field observer. In this highly intimate relationship the participating pollinators (typically birds and some kinds of bees) that visit such highly adapted flowers are easily deterred from their pursuit and, too fast, seek out alternative flowers. Even large native *Amegilla* (Plate 81) and *Xylocopa* bees are readily frightened; there must be a prolonged period of unmoving observation before the naturalist blends into the physical anonymity of the landscape, and the pollinator returns.

ULTRAVIOLET LIGHT AND PERCEPTION OF FLOWER COLOURS

'Individual variation is the substrate for evolution.'

(Briscoe and Chittka 2001)

The importance of flower colour in attracting pollinators generally has already been alluded to. Though predictions of what flower colours 'fit' with which pollinators are sometimes misplaced, nevertheless certain colours appear to select specific pollinator taxa. That is, specific colours can be considered agents by which the flower advertises its presence to desired pollinators such as birds or bats, and screens out or minimises its attractiveness to flower visitors that are minor, ecologically generalist pollinators or that do not undertake pollination (Kay *et al.* 2005). Equally, in the absence of any pronounced bright coloration (many small rainforest flowers are coloured cream, or are minute, inconspicuous and greenish-white as in *Glochidion ferdinandi* [Phyllanthaceae/ Euphorbiaceae]) (Plate 66), or where flowers are neutral in colour (e.g. where the perianth segments are not distinguishable in colour from the receptacle or gynoecium), the plant may be adapted to pollination by a broad taxonomic assemblage of pollinators. However, this statement should be considered with caution. While the latter assumption holds good for mass-flowering rainforest trees, shrubs and vines of the shaded understorey are also commonly pale in colour (e.g. as in the minute flowers of *Cleistanthus cunninghamii* [Phyllanthaceae/ Euphorbiaceae]). Large numbers of generalist pollinators do not congregate in such dim environments (many flower-visiting insects positively avoid them), and the plants are not gap-phase ones whose propagules have responded to the temporary intrusion of sunlight owing to damage and consequent opening of the canopy. Rather, they are understorey specialists that attract certain classes of pollinators (e.g. solitary bees not inhibited by shade) attuned to their diminutive pale flowers, or that recruit pollinators primarily by the release of floral fragrances. They are not necessarily reliant upon colour as a means of pollinator recruitment.

Flower colours can be considered as representing, or having a significant impact upon, the evolution of pollination mutualisms. Certain colours seem to target specific pollinator taxa, or at least elicit a particular recruitment response from them. Although individual flower colours may only loosely be considered attractive to particular groups of pollinators, colour can serve as a strong discriminating agent (Altshuler 2003; Cronk and Ojeda 2008); low UV white for example, is especially attractive to thrips (Kirk 1985a). Dyer *et al.* (2012) suggest that the long geographic isolation of the Australian continent has allowed an independent evolution (to that of plants in the Northern Hemisphere) of flower spectral signals that maximise colour discrimination by Hymenoptera, and that there may be significant fitness benefits for flowering plants that are readily discriminated by pollinators that are 'flower-constant', exhibiting some level of foraging fidelity.

In some instances there is variation in flower colour, the different colours holding true to isolated plant populations or along altitudinal gradients, or within a species there may be an adaptive overlap in colour forms. An example is the colonising and later climax forest tree *Metrosideros collina* (Myrtaceae) found at different altitudes in Hawaii and elsewhere in the Pacific (Carpenter 1976). The species has plants with yellow flowers (which are totally self-compatible though also pollinated by small insects, mainly bees) and red-flowered plants that are partially self-incompatible and exhibit greater adaptation to bird-pollination with increases in altitude. Birds perceive red (Shrestha *et al.* 2013), a faculty they share with humankind, and might claim a profound sensitivity to the many hues and saturations of red on the blossoms of rainforest trees. But while birds may be found foraging among the many red-flowering angiosperm species, the receptors of birds do not necessarily have an innate preferential sensitivity to that colour. In Australian landscapes, flower-frequenting birds may just as readily be found foraging among the many native and introduced flowers coloured cream or white, yellow or even green. The visual spectrum of birds typically comprises four spectral classes of single cones that contribute to the discrimination of colour (Dyer *et al.* 2012). Birds perceive colour over wavelengths ranging between 300 nm and 660 nm and in Neotropical forests bird-pollinated flowers have been shown to have a typical median reflectance greater than 585 nm; this is outside of the visual range of bees (300–550 nm) (Altshuler 2003; Kevan *et al.* 2001), a potential group of competitors for floral nectar in bird-adapted flowers. The human visual spectrum, by comparison, is in the range of 400–700 nm (Dafni 1992).

The eyes of insects possess a set of ultraviolet-, red- and blue- and green-sensitive photoreceptors that allow highly selective responses to floral displays (Briscoe and Chittka 2001; Peitsch *et al.* 1992; Willmer 2011). Bees recognise yellow, blue-green, blue and ultraviolet, but are blind or less sensitive to the orange-red band (Willmer 2011). They also lack brightness perception. Bees can preferentially discriminate between yellow over white flowers but in general cannot recognise true red, although red blossoms containing some blue will be recognised by bees as either blue or ultraviolet. Bee colour perception is considered to have evolved from a basal visual system that predates the evolution of angiosperms. Red receptors are more common in Lepidoptera, and it appears that receptors sensitive to red have evolved possibly as many as four times in that insect group. However, diurnal activity, as found in butterflies and some moths, is not necessarily a predictor of the presence of red-sensitive receptors. Among nymphalid butterflies (entirely day-active, though some species are crepuscular [such as *Melanitis leda*]) it appears that red receptors have been lost several times. Yet, overall, the evolution of red receptors in insects seems to be an important development that post-dates the ancestral suite of receptors that were sensitive to ultraviolet, blue and green (Briscoe and Chittka 2001). Clearly, a pollinator's ability to perceive floral colours in terms of an attracting signal strongly influences how the plant–pollinator relationship might progress or be constrained. Colours outside a pollinator's spectral range, no matter how resplendent to us, are of little or no use in effecting plant reproduction. However, the receptor capacity documented from some butterflies, dragonflies and Hymenoptera is among the broadest known in the animal kingdom (<300–>700 nm). The spectral range exhibited by insect photoreceptors can differ considerably between species, and there are instances of sexual dimorphism in receptor sensitivity. Individual insects may possess eyes in which different parts possess receptors of differing sensitivity, while species occupying quite distinct ecological niches may possess similar sets of receptors.

Ultraviolet (300–400 nm) recognition plays an important role as a discriminating factor in the pollination environment (Lunau *et al.* 2020), even if pollinators are unable to perceive it as a colour (Raguso 2004). Being able to *see* in the ultraviolet range was probably a faculty possessed by animals (e.g. aquatic invertebrates) that predated the terrestrial emergence of insects, and interestingly, examples of ultraviolet absorption are known from basal angiosperms. In *Cabomba* (Cabombaceae) the nectariferous parts of the petals are strongly ultraviolet absorbing, yet only weakly so in peripheral parts (Burr and Barthlott 1993). Though colours such as red may be invisible to a particular pollinator, as is the case for the majority of bees and some hawkmoths, sensitivity to ultraviolet (as demonstrated for the butterfly family Pieridae) attracts pollinators that would otherwise be lost to the flowering plant. The beguiling employment of ultraviolet nectar or floral guides can facilitate recruitment, by mimicry, of pollinators to otherwise-unrewarding flowers. For example, apparent behavioural entrainment to ultraviolet results in pollinating male *Trichocolletes* bees being attracted to the Australian orchid *Diuris maculata* (Indsto *et al.* 2006). This orchid offers no nectar reward (Beardsell *et al.* 1986) – its ultraviolet nectar signals closely resemble those of the unrelated pea-flower *Daviesia ulicifolia*, resulting in erroneous foraging by *Trichocolletes* and pollination of *Diuris*. Lunau *et al.* (2020) found that glossy floral areas are often positioned in proximity to UV-absorbing central flower parts and these produce a strong UV signal; and that the optical contrast produced by the glossiness of small smooth surface areas close to nectar holders represents a widespread yet overlooked floral cue that nectar-feeding flower visitors use to locate floral nectar. Thus such glossy surfaces can be categorised as nectar guides serving to direct UV sensitive flower visitors towards nectar.

Sensitivity to ultraviolet wavelengths has been found in different kinds of birds (once thought blind to ultraviolet), and ultraviolet-sensitive receptors are also known to occur in crustaceans, fishes, reptiles and amphibians. Even some mammals are able to see in the ultraviolet range. In their ability to perceive ultraviolet, some birds possess more than our 'blue', 'red' and 'green' primary colour arsenal. The colour sensitivity range of birds is potentially tetrachromatic, but humankind and bees share a colour vision that is trichromatic. In bees the perception of colour is skewed in favour of shorter wavelengths, likely best able to discriminate for wavelengths close to 400–500 nm (Dyer *et al.* 2012), thus they dwell in a visual world where ultraviolet is seen as a distinct colour. Honey bees, *Apis mellifera* (Plate 80), can detect a pronounced contrast between ultraviolet (344 nm) and green (544 nm) (Willmer 2011). It is unlikely that this sensitivity is their preserve alone, so it is reasonable to assume that rainforest flowers exhibiting ultraviolet reflectance may advertise to bees and other Hymenoptera more generally. For mass-flowering angiosperms competing for pollinators in an unspecialised pollinator environment, the preferential perception of one colour over another may have profound outcomes for pollen dispersal and plant reproductive success.

FLORAL REWARDS AND THE COMPOSITION OF NECTAR

Pollinators gain food from flowers. This mainly takes the form of nectar and pollen, less commonly resins and oils, but can also include flower structures such as petals, sepals and staminodes. Pollen is a food commonly consumed by beetles and thrips as well as vertebrates, and in its consumption a proportion of pollen available for transport and fertilisation is destroyed. Yet without it, a primary means of attracting pollinators would be lost.

The earliest records of pollen meals are from generalised polyneopteran insects (i.e. stoneflies, webspinners, zorapterans, walking-sticks, crickets, grasshoppers, cockroaches, mantises and termites) from the Permian with gymnosperm pollen preserved in their gut contents (e.g. Grimaldi and Engel 2005). Non-angiosperm (gymnospermous?) pollen has also been found in the guts of three different species of Xyledidae (Symphyta) described from the Late Cretaceous Baissa area of Eastern Siberia (see Krassilov *et al.* 2007). Thus a pollen-feeding pathway amongst early-derived invertebrates preceded the Tertiary radiation of the angiosperms.

Pollen and nectar are considered primary attractants and when pollen serves as an attractant the anthers are usually well exposed. It is often the reverse for nectar, for in gullet and bell-shaped blossoms the nectar-producing glands are hidden. Only in easily accessible entomophilous flowers (and the strobili of Gnetophyta/Gnetales) is nectar exposed. However, the availability of pollen, from dehiscent anthers, does not always occur simultaneously. In *Syzygium smithii* the anthers are present in two concentric ranks, the anthers of the outer longer rank opening first and the inner circle later (G. Williams 1995). Pollen is arguably the principal reward available to visitors of most Myrtoideae flowers (Lughadha and Proença 1996) but in several species of southern African *Eugenia*, and Australian *Decaspermum humile*, female flowers produce a non-viable pollen-like material which may represent an important resource to insect pollinators (Kevan and Lack 1985).

Pollen is sometimes the only floral resource offered to flower visitors (as in *Trimenia moorei* [Trimeniaceae] – Plate 73, and *Eidothea hardeniana* [Proteaceae] – Plate 68 [Bernhardt and Thien 1987; Bernhardt *et al.* 2003], and most buzz-pollinated flowers [Buchmann 1983]), and was likely to have been the principal reward offered to pollinators by Early Cretaceous angiosperms. Yet it is important to remember that pollen is a matured microspore functioning primarily as part of the plant's reproductive system. Nectar, pleasant-tasting as it might be, is just a sweet substance that attracts pollinators and serves no functional role (except as a possible antibacterial agent) in pollen grain germination and later fertilisation. Pollen occurs most commonly as single grains, or as composite clusters (tetrads, polyads etc.) such as those of *Acacia* (Fabaceae/Mimosaceae) and the waxy pollen aggregates, called pollinia (singular, pollinium), of orchids and Asclepiadaceae. The evolution of different types of pollen aggregation from single grain states has occurred independently at least 39 times in the history of angiosperms and may evolve when pollinators visit infrequently (Harder and Johnson 2008). Pollen is an important source of protein (certain grains contain over 60%), with other nutrients comprising nitrogen, amino acids, starches, sterols and lipids. But not all of the pollen body is digestible. Differences in digestibility may be attributable to variation in the porosity of the pollen wall as well as thickness and composition (Roulston and Cane 2000). The endangered gliding possum *Petaurus gracilis* is able to digest a high proportion (c. 80%) of pollen grains consumed (largely Myrtaceae) suggesting pollen is a major source of protein (Jackson 2001). The honey possum *Tarsipes rostratus* is able to digest high levels (90–100%) of *Banksia* pollen (Turner 1984) and the Queensland blossom bat *Syconycteris australis* has been recorded with more than 50% of *Banksia* pollen grains emptied by passing through the digestive tract (Law 1992), yet some flower visitors may gain little nutrient from ingested pollen (see Roulston and Cane 2000). The Australian New Holland honeyeater *Phylidonyris novaehollandiae* (Meliphagidae) eats pollen incidental to feeding on nectar, however, the contents of the pollen are not extracted during passage through the bird (Paton 1981). Worse, for the larvae of some bee species have failed to develop on particular pollens indicating that various pollen species can be chemically protected, and that animals utilising pollen as a food source, either as adults or larvae, might require physiological adaptations to overcome unfavourable chemical properties (Sedivy *et al.* 2011). Compounds responsible for the toxicity of certain pollen species possibly include alkaloids, mannose sugars and polyphenolics (Roulston and Cane 2000).

The pollen of flowering plants is formed, in general, by three main concentric layers (see Faegri *et al.* 1992) (oily pollenkitt is sometimes considered a fourth layer, but is not present in all angiosperm pollen species). The outer layer, normally the exine, is generally considered to be indigestible; undigested exines are found in the excreta of animals as diverse as beetles, birds and flower-visiting bats (e.g. Law 1992; Paton 1981). It is the exine that gives each pollen type a distinctive shape and sculpture, though individual grains can be transformed in appearance when dry or moist. The exine consists principally of the fatty acid derivative sporopollenin, which is resistant to water entry (Roulston and Cane 2000). Exines are highly resistant to decay and consequently, with the exception of very delicate exine types (e.g. Juncaceae), most exine structures fossilise very effectively, capable of staying fossilised in sediments for millions of years (Friis *et al.* 2011). This durability confronts animals that feed upon pollen with the problem of how to access the nutrients within. Animals must somehow physically break or pierce the pollen wall, utilise enzymes to dissolve or penetrate the pollen wall, or induce in the grain some form of germination or germination-like response (Roulston and Cane 2000).

The exine is one of the hardiest organic structures known, and can be treated with acids and bases (substances which react with an acid to form water and a salt) and subjected under specific conditions to temperatures as high as 300°C with little apparent detriment. Sporopollenin (formed by polymerisation of carotenes and related esters) is the secret of resistance to attack. Strength lies not simply in the thickness of the exine wall, nor does resistance rest with the amount of sporopollenin in the wall; the differential qualities of the chemical composition and degree of polymerisation of sporopollenin also play a role.

The intine, the second enveloping layer of which pollen is composed (though in some taxa the intine forms the outer of two-layered pollen types), is also likely indigestible. The intine consists of cellulose (though the percentage appears to vary dramatically between taxa), callose, pectine, proteins and enzymes (Faegri *et al.* 1992; Willmer 2011). Other substances may also be present. Only the innermost protoplast, the part which germinates upon the stigma, offers a food reward, but to claim this, the two outer layers must be pierced. This is quite common, though it is probable that much pollen ingested is incidental to the intake of nectar, as is likely the case with many birds.

For brood-rearing insects such as social apid and solitary bees, pollen is an essential food source for the developing young, and to specialist flower-visiting mammals, pollen is an important component in their dual nectar-pollen diet. Even sterile pollen can serve as a food source to flower visitors in general. High in protein and nitrogen, pollen also contains significant levels of starch, oils and fats. High levels of sugars may be present. Starches and oils represent the principal calorific resources in pollen. Since the essential role of pollen is to contribute to plant reproduction, it is to be expected that pollen contains compounds associated with the process of germination.

It is doubtful that pollen provides sufficient energy, of itself, to meet the energy requirements of larger flower-visiting insects. Bumblebees (*Bombus* spp.) are dependent on nectar intake periodically ingested during their pollen foraging cycles (Willmer 2011); while bumblebees are not native to Australia (though established as feral invaders [McQuillan and Hingston 1999]) large native bees such as the various *Xylocopa* species are likely to have similar energy demands. Large native bees may need to feed upon nectar from a source different from that where they collect pollen. Bees with long tongues that enable them to consume most or all of the nectar in deep tubular flowers may optimise energy returns (Johnson and Anderson 2010). However their food intake is achieved, and regardless of their size, bees (and all pollinators) must sustain a daily energy intake not less than that expended during foraging (positive energy balance) (see Higginson and Gilbert 2004), with longer foraging distances requiring more frequent or more extended nectar uptakes to meet the higher energy expenditure inherent in such foraging activity (Zurbuchen *et al.* 2010).

Pollinators, not just bees, that undertake accurate foraging strategies (i.e. make 'correct' or optimal resource decisions), such that energy returns are maximised, will potentially collect more nutrition per unit of time (Chittka *et al.* 2003; Dyer *et al.* 2012).

Where flowers simultaneously present pollen and nectar, as do many Myrtaceae, an energy benefit is potentially available as pollinators may not need to forage for independent sources of pollen and nectar. But some pollinators, for example butterflies that are restrained to a liquid diet since they possess no mouthparts capable of piercing pollen grains to access the protoplasm within, rely on nectar as the source of food. There are exceptions to this. Neotropical *Heliconius* butterflies are able to mix pollen grains that accumulate on the proboscis, with a clear liquid they exude (which is probably nectar [Gilbert 1972]). This mixture is agitated for several hours by repeatedly coiling and uncoiling the proboscis (Gilbert 1972; Penz and Krenn 2000). Following hydration, proteins and free amino acids are quickly released from the pollen (O'Brien *et al.* 2003). Under laboratory conditions certain pollens have been shown to release most of their proteins and amino acids within minutes of being incubated in sucrose solution (e.g. Linskens and Shrauwn 1969). The inclusion of pollen in the diet of *Heliconius* provides important nutrients (e.g. amino acids, peptides) extra to those obtained from nectar; the amino acids being used for egg production. A similar method of pollen feeding is exhibited by the Australian drosophilid *Scaptodrosophila flavohirta* (Nicolson 1994), and the syrphid fly *Cheilosia albitarsis* ingests both pollen and nectar from its host *Ranunculus repens* (Haslett 1983). As the pollen reaches the midgut most of the grains extrude their protoplasm in a manner like germination or pseudogermination (Haslett 1983).

Nectar is produced in specialised secretory structures – the floral nectaries. Like pollen, nectar is a major source of food for pollinators, and also attracts numerous non-pollinating flower visitors that are dependent upon nectar production. Owing to the importance of floral nectaries in pollination, the location of nectar secretion and presentation, combined with nectar quality and composition, are main factors in determining potential pollinators among nectar-feeding animals (Stpiczyńska *et al.* 2012).

A pollinator's sensitivity to variation in nectar reward can bias the way it responds to floral scent (Wright and Schiestl 2009). The ecologically specialised Western pygmy possum *Cercartetus concinnus* has been shown to preferentially select for hexose-rich *Eucalyptus rugosa* nectar when available (Morrant *et al.* 2010), and in that study mostly ignored *Eucalyptus diversifolia*; possibly because *E. diversifolia* nectar is proportionately much more rich in sucrose than other species. Such preferential selection poses questions as to the 'learning' pathway that leads nectivorous species to such strategic foraging decisions, and how plants might manipulate pollinator responses (Pyke *et al.* 2020).

However, nectar is generally absent in flowers that are wind-pollinated or attract by deception. Nectar can be a beguiling agent of death – when it takes a highly viscous form, numerous minute insects are held without prospect of escape. It is common to find on the flowers of *Schizomeria ovata* (Cunoniaceae), generally a small- to large-sized tree of subtropical and tropical Australian rainforests (Plate 25), many microhymenopterans snared in its sticky nectar (G. Williams 1995). Equally, nectar can be a toxic brew, the reported victims most commonly bees. In New Zealand the nectar of *Corynocarpus laevigatus* (Corynocarpaceae) is highly poisonous to introduced honey bees (Matheson and Reid 2011), the responsible toxic compound possibly being the glucoside karakin. Nectar palatability can serve to filter desired flower visitors (e.g. Johnson *et al.* 2006). In the open-flowered African milkweed *Pachycarpus grandiflorus* (Asclepiadaceae), the large quantity of nectar it produces is palatable to its desired pollinator (a wasp, *Hemipepsis*) but unpalatable to honey bees that also visit. *Pachycarpus* is specialised to *Hemipepsis* and in the absence of flower morphological adaptations, which might otherwise serve to filter visitors, it achieves specialisation by producing nectar of targeted palatability, in combination with cryptic flower coloration and scent, as a selective pollinator attractant (Shuttleworth and Johnson 2009). Amongst the carnivorous pitcher-plants, however, the presence of nectar, rather than flower colour and pattern, may be more important as an attractant (Bennett and Ellison 2009).

There are nectar-like substances exuded by gymnosperms (e.g. the so-called pollination or micropyle drop reported from various cycads). An evaluation of proteins in the pollination drop of gymnosperms strongly suggested that they function both to defend against pathogens and to promote pollen development (Wagner *et al.* 2007). Nectar-like exudates are also encountered in fungi, fern leaves, leaf bracts and the extrafloral nectaries (specialised nectar-excreting glands) of some plant phyllodes, leaves and stems (as in many *Acacia*, but also rainforest genera such as *Castanospermum, Dysoxylum, Glochidion, Melicope, Syzygium* and *Flagellaria* [e.g. Blüthgen and Reifenrath 2003]).

Extrafloral nectaries are recorded in more than 90 families of angiosperms and have been hypothesised to have evolved a secondary role, serving as decoys to keep ants away from floral nectar and away from flowers (Hölldobler and Wilson 1990; Willmer 2011). However, the abundance of ants commonly encountered at many Australian rainforest flowers, and the relative few numbers of rainforest plants with such structures, suggest little practical need in the rainforest environment for the deployment of such a resource expensive diversion. Examination of exudates from the extrafloral nectaries of the Brazilian *Sapium biglandulosum* (Euphorbiaceae) revealed a total sugar concentration of 32.5% (w/v) of which 38.1% was fructose, 43.7% glucose and 18.2% sucrose (Coutinho *et al.* 2010). Frustose, glucose and sucrose have also been identified in the extrafloral gland exudate of the tropical weedy species *Triumfetta semitriloba* (Tiliaceae) (Espolador Leitão *et al.* 2005).

But nectar in the strict sense is really a phenomenon of the angiosperms. In the evolutionary history of flowering plants nectar is likely to trace its history to some form of sugary exudate that had a role other than in pollination.

Considering the many alternative sources of nectar production unassociated with flowers, found in lower and higher plants, its production by floral nectaries has been viewed as an evolutionary development and condition derived from an ancestral state.

Nectar usually originates from the phloem, and nectaries are normally situated in proximity to the end of phloem strands (see Raven *et al.* 1992; Willmer 2011). However, there is variation in the construction of floral nectaries, suggesting numerous evolutionary origins. Many are formed by discrete glands and obvious apertures freely permitting nectar exudation, or are dense cell structures from which nectar diffuses through a thin-walled epidermis. In some floral nectaries the epidermal area from which nectar exudes is drawn out, or the epidermal cuticle may degrade to allow the escape of nectar. In *Cuttsia viburnea* and female flowers of dioecious *Diospyros australis,* the raised ovaries (superior ovaries) are surrounded by obvious, yellow, nectar-producing (nectariferous) glands (G. Williams 1995). In female flowers of dioecious *Euroschinus falcatus* a nectariferous gland surrounds the carpel (Plate 7), but nectar production by a similar-shaped gland surrounding male flowers seemingly does not occur. In many rainforest plants nectariferous glands are not readily identifiable, but in several instances, for example various tropical Fabaceae, the nectaries are whole flowers that serve no other purpose. Some floral exudates remain enigmatic, as in *Daphnandra micrantha* (Plate 47), whose flowers possess stigmatic papillae (but no obvious nectar-producing glands) which produce a clear gelatinous exudate on anthesis. An inexplicable occurrence is a highly viscous gel sometimes found coating the flower buds of *Elaeocarpus reticulatus* (Elaeocarpaceae) (G. Williams 1995); the gel's function is undetermined.

Wherever nectar originates in the flower's structure, its availability is constrained to some degree by the plant's overall physiological condition and daily cycles of anthesis and adaptation to particular pollinators. In addition, the age of the flower can have a significant effect on the amount of sugar produced. An increased availability of sugar has been found to correlate with younger inflorescences that have more flowers. However, particular episodes in the flower's life (e.g. initial anthesis, fertilisation, stigma receptivity and longevity) will overlay a level of daily autonomy in the production of nectar and its eventual cessation. In *Syzygium smithii* nectar is exuded on or shortly after anthesis and discrete nectar drops are found on the surface of the hypanthium. In *Acradenia euodiiformis* (Rutaceae) (Plate 80) nectar production commences before flowers are fully open and the greatest volume of nectar occurs before or at full opening, while in *Syzygium floribundum* (Plate 60) freshly opened flowers normally do not possess nectar (G. Williams 1995). Flowers of some species produce nectar irrespective of whether the perianth segments have been shed or pollinated flowers are senescent (as in *Acradenia euodiiformis*). In others, the loss of sepals and petals (*Tristaniopsis laurina* [Plate 100]) or general flower senescence (*Alphitonia excelsa* [Plate 89]) marks the cessation of nectar production. Even if flowers have been heavily damaged, their receptacles partly consumed and brought to dereliction by the ravages of herbivores, they may continue to produce nectar (as in *Syzygium smithii*).

Various kinds of sugars (a carbohydrate) are found in nectar – it is not a homogenous and universal soup. The most common sugar types present are fructose, glucose and sucrose, but sugar composition is sensitive to micro-organism impacts. The concentration of individual sugars within nectar varies, as does the overall dilution rate (Willmer 2011). Dilution rates vary with respect to the phylogenetic history and taxonomic affiliation of particular plant species, but daily events of heat and humidity, wind speed and solar radiation also affect sugar concentration. Rain can wash enough nectar from flowers that hours or even days may pass before recharging is complete, and in some species concealment of nectar by floral structures many be an adaptive means to prevent washing away or dilution by rain. It is not uncommon to note an absence of flower-visiting insects on the blossoms of rainforest Myrtaceae and other mass-flowering canopy taxa for lengthy periods after rainfall. But an absence of pollinators may at times be due to the reverse, where pollinator satiation has occurred and high levels of unused nectar have accumulated. Pollinator satiation can have important implications for plant reproductive outcomes, for example resulting in increased episodes of pollen transfer (as pollinators switch to an alternative food source) and subsequent out-crossing.

Sugar represents the major source of energy in nectar. The range of sugar concentrations reported from flowering plants typically varies from 25–75%, but may be as low as 5%. In bird-pollinated flowers the concentration of sugar is usually less than 20% but with typical ranges between 20–26% and extremes of 10–34% (Baker 1975; Cronk and Ojeda 2008). In Sumatran *Impatiens platypetala* (Balsaminaceae), pollinated by crepuscular hawkmoths, nectar contained 26% sugars, but three additional species, which were pollinated by day-active bees produced nectar of 34–39% sugar (Kato *et al.* 1991). The type of sugar can differ within individual plant families; Proteaceae generally have hexose-rich nectar but amongst *Banksia*, *Grevillea*, *Leucospermum* and *Protea* some of the species that appear to be more derived have sucrose-dominant nectar (Nicolson and Van Wyk 1998). The predominance of sugar types (e.g. sucrose, simple hexose sugars) has been found to differ in some plants with separate male and female flowers, and sugar composition is often associated with the kinds of anthophilous animals (bee, bird or bat) principally attracted to particular flowering plants (Baker and Baker 1990; Pyke *et al.* 2020). Bee-pollinated flowers, for example, characteristically have sucrose-rich or sucrose-dominant nectar (Cronk and Ojeda 2008). Even within closely related groups such as bees differential visitation may occur, depending on the types and composition of sugars present. For example, long-tongued bees have been noted to prefer nectar that is mainly constituted by sucrose. However, Perret *et al.* (2001), in a study relating nectar sugar composition to pollination syndromes in Sinningieae–Gesneriaceae, found that similarities of hummingbird and bee flowers contrasted with the presence of specific morphological traits associated with the two syndromes, indicating that plant–pollinator relationships rely on flower display rather than on nectar characteristics. In contrast, a distinct nectar chemistry was correlated with bat syndrome in which a particularly low sucrose production is responsible for hexose dominance.

Differences in concentration affect nectar viscosity, which in turn can have a significant impact on a given pollinator's ability to feed upon it (e.g. butterflies, which are restricted by their proboscis to imbibing less-viscous solutions), and thus acts to partition preferred or co-adapted pollinators. Flowers adapted to attract particular classes of pollinators may vary in volume produced and sugar concentration. Flowers adapted to bird pollination are known to produce larger quantities of nectar with lower concentrations of sugar, than flowers that are primarily bee-pollinated. This is likely an evolutionary response to the greater energy requirements of birds over that of individual insects (Johnson and Nicolson 2008). In Hawaiian *Metrosideros collina*, nectar produced at peak flow (which occurs during the earlier phase of flower opening) contains ~14% sucrose but mean concentrations of sucrose rise to ~32% in the last stages of nectar secretion (Carpenter 1976). The sucrose-rich nectar attracts insects and nectar of lower dilution is an apparent bird-pollination adaptation.

In addition to sugar, several other components can be found in nectar. These include amino acids (e.g. alanine and arginine) and essential oils, as well as small or trace amounts of alkaloids, antioxidants, lipids, pigments, polysaccharides, proteins and vitamins. The occurrence of several of these (e.g. protein) may be negligible. Cardiac glycosides may also be present. The amino acid content can change as a result of microorganism activity or alteration to pH, and considerable changes in pH levels may accompany accumulation of glucose (through the breakdown of sucrose). Amino acids and lipids have a calorific value and may serve as a direct source of energy. As well as being present in nectar (therefore originating from nectary exudates), lipids may also be secreted by separate oil glands. The presence of lipid films may help reduce the evaporation of nectar. Lipid presence has been considered to correlate slightly (though possibly significantly) with blossoms that are pollinated by bees and flies. The occurrence of small quantities of pigments in some nectars is of interest because these appear to serve as agents in the reflectance of ultraviolet, therefore facilitating a pollinator's ability to perceive the presence of nectar (enhanced nectar 'visibility') (Dafni 1992). Traces of antioxidants (e.g. ascorbic acid) serve to protect other nectar components from oxidation, and the role of small quantities of alkaloids relates to the deterrence of non-pollinating insects. The small polysaccharide component of nectars may contribute to higher viscosity, while the role of vitamins (typically riboflavin and thiamin) is uncertain but possibly as antioxidants. However, nectar chemistry can vary considerably, even between closely related plant species (Antoń *et al.* 2017).

Various species of certain angiosperm families (e.g. Melastomataceae, Orchidaceae and New World Maliphagaceae) possess small oil-secreting glands (elaiophores) that produce lipids, though little sugar. These lipids have been found to be chemically similar to lipids occurring in nectar. Elaiophores are found on different parts of the flower (e.g. on the epithelium, as trichome clusters and on anthers) and the oils they produce are attractive to several insects, notably tropical female bees of the tribe Anthophorini (Michener 2007). The female bees collect the oil in specially modified abdominal and tarsal (the jointed 'foot' at the end of each tibia) structures, mix it with pollen and provide the mixture to the brood cells upon which their larvae feed. Some male bees feed upon floral oil. Sometimes oil is used in nest construction, rather than as a food resource. Floral oils are a very high-energy source, greater than that of nectar. Oil extracts from various plants are similar: they are mainly saturated free fatty acids of like and sometimes identical composition. Other substances such as esters and paraffin compounds are present in floral oils, and further substances present in elaiophore exudates can include amino acids, carotenoids, glucose, glycosides, phenolics and saponins.

In addition to the primary food resources represented by pollen, nectar and oils, several secondary floral resources are available from various angiosperm species. These include various flower tissues such as staminodia and corolla segments (containing starches, sugars, lipids, proteins etc.), stigmatic exudates (sugars, lipids, amino acids and phenolics), resins (triterpenes) and perfumes (terpenes, and aminoid and aromatic compounds). Floral fragrances (perfumes) are sought by the males of Neotropical euglossine bees (Roubik 1989; Willmer 2011). They collect fragrances from orchids and aroids and store the perfume in the hind tibia. The function of this behaviour is not fully understood. Resins, as well as gums, are collected by female tropical bees of the tribes Anthidini, Euglossini and Meliponini; the bees use these in nest construction (Krahl *et al.* 2019). The resins also possess antibacterial properties and can inhibit the growth of fungi.

HEAT PRODUCTION IN ANGIOSPERMS

In addition to instances of thermogenesis in cycads, as already discussed, flowers that produce endogenous (internally generated) heat are known from only ~10 angiosperm families (Thien *et al.* 2000). Although most records are from basal families, heat production is well recorded from species of the more advanced Araceae (e.g. Maia *et al.* 2013). Reasons for such a function include an energy reward for pollinators, mimicry of odours attractive to faecal and carrion feeders, increased diffusion of carbon dioxide and volatilisation of chemicals (e.g. Kite and Hetterschield 1997; Seymour *et al.* 2003). Elevated levels of carbon dioxide in newly opened flowers may indicate resource profitability to foragers and elicit preferential visits by some taxa (e.g. hawkmoths), whilst an increase in the emission of volatile compounds has been shown to function to reduce herbivore attraction (Kessler and Baldwin 2001).

Classic examples of endothermic angiosperms are found within the tropical Rafflesiaceae, a family whose members are parasites living off their host plants. For example, the Oriental *Rafflesia tuan-mudae* maintains tissue temperatures of up to 6°C above the surrounding air temperature, while its relative *Rhizanthes lowii* achieves temperatures of 6–7°C above ambient (Patiño *et al.* 2000). Among the Neotropical Annonaceae, marked increases in floral temperature are associated with the diffusion of strong fragrances that attract respective pollinators. *Annona coriacea*, which is pollinated by large *Cyclocephala* scarab beetles, at dusk produces floral temperature peaks of ~34°C, ~15°C above ambient levels (Gottsberger 1999). Thermogenesis coupled with an adaptation to being pollinated by large beetles is well reported in the basal Nymphaeaceae; though there may be a broader taxonomic suite of pollinating insects involved (Hirthe and Porembski 2003; Seymour and Matthews 2006). In the thermogenic and strongly scent-emitting Neotropical understorey tree *Cymbopetalum brasiliense* (Annonaceae) however, Braun *et al.* (2011) did not observe any flower visitors during their study, and this led them to consider that fruit production was most likely via agamospermy rather than by self-pollination; an hypothesis seemingly at odds with the ecological purpose of endogenous heat production and the recruitment cue emittance of floral fragrances, both of which place high resource demands on the plant.

In the Araceae (as in species of Australian *Alocasia* and *Typhonium*) the inflorescence (the flower) is made up of an elongate spadix that consists of many small flowers. Partly enveloping the spadix there is normally a spathe (a large petal-like bract) (Plate 10), the base of which can trap small insects. In species with separate male and female flowers on the same plant the female flowers are nearer the base and the males are located towards the apex. The spadices of numerous species of Araceae are known to produce heat. In the Neotropical aroid *Taccarum ulei* (Araceae) beetles are attracted by the odouriferious inflorescences in the early evening of the first day of anthesis, during female phase. Emission of attractive volatiles is coupled with intense thermogenetic activity in the whole spadix, unlike other aroids in which certain zones of the spadix heat up; however, pollen release is not related to floral thermogenesis (Maia *et al.* 2013).

In the Australian aroid species *Alocasia brisbanensis* the spathe is white (Plates 9, 10). Although native *Tetragonula* bees (Plate 82), ants and various small flies (e.g. Neurochaetidae) are found upon the upper parts of the spadix the pollinators are mainly small beetles such as flattened staphylinids, nitidulids and weevils that are temporarily trapped, or out of preference gather, at the sheathed base of the spadix (Shaw and Cantrell 1982; G. Williams 1993). In the related Australian arum lily *Typhonium brownii*, pollinated by dung beetles (subfamily Scarabaeinae) (Monteith 1986), numerous tiny male and female flowers are located on the central unbranched spadix. Male flowers are located higher up the spadix and are separated from the lower cluster of female flowers by a dense collar of bristles. At the top of the spadix is a large sterile appendage, which becomes hot when the flower is emitting odours that are attractive to its beetle pollinators. The temperature rise of the appendage is greater than 15°C above ambient, significantly higher than that recorded for the better-known Rafflesiaceae and equalling that of *Annona coriacea*.

A second group of plants whose flowers create a temperature range markedly greater than that of the surrounding environment are those with large reflective surfaces. These do not generate heat but perform like small solar reflectors, capturing and enhancing the sun's heat. Such flowers benefit insects that bask upon them (Cooley 1995), and since insect pollinators need to allocate energy to maintain their own body temperature above the ambient temperature, floral heat might also be considered as an additional form of reward when other nutritional rewards are also available (Dyer *et al.* 2006). Dyer *et al.* (2006) considered that this may 'have implications for the evolution of specific floral structures and the connection between sensory cues, floral temperature and pollinator behaviour'. There are several records from the temperate and subarctic zones of the Northern Hemisphere, and indeed Hocking and Sharplin (1965) measured temperatures of up to 10.5°C, above air temperature inside flowers at Ellesmere Island, northern Canada. Flower temperature, and commensurate change in nectar temperature, has been shown to strongly influence pollinator foraging patterns. Shrestha *et al.* (2018b) found that bees preferred to collect warm nectar at cool temperatures but switched to cooler flowers at ambient temperatures above 30°C; though different flower responses may also constitute adaptations to local environmental conditions. Consequently, with respect to changing climatic conditions, flowers without a cooling mechanism may be less attractive to pollinators (Shrestha *et al.* 2018a).

However, it is difficult to imagine that flower visitors in tropical and subtropical rainforests are exposed to extreme cold where basking would greatly benefit daily activity patterns. Rather the reverse – midday temperatures can be quite hot and blossoms in the canopy suffer a reduction in abundance and variety of visitors, many of which seek shelter until the return of lower temperatures. Nevertheless, in tropical and subtropical forests basking behaviour is frequently exhibited by beetles. This might be because canopy-active beetles need to raise the temperature of their thoracic muscles for flight, and thus benefit from a period of basking (Crowson 1981).

A third group of flowers from which anthophilous insects may gain an energy benefit are those which provide shelter, providing an environment warmer than ambient temperatures (Dyer *et al.* 2006). Such microclimatic conditions might provide an opportunity to commence foraging on flowers sooner than other anthophiles. Flowers in this group are likely restricted to those with a depth effect in the perianth (tubular blossoms) or species whose flowers occur in dense clusters, allowing anthophilous insects to secrete themselves within the protection of the blossom mass. It is possibly for such comfort and protection that the larvae of some Australian species of the moth families (Ethmiidae and Hyblaeidae) build silken shelters on the flowers of rainforest trees and vines, from which they emerge to feed on adjacent foliage (Common 1990). In such instances the larvae are acting as herbivores, not as pollinators.

FLOWERING PLANTS AS BREEDING SITES FOR POLLINATORS

Some cycads (e.g. endemic *Macrozamia communis, Lepidozamia peroffskyana* and New World *Zamia furfuracea*) are known to provide breeding sites in which their pollinators are able to complete their life-cycles (e.g. Chadwick 1993; Hall *et al.* 2004). Likewise, the flowers of several angiosperms provide breeding sites for their pollinators (e.g. Ishida *et al.* 2009; Williams *et al.* 2001). Known examples include species that are considered primitive (*Idiospermum*) and those that are classified as occupying later stages of evolutionary development (*Macaranga*). Such sites have been viewed as rewarding the pollinators.

Several groups of pollinators are known to breed in the flowers of angiosperms. A number have been subject to a body of research undertaken by Shoko Sakai of the Smithsonian Tropical Research Institute in Panama, and by many other pollination ecologists worldwide. The three primary groups that Sakai (2002a, 2002b) recognised are those that parasitise plant ovules, parasites of pollen, and a group comprising pollinators which undergo larval development of young in decomposing flowers after pollination has been achieved. Two further groups are pollinators whose larvae feed upon nectar while developing within flowers, and those that meet and copulate on flowers (various buprestid, cantharid and melolonthine beetles). Other ecologists have recognised the intermediary role of flowers in the reproductive scheme. Although copulation is normally restricted to the time when they are active on flowers, pollinators generally complete their life cycles unassociated with flowers, depositing their eggs in bark, detritus or soil substrates.

In rainforests, including those of Australia, the best-known example of the first group is the mutualism existing between figs (*Ficus*, e.g. Plates 49, 50) and their fig wasp (Agaoninae-Agaonidae, Plate 87) pollinators (e.g. Lopez-Vaamonde *et al.* 2002; Wiebes 1979). Examples from exotic plants of non-rainforest habits include *Yucca* (Agavaceae) and *Lithophragma* (Saxifragaceae) by moths (Baker 1986; Thompson and Pellmyr 1992), and *Trollius* (Ranunculaceae) and *Pterostylis* (Orchidaceae) by flies (Pellmyr 1989; Thalwitzer *et al.* 2018). In *Yucca* the mutualism is known to be mediated by chemical cues [Svensson *et al.* 2011]). Chemical recruitment cues have also been postulated for *Pterostylis* (Thalwitzer *et al.* 2018). Obligate mutualisms have been recorded between the tropical and subtropical tree genus *Glochidion* (Phyllanthaceae/Euphorbiaceae) and small moths of the genus *Epicephala* (Gracillariidae) that are its pollinators, but whose larvae are also seed predators (Henderson *et al.* 2019; Kato *et al.* 2003; Kawakita *et al.* 2004). Species of *Epicephala* are also considered to be pollinators of *Breynia oblongifolia* (Finch *et al.* 2018). For ovule parasites, the incidence of pollinator specificity to the host is mainly at the one-to-one level (as in *Yucca* and *Ficus*), but this highly co-evolved category of breeding site relationship represents only a small proportion of overall records for the three groups. In the species of *Glochidion* that have been studied, nocturnally active female *Epicephala* moths deposit pollen onto the stigma and use a long ovipositor to lay their eggs into the style. The structure of the style is highly specialised, with the stigma hidden in a pit at its tip. The specialised stylar formation, and the nature of the moth's morphological and behavioural adaptations for oviposition, have been conjectured as driving speciation in both *Glochidion* and *Epicephala*. However, co-pollinator presence and predator dynamics (e.g. occurrence, abundance, daily co-activity patterns) can affect relationship outcomes. In *Lithophragma*, a suite of pollinators includes species that are not ovule parasites, so the latter are not necessary to achieve pollination (Pellmyr and Thompson 1992). Thus no highly specialised co-dependency between its moth pollinator and ovule parasite (*Greya*) has evolved, unlike in *Yucca* and *Ficus*, which is generally dependent upon a single pollinator taxon (though some pollinator taxa are recorded as pollinators of more than one species of fig).

In the second group (the pollen parasites), larval and adult parasites eat pollen directly from the flowers. This includes pollen that is still held upon the anthers or that has been deposited as pollen dust on surrounding petals or sepals. This group of pollinators is almost exclusively composed of thrips (Sakai 2002a), and some groups of small beetles (e.g. Curculionidae-Acalyptini/Derelomini, Nitidulidae-*Brachypeplus*, and Staphylinidae on *Alocasia brisbanensis* in subtropical rainforest [G. Williams 2020]), though many of the thrips species recorded from flowers are not necessarily pollinators. Thrips observed on flowers fall into two subcategories – those that consume a percentage of the flower's available pollen but serve as potential pollinators (even if the number of grains carried by each thrips individual is few), and those that feed on pollen but do not necessarily undertake flight patterns or exhibit behaviour that results in some level of pollination for the individual plant or plant species. Other insects, such as small chrysomelid beetles, also fall into the category of insects that can serve as pollinators (e.g. Annonaceae, Achariaceae) and are consumers of displaced pollen. These include several chrysomelid genera (e.g. *Crepidodera, Ditriopidus, Monolepta, Rhyparida*) commonly encountered on the flowers of Australian subtropical rainforest trees (G. Williams 2020).

Thrips (*Neoheegeria*) are pollinators of dioecious *Macaranga hullettii* (Euphorbiaceae) in West Malaya (Moog *et al.* 2002), to which they are apparently attracted primarily by olfactory cues, and breed in the floral bracteoles, these having openings along their margins that allow entry only by small insects. In this instance adults and larval instars of both sexes feed on nectar within hairy patches on the adaxial surface of the floral bracteoles. In previous cases of thrips pollination pollen and or floral tissues have been recorded as rewards, but not nectar or nectar-like substances.

Historically, thrips associated with flowers have been disregarded as fulfilling any substantial role as pollinators or have been regarded as flower pests and pollen predators, with no level of beneficial symbiotic parasitism. However, pollination by thrips is now recognised in numerous angiosperm families including Araceae, Dipterocarpaceae, Euphorbiaceae, Lecythidaceae, Moraceae and Myristicaceae. Thrips are also pollinators of the basal angiosperm families Annonaceae, Calycanthaceae, Monimiaceae and Winteraceae, and are commonly found on horticultural specimens of *Magnolia*, where they may be the only flower visitors. Thrips have fast reproduction rates, often reaching the adult stage in less than two weeks (e.g. *Neoheegeria* in 16–18 days [Moog *et al.* 2002]). Consequently they can establish very large populations in a single season, from a small founding population. In Australian rainforests there are numerous plant taxa with enclosed flowers that provide sheltered spaces for thrips, the deposition of their eggs and later larval development (though their mobile immature stages are more correctly wingless nymphs that resemble the adults, rather than larvae). They include species of *Myrsine, Breynia, Wilkiea, Pittosporum, Idiospermum, Anopterus* and members of the Lauraceae; though moths (*Epicephala* – Gracillariidae) are also recorded as pollinators of *Breynia oblongifolia* ([Finch *et al.* 2018). In some instances thrips are able to complete their life-cycles within the flower while it is still upon the plant, in others eggs and immature thrips are carried to the ground when the flowers are shed. In *Pittosporum revolutum*, the pollinators include species of nitidulid beetles (G. Williams 2020). These live deep within the protective tubular perianth, where they consume some of the pollen but also transfer pollen between flowers of the same and different plants.

Plants whose flowers provide post-pollination breeding sites in shed and decomposing flowers (Sakai 2002a) are a diverse group. However, it is not always possible to clearly differentiate between insect pollinators that occupied flowers before abscission and those that entered shed inflorescences (e.g. *Brachychiton acerifolius* and *Brachychiton discolor* [Malvaceae/Sterculiaceae], Plates 42, 43) when these litter the forest floor below the parent plants (see plate in G. Williams 1993). There is a further division – some shed flowers retain the immature pollinators throughout their full development, while in others the immature stages leave the fallen flower and enter the upper part of the soil profile to complete their life cycle. In *Hibiscus splendens* and *Hibiscus heterophyllus* the pollinators are small beetles (usually Nitidulidae) and drosophilid flies (e.g. *Scaptodrosophila hibisci*) (Plate 45), even though the flowers are individually very large. Flowers open sequentially, and normally close after being unfurled for a single day. They later fall to the ground, carrying some of the nitidulids remaining within. The location of the final maturation of their immature stages, and emergence as adults, is not fully understood. *Scaptodrosophila hibisci* and related drosophilids also breed within various *Hibiscus* flowers (McEvey and Barker 2001), but for *Scaptodrosophila hibisci* the host plant association does not appear to extend to other plant genera. Drosophilids also use the inflorescences of several *Alocasia* species (Araceae) as breeding sites, as in the case of *Alocasia hollrungii* from Papua New Guinea (Okada and Carson 1980). However, they are normally not pollinators (being unassociated with the female part of the floret-bearing spadix) and the flower body is not shed upon pollination and fertilisation. Instead, the spadix is retained as seeds develop.

In Eupomatiaceae, the floral segments fall after fertilisation as a single unit (the synandrium). Pioneering studies on the pollination ecology of the Eupomatiaceae were undertaken in the 19th century by Alexander Hamilton (1897), with detailed observations of a detailed nature far in advance of his time. The Eupomatiaceae are pollinated solely by weevils of the genus *Elleschodes* (Plate 35). These deposit their eggs within the synandria, from which the larval weevils emerge following abscission then enter the soil to complete their development. Other flowers in which pollinating insects breed following pollination include those of exotic *Siparuna* (Monimiaceae/Siparunaceae), *Aristolochia* (Aristolochiaceae), *Balanophora* (Balanophoraceae), *Artocarpus* (Moraceae) and several palms (Feil 1992; Kawakita and Kato 2002; Sakai 2002a, 2002b; Sakai *et al.* 2000). The West African root parasite *Thonningia sanguinea* (Balanophoraceae) is pollinated by a species of *Morellia* (Muscidae) (Goto *et al.* 2012). Adult *Morellia* flies lay their eggs in *Thonningia sanguinea* inflorescences but the latter emerging and developing larvae feed only on the decaying inflorescence vegetative tissue. However, non-pollinating insects also breed in flowers and may be closely related taxonomically to the true pollinators. The zygomorphic flowers of exotic *Aristolochia* vines are pollinated by numerous Diptera, including the families Drosophilidae, Phoridae, Chloropidae, Syrphidae, Milichiidae, Agromyzidae, Sphaeroceridae and Anthomyiidae (Bänziger and Disney 2006; Sakai 2002a). The related Australian subtropical rainforest species *Pararistolochia praevenosa* (Plate 15), a foodplant of the threatened birdwing butterfly *Ornithoptera richmondia*, is pollinated by species of *Forcipomyia*, small midges of the family Ceratopogonidae (Nematocera) (Sands *et al.* 1997) ('Nematocera' are now considered an un-natural grouping but are retained here for convenience). The flowers of Panamanian rainforest *Aristolochia maxima* and *Aristolochia inflata* last about two days, being functionally female on the first day and male on the second. On the second day the perianth of both species undergoes abscission; their pollinators (drosophilid and phorid flies respectively) are carried to the ground and breed within the fallen flowers (Sakai 2002a).

An unusual mutualism is found in *Artocarpus integer* (Moraceae) from Sarawak, in which the plant and its pollinating gall midges (Cecidomyiidae) participate in a 3-way mutualism involving a species of *Choanephora* fungus (Mucorales, Zygomycetes) (Sakai 2002a). The midges are attracted by an unpleasant odour emitted by both male and female flowers but only stay on male flowers, where they lay their eggs. The fungus grows on the plant's male flowers, and the midge larvae feed upon the developing fungal mycelia. The larvae pupate within the *Artocarpus* inflorescence and the emerging adults move to other flowers.

A further twist in the role of flowers as feeding and breeding sites is the vectoring of yeasts by flower-visiting insects (see Basukriadi *et al.* 2010; Pimentel *et al.* 2005), and these can induce changes in the nature of nectar sugar, reducing the chemical control of the plant over this important floral trait (de Vega and Herrera 2013). Nectar-inhabiting microorganisms produce volatile compounds which can differentially affect bee preferences, with the presence of microbes providing volatile cues that may influence plant–pollinator interactions more generally (Rering *et al.* 2018). Ants, nitidulid beetles and drosophilid flies are known yeast vectors, but pollinating bees can also carry yeasts; as in the Asian *Apis cerana* from which associated yeasts include *Aureobasidium, Candida, Metshnikowia, Kodamaea, Yarrowia* and species of Dothoriaceae. However, nectar-collecting bees and pollen-collecting bees may possess different and specific yeast communities (Basukriadi *et al.* 2010). Beetles and flies vector highly specific yeast communities to flowers that then serve as food for their larvae, though in some instances a particular yeast species may be carried by bees, flies and beetles (*Candida restingae* isolated from *Hylocereus costaricensis* [Cactaceae] in Costa Rica [Lachance *et al.* 2001]). The establishment of the yeast community generally occurs within several hours, with yeast cells eventually distributed over the (usually inner) surface of the corolla. Concentration of yeast cells can also occur, for example in proximity to floral nectaries. In this example flowers provide a setting that is apparently conducive to the constitution of a yeast community, but the insects are the sole founders of their children's meal.

ATTRACTION OF THE COMELY SHAPE: ORCHID FLOWERS AND BARREN ILLUSION

The flowers of some rainforest plants attract pollinators by deceitful fragrances and odour. Others offer false and beguiling form and colour as a lure. Such methods of deceitful attraction are found most notably among the Orchidaceae (Pemberton 2013; Willmer 2011), in which family the phenomenon of pseudocopulation by male tynnid wasps and ants has been recorded (Peakall *et al.* 1987, 2010). The pseudocopulation pollination syndrome involves the attraction of male insects (usually of the orders Hymenoptera and Diptera, and occasionally Coleoptera) that are deceived by the flower's shape or fragrance and are stimulated into sexual behaviour.

Attributes that singly or in combination characterise sexually deceptive orchids are the lack of nectar and collectable pollen rewards, flowers that tend to be similarly coloured to the insects they are seeking to attract, flowers that emit fragrance stimuli (kairomones) that apparently mimic the pheromones of female insects, and a flower labellum that mimics (in shape and colour etc.) the female. All this elicits copulation-like behaviour by the male. Orchids employing these adaptations usually participate in a highly specialised pollinator relationship, often species-specific or involving closely related pollinator taxa. Because of this adaptive specialisation there is a relatively pronounced level of reproductive isolation (being pollinated by a single species), and even orchids that are morphologically difficult to distinguish on the basis of gross structure and form may be reproductively isolated due to subtle variation in kairomone chemistry.

Among the Australian orchids there are several instances of flowers that physically mimic their wasp pollinators, luring them by shape and a false hope of copulation with a mate. The wasp unwittingly finds partnership only with the orchid flower, and only the orchid achieves reproduction. Australian orchids known to participate in pollination strategies based upon sexual deceit are in open vegetation habitats usually removed from rainforests (e.g. Indsto *et al.* 2006; Peakall 1990; Reiter *et al.* 2019). Typical orchid genera are *Caladenia* and *Chiloglottis*, which lure males of the wasp family Thynnidae (e.g. *Neozeleboria, Zaspilothynnus*), though these and other terrestrial orchids are frequently encountered on rainforest margins and sunlit gaps below damaged canopies and along stream lines. Elsewhere, pseudocopulation strategies are known in *Ophrys* (in Europe), which is pollinated by bees and scoliid and sphecid wasps, and in *Trichoceros* and *Trigonidium* (in Central and South America) which are pollinated by tachinid flies and *Plebeia* bees (as '*Trigona*') respectively (Singer 2002; Willmer 2011). *Trigonidium obtusum* is a sexually mimicking orchid which is pollinated by sexually excited *Plebeia droryana* bees (pseudocopulation). *Plebeia droryana* is a small bee, seemingly too small to be considered the likely pollinator without close observation of its behaviour (Singer 2002).

The promise of food, where none exists, is also a common strategy employed by the Orchidaceae. In the epiphytic and deceptive lady slipper orchid *Paphiopedilum villosum*, from the rainforests of northern Thailand, pollinators are syrphid flies (e.g. species of *Episyrphus, Betasyrphus* and *Syrphus*) lured by the expectation of finding food in the orchid's barren flowers, which do not produce nectar (Bänziger 2002). Long-distance attraction is thought to be due to the urine-like odour emitted by the plant, with close-range attraction possibly to shining staminodes that may mimic droplets of honeydew. Upon each staminode is a slippery wart-like perch; flying towards and landing upon these, the flies fall into the pouched flower structure. The only exit is past the stigma and thus contact with, and removal of, sticky pollen. There is no food reward within the flower, just a cruel tumble and a difficult extrication.

Though the example of *Paphiopedilum villosum* is not an Australian one, it is mentioned so as to highlight the highly specialised co-adaptive relationships between plants and insect pollinators in the rainforest Orchidaceae, which field ecologists may also discover more widely in Australian rainforest communities. In the synchronously mass flowering Australian wet forest orchid *Dendrobium speciosum* potential pollinators (bees) are attracted by the sweetly aromatic large normally cream-coloured inflorescence displays (Slater and Calder 1988). The flowers are highly fragrant, and though hermaphroditic they do not produce nectar, the attraction being one of deceit. Nectar-seeking insects are guided to the central reproductive area of the flower by the colour gradation of the perianth. This includes an area of high ultraviolet reflection near the centre, and a bright yellow ridge along the labellum. A tube formed by the labellum and column further directs potential pollinators.

Such spectacular relationships are the exception rather than the rule among flowering plants and their pollinators, and are ecologically dangerous if one part of the relationship is more vulnerable than the other. *Paphiopedilum villosum*, cited earlier, is an example of the threat to interdependencies posed by the collection of rare plants. Hans Bänziger of Chiang Mai University, who has studied the pollination ecology of many Thai orchids, and noted the threat of extinction through habitat destruction and poaching, has to undertake fieldwork with the greatest secrecy lest his plants be discovered and collected. Similarly, Bower *et al.* (2015), in their study of the Australian Tuncurry Midge orchid (*Genoplesium littorale*), a threatened species greatly restricted in distribution in New South Wales, were obliged to mask the location of their study site.

The 'emergence' of orchids

The Orchidaceae are one of the two largest plant families (the other being the Asteraceae/Compositae), with ~25 000 species worldwide, of which ~700 species are recorded from Australia. The family includes taxa that in Australian rainforests and wet sclerophyll forest are terrestrial (e.g. *Calanthe* and *Pterostylis*), epiphytic (*Bulbophyllum, Dendrobium* and *Rhinerrhiza*, saprophytic (deriving nourishment from dead animals) (*Rhizanthella* and *Epigonium*), lithophytic/epilithic (growing on rocks) (*Oberonia* and *Sarcochilus*) and a few species that are climbers (*Erythrorchis* and *Pseudovanilla*). Although the Orchidaceae exhibit highly specialised flower structures and pollination adaptations their fossil record is poor and provides little insight into their evolutionary history. The most recent common ancestor of living orchids is hypothesised to have been present in the Late Cretaceous. The great diversification of the family occurred soon after the mass extinction event at the K/T (K-Pg) boundary. A fossil pollen aggregate (pollinarium, pollinium) of *Meliorchis caribea* has been found attached to an extinct stingless bee in Miocene amber (c. 15–20 mya) collected in the Dominican Republic (Ramírez *et al.* 2007).

FLOWERING PLANTS THAT MIMIC DEATH

'The taste and the smell were inextricably confounded,
and became one sentiment, abnormal and intense.
The rose-water with which your tenderness had moistened my lips to the last,
affected me with sweet fancies of flowers – fantastic flowers,
far more lovely than any of old Earth,
but whose prototypes we have blooming around us.'

The Colloquy of Monos and Una (Edgar Allen Poe)

If the 19th century American writer Edgar Allan Poe had a hand in the design of flowers and floral syndromes this section would be his domain – one of grim allurement by plants whose very lives are fundamentally embedded with those of unwilling hosts, from whom they extract sustenance and support. Within this group of plants reside the huge short-lived flowers of parasitic *Rafflesia* (Rafflesiaceae), a genus of Oriental distribution, whose flowers are renowned for a stench akin to that of rotting flesh; and to which calliphorid flies are attracted. Though not producing nectar, *Rafflesia* does offer food resources to the attending Calliphoridae flies in the form of tissue secretions and excess pollen.

Rafflesia is not alone in its peculiar pollinator recruitment strategy. The related *Rhizanthes infanticida* from southern Thailand goes a step further in the world of deathly and perverse deception (Bänziger and Hansen 2000). As with other Rafflesiaceae, it is a parasite (of *Tetrastigma* [Vitaceae]) but unlike *Rafflesia* does not emit any obvious odour of decay. Its trick is to apparently mimic a mammalian carcass or parts thereof – the dense mat of hairs that cover the flower resemble mammalian fur and the overall circular flower structure is shaped like an orifice. Again it is carrion flies, mainly *Lucilia* and *Chrysomya*, which are the pollinators of *Rhizanthes*. They settle on the flower and pollen is smeared upon the flies' backs. This clots and hardens. When the flies are attracted to female flowers the pollen mass is reliquefied and collected by fluid-soaked papillae on the surface of the stigma. The pollinating calliphorids lay many eggs within the flowers of *Rhizanthes infanticida*, but the emerging fly larvae are doomed to die of starvation and predation by ants. There is nothing within the floral incubus for them to sup upon and complete their life cycles. In luring its gravid pollinators by deceit, *Rhizanthes infanticida* conspires in the murder of new-born children, as its name implies.

Flesh flies (Sarcophagidae) are also attracted to the flowers of *Rhizanthes infanticida* but are not its pollinators; they are redundant. Instead, flesh flies play the role of pollinator on another member of the Rafflesiaceae, *Sapria ram*. *Sapria ram* occurs in Thailand and is a spectacular brilliant-pink parasitic angiosperm, the inner surfaces of its splayed urn-shaped flowers flecked with white blotches. As in *Rafflesia*, *Sapria ram* emits an odour mainly evoking that of a decaying corpse, and to a lesser degree like that of faeces (Bänziger and Pape 2004). Flies of the genus *Sarcophaga* are attracted to the odorous flowers, either (depending on the species) in the hope of finding a putrid carcass or excrement on which they can oviposit, or in expectation of finding food. But as with *Rafflesia* and *Rhizanthes*, *Sapria ram* is a deceiver that provokes unfulfilled desire; a chimera, which by brood site or food deception, solicits from *Sarcophaga* the function of pollinator.

The Rafflesiaceae do not occur in Australia, and though we are not without flowering plants that are parasites or semi-parasites attached to the roots of hosts (e.g. *Exocarpos*, *Leptomeria*, *Santalum* [Santalaceae], *Nuytsia* [Loranthaceae]) these are not rainforest species. However, Calliphoridae and Sarcophagidae do occur as pollinators on many mass-flowering rainforest trees. In subtropical rainforest several large metallic coloured species of *Amenia* and *Paramenia* (Calliphoridae) occur, and characteristic anthophilous calliphorid genera also include *Chlororhinia* and *Stomorhina*[1] (Plate 98) (G. Williams 2020). However, the life histories of these, and their sarcophagid cohorts, are poorly documented.

At least two other families of angiosperms include species whose flowers mimic the dead – the Austrobaileyaceae and Aristolochiaceae. The protogynous flowers of *Austrobaileya scandens*, a species that grows in low-light forest understorey habitats (Feild *et al.* 2003), smell like decaying fish, the smell being produced by the stamens and inner staminodes, and this brings visiting flies to oviposit on them (Endress 2001). The Aristolochiaceae are a family with ~400 species found in the world's tropical and subtropical regions (Gonzalez and Poncy 1999), extending southwards in Australia to the rainforests of northern New South Wales (Plate 15) (Harden *et al.* 2007). Flowers of *Aristolochia* are protogynous and their pollinators are mainly small brachyceran flies (e.g. Bänziger and Disney 2006). Pollinators small enough to fully enter the flowers are trapped during the first day when plants are in female phase and anthers are closed.

1. Both genera previously Calliphoridae-Rhiniinae, now Rhiniidae. See also Table A1.3, and caption 98 (p. 218).

They are released (pollen emplaced upon their thorax) on the second day when the stigmas are no longer receptive and anthers are dehiscent. In northern Thailand is the cauliflorous *Aristolochia baenzigeri*, a species that tempts its pollinators by smell. In freshly opened flowers the floral odour is weak but as flowers draw near the end of their life their odour becomes more discernible and unpleasant. Newly fallen flowers smell offensive, distinctly cadaveric. Pollinators of *Aristolochia baenzigeri* are mainly Phoridae (e.g. *Dohrniphora cornuta*), and while larger Sarcophagidae are attracted to flowers they are unable to enter the depths of the flower structure where the stigma is hidden (Bänziger and Disney 2006). Phorids approach flowers, orienting along the scent plume until visual perception provides short-range alignment to the flower. Upon landing on the slippery inner surface of the flower their grip is lost and they tumble into the floral tube. They escape once the flower falls to the ground: the flower's horizontal position upon the forest floor is no longer an obstacle to freedom and the flies are finally able to gain purchase within the previously slippery chamber. The attraction of *Aristolochia baenzigeri* flowers to pollinators is not nectar but the promise of products of decay, upon which they or their young might feast. As with *Sapria* and its kin, no meal of animal decomposition awaits.

Although phorid flies are unrecorded from Australian species of Aristolochiaceae, they are part of the broad diversity of brachyceran flies found on rainforest flowers with simple open actinomorphic flowers. In subtropical rainforests Phoridae have been recorded from Myrtaceae (*Syzygium smithii*, *Syzygium floribundum* and *Tristaniopsis laurina*), Sapindaceae (*Alectryon coriaceus* and *Guioa semiglauca*), Rhamnaceae (*Alphitonia excelsa*), Escalloniaceae/Rousseaceae (*Cuttsia viburnea*), Anacardiaceae (*Euroschinus falcatus*), Cunoniaceae (*Schizomeria ovata*) and Flacourtiaceae/Salicaceae (*Scolopia braunii*) (G. Williams 1995). They are associated with the general guild of small insects (pollinators or not) that, *en masse*, attend to flowers.

DECIDUOUSNESS AND ITS BENEFITS TO POLLINATION

Worldwide, foliage deciduousness in angiosperms is believed to have arisen in tropical species as a response to periodic drought, to which rainforest trees have varying tolerances (Choat *et al.* 2012; González-Varo *et* al. 2013). From these warm areas, the condition of inherited deciduousness mainly spread north to regions where intense winter cold seasonally locks up available water. The pronounced leaf drop of trees in these regions may be related to this lack of water, reduced rainfall during the winter months and the vulnerability of broad leaves to frost damage. In a phenological study in secondary forest in West Sumatra, Chairul *et al.* (2010) noted that smaller trees shed leaves simultaneously during drought, and larger trees had highly synchronised leaf fall during the peak shedding period. The deciduous habit may be particularly characteristic of trees occupying the upper canopy (e.g. Justiniano and Fredericksen 2000), suggesting leaf loss may be related to water stress (Archibold 1995), but Wright and Cornejo (1990) found that in Panamanian mature rainforest (for trees, saplings and lianas) plant water status alone did not control leaf fall (nor cue flowering), though they cautioned that water status may impact differentially on forest life forms (i.e. trees, shrubs, lianas, herbs).

Almost all trees found in tropical savannah communities are deciduous, losing their leaves during the dry season and commencing flowering while still leafless; the bottle-trunked boab tree, *Adansonia gregorii* (Malvaceae/Bombacaceae), of north-western Australia, being a conspicuous deciduous tree of savannah woodlands (Plate 116). In northern Australian vine forests (often nested in a savannah matrix), especially coastal thickets, over 90% of canopy trees may be deciduous (Russell-Smith and Dunlop 1987), however, the Southern Hemisphere as such has little temperate deciduous forest – though such forests are extensive in the Northern Hemisphere where, as with savannah-inhabiting trees, flower production before leaf growth in many deciduous tree species is a well known phenomenon.

But relatively few Australian rainforest angiosperms are truly deciduous; although heavy leaf fall may be seasonally experienced among normally evergreen species in response to drought conditions (Melick 1990), and can be severe, threatening mortality, in species such as *Glochidion ferdinandi*, *Cryptocarya glaucescens* and *Archontophoenix cunninghamiana* (Plate 11) (G. Williams pers. obs.). However, there are several deciduous or semi-deciduous tree species. In subtropical rainforests these include *Brachychiton discolor*, *Cassia marksiana*, *Croton insularis* (Plate 34), *Ehretia acuminata*, *Ficus henneana*, *Flindersia australis*, *Gyrocarpus americanus* and *Toona ciliata*, the latter playing a key role in the exploration of rainforest because its timber is valuable.

Such instances relate to leaves and their disposal; although some species retain floral structures long into fruit maturation (e.g. stamens and staminodes, as in *Idiospermum*), deciduousness can also relate to the shedding of anthers and stigmas. Among Australian rainforest trees this phenomenon has received little attention and record. Loss of anthers is to be expected in the female phase of bisexual flowers and extreme examples of anther shedding are documented from *Impatiens* (Balsaminaceae) (see Schemske 1978; Willmer 2011), a tropical and subtropical genus not native to Australia, though introduced and now widely naturalised. Ready abscission of numerous anthers is also a common phenomenon in bisexual flowers following pollination, as in *Syzygium floribundum* and *Tristaniopsis laurina* (in the latter the stamens are fused clusters which are shed as whole units) (G. Williams 1995). In some taxa the style is shed before the male phase, as recorded for members of the Annonaceae. In Annonaceae, shedding of individual petals allows the release of trapped pollinators, allowing them to fly to alternate receptive flowers (Neotropical *Annona crassiflora, A. coriacea*) (Gottsberger 1999). In species in which the whole perianth unit is lost following fertilisation (Australian *Hibiscus splendens, Pittosporum revolutum, Eupomatia laurina, Archontophoenix cunninghamiana* and *Diospyros australis*, or the Neotropical genus *Chorisia* [Bombacaceae]), any redundant stigmas (or ornate stigmatic structures) and any residual attached anthers are shed, their reproductive function having been served or the opportunity for fertilisation expended.

Flower production and anthesis, before the onset of new seasonal foliage growth, is common in non-deciduous rainforest angiosperms. Flowering before onset of foliage production in deciduous rainforest trees is less commonly recorded but examples include *Brachychiton acerifolius* (Malvaceae/Sterculiaceae), whose bright red flowers are produced in spring, usually on leafless branches (Plate 42). In the monsoon forests and vine thickets of northern Australia resides the deciduous *Cochlospermum gillivraei* (Cochlospermaceae), whose large yellow flowers are normally borne June to September (again, usually when trees are fully leafless), as well as deciduous *Bombax ceiba* (Malvaceae) the red flowers of which appear in spring before new leaf flush (Beard 1990). Also from Australia, and encountered in a diversity of vegetation communities from dry rainforest to desert, is *Erythrina vespertilio* (Fabaceae) (Boland *et al.* 1984). This frequently deciduous species often produces conspicuous red-orange flowers before foliage is present. *Erythrina vespertilio, Bombax ceiba* and *Cochlospermum gillivraei* behave like typical flowering savannah angiosperms recorded elsewhere in the world. The tree *Grevillea robusta*, restricted to southern Queensland and northern New South Wales, is partly deciduous, being nearly leafless before the blooming of its massed orange-yellow flowers.

Deciduousness has inherent costs. These are incurred through loss of carbon and other nutrients due to leaf fall, and the inability to fix carbon during a substantial part of the year. The benefits include the avoidance of foliar drought and a reduction in the requirement to maintain more extensive root systems and the avoidance of sustained herbivore attack throughout the dry season (see Atwell *et al.* 1999). So what are the benefits to flowering before the onset of new foliage in rainforest trees? In closely spaced Northern Hemisphere trees where wind-pollination takes place, the benefits are obvious – leaves would be an obstacle to efficient pollen transfer to receptive flowers. However, populations of deciduous trees in Australian rainforests are usually well separated, as in all floristically complex rainforests. One possible answer is that deciduous rainforest trees, when flowering in the absence of leaves, conspicuously advertise their presence and so enhance their ability to attract pollinators over greater distances.

4 Introduction to breeding systems

INFLUENCE OF BREEDING SYSTEMS

Many tropical and subtropical trees and shrubs are characterised by wide geographical and altitudinal distributions, but at individual sites species often have low population densities. Due to the fragmentation and spatial reduction of rainforests, especially lowland tropical and subtropical rainforests, the local populations of many species are now very small, in some cases reduced to single individuals per stand (Fox *et al.* 1997; Hobbs and Yates 2003; Neal *et al.* 2010).

Knowledge of the reproductive biology/breeding system of species is an important prerequisite to understanding their resilience to environmental change (extinction risk), and changes in population density and pollinators (Gauli *et al.* 2013; Sjöstrom and Gross 2006). This is a scenario not restricted to rainforest plants, and although mating systems of plants may vary depending on their pollination biology and life history, as plant populations of any vegetation community get smaller there is likely a progressive trend towards increased inbreeding, smaller effective sizes of pollen pools, and greater variation in out-crossing. Whether a species has the ability to self-fertilise and whether there is plasticity in requirements for pollinators (i.e. numerous alternative species can fulfil the role) are particularly important to a species' ability to successfully colonise new sites, to survive in highly disturbed remnants, and is highly relevant to the management and preservation of remnant rainforest stands. Rainforest plants that cannot reproduce vegetatively, and obligate out-crossers that are incapable of producing seed by selfing, are unable to establish colonies from single recruits. Where such populations are reduced to single individuals and are isolated from external pollen sources, the species are functionally extinct within individual remnants.

The requirement for a pollination vector is dependent upon the particular breeding system of a plant species. Essentially, there are three main categories of breeding systems in plants (though overall angiosperm pollination strategies comprise a continuum). These are represented by obligate out-crossing species (dioecious and self-incompatible hermaphroditic species), facultative out-crossers (monoecious, gynomonoecious, protandrous and protogynous species) including some self-compatible hermaphrodites, and facultative inbreeders (self-compatible and apomictic plants) (Gross 2005; House 1985). In the third group are some apomicts (discussed below), in which inbreeding is obligatory. Self-pollinating plants are less reliant on pollinators, however, pollinators may remain heavily dependent on the plant for pollen and/or nectar (Spigler and Kalisz 2017).

Approximately 50% of flowering plants show some level of self-incompatibility (Gibbs 2014), with obligate out-crossing, due to self-incompatibility, being reported from more than 100 angiosperm families and estimated to occur in ~39% of species (Igic *et al.* 2008). However, self-incompatibility frequently has been lost during the evolution of flowering plants; and there is no evidence that once this has been lost it has been regained (Igic *et al.* 2008). For individuals of obligate out-crossing species within plant populations that are widely separated from one another, inadequate receipt of pollen and commensurate limitation of female success (fruit- and seed-set) may be common. Studies in various foreign rainforests have indicated that out-crossing generally predominates (e.g. Murawski *et al.* 1994); in Brazilian tropical dry forest 83% of all growth forms were hermaphrodites but obligate self-incompatibility occurred in more than 60% of the 147 species studied (Machado *et al.* 2006). However, the spatial isolation of trees may influence the degree of out-breeding and fecundity (lower levels of fertilisation and less seeds produced) (House 1985, 1992). This is the situation for Australian subtropical rainforests, but many bisexual species retain some capacity for self-pollination (Table 4.1). This ability may be important for the survival of biodiversity in the face of continuing disturbance to rainforest, and may explain rainforest and species survival during periods of naturally rapid environmental change. Historically, however, the level of self-incompatibility and requirement for out-crossing in tropical rainforest trees has been a contentious topic. Fedorov, discussing in 1966 the structure of tropical rainforest and speciation in the humid tropics, suggested that out-crossing was rare (due to the spatial isolation of species in complex tropical rainforests) and that self-compatibility was general among rainforest trees. Corner, in 1954, argued that selfing is the tendency of all trees even though it will tend to reduce heterozygosity in populations. Ashton (1969) partly summarised Fedorov's case, taking into account the low densities of rainforest species, as 'given the lack of marked seasonality which leads to irregularity and lack of coincidence in flowering' and assuming that pollen dispersal in temperate and tropical regions was equally efficient, 'then the effective populations in tropical tree species must contain on average fewer individuals than their temperate counterparts, and indeed, self-fertilisation must be the rule'. He later argued that, if Fedorov (1966) and Corner (1954) were correct, dioecism and other breeding systems promoting out-crossing would be eliminated. However, Ashton (1969) showed that contrary to this prediction and based upon his own field investigations, that a significant proportion (26%) of tree species in a Sarawak mixed dipterocarp forest study site were dioecious, and thus obligate out-crossers. He argued on the basis of plant sexual systems, as indicated by floral morphology, but had no data on the potential for self-compatibility in non-dioecious species.

The majority of woody tropical rainforest plant species which have been studied have adaptations for out-crossing (Sands and House 1990), although the rate of out-crossing in populations and fecundity levels in individuals may depend upon the density of trees in flower (House 1992, 1993). A reduction in population density and fecundity levels due to habitat fragmentation and consequent increased distances between obligate out-crossed rainforest plants may threaten the long-term viability of local populations. Thus an ability to self-fertilise reduces vulnerability to extinction. Out-crossing is often obligatory (e.g. in dioecious species such as the Australian *Neolitsea dealbata* [Lauraceae]) and self-incompatibility in tropical trees, especially pioneer species, has been reported as widespread throughout the world's rainforests (Dayanandan *et al.* 1990; Perry and Starrett 1980). However, because of the occurrence in rainforests of populations and individuals with intermediate levels of self-incompatibility categorisation of the expression of self-incompatibility of a species may prove complex.

Low or variable levels of self-compatibility may occur in otherwise out-crossing species (e.g. Appanah 1982; Crome and Irvine 1986) and this may result in successful pollinations where cross-pollination is unpredictable or pollinators are absent. Self-compatibility may result in reduced offspring vigour but, where species are vulnerable to or undergo frequent colonisation and extinction cycles, as do many short-lived rainforest pioneers (e.g. *Commersonia fraseri* – Malvaceae/Sterculiaceae [Plate 44]), levels of self-compatibility in an otherwise predominantly self-incompatible breeding system would be advantageous for long-term species survivorship. Among cosexual species (those with bisexual or monoecious breeding systems; such as the ephemeral monoecious vine *Diplocyclos palmatus* (Cucurbitaceae [Plate 23]) out-crossing may be obligatory because of self-incompatibility. However, if self-pollination can occur there is the possibility of seed-set in the absence of pollination vectors, or where the plant population size is so small as not to attract pollinators. Environmental conditions can affect the degree of self-compatibility, for example, many tree species in Neotropical montane forests are self-compatible (Bawa 1990). The frequently cold and wet montane climates, which limit pollinator activity, may explain this preponderance of self-compatibility.

There have been few studies of pollination systems in Australian rainforest species but, as a broad generalisation based on floral morphology and patterns of anthesis, it appears that many species are either facultative or obligate out-crossers (see Tables 4.1, 4.2). Some of these species are dioecious (*Rhodosphaera rhodanthema* [Anacardiaceae] (Plate 95), *Drypetes deplanchei* [Putranjivaceae/Euphorbiaceae] and *Citronella moorei* [Icacinaceae/Cardiopteridaceae]) but descriptions of floral structure in 'Floras' and botanical handbooks may be inaccurate and convey misleading impressions of possible breeding systems. For example, *Wilkiea huegeliana* (Monimiaceae) (Plate 48), though reported to be dioecious, is frequently found to be monoecious and *Euroschinus falcatus* (Anacardiaceae) (Plate 7) is reported as polygamous but is frequently dioecious, at least in those New South Wales populations that have been studied (G. Williams 1995; Williams *et al.* 2001).

A high proportion (>50%) of trees in Australian tropical and subtropical rainforests are hermaphrodites but their capacity to self-pollinate (and the degree of self-compatibility) is largely unstudied. In the Australian tropical rainforest herb *Hornstedtia scottiana* (Zingiberaceae) self-pollination is effected (almost 100% of the time) by the rapid elongation of the corolla tube should flowers not be otherwise pollinated by their bird (Melphagidae) pollinators (Ippolito and Armstrong 1993). This delayed self-pollination strategy is characteristic of self-compatible species which initially offer opportunity for out-crossing (Willmer 2011). However, data on the

Table 4.1. Self-compatibility in Australian lowland subtropical rainforest species (after Adam and Williams 2001).

Family/species	%	Family/species	%
Anacardiaceae		**Monimiaceae/ Atherospermataceae**	
Euroschinus falcatus	0.6 (20.1)	*Daphnandra micrantha*	2.1 (7.1)
Celastraceae		**Myrsinaceae/ Primulaceae**	
Denhamia celastroides	0 (7.2)	*Myrsine howittiana*	0 (0)a
Ebenaceae		**Myrtaceae**	
Diospyros australis	2 (24.6)	*Archirhodomyrtus beckleri*	0 (31.3)
Elaeocarpaceae		*Rhodomyrtus psidioides*	0 (2.2)
Elaeocarpus reticulatus	36.9 (16.1)	*Syzygium floribundum*	0.8 (20.1)
Ericaceae/Epacridaceae		*Syzygium smithii*	3.6 (4)
Trochocarpa laurina	0 (9.1)	*Tristaniopsis laurina*	0 (19.1)
Escalloniaceae/Rousseaceae		**Pittosporaceae**	
Abrophyllum ornans	16.1–29.4 (93.9)	*Hymenosporum flavum*	0 (2.1)
Cuttsia viburnea	0 8 (2.9)	**Rhamnaceae**	
Eupomatiaceae		*Alphitonia excelsa*	0 (5.4)
Eupomatia laurina	9.7 (59.1)	**Rutaceae**	
Flacourtiaceae/Salicaceae		*Acradenia euodiiformis*	10.3 (4)
Scolopia braunii	17.8 (28)	*Acronychia imperforata*	0 (0)b
Lamiaceae/Verbenaceae		*Geijera salicifolia*	0 (8.9)
Clerodendrum tomentosum	0.5 (1.2)	**Sapindaceae**	
Lauraceae		*Alectryon coriaceus*	0 (0–7.8)
Cryptocarya glaucescens	18.1 (24.7)	*Cupaniopsis anacardioides*	9.2 (19.8)
Cryptocarya rigida	0 (3.8)	*Guioa semiglauca*	0.8 (8.9)
Malvaceae		*Mischocarpus pyriformis*	0 (9.7)
Hibiscus splendens	9.1 (29)	**Winteraceae**	
		Tasmannia insipida	58.7 (97.7)

Unenclosed (%) values represent percentage of developing fruit from inflorescences at conclusion of flowering in which pollinators were excluded; values in parentheses represent percentage of developing fruit from open pollinated flowers to which pollinators were given unobstructed access. All values are indicative only. Lower or zero values for some open pollination results illustrate the uncertainty of the pollination environment.
a: Species represented by only one tree in population, suggesting it is an obligate outcrosser.
b: No plants in population set seed in year of study.

incidence of self-compatibility in Australian rainforest trees is sparse; that which are available indicate a high frequency of selfing potential among the rainforest tree and shrub flora even if the level of selfing for individual species is low. Several Australian subtropical rainforest trees and shrubs exhibit relatively high levels of selfing (*Abrophyllum ornans, Acradenia euodiiformis, Cupaniopsis anacardioides, Cryptocarya glaucescens, Endiandra discolor, Elaeocarpus reticulatus, Davidsonia jerseyana, Eupomatia laurina, Hibiscus splendens, Scolopia braunii, Hicksbeachia pinnatifolia* and *Tasmannia insipida*) (see Table 4.1; Adam and Williams 2001). Those cited are predominantly pioneer or edge species. Reproductively, *Eupomatia laurina* (Plates 35, 36) and *Hibiscus splendens* (Plate 45) (and possibly *Hicksbeachia pinnatifolia* [Proteaceae]) are specialist species, pollinated by one or a small number of pollinators, and their ability to self-pollinate may have important consequences for the maintenance of populations in the absence of their particular pollinators. Yet some species do not demonstrate any apparent capacity for selfing. These also include pioneers (e.g. *Alphitonia excelsa* and *Rhodomyrtus psidioides*). In *Alphitonia excelsa* self-incompatibility may be a consequence of protandry (where male sexual maturity precedes that of female receptivity within a bisexual flower) and of physical changes as the stamens, encapsulated by the petals, are carried away from the developing gynoecium as the petals recurve (Williams and Adam 2001).

Within a family, species exhibit different behaviour (see Table 4.1); compare *Cryptocarya glaucescens* and *Cryptocarya rigida, Acradenia euodiiformis* and *Acronychia imperforata, Geijera salicifolia*). Within the Escalloniaceae/Rousseaceae *Abrophyllum ornans* (Plate 30) exhibits high levels of self-compatibility but *Cuttsia viburnea* (Plate 31) is apparently an obligate out-breeder, being self-incompatible.

There may be variation in the reproductive success of species both within and between years. This is a common phenomenon among rainforest trees and is caused by numerous influences, such as seasonal climatic events, plant physiology and variations in pollinator visitation. For some species (e.g. *Acronychia imperforata* [Plate 83] and *Rhodomyrtus psidioides* [Plate 58] – the latter species apparently now functionally extinct in nature) local populations may seasonally fail to set seed, while in the same rainforest stand others (e.g. *Abrophyllum ornans, Alphitonia excelsa, Cuttsia viburnea, Syzygium floribundum, Syzygium smithii, Tasmannia insipida*) may set large numbers of seed every year, regardless of climatic vagaries (G. Williams 1995). While continued selfing by self-compatible hermaphrodites may result in a decline in offspring quality, selfing may permit the local survival of species in fragmented stands where individual populations are severely reduced.

APOMIXIS AND COPPICING: LIFE WITHOUT SEX

Pollination is an essential process in sexual reproduction, for the movement of pollen between plants, in addition to the dispersal of seeds, affects the genetic variation of populations. Restriction in pollen movement reduces genetic variation within populations, while increases in genetic variation (heterozygosity) are achieved by out-crossing. There are definite benefits to a plant in 'sleeping around' to avoid the dangers of inbreeding (through loss of genetic variability because of the accentuated threat of homozygosity in isolated and fragmented populations). Avoiding extinction in fragmented rainforest remnants is one of them, such as a slip towards subtle loss of genetic diversity (recorded for fragmented *Elaeocarpus grandis* [Elaeocarpaceae] populations in northern New South Wales rainforest) (Rossetto *et al.* 2004; see also Whitehead *et al.* 2019 for a discussion of genetic differentiation in isolated *Elaeocarpus reticulatus*), and in cases where there is very little genetic diversity within and among populations, as is the case with the rare Australian *Syzygium paniculatum* (Myrtaceae), species may be susceptible to extreme stochastic events (Thurlby *et. al.* 2012).

Sexual reproduction, as well as allowing the expression of genetic variation, results in the production of fruits and seeds (dispersal units) which permit a species to colonise new locations, or re-establish in areas from which rainforest may have previously contracted (Rossetto *et al.* 2015). However, many plants have a capacity for vegetative, or asexual, reproduction (apomixis), such as from stolons (e.g. *Pollia* [Commelinaceae]) (Plates 21, 22) and rhizomes (*Hornstedtia* and *Alpinia* [Zingiberaceae]) that progressively creep below or just upon the soil surface, or from bulbs and more rarely bulbils (small bulbs that arise from leaf axils or inflorescences) (e.g. Ippolito and Armstrong 1993; Williams and Walker 2003). From stolons and rhizomes stems and leaves arise, from each node a new plant can establish and invade adjacent areas. Numerous grasses, and weeds such as blackberries, are very good at it. While this permits long-term maintenance of the population without sexual reproduction, vegetative (apomictic) reproduction, except in agamospermy, rarely provides a long-distance dispersal mechanism and does not permit the expression of genetic variation. Only clones are born, each genetically identical to the parent plant. An interesting example of this is the rare *Lomatia tasmanica* (Proteaceae), a red-flowered shrub – small tree (up to 8 m in height) restricted to south-western Tasmania (Lynch *et al.* 1998). Only one population occurs, this some 300 plants growing along stream gullies in rainforest; an area less than 1.2 km long. It flowers in the austral summer but bears neither fruit nor seeds. It has three sets of chromosones (triploid) and is therefore sterile. Reproduction occurs only vegetatively. When branches fall these produce new roots, each branch establishing a new plant that is genetically identical to its parent.

Vegetative reproduction is employed by some herbs (as in *Viola* [Violaceae], a genus also known for its ability to self-pollinate), and is frequently encountered among epiphytic and lithophytic rainforest ferns. Vegetative growth, in the form of basal coppicing of stems, without the inducement of disturbance, is exhibited by many Australian sclerophyllous flowering plants such as eucalypts and banksias, and even several gymnosperms, such as *Wollemia nobilis*, employ stem coppicing as an asexual reproductive strategy (Zimmer *et al.* 2015). However, vegetative regrowth is also an important part of the regeneration strategies used by many plants to respond to disturbance, for example fire and episodes of volcanism (Abe *et al.* 2013; Hill and Read 1984; Ondei *et al.* 2016). Such a response is seen less commonly among rainforest angiosperms, if only because these communities were historically considered to be buffered by humid, moist conditions (in the absence of extreme weather events), and so would carry fire (except at their margins [e.g. see Plate 148]) only in times of pronounced drought (e.g. Fensham *et al.* 2003; Miller and Murphy 2017).

Whilst fire in tropical, subtropical and temperate rainforest can change the relative abundance of individual species, and indeed the extent, distribution and vulnerability to alien plant invasion of rainforest (Plates 144, 145, 148) (e.g. Brown *et al.* 2017; Cameron 1992; Gentle and Duggin 1997), many Australian rainforest species studied appear to have potential to resprout (Williams *et al.* 2012); though for some species this capacity may be weak (e.g. *Pittosporum undulatum* [Pittosporaceae] [McMahon 1987]), and is at least partly dependent on fire intensity. The rainforest trees *Bridelia exaltata*, *Drypetes deplanchei Eupomatia laurina*, *Endiandra floydii* (DEC 2004c), *Mischocarpus pyriformis*, *Nothofagus moorei* and *Nothofagus cunninghamii*, *Schizomeria ovata* and *Syzygium oleosum*, as examples, use coppice growth to recover from low- to medium-intensity fire, or to regenerate following storm throw (Bowman 1991; Chapter 9– Table 9.1, Plate 150). Pioneer rainforest species in particular can exhibit significant vegetative regeneration responses to low and moderate intensity fires (Table 9.1); as with *Alstonia muelleriana* (Apocynaceae) in north-eastern Australia (P. Williams 2000), *Commersonia fraseri* (Malvaceae/Sterculiaceae) in lowland forests generally, *Cupaniopsis anacardioides* (Sapindaceae) in littoral rainforest (G. Williams pers. obs.), and *Syzygium smithii* in lowland subtropical rainforest; the latter being resistant to irregular-occurring fires because it develops lignotubers which confer a degree of coppicing (Adam 1992; Ashton and Frankenberg 1976). Plant age/size strongly influences survival outcomes, but intense fires (their impact exacerbated by the absence of post-fire rainfall), however, can result in catastrophic destruction of stands or loss of individual species within stands (Van Nieuwstadt and Sheil 2005).

The nature of resprouting, in rainforests worldwide, can vary. For example, in the aftermath of an Amazonian rainforest fire, of the species that resprouted 46% were subterranean sprouters, 27% were epicormic resprouters and 15% exhibited both methods of reproduction (Kauffman (1991), though the percentage of individuals that survived by resprouting varied among tree species. Age classes may explain the different forms of resprouting though some families, by their nature, may be less capable of recovery by resprouting (indicated for Dipterocarpaceae and Euphorbiaceae in Malaysian rainforest). Species that are less adapted to resprout post-fire are likely to be vulnerable to frequent fire cycles, as these plants are heavily dependent on seed stock for replenishment. Frequent fires progressively diminish surviving parent trees and their ability to set sufficient seed and replenish seedling stocks.

Even-aged (over-mature) rainforest trees of many species are able to cling tenaciously to life by the vigorous production of suckers from ever-collapsing and senescent parent tree trunks, for example in Australian *Doryphora sassafras* (Monimiaceae/Atherospermacae) and *Eucryphia moorei* (Cunoniaceae/Eucryphiaceae) where coppice growth can arise from a swollen stem base that resembles a lignotuber (Adam 1992). In *Doryphora sassafras* and

Eucryphia moorei the stem base can survive the death of the parent stem, the original tree being able to vegetatively reproduce. In similar fashion *Nothofagus moorei*, in the Barrington Tops, Werrikimbe and Border Ranges national parks, northern New South Wales, senescent populations are sustained by vegetative growth arising from the bases of aging parent stems. The endangered *Eucryphia jinksii* (Plate 24), a species with a highly restricted geographical and habitat distribution in south-east Queensland and north-eastern New South Wales, also produces coppice growth. In such species vegetative reproduction can compensate for lack of seedling regeneration on the forest floor and can result in ecologically stable, though genetically conservative, communities (Taylor *et al.* 2005). However, vegetative reproduction by resprouting may also favour long-term persistence of genetic variation in relatively large populations, as Acosta *et al.* (2012) considered for *Nothofagus antarctica* (predominantly a resprouter) in Chile and Argentina.

Root suckering is also a means of vegetative establishment, with aerial growth initiated at various distances from the parent plant (common in Australian *Melia azedarach* and *Glochidion ferdinandi*), and is an effective invasion mechanism in unoccupied areas following canopy disturbance or superficial clearance. Attempted survival by resort to suckering from root stock is a strategy observed in *Rhodomyrtus psidioides* (Myrtaceae) (Plate 149) following death of the parent trees caused by the fungal pathogen 'Myrtle rust'(*Austropuccinia psidii*) (G. Williams 2018). Although suckering is exhibited by several widespread species, such as *Atalaya hemiglauca* (Sapindaceae), *Glochidion ferdinandi*, and *Commersonia fraseri*, reproduction from root stock (sometimes following soil disturbance) is also a means by which rare and threatened subtropical rainforest species can maintain or extend population size, and has been recorded for *Elaeocarpus williamsianus* (Elaeocarpaceae), *Randia moorei* (Rubiaceae), *Davidsonia jerseyana* and *Davidsonia johnsonii* [Davidsoniaceae/Cunoniaceae] (DEC 2003, 2004a, 2004b, 2004d; Rossetto *et al.* 2003).

It is not uncommon to encounter whole trees (e.g. *Aphananthe philippinensis* [Cannabaceae/Ulmaceae], *Syzygium floribundum* [Myrtaceae]) that have collapsed, presumably through an unfortunate combination of insufficient root anchoring in shallow soils and the impact of strong winds, yet by the retention of functioning roots have sent adventitious vertical stems from their fallen trunks and, in so doing, survive and seemingly prosper. Such renewed aerial growth may depend on the possession of epicormic buds or lignotubers (outgrowths developed from accessory buds), known from many families throughout the world's ecosystems. Classically, within Australian landscapes, lignotubers are most notably associated with eucalypts and other woody members of fire-prone environments. Lignotubers and epicormic buds are not unknown among rainforest trees; both angiosperms (e.g. *Doryphora sassafras, Elaeocarpus reticulatus, Notelaea longifolia*), *Syzygium smithii*, and gymnosperms (*Wollemia nobilis*) have them (e.g. Ashton and Frankenberg 1976; Lacey and Jahnke 1984, Zimmer *et al.* 2015). In the Northern Territory the myrtle *Allosyncarpia ternata* (Myrtaceae) is an endemic tree that can occur as a dominant canopy species in spatially discrete pockets of monsoonal rainforest found within a matrix of eucalypt-dominated savannah (Fordyce *et al.* 2000). *Allosyncarpia ternata* is a slow-growing tree that takes up to 20 years to reach maturity and has a wide ecological range, being encountered in sheltered riverine zones as well as exposed drought-prone hills. Following damage or death of aerial parts, the species is able to regenerate by resprouting from lignotubers; the ability of these to store starch may be important for the survival of *Allosyncarpia ternata* seedlings.

However, reliance on vegetative reproductive strategies may only allow tenuous release from reproduction by sexual means. In the threatened New South Wales rainforest tree *Fontainea oraria* (Euphorbiaceae), a species with a very low population, some adult trees have been observed to coppice but otherwise the species does not appear to be able to reproduce vegetatively (DEC 2004e). Some species also appear to exhibit differential vegetative responses to threat events. The small endangered rainforest tree *Rhodamnia rubescens* (Myrtaceae) (Plate 57) has been observed to coppice following fire (Adam and Williams 2001) but was not observed to coppice following death of most plants induced by Myrtle rust at Lorien Wildlife Refuge, northern New South Wales (G. Williams pers. obs.).

Although regeneration by vegetative regrowth is an important adaptation to survival in an uncertain ecological landscape, it represents the final loss of sexual reproduction. One step above such an asexual survival strategy is that of apomixis via agamospermy. In plants which exhibit agamospermy, seeds may be formed without the requirement for pollination. The embryo is formed from the female gamete without fertilisation taking place, thus meiosis does not occur. Consequentially, a dispersal unit is produced but genetic variation is limited, if expressed at all. In some agamospermous species pollination is still required even though fertilisation does not take place. This is because the growth of the redundant pollen tube leads to the production of plant hormones and other phytochemicals involved in the stimulation of fruit growth. Eliott *et al.* (2014) report that tropical Queensland *Davidsonia pruriens* is likely apomictic, however, worldwide, the first documented example of agamospermy was from a widespread Australian subtropical rainforest tree, *Alchornea ilicifolia* (Euphorbiaceae) (Briggs and Walters 1984). Obligatory apomixis results in distinctive local clonal populations, sometimes recognised taxonomically as microspecies, but local variation in *Alchornea* awaits study.

But for the majority of rainforest species sexual reproduction (perhaps augmented by an ability for vegetative regrowth) is the norm, and thus pollination is a necessary part of the life-cycle of most species.

DIOECY: SEPARATION AS AN EXAMPLE OF OBLIGATE OUT-CROSSING

Dioecy is the phenomenon where individual plants of a particular species are either male or female. Consequently, the male and female flowers in dioecious species are only borne on separate plants, which may be widely separated from one another. This is distinct from plants that are monoecious, where individual plants of a species carry both female and male flowers. Be the species dioecious or monoecious, the two flower conditions are referred to as being *unisexual* or *cosexual*. However, the great majority (some 90%) of flowering plants produce 'perfect' (hermaphroditic/bisexual) flowers with stamens (thus staminate) and carpels (pistillate) (Ainsworth 2000).

The evolution of dioecy represents one of the major transitions in flowering plants and has evolved independently in several groups from a hermaphroditic or monoecious condition (Barrett 2008; Charlesworth and Charlesworth 1978). Though of recent evolutionary origin, the dioecious condition is uncommon yet widespread within higher order plant taxa, and occurs in ~4–6% of the >250 000 known species, suggesting it has arisen many times during the evolution of flowering plants (Negrutiu *et al.* 2001; Renner and Ricklefs 1995).

There are benefits in both unisexual conditions, and perceived disadvantages. Unless able to reproduce by the faculty of agamospermy, female dioecious plants are wholly dependent upon pollen transfer from separate male plants for sexual reproduction to occur. Monoecious species may be swamped by their own pollen, which may adversely affect fecundity and breeding success in the crowded rainforest environment. Interestingly, Sjöstrom and Gross (2006), in a countrywide exploration of four life history characters in the Australian flora, highlighted (dependent on life form) that unisexual breeding systems can be significantly and positively associated with endangered species, compared with non-threatened species.

By requiring the transfer of pollen between male and female plants dioecy, therefore, is a mechanism which promotes out-crossing. In the Australian tropical and subtropical dioecious trees *Diospyros pentamera* (Ebenaceae), *Litsea leefeana* and *Neolitsea dealbata* (Lauraceae) pollination is achieved mainly by Diptera, Hymenoptera and Coleoptera that are no larger in size than 6 mm long (House 1985). These tiny pollinating insects carry pollen from male (staminate) to female (pistillate) trees of the three species, though the pollen loads carried are not totally constituted of pollen of each species. The insects generally carry mixed pollen loads and are non-specialist pollinators that forage at a range of flowering rainforest trees – basically, anything available at the time. No apparent expression of fidelity. This absence of pollen load fidelity suggests that dioecy, as a sexual reproductive strategy, does not restrict pollen transport to host-specific pollinators and pollen flow between conspecific plants. Senarath (2008) found that in the Sri Lankan flora dioecy (along with monoecy) is significantly associated with plant growth form, fruit type, seed number and flower size, with dioecious species favoured by trees and small flowers. In addition there was a further significant association between dioecy and endemism. However, dioecy is not restricted to trees and shrubs, being found in numerous vines (lianas) and herbs (e.g. Higham and McQuillan 2000; Senarath 2008). In the dioecious individuals of the Australian rainforest and wet sclerophyll forest edge vine *Clematis glycinoides* (Ranunculaceae), as with *Diospyros pentamera*, *Litsea leefeana* and *Neolitsea dealbata*, flower visitors are small insects, mostly flies such as lauxaniids, muscids, calliphorids and syrphids, with occasional visits from other insects (e.g. vespid wasps and *Tetragonula* bees) (G. Williams pers. obs.). Many dioecious species have flowers characteristic of the general entomophilous pollination syndrome. They are simple in structure, usually small and accessible to all (Vamosi and Otto 2002). This appears to be the case regardless of the vegetation type that they inhabit; though the apparent pollinator syndrome indicated by dioecious plants may be misleading, as Higham and McQuillan (2000) found for the Australian endemic *Cyathodes divaricata* (Ericaceae/Epacridaceae), in which the small white flowers conformed to that suggestive of insect pollination, but which are actually pollinated by meliphagid honeyeaters.

Though the dominant theme among dioecious plants is one of floral drabness, in species that are pollinated by animals (rather than those that are wind-pollinated) male plants may be more likely to evolve more showy flowers than females (Eckhart 1999). This floral dimorphism may place showy dioecious species at risk of extinction during periods of lower pollinator abundance because pollinators may demonstrate marked preference in their visits to male plants, and this may explain the prevalence of small-sized flowers and the high incidence of wind pollination in dioecious species, as well as the relative rarity of dioecious to bisexual species (Vamosi and Otto 2002).

Some angiosperms may be functionally dioecious. For example, the endemic North Queensland palm *Normanbya normanbyi* (Plate 14) is monoecious but has flowers that are temporarily dioecious, and some androdioecious plants (with males and hermaphrodites) possess hermaphrodites with non-functional pollen (e.g. Kolesik *et al.* 2006). Androdioecy is quite rare. In the case of the usually dioecious Samoan liana *Freycinetia reineckei* (Pandanaceae), pollinated by the fruit-bat *Pteropus samoensis*, the dioecious condition is apparently maintained through the damaging impact of its pollinator, with Cox *et al.* (1991) demonstrating that the very few androdioecious plants produced were more likely to have their male and hermaphroditic flower spikes damaged by the pollinator than were dioecious ones within the population. Thus, dioecy in *Freycinetia reineckei* was likely to be the only condition given successful expression. Although *Pteropus samoensis* was exerting an evolutionary effect on *Freycinetia reineckei* populations by pushing it constantly in the direction of dioecy, the plant–pollinator relationship is asymmetrical since the pollinator feeds on a variety of plants. There is no reciprocal evolutionary influence by the plant. In thrips-pollinated *Castilla elastica* (Moraceae) growing in Panama (Sakai 2001), the androdioecious condition is thought to have evolved from a prior dioecious state, with other species in the genus being androdioecious (*Castilla ulei*) or strictly dioecious (*Castilla tunu*). In *Castilla elastica* male and female flowers on cosexual plants are not borne on the same inflorescence. Androdioecy in the species has a further expression in addition to that of the breeding system and structural separation of flower sexes on cosexual plants – flowers on male-only individuals are laterally broadened with elongate slits along the apex (allowing entry of thrips). Flowers on cosexual plants are urn-shaped, reminiscent of those found on Australian *Wilkiea* (Monimiaceae) (Plate 48), with access only through a small apical ostiole again similar to that of *Wilkiea* but also paralleling the restrictive pollinator morphology of *Ficus*, whose family Moraceae *Castilla* shares.

Although dioecy is generally uncommon, the incidence is higher among trees (Grant *et al.* 2017; Schlessman *et al.* 2014) than among herbs and is higher in certain habitats, notably rainforest and on oceanic islands (Weller *et al.* 1998). Dioecy is strikingly high among the New Zealand woody flora. Gender dimorphism is recorded in 83 indigenous genera of New Zealand seed plants, 65 genera being dioecious and a further 26 genera gynodioecious; though significantly only 17 of these are estimated to have arisen autochthonously (Newstrom and Robertson 2005; Webb *et al.* 1999). Thus there is an inherently high level of dependence on pollinators. New Zealand's flora is a combination of both relict and dispersed taxa, a large percentage of which is endemic (>80%), but the available suite of pollinators is a depauperate one (Webb and Kelly 1993). The high level of obligate out-crossing among indigenous plants would suggest that an impoverished pollinator fauna is not necessarily disadvantageous.

The incidence of dioecious plants in rainforests, given the spatial and floristic complexity of these communities, would be expected to be low. This does not appear to be the situation. For example, in isolated Northern Territory monsoonal rainforest patches a significant proportion of dioecious tree species have been recorded (19%, or 16 out of 84 species recorded) (Russell-Smith and Lee 1992), and in North Queensland dry rainforest, occurring in a matrix of savannah woodland, the level of dioecy was 21% with pollen-ovule ratios of most species (91 of 97 studied [Hansman 2001]) indicating obligate outbreeding; though in Australia's tropics overall, the incidence of monoecy is greater than that of dioecy (Gross 2005), the Euphorbiaceae and Sapindaceae contributing ~50% of the monoecious taxa. Similar average proportions (18%, but ranging to higher than 30%) of dioecious trees occur in New South Wales lowland subtropical rainforests (see Table 4.2, and Appendix 1), Jamaican montane forest

(21%), Costa Rican tropical lowland rainforest (23%), Madagascan littoral forest (18.4%) and Venezualan montane forest (31%) (Bawa 1990; Vary *et al.* 2011; Williams and Adam 2001). However, the proportion of dioecious species can differ in relation to elevation (Lin *et al.* 2019). A high incidence of dioecy (~20%) is also noted for New Caledonia's angiosperms, with the dioecious condition overrepresented among species that were woody, have plain flowers, have fleshy fruit, occur in rainforest, or are endemic (Schlessman *et al.* 2014). Schlessman *et al.* (2014) considered that 'successful dispersal and establishment of already dioecious colonists and autochthonous speciation of dioecious lineages are primarily responsible for the high incidence of dioecy in New Caledonia' and that the 'associations of dioecy with rainforest habitat and endemism appear to result from autochthonous speciation of dioecious lineages'.

The various figures cited above indicate that ~20% of species in different types of rainforests might be predicted to be dioecious (Adam and Williams 2001), the actual figure skewing a little (though sometimes substantially) either way from this value. However, the figure can be higher. In the sandy, resource-poor, coastal plain ('restingia') vegetation of south-east Brazil a high percentage (35%) of dominant woody plants are dioecious (Matallana *et al.* 2005) and in Nigeria 40% of trees have been reported as dioecious (Archibold 1995). In a montane sclerophyllous forest in central Chile the frequency for all life forms was 9% but 57% of the trees were reported as dioecious (Arroyo and Uslar 1993). For other growth forms, however, the dioecious condition was recorded in 17% of shrubs, 2% in perennial herbs, but absent for annual herbs.

Dioecy is strongly correlated with woodiness (Schlessman *et al.* 2014), fleshy fruit (Schlessman *et al.* 2014) and pollination by small insects (Grant *et al.* 2017), and this exceptional incidence is argued as attributable to ecological traits related to long-distance dispersal, ecological vigour and possibly vegetative reproduction. Though there are examples of the incidence of overall dioecy being generally as low as 3–7% (e.g. in Sri Lankan forests [Senarath 2008] and Brazilian tropical dry forest [Machado *et al.* 2006]), which might be explained by low incidence of small generalist flowers (Machado *et al.* 2006) or by aggregating all growth forms (Arroyo and Uslar 1993), nevertheless it is unclear why there should be relatively similar levels of dioecious rainforest floras in communities with different histories and different contemporaneous environmental conditions.

Table 4.2: Some dioecious and partly dioecious (sometimes monoecious or bisexual) tree and shrub species in Australian subtropical rainforests (after Harden 1990, 1991, 1992).

Family/species	Family/species
Anacardiaceae	**Moraceae**
Euroschinus falcatus	*Ficus coronata*
Rhodosphaera rhodanthema	*F. fraseri*
Celastraceae	*F. macrophylla*
Elaeodendron australe	*F. henneana*
Ebenaceae	*F. obliqua*
Diospyros australis	*F. rubiginosa*
D. pentamera	*F. watkinsiana*
Euphorbiaceae	*Streblus brunonianus*
Alchornea ilicifolia	**Nothofagaceae**
Claoxylon australe	*Nothofagus moorei*
Croton insularis	**Putranjivaceae/ Euphorbiaceae**
Croton verreauxii	*Drypetes deplanchei*
Fontainea australis	**Sapindaceae**
F. oraria	*Atalaya salicifolia*
Macaranga tanarius	*A. multiflora*
Mallotus claoxyloides	*Castanospora alphandii*
M. discolor	*Cupaniopsis anacardioides*
M. philippensis	*C. flagelliformis*
Icacinaceae/ Cardiopteridaceae	*C. foveolata*
Citronella moorei	*C. newmani*
Lauraceae	*C. parvifolia*
Litsea australis	*C. serrata*
L. reticulata	*Diploglottis australis*
Neolitsea australiensis	*Harpullia alata*
N. dealbata	*H. hillii,*
Malvaceae/Sterculiaceae	*H. pendula*
Argyrodendron actinophyllum	*Jagera pseudorhus*
A. trifoliatum	
Meliaceae	**Urticaceae**
Dysoxylum fraserianum	*Dendrocnide excelsa*
D. rufum	*D. moroides*
Melia azedarach	*D. photinophylla*
Owenia cepiodora	
Toona ciliata	**Winteraceae**
Monimiaceae	*Tasmannia glaucifolia*
Hedycarya angustifolia	*T.insipida*
Wilkiea huegeliana	*T. lanceolata*
Wilkiea macrophylla	*T. purpurascens*
Monimiaceae/ Atherospermataceae	*T. stipata*
Atherosperma moschatum	

Dioecious species would appear to be at a particular disadvantage because reproductive success is linked to the receipt of pollen from male plants. Clearly, male plants alone cannot initiate foundation events, so colonisation episodes should select against dioecious (and monoecious) species. In rainforests, dioecy has not proven an obstacle to success. The evolutionary explanation for the high incidence of dioecy in rainforest is unclear, and various hypotheses have been advanced. It has been argued that female flowers, with no investment in male function, may be able to make a high investment in female function, and that in the rainforest environment there is an advantage to producing large seeds (which many dioecious species do). Within rainforest, individuals of a species often occur at a low density with large distances between them. In consequence, self-compatible hermaphrodites may be largely self-pollinated, with a loss of vigour and viability. If dioecious variants arise in such species, rare long-distance cross-pollination may result in higher-quality offspring and hence a long-term evolutionary advantage, despite lower fecundity, over self-compatible hermaphrodites.

Regardless of the original advantages of dioecy, in rainforest stands that are very fragmented the consequence is a high risk of local extinction for dioecious taxa, and little capacity for successful colonisation of new sites. Where only male trees grow, the species is already functionally extinct unless female trees are recruited or artificially reintroduced.

PROTOGYNY AND PROTANDRY: SEGREGATION OF SEXUAL FUNCTION

Dioecy is about separation in space. The seemingly arcane terms 'protogyny' and 'protandry' are used to describe the temporal separation of female and male function in bisexual flowers, the two phenomena grouped in the broader term 'dichogamy'. Such functions impede the likelihood of self-pollination in self-compatible species, however, the occurrence of protogyny and protandry appears to correlate with several ecological conditions such as latitude and altitude, and also mating system, pollination mode and flower size (Willmer 2011). Dichogamous traits may be partly explained by plant phylogeny, or might be the product of convergent evolution rather than artefacts of phylogenetic relatedness (Sargent and Otto 2004)

Protogyny is common in basal angiosperms that have bisexual flowers (i.e. all magnoliids, basal monocots and eudicots [Endress 2001]). Examples are encountered in the widely distributed Australian rainforest trees *Scolopia braunii* (Flacourtiaceae/Salicaceae), *Syzygium smithii* (Myrtaceae), *Daphnandra micrantha* (Monimiaceae/Atherospermataceae), *Eupomatia laurina* (Eupomatiaceae), *Idiospermum australiense* (Calycanthaceae) and *Pouteria australis* (Sapotaceae) (Armstrong and Irvine 1990; Endress 2003; Hamilton 1897; G. Williams 1995; Worboys and Jackes 2005), and also the vines *Trimenia moorei*, in which the stigma is exposed before the opening of the stamens, and *Austrobaileya scandens* which also is bisexual though self-incompatible (Endress 2001). The Eupomatiaceae are strongy protogynous (see discussion in '*Elleschodes* and *Eupomatia*' Chapter 8). The Australian species *Pandorea pandorana* (Bignoniaceae) (Plate 19), a vine of rainforest and vine thickets, is protogynous but in the related *Pandorea jasminoides* and *Pandorea baileyana* anther dehiscence and stigma receptivity occur at the same time (James and Knox 1993).

Protandrous examples include *Alphitonia excelsa* (Rhamnaceae) (Plate 89), a species commonly encountered on the margins of littoral rainforests (Williams and Adam 2001), the endemic North Queensland palm *Normanbya normanbyi* (Arecaceae) (Plate 14) (Inkrot *et al.* 2007), and the leatherwoods *Eucryphia lucida* and *Eucryphia milliganii* (Cunoniaceae/Eucryphiaceae), species endemic to Tasmania (Taylor and Hill 1996). All Australian Proteaceae are hermaphroditic and protandrous (Collins and Rebelo 1987). In the palm *Normanbya normanbyi* the staminate phase lasts 40 days with the later pistillate stage lasting about two weeks. The breeding systems of some rainforest species may exhibit a trend to protogyny or protandry but without complete or certain expression of either; for example, the North Queensland rainforest tree *Syzygium cormiflorum* (Myrtaceae) (Crome and Irvine 1986) may be partially protandrous. In such instances we can often only guess in which evolutionary direction the breeding system expression is progressing, or whether the dual condition is static. Maybe someone will look back through time and note any direction unequivocally evident.

In protandry, pollen is shed by the anthers before the stigma becoming receptive. This is often achieved by the physical immaturity of the stigma, full growth and receptivity being constrained until anther dehiscence. However, in *Commersonia* central staminodes are initially incurved and cover the stigmas, and then, when each stigma is receptive, the staminodes recurve to cover post-dehiscent anthers (Wilkins and Whitlock 2011).

The reverse of the protandrous function is that of protogyny. Here the stigma matures and is receptive before maturation of the anthers and subsequent release of pollen. Flowers of the basal angiosperm *Idiospermum australiense* are protogynous, but some populations are andromonoecious (where inflorescences include bisexual flowers and some that are functionally male) and others are hermaphroditic (Worboys and Jackes 2005). In *Daphnandra micrantha* (Plate 47) anther dehiscence may occur more than two days after initial opening of the flower so that a relatively long period of time can elapse, allowing the stigma to receive foreign pollen (G. Williams 1995). Some species appear to be only slightly protogynous, such as *Cuttsia viburnea* (Plate 31) and *Acradenia euodiiformis* (Plate 80). The effectiveness of the protogynous function is increased where the stigma is exerted beyond the perianth before the numerous anthers have been exposed (as in *Scolopia braunii*), these being held for a time by the unopened corolla. But by whichever method protogyny is expressed, in general it is a surer guard than protandry against the possibility of self-pollination or contamination by a plant's own incompatible pollen. And though, in the absence of a better choice, self-pollination in self-compatible species is not necessarily a bad thing, out-crossing is the preferred strategy because it results in the heterozygous blending of genetic attributes from different parent plants.

Comparative studies undertaken at the species level indicate protandrous flowering plants tend to be pollinated by bees or flies, but in protogynous species there is a tendency to wind or beetle pollination. This has been taken to suggest that a functional role for pollinators exists as a driving mechanism in the evolution of dichogamy. The mode of pollination (i.e. biotic or abiotic) may not affect transitions between protandrous or protogynous states, rather rates of transition between animal-mediated and abiotic pollination may depend on whether a species is protandrous or protogynous (Sargent and Otto 2004).

Multiple threats from invasive species – a diversity of enemies

Invasive species, both plant and animal, can pose long-term threats to native plant–pollinator mutualisms, plant breeding systems, and the recruitment and survival of individual native plants species within rainforest stands (Arroyo-Correa *et al.* 2019). Invasion is facilitated where alien plants and animals are able to act in liaison (e.g. as in Gross *et al.* 2017), with the facultative interaction between non-indigenous species gaining recognition as a major driver of plant invasion success (Simpson *et al.* 2005).

Observations of invasive flower-frequenting animals in Australian rainforests are generally restricted to the honey bee *Apis mellifera*. This 'super-generalist' pollinator will enter both disturbed and undisturbed rainforest stands to forage, and and can take residence in tree hollows. *Apis mellifera* has been recorded from numerous mass-flowering trees and shrubs, but its impact on the reproductive ecology of rainforest plants in Australia is little studied. Although it competes with native anthophilous species for floral resources (e.g. Gilpin 2017; Williams and Adam 1997), with Hermansen *et al.* (2014b) suggesting it may have displaced native pollinators of the mangrove *Avicennia marina*, its impact to the detriment of native plant sexual reproduction in Australian rainforests has not been clearly demonstrated. However, *Apis mellifera* is known to forage at a wide taxonomic range of highly invasive plants in Australian rainforests, such as *Asparagus* spp. (Asparagaceae) and *Chrysanthemoides monilifera* (Asteraceae), in littoral rainforest, *Ligustrum sinense* (Oleaceae) in riparian rainforest, *Anredera cordifolia* (Basellaceae) in lowland floodplain rainforest, and *Lantana camara* (Verbenaceae) extensively in eastern mainland Australia, with consequent *Apis mellifera*-induced pollination promoting high levels of seed production and further weed infestation.

However, five additional exotic bees, *Bombus terrestris, Apis cerana, Megachile rotundata, Afranthidium repititum* and *Seladonia hotoni* are resident in Australia (T. Houston 2018). The large European bumblebee *Bombus terrestris* (Apidae) was introduced to Tasmania in 1992 where it is now widely established, spreading to numerous vegetation formations from sea level to elevations above 1400m. It is a generalist feeder (Moore and Gross 2012) that forages at both native and exotic plants, and has been recorded competing with, and displacing, native megachilid bees (McQuillan and Hingston 1999). *Bombus terrestris* has the potential to spread widely throughout temperate mainland Australia, with the likelihood of impacting on native plants not previously known to be visited by exotic bees (Kingston and McQuillan 1998). The foraging activities of *Apis mellifera* and *Bombus terrestris* threaten the declining swift parrot *Lathamus discolor* (Hingston and Wotherspoon 2017), which feeds largely on nectar and pollen of eucalypts. Hingston and Wotherspoon found that *Bombus terrestris* and *Apis mellifera* compete with swift parrots for the nectar of flowering *Eucalyptus globulus*, the two bees consuming most of the available nectar in their exposed study blossoms. *Apis mellifera* was the most common species, but

Bombus terrestris had greater *per capita* rates of consumption. The introduced bee, *Apis cerana*, was recorded from Cairns, North Queensland, in 2007. Attempts to eradicate *Apis cerana* in Australia have failed. Widely distributed in Asia and the south-west Pacific, the species is resilient to different environmental conditions (Koetz 2013), and given its ability to forage during cool months at high latitudes in the Northern Hemisphere (Kato 2000) *Apis cerana* may be able to invade regions and ecosystems far beyond its currently known Australian distribution. *Megachile rotundata* is a native to Europe and was introduced in attempts to establish it as a pollinator of lucerne crops. It appears restricted to areas where lucerne is farmed. *Afranthidium repititum* is a native of South Africa and is established in Queensland, New South Wales and one record from South Australia. *Seladonia hotoni*, a native of southern Africa, was first detected in the Upper Hunter Valley region of New South Wales in 2004. It visits flowers of introduced plants, including the noxious weed *Galenia pubescens* (T. Houston 2018).

Though currently unexplored, exotic ants may also pose a threat to the reproductive ecology of Australian rainforest plants, this through their interference with pollinators (as has befallen the threatened Mauritian gecko-pollinated plant *Roussea cepediana* (Rousseaceae) (Hansen and Müller 2009). A precautionary case is the Yellow crazy ant *Anoplolepsis gracilipes* (Lowe *et al.* 2000), potentially capable of inhabiting Australian ecosystems from northern Western Australia, south to parts of inland and coastal New South Wales. Loss of biodiversity and ecosystem integrity followed its invasion of Christmas Island, Indian Ocean, resulting in listing there as a 'key threatening process' under the *Environmental Protection and Biodiversity Conservation Act, 1999.* A second potentially threatening species is *Wasmannia auropunctata*, commonly called the Electric or Little fire ant. This has a limited known occurrence in northern Queensland, but has displaced other ants, or caused a decline in invertebrate populations, wherever it has been introduced.

In Australia cattle, feral Asian water buffalo (*Bubalus bubalus*) and pigs have been implicated as agents of change to rainforest structure and floristic composition, and as agents of exotic plant infestation (e.g. Gentle and Duggin 1997), to which native pollinators will variously respond. In the Forty Mile Scrub National Park, North Queensland, rooting of soil by pigs has been suggested to cause tree death and subsequent increased light penetration in dry rainforest; this favouring the spread and dense infestation of *Lantana camara* (Fensham *et al.* 1994). *Lantana camara* is widespread in eastern Australia, forming dense infestations in logged and otherwise disturbed mesic forests of northern New South Wales and south-eastern Queensland. It will also invade seemingly little-disturbed rainforest stands, as at Camels Hump Nature Reserve, northern New South Wales; and in such instances lantana can add significantly to the fuel load (and fire potential) in the understorey. However, *Lantana camara* provides a nectar source for numerous native butterflies (e.g. *Graphium agamemnon, G. eurypylus, G. macfarlanei, G. macleayanum, G. sarpedon, Ornithoptera priamus, O. richmondia, Pachlioptera polydorus, Papilio ulysses, Belenois java* [Plate 114], many Hesperiidae) but the ecological implications of their diversion from co-flowering rainforest trees and shrubs is generally unknown.

Native animals can also be functionally invasive. These are species, vertebrate and invertebrate, from adjacent non-rainforest communities that forage on rainforest margins, or owing to the narrow or small patch size of a stand (where no effective 'core' area remains) can penetrate past the rainforest boundary. Where these are species that forage upon floral resources, and in so doing competing with resident faunas, their movements might be such that they perform minimal or no pollination service, yet deplete available sources of pollen and nectar. Flying-foxes can be highly destructive to small rainforest stands (as at Maclean and Wingham Brush Nature Reserve, northern New South Wales) (Plate 142), destroying the canopy, facilitating the invasion of smothering alien plants, and so driving the breakdown of plant reproduction networks. In similar fashion, some native plants can pose a threat to pollination networks, and the floristic composition and structure of stands more generally; for example the rampant vine *Cissus antarctica* (Vitaceae) and the small rainforest tree *Pittosporum undulatum* (Pittosporaceae). *Pittosporum undulatum* has a wide distribution extending from eastern Victoria north to Kaputar National Park, Kroombit Tops and Carnarvon National Park (Harden *et al.* 2006), but is now naturalised beyond this range; and is capable of vigorously invading remnant native bushland and small rainforest stands. *Cissus antarctica* is widespread and can be highly invasive and persistant in mesic forests that have been disturbed by logging, storm damage and fire (Plates 143, 148).

Alien and native plants in rainforests will compete to some degree for pollinators from the same resident pollinator network, and this can cause disruption to native plant–animal interactions. Shade is a major factor that usually limits exotic plant invasion of rainforest, consequently physical disturbance, such as the structural damage caused by periodic storms, extreme penetrating fire events, logging or natural tree fall can promote invasions by exotic plants that respond to increased light levels on the forest floor (Plate 145). However, alien plants which are able to develop shade tolerance, as Godoy *et al.* (2011) noted for invasive *Prunella vulgaris* in Neotropical temperate evergreen forest, and as *Lantana camara* and various Asparagaceae (Plate 147) exhibit in some eastern Australian forests, pose a potential threat to seemingly undisturbed rainforests. The small, often narrow, size of floodplain and littoral rainforest remnants (subject to high light levels) makes them especially vulnerable to invasion by a large suite of aggressive and persistent exotic plants (e.g. *Asparagus aethiopicus* [Asparagaceae], *Chrysanthemoides monilifera* [Asteraceae], *Gloriosa superba* [Colchicaceae], *Anredera cordifolia* [Basellaceae], *Dolichandra unguis-cati* [Bignoniaceae], *Tradescantia fluminensis* [Commelinaceae]). Vulnerability to invasion, especially by *Chrysanthemoides monilifera*, is exacerbated by fire. Small hinterland rainforests (e.g. narrow lowland riparian stands) are vulnerable to canopy invasion and replacement by profusely seeding *Cinnamomun camphora* (Lauraceae) and *Ligustrum lucidum* (Oleaceae). Their seeds are dispersed by native birds, as are those of the highly invasive *Hedychium gardnerianum* (Zingiberaceae), a large clumping herb native of India and increasingly naturalised in several hinterland mesic forests (and capable of establishing in fully shaded rainforest understoreys). Although it has been argued that *Cinnamomum camphora* should be retained, at least as an interim measure, during rainforest restoration projects (Neilan *et al.* 2006), trees can live for centuries, over which time they can suppress native plant recuits (Kanowski *et al.* 2009).

Periodic cyclones in Tropical Queensland (e.g. Turton 2008) result in extensive foliage and crown loss and widespread tree mortality, which exposes individual stands to alien plant invasion, but the broad ecological consequences (including impacts on pollination services) of invasion following cyclone, storm and fire damage remain poorly understood. In a study of the dynamics of invasive plant recruitment in tropical Queensland rainforest following Cyclone Larry (occurring in 2006), Murphy *et al.* (2008) found that invasive species may comprise a transient component of the regenerating flora in early regenerative stages, but with a persistent aggressive suite of species that can lead to the smothering of tree crowns and the forest floor. These can have long-term effects on the successional trajectory of the rainforest and its future composition and structure (Murphy *et al.* 2008), and though subsequent impacts on pre-existing ecological interactions can be predicted in broad terms, they cannot be foretold in detail.

Small rainforest remnants, regrowth forests, and riparian forests have much lower levels of resistance to high energy winds due to patch size and edge effects. Topography can influence storm impacts (Negrón-Juárez *et al.* 2014), such that the nature and degree of damage can be patchy, ranging from severe to localised, but rainforests that are subject to frequent cyclonic events may suffer canopy lowering and changes to species composition; to which surviving native pollinators, and anthophilous animal species more generally, respond. In wet sclerophyll forest logging operations can result in heavy invasion of 'secondary' native vines (e.g. *Cissus antarctica* [Vitaceae]) which suppress regeneration of prior species composition and forest structure (Plates 143, 144). Although this may not result in a gross reduction in floral resources, a reduction in the diversity of flowering species can result in temporal gaps in the seasonal sequence of available nectar and pollen, to which the anthophilous fauna might have otherwise attended.

Invasive pathogens, such as *Phytophthora cinnamoni* (e.g. invasive in the Barrington Tops, northern New South Wales), *Chalara australis* (which threatens *Nothofagus cunninghamii* in southern cool temperate rainforests [Packham *et al.* 2008]), and more recently the widely established Myrtle rust (*Austropuccinia psidii*), also constitute a high level threat to Australian plant species and plant communities; Myrtle rust posing a threat at the landscape scale. Myrtle rust is able to re-infect myrtaceous species in post-fire environments, with infections of coppicing rainforest Myrtaceae causing subsequent flow-on effects to species survival and community dynamics.

Plate 1. Emergent *Araucaria cunninghamii* (Araucariaceae)

Plate 4. Pollinating *Tranes* weevils on male *Lepidozamia peroffskyana* cone

Plate 2. *Macrozamia* sp. (Zamiaceae) in dry rainforest

Plate 5. *Drymophila moorei* (Alstroemeriaceae/Luzuriagaceae)

Plate 3. Female cone, *Lepidozamia peroffskyana* (Zamiaceae)

Plate 6. *Crinum pedunculatum* (Amarylliidaceae/Liliaceae)

Plate 7. *Euroschinus falcatus* (Anacardiaceae)

Plate 8. *Melodinus australis* (Apocynaceae)

Plate 9. *Alocasia brisbanensis* (Araceae)

Plate 10. *Alocasia brisbanensis* (Araceae), showing spathe

Plate 11. *Archontophoenix cunninghamiana* (Arecaceae) in subtropical rainforest

Plate 12. *Calamus caryotoides* (Arecaceae)

Plate 13. *Linospadix monostachyos* (Arecaceae)

Plate 14. *Normanbya normanbyi* (Arecaceae)

Plate 15. *Pararistolochia praevenosa* (Aristolochiaceae)

Plate 16. *Cordyline stricta* (Asparagaceae)

Plate 17. *Neoastelia spectabilis* (Asteliaceae)

Plate 18. *Balanophora fungosa* (Balanophoraceae)

Plate 19. *Pandorea pandorana* (Bignoniaceae)

Plate 20. *Capparis arborea* (Capparaceae)

Plate 21. *Pollia crispata* (Commelinaceae)

Plate 22. *Pollia crispata* in subtropical rainforest understorey

Plate 23. *Diplocyclos palmatus* (Cucurbitaceae)

Plate 24. *Eucryphia jinksii* (Cunoniaceae/Eucryphiaceae)

Plate 25. *Schizomeria ovata* (Cunoniaceae)

Plate 26. *Vesselowskya ?rubifolia* (Cunoniaceae)

Plate 27. *Doryanthes palmeri* (Doryanthaceae)

Plate 28. *Diospyros australis* (Ebenaceae)

Plate 29. *Elaeocarpus holopetalus* (Elaeocarpaceae)

Plate 30. *Abrophyllum ornans* (Escalloniaceae/Rousseaceae)

Plate 31. *Cuttsia viburnea* (Escalloniaceae/Rousseaceae)

Plate 32. *Aleurites moluccana* (Euphorbiaceae)

Plate 33. *Baloghia inophylla* (Euphorbiaceae)

Plate 34. *Croton insularis* (Euphorbiaceae)

Plate 35. *Eupomatia laurina* (Eupomatiaceae) and *Elleschodes* weevil

Plate 36. *Eupomatia laurina* with operculum

Plate 37. *Cynometra lenticellata* (Fabaceae)

Plate 40. *Endiandra muelleri* (Lauraceae)

Plate 38. *Cynometra lenticellata* (Fabaceae), showing flower detail

Plate 41. *Argyrodendron actinophyllum* (Malvaceae/ Sterculiaceae)

Plate 39. *Clerodendrum tomentosum* (Lamiaceae/Verbenaceae)

Plate 42. *Brachychiton acerifolius* (Malvaceae/Sterculiaceae)

Plate 43. *Brachychiton discolor* (Malvaceae/Sterculiaceae)

Plate 44. *Commersonia fraseri* (Malvaceae/Sterculiaceae)

Plate 45. *Hibiscus splendens* (Malvaceae)

Plate 46. *Synoum glandulosum* (Meliaceae)

Plate 47. *Daphnandra micrantha* (Monimiaceae/Atherospermataceae)

Plate 48. *Wilkiea huegeliana* (Monimiaceae)

Plate 49. *Ficus coronata* (Moraceae)

Plate 52. *Trophis scandens* (Moraceae)

Plate 50. *Ficus henneana* (Moraceae)

Plate 53. *Myrsine variabilis* (Myrsinaceae/Primulaceae)

Plate 51. *Streblus brunonianus* (Moraceae)

Plate 54. *Archirhodomyrtus beckleri* (Myrtaceae)

Plate 55. *Backhousia myrtifolia* (Myrtaceae)

Plate 56. *Metrosideros* sp. (Myrtaceae)

Plate 57. *Rhodamnia rubescens* (Myrtaceae)

Plate 58. *Rhodomyrtus psidioides* (Myrtaceae)

Plate 59. *Syzygium cormiflorum* (Myrtaceae)

Plate 60. *Syzygium floribundum* (Myrtaceae)

Plate 61. *Xanthostemon chrysanthus* (Myrtaceae)

Plate 62. *Olea paniculata* (Oleaceae)

Plate 63. *Pseudovanilla foliata* (Orchidaceae)

Plate 64. *Rhinerrhiza divitiflora* (Orchidaceae)

Plate 65. *Breynia oblongifolia* (Phyllanthaceae/Euphorbiaceae)

Plate 66. *Glochidion ferdinandi* (Phyllanthaceae/Euphorbiaceae)

Plate 67. *Banksia integrifolia* (Proteaceae)

Plate 68. *Eidothea hardeniana* (Proteaceae)

Plate 69. *Stenocarpus sinuatus* (Proteaceae)

Plate 70. *Ripogonum discolor* (Ripogonaceae)

Plate 71. *Guioa semiglauca* (Sapindaceae)

Plate 72. *Solanum aviculare* (Solonaceae)

Plate 73. *Trimenia moorei* (Trimeniaceae)

Plate 74. *Elatostema ?reticulatum* (Urticaceae)

Plate 75. *Cayratia clematidea* (Vitaceae)

Plate 76. *Cissus antarctica* (Vitaceae)

Plate 77. *Cissus sterculifolia* (Vitaceae)

Plate 78. *Tasmannia insipida* (Winteraceae)

Plate 79. *Alpinia caerulea* (Zingiberaceae)

Plate 80. *Apis mellifera* foraging on *Acradenia euodiiformis* (Rutaceae)

Plate 81. *Amegilla bombiformis* (Apidae) sonicating *Senna acclinis* (Fabaceae)

Plate 82. *Tetragonula carbonaria* (Apidae)

Plate 83. *Leioproctus flavomaculatus* (Colletidae–Colletinae) on *Acronychia imperforata* (Rutaceae)

Plate 84. *Amphylaeus nubilosellus* (Colletidae–Hylaeinae)

Plate 85. *Megachile pictiventris* (Megachilidae)

Plate 88. *Piyuma* sp. (Crabronidae)

Plate 86. *Lasioglossum* sp. (Halictidae)

Plate 89. *Anthobosca signata* (Thynnidae) on *Alphitonia excelsa* (Rhamnaceae)

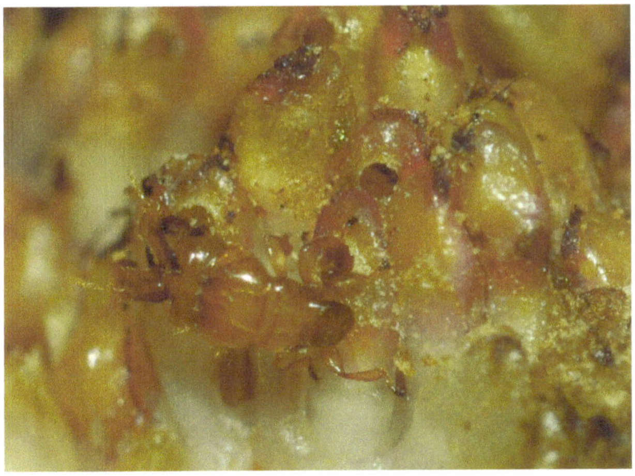

Plate 87. Male 'fig wasp' (Agaonidae–Agaoninae)

Plate 90. *Dimorphothynnus dimidiatus* (Thynnidae) on *Elaeocarpus obovatus* (Elaeocarpaceae)

Plate 91. *Rhagigaster ?mutatus* and *Gasterupton ?pallidicus* foraging on *Cayratia clematidea* (Vitaceae)

Plate 92. *Deuterodiscoelius ephippium* (Vespidae) foraging on *Trochocarpa laurina* (Ericaceae/Epacridaceae)

Plate 93. *Myrmecia nigrocincta* (Formicidae) on flowering *Alectryon subcinereus* (Sapindaceae)

Plate 94. *Polyrhachis ammon* (Formicidae) on *Alectryon coriaceus* (Sapindaceae)

Plate 95. *Mesophysa flavipes* (Acroceridae) on *Rhodosphaera rhodanthema* (Anacardiaceae)

Plate 96. Female *Bibio imitator* (Bibionidae) on *Avicennia marina* (Acanthaceae/Avicenniaceae)

Plate 97. *Geron* sp. (Bombyliidae) on *Alectryon subcinereus* (Sapindaceae)

Plate 98. *Stomorhina* sp. (Calliphoridae)

Plate 99. *Trichophthalma costalis* (Nemestrinidae)

Plate 100. *Calodema regale* (Buprestidae) on *Tristaniopsis laurina* (Myrtaceae)

Plate 101. *Castiarina rollei* (Buprestidae)

Plate 102. *Metaxymopha gloriosa* and *M. hauseri* (Buprestidae) on *Buckinghamia celsissima* (Proteaceae)

Plate 103. *Distichocera superba* (Cerambycidae) on *Syzygium floribundum* (Myrtaceae)

Plate 106. *Amata* sp. (Erebidae–Arctiinae) feeding on *Acronychia oblongifolia*

Plate 104. *Tropocalymma dimidiata* (Cerambycidae) on *Acronychia imperforata* (Rutaceae)

Plate 107. *Oxyodes tricolor* (Noctuidae) larvae feeding on *Arytera divaricata* (Sapindaceae) inflorescences

Plate 105. *Glycyphana brunnipes* (Scarabaeidae) on *Avicennia marina* (Acanthaceae/Avicenniaceae)

Plate 108. *Hypocysta metirius* (Nymphalidae) on *Trochocarpa laurina* (Ericaceae/Epacridaceae)

Plate 109. *Appias paulina* (Pieridae) on *Cayratia clematidea* (Vitaceae)

Plate 110. *Oncopeltus sordidus* (Lygaeidae) on *Avicennia marina* (Acanthaceae/Avicenniaceae)

Plate 111. *Pristhesancus plagipennis* (Reduviidae)

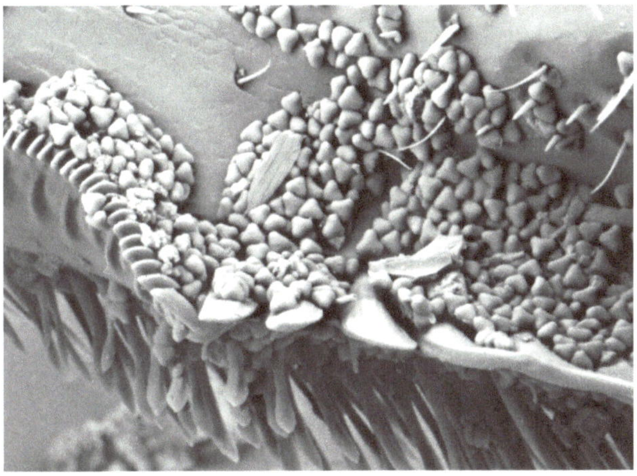

Plate 112. Pollen comb on *Anthobosca signata* (Thynnidae)

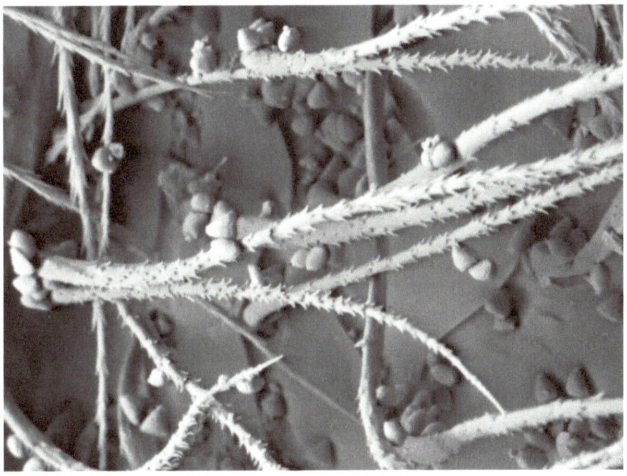

Plate 113. Thoracic setae, *Bisallardiana gymnopleura* (Scarabaeidae-Cetoniinae)

Plate 114. *Belenois java* (Pieridae) feeding on *Lantana camara*

Plate 115. Emergent *Araucaria* in low vine thicket, central Queensland

Plate 116. *Adansonia gregorii* (Malvaceae)

Plate 117. Rainforest mosaic within eucalypt forest matrix

Plate 118. Dry rainforest dominated by *Backhousia sciadophora* (Myrtaceae)

Plate 119. Rainforest understorey, Werrikimbe National Park

Plate 120. Subtropical rainforest-wet sclerophyll forest margin

Plate 121. Subtropical rainforest, League Scrub, northern New South Wales

Plate 122. Understorey *Nothofagus moorei*–cool temperate rainforest

Plate 123. *Nothofagus moorei* (Nothofagaceae)-dominated canopy, Barrington Tops

Plate 124. Warm temperate rainforest, Boorganna Nature Reserve

Plate 125. Cool temperate rainforest understorey, Otway Ranges

Plate 126. Cool temperate rainforest canopy, Otway Ranges

Plate 127. Cool temperate rainforest, 'Balfour Track', Tasmania

Plate 128. Victoria Pass, Tasmania

Plate 129. Tahune Forest Reserve, Tasmania

Plate 130. *Livistona australis*, *Eucalyptus grandis* in wet sclerophyll forest

Plate 131. Gallery rainforest

Plate 132. *Avicennia marina* (Acanthaceae/Avicenniaceae)-dominated mangrove forest

Plate 133. Littoral rainforest, Redhead, northern New South Wales

Plate 134. Littoral rainforest understorey, Seal Rocks, New South Wales

Plate 135. Canopy, Wooroonooran National Park, North Queensland

Plate 136. Understorey, Tully Gorge, North Queensland

Plate 137. Understorey, lowland rainforest, Mission Beach, North Queensland

Plate 138. Upland tropical rainforest, Bellenden Ker, North Queensland

Plate 139. Vine thicket pockets, Katherine River, Northern Territory

Plate 140. *Brachychiton rupestris* (Malvaceae/Sterculiaceae) in vine thicket remnant

Plate 141. Canopy gap, Victoria Park Nature Reserve, northern New South Wales

Plate 142. Damage to canopy by flying-foxes, Maclean, northern New South Wales

Plate 143. *Cissus antarctica* infestation

Plate 144. *Argyrodendron actinophyllum* (Malvaceae/ Sterculiaceae) in burnt mesic forest gully

Plate 145. Weed invasion following fire in dry rainforest

Plate 146. Dry rainforest prior to fire and weed invasion

Plate 147. *Asparagus aethiopicus* (Asparagaceae) infestation in littoral rainforest

Plate 148. Fire damage to littoral rainforest

Plate 149. 'Myrtle rust' re-infection of *Rhodomyrtus psidioides*

Plate 150. Coppicing *Mischocarpus pyriformis* following fire

5 Spatial and temporal structure of rainforest: general mechanisms that influence pollination and reproductive ecology

> *'Chance decides where the pattern starts; the nature of the system governs the sort of pattern that is elaborated on the random foundations. In this sense all such systems are unpredictable in detail, describable in general terms.'*
>
> *(Canny 1981)*

Cycles of day and night, and the environmental conditions and flowering events that are attendant or attuned to them, are primary drivers of pollinator activity. Most angiosperms flower during the day, however night, and the periods of dusk and dawn that precede and follow, have abiotic features that markedly differ from those during the day. Illumination, wind speed, turbulence and temperatures are lower, but humidity may be higher at night (Borges *et al.* 2016). These environmental conditions influence pollinator patterns and duration of foraging, and the kinds of pollinators that are active. For example, Xing *et al.* (2016), in a study of how butterflies might respond to microclimate in Australian tropical forests, highlighted how species responded differently to warming based on ecophysiological traits and how thermal refuges might emerge at microclimate and habitat scales. In general, they observed that larger and darker individuals were more prevalent in cool, closed canopy rainforests than in immediately adjacent or hotter open woodlands. Darker and larger butterflies preferred to be active in the shade and during crepuscular hours, while smaller and lighter coloured butterflies were more active in the sun and during midday. Their results demonstrated a thermal consequence of butterfly morphology across habitats and how environmental factors at the microhabitat scale may affect species distribution based on these traits. Such responses have ramifications for how flowering plants may be visited, and how pollen might be distributed within rainforest stands, and has particular relevance to pollinator foraging behaviour in highly disturbed remnants (e.g. those with large boundary to 'core' ratios) where little refuge from temperature impacts is offered.

In addition, the effectiveness and availability of particular pollinators is also influenced by the structure of the rainforest. Rainforests, especially those at lower latitudes, are frequently structurally and floristically diverse (see Plates 135, 136, 137), with numerous species occupying, and often restricted to, particular vertical layers from the canopy to the forest floor (strata). These layers are sometimes highly stratified but more commonly there is a broad overlap, the boundaries at best loosely defined. There are exceptions to the rule of diversity but, for many tree species in complex rainforests, mature individuals are scattered and there are difficulties in the transfer of pollen from one to another. If a pollinator is highly specialised, how does it find individuals of a particular species? If the pollinator is a generalist, seeking out whatever plant is in flower with little or no apparent discrimination, most pollen may be transported to other species and wasted, with no reproductive benefit.

In addition to the spatial discontinuity of flowers (resources) the temporal availability of resources in rainforest plant communities is normally patchy. That is, they are clumped or broadly scattered, rather than constant and uniform. This temporal heterogeneity, or mosaic, of resource availability also influences pollination. Pollinator populations may fluctuate in response to changes in resource availability and in response to various environmental factors. If there are insufficient resources over single or multiple seasons, due to severe or ongoing environmental perturbations (such as prolonged droughts, shifts in climatic patterns and associated flowering disruption, cyclone damage), the populations of individual pollinators, or even whole suites of taxonomically related pollinators, may crash. The long life-span of many rainforest trees and the large number of flowers that they can produce (in excess of that required to achieve adequate seed-set) may counter fluctuations in pollinator availability. It remains to be tested whether such a strategy can be sustained in the face of rapid climate change and rapid forest clearance (deforestation) and fragmentation, which also entails loss of pollinators over time due to inadequate home ranges and habitat niche loss.

PHENOLOGY: RECURRENCE OF THE FLOWERING PHENOMENON

Plant phenological events encompass flower bud production, length of flowering, fruiting and germination of seeds, and the timing of leaf production and abscission. Such events impact widely upon rainforests, as well as other plant communities, modifying the movements and feeding behaviour of pollinators, herbivores and frugivores as they respond to the diverse mosaics of resource availability that phenological episodes determine (e.g. Basilio *et al.* 2006). Thus visitor assemblages may vary widely over time as a consequence of the differences in plant phenology and the idiosyncratic and contextual effect of individual pollinators.

Climate change – some considerations

The implications of climate change – global warming are diverse, and impact with respect to individual species and species mutualisms, as well as stand structure and size, exposure within landscapes, and rainfall patterns (e.g. Laidlaw *et al.* 2011; NESP Earth Systems and Climate Change Hub 2019). Climate change is predicted to cause spatial and temporal mismatches between pollinators and their foodplants due to differential shifts in the distribution of species and the phenological patterns of interacting species. Mismatches can cause pollen limitation to plants and gaps in food resources to pollinators – thus the more generalist the relationship the more potentially resilient/adaptive the interaction under climate change (González-Varo *et al.* 2013).

Changes to temperature and rainfall patterns can ultimately lead to a reduction or shift in distribution in species, as well as changes in the composition of the community. Plant species occurring at lower (tropical) latitudes may be especially vulnerable due to a reduced ability to cope with warming; this owing to a lesser capacity to adjust to warmer temperatures because they have adapted to stable year-round thermal conditions (Crous 2019). For such plants warmer temperatures can result in reduced plant growth and this affects seed production and dispersal leading to less seedling establishment and forest dieback/retreat at larger scales. However, cascading extinctions under climate change are considered more likely to trigger animal extinctions than plant extinctions (Schleuning *et al.* 2016), with increasing temperature potentially resulting in significant reduction or complete loss of core environment of regionally endemic vertebrates (Williams *et al.* 2003). Nevertheless associated drought conditions may induce mortality with both catastrophic plant hydraulic failure and gradual carbon starvation being plausable (Choat *et al.* 2012). Choat *et al.* (2012) highlight that drought stress creates trapped gas emboli (effectively 'clots') in the water transport system, which reduce the ability of plants to supply water to leaves for photosynthetic gas exchange (Konrad and Roth-Nebelsick 2003). This results ultimately in dessication and death.

Thus shifts in rainfall patterns, and increasing temperatures associated with climate change, are likely to cause widespread species loss and overall rainforest declines in regions where droughts are predicted to increase in duration and severity.

Populations of many Australian subtropical rainforest trees, such as members of the families Myrtaceae and Sapindaceae, are highly synchronous in their flowering (G. Williams 1995). Synchronous 'mast' flowering is also exhibited by species such as *Argyrodendron actinophyllum* (Plate 41) and *Brachychiton acerifolius* (Plate 42), their respective tall white and red-flowered canopies scattered across the landscape. Synchronous flowering promotes conspecific pollen exchange within populations. These can be gender-specific in individual species, as with the female flowers of dioecious *Diospyros australis*, or occur within whole populations (as in *Eupomatia laurina*). Flowering synchrony is thought to be strongly controlled by environmental conditions (Newstrom *et al.* 1994; Rathcke and Lacey 1985) and the ability to flower when potential pollinators are also abundant is generally considered to be strongly selected for, resulting in interspecific synchronisation of flowering events in tropical rainforests (Sands and House 1990).

However, there are many examples of peak pollinator activity and sequential or differential interspecific flowering, which may indicate selection pressure to compete for pollinators or selection against hybridisation. Examples of subtropical rainforest trees flowering out of synchrony with peak pollinator densities include *Synoum glandulosum* (Plate 46) and *Dysoxylum fraserianum* (Meliaceae), *Backhousia sciadophora* (Myrtaceae), *Cupaniopsis anacardioides*, *Mischocarpus pyriformis* (Sapindaceae), *Orites excelsa* (Proteaceae), *Polyosma cunninghamii* (Polyosmaceae/Grossulariaceae), *Doryphora sassafras* (Monimiaceae/Atherospermataceae), *Flindersia bennettiana* and *Geijera salicifolia* (Rutaceae) (Boland *et al.* 1984; G. Williams 1995). These generally flower during cooler months when the diversity of pollinators is reduced and individual populations of pollinators are less numerous. Flowering of *Synoum glandulosum*, *Mischocarpus pyriformis* and *Backhousia sciadophora*, as well as other species, in subtropical latitudes regularly corresponds with the winter months (June–August) of the Southern Hemisphere, when day temperatures are predictably less than 20°C and night temperatures reliably less than 10°C, with few insects of any kind and size noticeably active on inflorescences. Certainly the diversity of potential insect pollinators plummets during this period, but large numbers at least of nematoceran flies and moths are still active during cooler months. In studies that I and Paul Adam undertook to investigate the possible pollinators of *Daphnandra micrantha* (Plate 47) in New South Wales lowland subtropical rainforest, nematocerans (43–67% of individuals, represented by more than 55 species, over two months sampling) and moths (7–14%, >20 species) greatly outnumbered all other insect groups collected in intercept trap samples collected during late winter and early spring (August, September) (G. Williams 1995). Other taxa collected were mainly small Collembola, book-lice (Psocoptera), brachyceran flies, beetles, wasps and ants. Although these include many insects that do not usually frequent flowers, it demonstrates that a pool of potential pollinators is available throughout the year in southern rainforests and that the role of nematoceran flies and small moths as pollinators may be greatly underestimated – being small in size and active mainly at night is a recipe for being overlooked.

There can be considerable variation in the flowering patterns of rainforest trees and shrubs, expressed heterogeneously by populations and individuals (Brown and Hopkins 1996). Flowering patterns can vary over altitudinal and geographic ranges and within rainforest strata. The variation in flowering events recorded in Australian rainforest is broadly in accord with the heterogeneous flowering patterns in rainforests elsewhere. Microgeographic (between-plot) variation in flowering can be great; an order of 10- to 100-fold has been recorded for Philippines primary rainforest (Heideman 1989). Variation in flowering will be reflected later in variation in the availability of fruit, although seasonal fluctuations in the proportion of fruit-set may not mirror oscillations in flower production.

Solar irradiance and seasonal climatic patterns of temperature, rainfall and subsequent soil moisture availability are generally considered to strongly influence flowering episodes in rainforest trees. Amongst South American Arecaceae flowering may peak during the rainy season (e.g. *Orbignya phalerata*, Brazil [Anderson *et al.* 1988]) or be restricted to the dry season (e.g. *Chaemaedorea pinnatifrons*, *Wendlandiella* sp., Peru [Listabarth 1993]). In South America there is a considerable body of folklore associated with enigmatic rodent population outbreaks, 'ratadas', that may follow widespread episodes of bamboo (*Chusquea*) flowering. Synchronous mass-flowering events are also recorded by Asian bamboo species (e.g. *Phyllostachys bambusoides*), but most species flower infrequently. Causes of mass-flowering (sometimes called general flowering) events by rainforest trees at the community level in tropical rainforest have included a drop of 2–3°C in minium daily temperature, and increases in sunshine (Sakai *et al.* 1999b; Van Schaik 1986). Ushio *et al.* (2019) found that general flowering in Bornean lowland rainforests has been found to be induced synergistically, rather than independently, by cool air temperature and drought.

However, hypotheses that suggest associations with El Niño and La Niña events often have been inconclusive – mass flowering in Sarawak rainforest has occurred in El Niño as well as La Niña episodes, and in adjacent peninsular Malaysia mass flowering appears to take place in normal and La Niña years (Ashton *et al.* 1988). Inouye *et al.* (2002) found significant reduction in flower production of *Delphinium barbeyi* populations in North American montane meadows were associated with La Niña episodes, and Wright and Calderón (2006) record an increase in flower and seed production during El Niño events in an old growth tropical moist forest in Panama. Among individual Australian subtropical rainforest plants there are cyclic patterns that are difficult to explain, such as that of certain *Eupomatia laurina* populations that predictably flower every two years (G. Williams pers. obs.). Nevertheless more adverse impacts have attended an El Niño onset. In a study of logged and undisturbed forest in East Kalimantan, Indonesia, (Slik 2004) extreme droughts, such as those that are associated with El Niño events were shown to affect the floristic composition and species richness of tropical rainforests; there was also a disproportionate mortality of certain tree species groups and size classes, and where disturbance affected light environments in the understorey there was reduced recruitment and growth of small and immature trees. Sixty per cent of the Way Kambas National Park, Sumatra, burnt during the 1997 El Niño drought, the El Niño event dramatically exacerbating ongoing man-induced fire impacts on native forests in Indonesia, these stemming from competing land use activities in Kalimantan and Sumatra (Byron and Shepherd 1998). Catastrophic fires (their impact exacerbated by subsequent prolonged drought) in northern New South Wales in late 2019 obliterated entire forested landscapes, such that whole suites of plant communities and their attendent pollinator faunas were extirpated; the wholesale loss of floral resources immediate, and their reconstituiton dependent on uncertain recolonisation by distant populations, or from residual communities surviving in isolated unburnt rainforest pockets.

Fire – an ever present and increasing threat to rainforest

Fire drives global biodiversity patterns and affects plant–pollinator interactions (Banza *et al.* 2019). Fire is a major limiting factor in the stability and advance of Australian rainforest boundaries (Little *et al.* 2012; Russell-Smith and Dunlop 1987) and represents an increasing threat to the survival of spatially small stands, or where rainforest exists as narrow bands or mosaics within highly flammable vegetation types (Plate 117). Fires are expected to become more widespread and prevalent worldwide under climate change (Jolly *et al.* 2015). Their impact is exacerbated where drought conditions extend past fire events. Fire can alter the relative abundance, composition and distribution of plant species to which plant–pollinator interactions respond at multiple spatial scales (Kelly *et al.* 2015). Various plant traits that may mediate flowering responses to fire include growth form, phenology and bud location, seasonal changes in bud exposure and responses to bud damage. However, predicting the ecological consequences of fire requires knowledge of relationships between fire regimes and biota (Brown *et al.* 2017). As an example plant density can act as a mechanism by which fire regimes indirectly impact floral visitation; where species regeneration in post-fire environments may result in increased (or reduced) abundance and proximity to which pollinators respond (Van Nuland *et al.* 2013). Post-fire flowering (reproductive maturity) can precede the recovery of pollinators (and so pollination), by several, if not many, years (Geerts *et al.* 2012); with the rate of pollinator recovery depending on fire characteristics such as size and patchiness. Thus the impact of subsequent fires that occur before significant reproduction (which can cause local extinctions), might be mediated by spatial patterns of fire (Watson *et al.* 2012). Fire also promotes invasion by alien plants, such as the southern African bitou bush *Chrysanthemoides monilifera* (Asteraceae) in littoral rainforest, disturbance to canopy cover allowing penetration of higher light levels to the forest floor, to which alien plant species respond (Plate 148). Post-fire invasion by grasses can also threaten rainforest stands (and plant reproduction processes), especially those of vine thickets, owing to the increased fuel loads that these represent (McDonald 2010).

Tropical climates are relatively seasonal (normally markedly wet or dry), so the significance of abiotic factors on the production of flowers in subtropical and temperate rainforests (where seasonality is more diverse or subtle, as there are no obvious wet and dry seasons) may be different. In northern Australian monsoon forests, major peaks in flowering events coincide with leaf flush (Hopkins and Graham 1989). In wet monsoon forest this occurs before the commencement of the wet season, but in dry monsoon forest leaf flush coincides with the beginning of the wet season (Bach 2002). Flowering peaks are often associated with drought or, for riverine forests, low river levels. The function of plant water status, as a physiological mechanism initiating onset of flowering, is unclear (Wright and Cornejo 1990) particularly where trees may shed leaves, produce new leaf flush, and initiate inflorescence and fruit production in any month. Borchert (1983) suggests that physiological factors that may be related to seasonally induced changes in water status may trigger phenological events, such as leaf drop, and the timing of such events through interrelated physiological mechanisms could influence other phenological events such as the initiation of flowering.

In East African riverine trees there is marked temporal and spatial variation in flowering, with flowering apparently triggered by low river levels rather than local reductions in rainfall (Kinnaird 1992). In species-rich tropical and subtropical rainforest, numerous trees in study plots are often represented by a single individual and interpretation of phenological data, be it of flowering or leaf production, can be fraught with the possibility of inaccurate conclusions based upon outcomes from an unrepresentative or inadequate sample population.

In North Queensland 'dry' rainforest most canopy trees tend to flower in the wet season (Hansman 2001), but in South-east Asian tropical monsoon forest the number of species in flower peaks at the end of the dry season (Gardner *et al.* 2000), with dominant species flowering annually. This is opposed to the supra-annual general flowering that occurs in South-east Asian tropical rainforest (Kato *et al.* 2008). Among the Fagaceae (e.g. *Castanopsis*, *Lithocarpus*) in Borneo flowering occurs during the dry season; for entomophilous species this maximises pollination as this is the season in which insect populations peak (Kaul *et al.*1986). Flowering of the woody species (4 trees, 6 shrubs, 3 vines), studied by Nayak and Davidar (2010) in endangered tropical dry evergreen forest remnants (0.5–~10 ha) in peninsula India, occurred during the dry season (January–July). However, in Bolivian dry forest flowering is bimodal, with a major peak at the end of the dry season and a minor peak during the wet season (Justiniano and Fredericksen 2000).

In Malaysian lowland dipterocarp forests (those dominated by Dipterocarpaceae) several patterns have been noted among flowering trees. There are those that are periodic but regular flowering species, those that predictably flower each year, and those that flower supra-annually (Yap and Chan 1990). In New Guinea lowland tropical hill forest (Brown and Hopkins 1996) the total number of species and trees flowering (as well as numbers of available flowers) fluctuated during the year, decreasing in the dry season and increasing during the wet season. There was also a wide range of temporal variation within and among tree species in the length and timing of the flowering period, the degree of synchronicity amongst conspecifics in flower, and the percentage of each population that was in flower each year. Birds responded differentially to these events. Resident birds were correlated with synchronously flowering trees and smaller blooming displays per tree but nomadic birds were correlated with trees that flowered asynchronously and seemed to range in search of resource-rich, massively blooming trees that were unpredictable in their flowering episodes.

Many Neotropical tree species flower each year, although numbers of trees flowering within populations, and numbers of flowers produced by individual trees, may vary between years (e.g. in *Cavanillesia platanifolia* [Bombacaceae] [Murawski *et al.* 1990]). Intuitively, this is probably the norm for most species irrespective of the rainforest community in which flowering events occur. Continuous flowering in the strict sense is uncommon. Rainforest trees that flower over extended periods, or continuously, need to partition flowers and developing fruit so that pollination and dispersal functions do not interfere with each other. In continuously flowering individuals of the lowland Central American tree *Muntingia calabura* (Elaeocarpaceae), the elongation and erection of the pedicel raises the flower buds above branches where they are pollinated by bees (C. Webb 1984). Later, the developing fruit are lowered, concealed from predators below leaves and finally dispersed by bats. For the majority of woody rainforest flowering plants, considered as individual species, the possibility of interference between pollination and dispersal vectors does not arise because fruit maturation and dispersal occur after the cessation of flowering. However, where different species flower and fruit simultaneously, interference may occur.

Contrary to flowering patterns, continuous patterns of leaf production are common in tropical forests (Frankie *et al.* 1974). Patterns of leaf production in tropical rainforests are often indirectly related to rainfall (Wolda 1978), although individual responses by species to the same environmental cue may differ and patterns of leaf and flower production in closely related species (e.g. *Shorea* spp. [Asian Dipterocarpaceae], Neotropical Moraceae) can be widely divergent (Dayanandan *et al.* 1990; Milton 1991). Similar divergences occur among closely related Australian subtropical and tropical species, such as various members of the genus *Syzygium* (Myrtoideae/ Myrtaceae). Worldwide the subfamily Myrtoideae demonstrates a tendency to flower at the dry to rainy season transition (Lughadha and Proença 1996), however, individual species can exhibit distinct annual patterns of flowering. In North Queensland rainforest *Syzygium tierneyanum* flowers in January but *Syzygium cormiflorum* has a flowering period that can last from July to mid-November (Crome and Irvine 1986; Hopper 1980). Other climatic factors may also influence flowering, with severe and prolonged drought acting to suppress flowering in stands across the whole plant community, with consequent loss of floral resources to flower-dependent animals. In the temperate rainforests of Tasmania, delays in flowering of *Eucryphia lucida* and *Atherosperma moschatum* can be related to slightly lower temperatures in the growing season (Read 1989). Different species are likely to respond differentially to environmental cues, so that year-to-year variation in flowering patterns are unlikely to be in phase for all species.

Extreme storm events may also affect flowering events. Hopkins and Graham (1987) studied the impact of Cyclone Winifred on flowering patterns in northern Australia, and noted that ~80% of smaller trees and shrubs along a 4 km transect in cyclone-damaged forest were in flower. They suggested that such widespread, episodic, environmental impacts may effect the synchronisation of flowering of a larger number of species than would normally take place, and that long-term changes in the species composition of damaged forests as a result of coincidences of seed availability and regenerative space may occur.

Life cycles of phytophagous insects are generally synchronised to the availability of food resources, to lowered chemical defences or increases in palatability of foliage, with flower and leaf production and adult insect emergence probably cued by similar environmental triggers. Although there may be a lag in pollinator response to the availability of flowers, in both Neotropical lowland rainforests and tropical Queensland montane rainforest peaks in flowering and insect numbers generally coincide (Sands and House 1990), though interactions can fluctuate monthly (Basilio *et al.* 2006). However, in Australian tropical and subtropical rainforest, and in sclerophyllous communities dominated by Myrtaceae, mass-flowering events are not always paralleled by high numbers and diversity of insects, a phenomenon well known to field naturalists and entomologists. This suggests that at least some entomophilous pollination systems may be pollinator- rather than resource-limited in some years. Unreliability in the presence of pollinators may act as selection pressure in favour of self-compatibility and or broad flexibility in potential pollinators. For long-lived rainforest trees, there may not be strong selection in favour of frequent reproduction. In shorter-lived pioneer species in Australian subtropical rainforest, frequent flowering is usually predictable (e.g. G. Williams 1995), suggesting that regularity of reproduction is favoured or normal.

The extinction of insects?

Insects, and invertebrates more generally, play a central role in ecosystem function, but are understudied and poorly appreciated. However, in recent years extinction events and perceived declines in insect biomass, abundance and diversity at a global scale have been highlighted (e.g. Dirzo *et al.* 2014; Samways *et al.* 2020; Sánchez-Bayo and Wyckhuysb 2019). In particular, evidence of the risk of pollinator extinction (not solely that of insects), and threats to the ecosystem services they provide to mankind, is compelling and has received increasing attention (e.g. Powney *et al.* 2019; Thomann *et al.* 2013). Identified threats to pollinators, and invertebrate species and populations more generally, include deforestation, loss of microhabitats, pesticide use, pathogens, climate change and alien species (Vanbergen 2013). Extreme and recurring drought, asynchronous flowering patterns and loss of plant resources, also impact on species. Outcomes of pollinator decline include pollen limitation and a reduction in plant fitness, threatening natural plant populations with extinction and agricultural crops with reduced pollination services and subsequent lower production yields.

In Australian subtropical rainforest, most tree species flower between September–October and March (Boulter *et al.* 2011; Floyd 2007). There is, however, specific overlap with, and divergence from, this general flowering period. Most Myrtaceae flower (at different amplitudes) annually although the onset of flowering may vary between years. Flowering episodes in some species (e.g. *Acradenia euodiiformis* [Rutaceae], *Guioa semiglauca* [Sapindaceae], and *Dysoxylum fraserianum* [Meliaceae]) are separated by several years in which no or few flowers are produced. Phenological patterns within families may vary; for example, within the Rutaceae some species flower unpredictably (e.g. *Acradenia euodiiformis* [Plate 80]), flower annually with relative constancy of a flower crop (e.g. *Acronychia oblongifolia* [Plate 106]) or generally flower annually but with frequent marked variation in numbers of flowers produced (e.g. *Acronychia imperforata* [Plate 83]) (G. Williams 1995). Related species within genera may also vary with respect to season. Again in the Rutaceae, the genus *Flindersia* includes late spring- and summer-flowering species (*Flindersia schottiana* and *Flindersia xanthoxyla*), species whose flowering events are primarily restricted to spring (*Flindersia australis* and *Flindersia collina*) and autumn – winter-flowering species (*Flindersia bennettiana*).

Distinct flowering strategies are also encountered in Australian tropical rainforest trees. In North Queensland populations of *Neolitsea dealbata* and *Litsea leefeana* (Lauraceae), flowering occurs over periods of three to four weeks and four to five weeks respectively (periods similar to those noted for many Australian subtropical rainforest trees). However, in *Diospyros pentamera* (Ebenaceae) flowering takes place over a greatly extended period of more than 15 weeks (House 1985). These are periods of flowering length that relate to the flowering phenology of species populations, not flower life-spans on individual trees nor the phenological patterns of unisexual flowers. The latter exhibit a further degree of phenological complexity. The male flowers of *Guioa semiglauca* (Sapindaceae) commence anthesis approximately one to three weeks before the opening of female flowers, the female flowers of dioecious *Diospyros australis* (Plate 28) and *Euroschinus falcatus* (Plate 7) open synchronously or as an episodic flush within a longer duration of male flowering. Though male flowers of *Euroschinus falcatus* commence opening before that of females, bud burst on individual panicles is more prolonged with fewer florets (little flowers) available at any one time. Maximum opening of male flowers occurs at the end of the flowering period (G. Williams 1995). This may 'train' pollinators and mobilise pollen before availability of female flowers. In the rare and endangered dioecious shrub *Gardenia actinocarpa* (Rubiaceae), in North Queensland lowland rainforest, flowering can extend over six months, with length of flowering correlated with plant size in both male and female plants (Osunkoya 1999). Male *Gardenia actinocarpa* plants commence flowering before females, produce more flowers and flower longer than females. The populations of many subtropical rainforest species in Australia (e.g. *Rhodomyrtus psidioides* [Plate 58] and *Alpinia caerulea* [Plate 79]) flower synchronously, but the flowers of individual plants open sequentially. In several species there is a conspicuous sequential opening of flowers along the length of the flower spike (e.g. *Gymnostachys anceps*).

Species groups that exhibit different phenological patterns (e.g. annually constant, supra-seasonal absence of flowering) rarely suggest a distinct association with the dynamics of rainforest succession, pollination and dispersal ecology or breeding systems. These groups represent markedly heterogeneous assemblages of pioneer, 'climax' (a misnomer, as rainforests are usually in a state of change relating to floristic composition and abundance) and later successional stage species from diverse families. They may express varying flowering patterns between sites, such that there is an absence of local and regional predictability as well as lack of constancy within any rainforest stand.

Subtropical rainforest trees can be grouped into four broadly defined spatial and temporal heterogeneous flowering: species that generally flower each year at all stands (although the length of the flowering season within populations may vary between years) and in which all mature individuals normally flower intensively (e.g. *Commersonia fraseri* and *Hibiscus splendens*); species that generally flower intensively each year at each stand but with occasional years in which trees may not flower, or a few trees flower poorly (e.g. *Rhodamnia rubescens*, *Tristaniopsis laurina*, *Syzygium floribundum*, *Cryptocarya glaucescens*, *Mischocarpus pyriformis*, *Cuttsia viburnea*, *Guilfoylia monostylis* and *Alphitonia excelsa*); species that habitually flower with fluctuating intensity at different sites and fail to flower at some sites in some years (e.g. *Rhodomyrtus psidioides*, *Syzygium smithii*, *Elaeocarpus obovatus*, *Streblus brunonianus*, *Cryptocarya rigida* and *Wilkiea huegeliana*); and trees that do not flower regularly at any stand, and in which regional flowering by populations or individuals is often episodic and conspicuous (e.g. *Brachychiton acerifolius*, *Gmelina leichhardtii*, *Argyrodendron actinophyllum* and *Doryphora sassafras*) (G. Williams 1995). However, in situations of drought-induced extreme water stress flowering events may be unpredictable, with many species failing to flower (even though buds may be at first produced).

Although seasonal variation in the intensity and numbers of species flowering may have several explanations, the spatial and temporal heterogeneous patterns of rainforest flowering (and resource availability) must strongly influence pollinator recruitment and distribution, and the establishment, retention and nature of pollination mutualisms. Where pollinator abundance is lowered as a result of a reduction in floral resources there can be pollination limitation, leading to an increased risk of pollination failure for plant species existing at low population levels or that possess reproductive traits and strategies (e.g. dioecious species) that place them at a higher risk of localised extinction (Wilcock and Neiland 2002). We should not forget that rainforests typically are composed of tree species that are predictably present in individual stands (matrix or core species), as well as species whose presence within stands cannot be forecast (satellites). Such patterns do not mean that distribution is random, although in some cases it might be, or might be explained by species-specific responses to gradients in water availability and altitude (Balvanera *et al.* 2011; Boulter *et al.* 2011) rather than chance recruitment/extinction events or physiological predilection. This mosaic-like floristic composition imposes a landscape-scale layer of spatial heterogeneity upon the seasonal availability of flowers to pollinators, which in turn can influence the establishment of plant–pollinator relationships.

LENGTH OF FLOWERING LIFE

Other factors which are important to pollination in Australian rainforest trees include the length of life of individual flowers, the pattern of opening of flowers within inflorescences, and the positioning of inflorescences. Even when individual flowers are inconspicuous their positioning within inflorescences may make spectacular displays, while the length of time inflorescences are displayed may determine whether a particular pollinator becomes conditioned to visiting a species.

The longevity (the period between opening and senescence) of individual rainforest flowers has rarely been investigated although the life span of flowers could have a strong selective influence on pollinator visits and could affect the plant's reproductive energy budget. There is general correlation of increased flower longevity with increased altitude, and longevity is greater in more mesic habitats (such as rainforest) (Primack 1985). However, ecological differences between adjoining communities, while sufficient to alter floristic composition, do not necessarily alter mean floral longevity. There is evidence (for example, from studies in Costa Rican cloud forest and semi-deciduous forest) to suggest that strong phylogenetic constraints operate at the family level to limit the evolution of flower longevity (Stratton 1989). Factors that may promote a longer flower life span include a scarcity of pollinators, unpredictable visitation by pollinators, and pollen limitation in plants with many ovules per flower requiring longer periods to achieve maximum seed-set. Selection for male function in hermaphroditic flowers has been suggested as an important influencing factor, however, greater flower duration has been recorded in Neotropical dioecious female plants and in the female phase of dichogamous bisexual flowers.

Flowers in high-altitude tropical forests last an average two to eight days longer than those in tropical lowland rainforest, the longer life span possibly consistent with the notion of unpredictability in pollination at higher altitudes (Bawa 1990). Flower longevity is also related to structure; harsh environments might select for stronger structure, size and form in flowers which, coincidentally, live longer. Australia has several localities in the tropics and subtropics where there are continuous sequences of rainforest from lowland valleys to mountain tops. Studies of flower longevity and flowering behaviour along these gradients would be of considerable interest.

Although there are few data on the longevity of individual flowers on rainforest plants in general, limited studies on the life spans (i.e. receptivity) of Australian subtropical rainforest flowers reveal considerable variation. For example, flowers of *Rhodomyrtus psidioides*, *Tristaniopsis laurina*, *Syzygium floribundum*, *Syzygium smithii*, *Alphitonia excelsa*, *Acradenia euodiiformis*, *Guoia semiglauca* and *Cuttsia viburnea* have average life spans of five to seven days, those of *Daphnandra micrantha* last approximately four days). *Diospyros australis* (Plate 28) and *Eupomatia laurina* (Plates 35, 36) persist for only a single day before being shed as a single unit

(G. Williams 1995), however, the fused floral bodies of *Eupomatia bennettii* are 2-day flowers (Endress 2003. Female flowers of *Wilkiea huegeliana* (Plate 48) on average last longer than those of males (mean longevity 12 compared to 8 days), but male flowers of *Euroschinus falcatus* outlast those of females (mean longevity more than eight days as opposed to approximately four) (G. Williams 1995). Individual male flowers of the dioecious Australian rainforest tree *Fontainea picrosperma* (Euphorbiaceae) open for 1–2 days, with male inflorescences open for a maximum of 14 days, and that of females 11 days (Grant *et al.* 2017). Duration of anthesis in the tropical Queensland vine *Austrobaileya scandens* is about five days (Endress 2001). Individual flowers of *Syzygium sayeri* and *Eupomatia barbata*, both tropical Queensland species, last for three and one day respectively (Boulter *et al.* 2005; Endress 2003). These are all shorter life spans than the 10–16 day life span recorded for the flowers of protogynous *Idiospermum australiense* in the rainforests of tropical northern Queensland, during which female and male expression of function is reassigned and spatially separated (Worboys and Jackes 2005).

The primary factors underlying flower longevity are not fully understood but are likely influenced by individual breeding systems and seasonal environmental conditions; episodes of water stress often result in pronounced flower abortion. High levels of inexplicable spontaneous abortion in fertilised flowers can occur, as recorded for *Syzygium cormiflorum* (Plate 59) in tropical Queensland (Crome and Irvine 1986). Pollination and fertilisation success and anther dehiscence are also strong determinants of how long an individual flower will be retained (or continue to be receptive or possess viable pollen) before abscission. Even when flowers have shed perianth segments, the flower may remain functionally capable of pollination or pollen presentation (*Syzygium smithii* loses its petals on the first day of anthesis) or continue to produce nectar (*Acradenia euodiiformis*) (G. Williams 1995). For most flowers studied, nectar production does not continue once they have become senescent.

FOREST STRATA AND SYNUSIAE

Structural complexity and floristic composition vary at any point in a forest, both horizontally and vertically, with the stage of the regeneration cycle and within the succession sequence (Hopkins 1990). Rainforest is constantly in a state of change (even though apparently stable to the fleeting appraisals of human life spans) or is responding to localised disruption such as tree fall (Plate 141) and storm damage (cyclones cause disturbance at the landscape scale, resulting in impacts to the populations of individual species as well as the community level). Flowering behaviour and pollination requirements are likely to vary in a multidimensional temporal–spatial mosaic reflecting the regeneration cycle (Adam 1992).

The understorey/subcanopy plant layers (strata) (e.g. Plates 118, 119, 120, 122, 124, 125, 134, 136, 137, 146) in rainforest usually intergrade, and so recognition of distinct vertical layers is often more in the eye of the beholder than real. Nevertheless, change in forest height can result in associated changes to floral species composition, temperature and light regimes, foliage density and resource abundance to which animals respond (Ramírez *et al.* 2016; Stork *et al.* 2016), such that recognition of emergent, canopy and understorey layers may be a useful level of subdivision. Recognition of specialised synusiae may also be useful, since the patterns of flowering behaviour may differ between strata and synusiae in the one forest. These different flowering patterns and different microclimatic conditions may mean that different pollination mechanisms are favoured at different positions within the forest structure, or even within the varying height of individual trees.

In Malaysian lowland dipterocarp forests three stratified layers, or synusiae, have been recognised (Appanah 1990). Within these there is differential production of flowers, to which pollinators and other animals respond (Kato 1996). The strata comprise a canopy layer in which flowering trees hold their flowers above the foliage, a second synusia in which flowers are borne within or below the canopy foliage (this includes species whose flowers are carried on trunks [cauliflory, cauliflorous] or lateral branches [ramiflory, ramiflorous]), and a ground-layer synusia composed of plants that flower or grow close to the forest floor. In Australian subtropical rainforest ramiflory is an uncommon strategy; *Myrsine variabilis* (Plate 53) and *Myrsine howittiana* (Myrsinaceae/Primulaceae) are examples. Flowers (within the 'fig' synconium) of *Ficus coronata* (Moraceae) are commonly cauliflorous (Plate 49), but often ramiflorous on large trees. Flowers of *Stenocarpus sinuatus* (Plate 69) are normally terminal, but may also be ramiflorous. Flowering epiphytes such as orchids occupy a range of vertical strata and forest gaps are occupied by transient species (frequently shrubs, herbs) (e.g. *Alocasia brisbanensis*, *Pollia crispata*) (Plates 9, 22) that are eliminated as the gap closes, vines that later occupy the upper canopy, and tree species that may finally grow to form part of the canopy or survive as suppressed elements in the dimly lit subcanopy layer (Plate 141).

Such a broad structural characterisation occurs in Australian tropical and complex subtropical rainforests. The recognition of stratified synusiae adds a vertical dimension to pollinator and resource selection in the rainforest community to which insects (e.g. bees and wasps) and vertebrates commonly respond (e.g. Janzen 1971; Sobek *et al.* 2009); though, interestingly, Morris and Sinclair (2015) found no significant differences in insect abundance, diversity or parasitism rate between canopy and subcanopy strata in an Australian subtropical rainforest, though they did find that food web structure significantly changed with elevation. This notwithstanding, it is in the below-canopy synusiae that many instances of close plant–pollinator mutualisms have been discovered. However, pollinator responses to floral resources within synusiae also have a behavioural basis. For example, it has been noted that, in Neotropical lowland forests, approximate stability (fidelity of a pollinator to synusiae) in the distribution and type of pollinator exists between understorey and canopy and is reflected in flowering strategies (Perry and Starrett 1980). Pollination of understorey plants is dominated by insects exhibiting trap-lining syndromes (pollinators undertake long-range foraging of a relatively constant spatial and temporal pattern), while mass-flowering dominates the canopy layer. The two strategies may be related to pollinator density, and 'energetics' (caloric reward and energy cost of foraging) probably serves as a causal mechanism determining the vertical distribution of pattern of strategies (see Kearns and Inouye 1993; Willmer 2011). In the case of sparsely distributed rainforest plants, trap-lining might provide a mechanism for pollination but the distribution of plants can be explained in many ways unrelated to pollinator requirements, such as chance dispersal and founding events and availability of appropriate microhabitats (Harrington *et al.* 1997).

Although the plant members of various tropical forest synusiae possess broadly defined characteristics of floral morphology, colour and scent cues, structure, floral rewards and breeding systems, the restriction of pollinators to the various rainforest strata is not necessarily obligate. For example, in South-east Asia, cauliflorous trees may recruit pollinating beetles and bees from the understorey synusia, and bats and birds from the upper canopy layer descend to forage on ramiflorous *Durio malaccensis* and *Ganua* species (Appanah 1990). Likewise, in Panamanian lowland rainforest, canopy-foraging *Melipona* bees descend to forage *en masse* on synchronously mass-flowering understorey shrubs (Augspurger 1980). Under conditions of low understorey flower resources, however, *Melipona* bees only occasionally forage in the understorey. Mass-flowering of individual plants is required to initiate foraging responses by *Melipona* colonies.

6 Australian vegetation history and its influence on plant–pollinator relationships

The Australian continent has been separated from other major continental landmasses for ~34 million years, and this temporal length of isolation, and the north-west drift of the continent, has resulted in a high level of plant endemism; ~6% of families, 22% of genera and 80% of species being endemic, with many lineages tracing their ancestry to the Palaeogene and Cretaceous (Dyer *et al.* 2012; Hill *et al.* 1999).

The modern Australian vegetation, at least in the south-east, has its genesis in the Late Cretaceous (c. 98–65 mya), being at that time part of the connected landmass additionally comprised by what is now South America, Western Antarctica and New Zealand, and was an environment that is totally foreign to that of today (R. Hill 2004). The present species composition of Australia's rainforests reflects a distinctive east Gondwanan, as well as an Asian, ancestry (Crayn *et al.* 2006; Delmas *et al.* 2020), with vegetation more generally reflecting episodes of extinction and adaptation due to climatic change occurring during the Late Tertiary and Quaternary (McKenzie 2002), as well as 38 000–45 000 years or more of Indigenous Australian burning practices (Benson 1991; see Russell-Smith and Dunlop 1987 for discussion of the impact of burning practices on northern Australian monsoon forests); though fire-adapted Gondwanan angiosperms (in Australia; and southern Africa) had already appeared ~60 million years ago (Carpenter *et al.* 2015; Lamont and He 2012; also see Bond and Midgley 2012; Bond and Scott 2010 for discussion of fire and the spread of flowering plants in the Cretaceous). However, Rule *et al.* (2012) suggest that the arrival of humans led to the extinction of the Australian megafauna and this triggered the replacement of mixed rainforest by sclerophyll vegetation – this being achieved by direct impacts on vegetation through relaxed herbivore pressure and an increase in the incidence of fire in the landscape. They contend that this ecosystem shift equalled any effect of climate change over the last glacial cycle, indicating 'the magnitude of changes that may have followed megafaunal extinction in the world'.

For much of the Tertiary rainforest was the predominant vegetation over most of Australia (Truswell 1990; though see Carpenter *et al.* 2015 for a consideration of the extent of a Late Cretaceous open sclerophyll habitat). With the onset of drier conditions in the Late Tertiary, rainforest retreated in the face of the spread of sclerophyll vegetation (Byrne *et al.* 2011), with instances of of dry-adapted taxa radiating from rainforest tree progenitors during this period (Crayn *et al.* 2006). During the Quaternary there have been substantial fluctuations in the extent of rainforest. Even within what is now the most extensive contiguous rainforest in Australia – north-east Queensland – the rainforest was fragmented and greatly reduced in area during dry (glacial) periods. About 6000 years ago, wet sclerophyll forest and *Nothofagus* had reached their maximum postglacial extent in the south-central highlands of Victoria and underwent subsequent retreat to their present range (McKenzie 2002). Although shared predispositions could have served to retain some plant–animal linkages, fluctuations in the extent and nature of rainforests may have reduced opportunities for the establishment or retention of complex interdependencies and strong co-evolutionary mutualisms.

Plants and pollinators participating in obligate pollination mutualisms will be at risk if they respond differentially, or lack the ability to respond, to climatic and environmental fluctuations over evolutionary time spans. Communities in which obligate mutualisms are abundant may therefore be sensitive to instability. Over shorter time scales (decades and centuries) the spatial and floristic complexity of plant distribution in rainforest can be initiated by chance events of canopy disturbance, fortuitous seed dispersal and canopy establishment (Floyd 1990; Xu and Li 2002). The resulting stochastic instability within the trophic structure in rainforest communities may further preclude the establishment and maintenance of tightly circumscribed plant–pollinator relationships.

In light of this temporal variability in resource availability, flexibility in the pollination strategy of rainforest plants – what might initially be considered a generalist strategy – may really be a specialised adaptation rather than an indication of lack of specialisation.

PLANT–POLLINATOR INTERACTIONS

Flowers of individual species in the rainforest are, from the viewpoint of the potential pollinator, a resource which is variably abundant in space and time because of the lack of synchrony in flowering between and within populations (Newstrom *et al.* 1994). Some species may flower each year, although intra-seasonal length of flowering and the numbers of flowers produced may vary. Flowering by regional populations of other species can be episodic, with flowering events separated by several or more years. From ecological and evolutionary perspectives, this distribution of resources suggests that opportunities for closely co-adapted plant–pollinator relationships would be rare. Most pollination would be effected by pollinators which, although they might be specialised (visit particular flower types or have selective and efficient foraging patterns), are not closely tied to a particular plant species. To test this hypothesis, it is necessary to identify the pollination vectors.

This is not an easy task. Many statements regarding pollination are often generalised inferences drawn from flower structure (which can be misleading) and behaviour. Certain syndromes of floral characteristics have been correlated with particular modes of pollination (Faegri and van der Pijl 1979; Fenster *et al.* 2004). While a

Wet sclerophyll forest – a transitional rainforest type

Australian rainforests frequently occur as spatially discrete 'closed canopy' plant communities within otherwise non-rainforest landscapes (Plate 117). However, rainforests are often bounded by mesic forests whose overarching canopies are dominated by dry-fruited eucalypts and related genera (e.g. *Lophostemon*). These so-called 'wet sclerophyll forests' (also known as 'tall open forest') exist under conditions of high annual rainfall, and can occur as narrow transitional zones between rainforest and more open, xeric-adapted, plant communities. The immediate boundaries between wet sclerophyll forest and rainforest are often floristically dynamic, rarely with well defined lines of demarcation (such that in effect no distinct boundary may be drawn), with the transitional (ecotonal) nature of the canopy and subcanopy flora allowing for a rich ecological diversity of plant–pollinator relationships (Plate 120). Wet sclerophyll forests are widely represented in the coastal and associated hinterland regions of south-eastern Australia, in addition to Tasmania and the extreme south-western corner of Western Australia. More limited occurrences are to be found in north-eastern Queensland. They are characterised by tall eucalypts (e.g. *Eucalyptus grandis, E. dunnii, E. microcorys, E. pilularis, E. saligna, Corymbia variegata*) with a subcanopy stratum constituted of rainforest species. This substratum may comprise a dense 'closed' assemblage of establishing trees and shrubs, but these often exhibiting floristically impoverished patterns along gradients that are influenced by increasing exposure, and reduced humidity, soil fertility and soil moisture. Unlike rainforest wet sclerophyll forests can generate very high fuel loads (Plate 130). Fire is a strong limiting determinant of wet sclerophyll forest – rainforest boundaries, their frequency and intensity restricting the expansion of rainforest. But in the absence of fire eucalypts within wet sclerophyll forest will progressively fail to regenerate; the formation evolving to that of pure rainforest (Plate 120) (Adam 1992).

particular syndrome may not be an absolutely reliable indication of pollination mechanisms, it can be a useful predictor of biotic pollination vectors at the higher taxonomic level. It cannot indicate the specific identity of the agent. An example of the uncertainty of predicting specific agents is offered by *Cayratia clematidea* (Vitaceae) (Plate 75), a weakly climbing native vine commonly encountered on the margins of subtropical rainforests of New South Wales. The composition of the attendant anthophilous fauna can exhibit major spatial differences, both with regard to the nature of the participating species and that of their morphological adaptation to feeding from floral resources. During March–April 2020 flowering *Cayratia clematidea* plants growing on the margin of a fire-impacted littoral rainforest near Harrington (northern New South Wales) were visited commonly by wasps of the families Vespidae, Gasteruptiidae and Thynnidae, and tachind and syrphid flies; all short-tongued species. However, during the same period, flowering *Cayratia clematidea* vines growing profusely on the margin of a fire-impacted dry rainforest, in Kiwarrak State Forest to the south-west (Plate 145), were visited exclusively by large numbers of long-tongued pierid butterflies (predominantly *Appias paulina*, 109) (G. Williams pers. obs.). In each case plants developed a high proportion of fruit (>50%) from available flowers.

Recognition of specific pollination syndromes requires identification of visitors to flowers, from direct observation or sampling, and assessment of the efficiency of pollen transfer and subsequent fertilisation. The factors are difficult to assess without experimental manipulation and it is easy to reach misleading conclusions from databased on the abundance, frequency and conspicuousness of visitors. In any discussion of the dispersal of pollen by pollinators, a distinction needs to be made between the capacity of the pollinator to *transport* pollen and its ability to *transfer* pollen, resulting in placement of pollen on an appropriate stigmatic surface. *Transport* can be relatively easily assessed, *transfer* is much more difficult to demonstrate.

FACTORS AFFECTING MOVEMENT AND RECRUITMENT OF POLLINATORS

Heinrich (1975) suggested that the massive peak-phase blossoming of mass-flowering trees could attract large numbers of pollinators that would deplete the energy resource provided by flowers. The pollinators would be obliged to move to other conspecific trees, thus promoting pollen transfer between plants and achieving out-crossing, which primarily took place during this peak flowering stage. Carpenter (1976) studied the pollination of the Hawaiian myrtle, *Metrosideros collina*, and hypothesised that out-crossing took place in early- and late-phase flowering of the study plants and that self-pollination dominated the peak phase of flower availability. In contrast Stephenson (1982) found that out-crossing predominated in the late, nectar-depleted flowering phase of North American *Catalpa speciosa* and that a large proportion of the receptive flowers available in the peak flowering phase remained unpollinated. This suggested that the flowers had retained their nectar reward. Stephenson considered that early- and peak-phase flowering served to recruit then condition pollinators to that species, with inter-tree movements increasing as conditioned insects feed on the resource-diminishing late-phase *Catalpa* trees. Differences in these predicted out-crossing and recruitment hypotheses may be explained by differences in the breeding systems of the plants studied. The mechanisms by which the insects locate flower resources then remain at rewarding resources, or continue resource searches elsewhere, can be partly explained by simple neural responses.

However, factors which influence the movement of pollinators between plants may be both biological and environmental.

Biological factors promoting pollen transfer between sparse- and mass-flowering plants include temporal nectar flow pulses between sexes in dioecious species, periodic asynchronous cessation of flowering in male trees, increased female flower longevity, nectar satiation-induced out-crossing, dilute nectars forcing pollinators (birds) to forage more widely to satisfy energy requirements (and thus promote out-crossing), insect density-induced disturbance, territory (resource) defence and disturbance by insectivorous birds that induce other pollinators to disperse between flowering plants. For example, in southern Chile the tree *Embothrium coccineum* occurs in forest fragments and pastures, also a common scenario for Australian subtropical rainforest trees. The trees, which are a resource territory to their pollinators, are often defended by a pollinating flycatcher, *Elaenia albiceps* (Devoto *et al.* 2006). However, *Elaenia* often switches from nectar-feeding to feeding on succulent fruit and insects. The species mirrors the feeding habits of Australian meliphagid honeyeaters. Where

Embothrium trees are undefended there is less probability of pollen being transferred between trees because birds remain for longer periods of time in individual trees, where they may forage at more flowers.

Circadian rhythms provide a daily 'pulse cycle' to the activity patterns of pollinators on flowers, with the movements of diurnal, crepuscular and nocturnal species facilitating some level of intra- and inter-population carriage of pollen. An additional, largely overlooked, mechanism for pollen movement is the influence of insect pheromones. In relation to sexual aggregation, pheromones attract males to conspecific females on flowering trees and the reproductively cued movement of males results in transference of pollen within single trees and within single tree populations. Such intra- and inter-tree pollen transfer is probably significant at least in crepuscular and fully nocturnal active moths and in some diurnal anthophilous beetles, especially in the subfamilies Melolonthinae (Scarabaeidae) and Cetoniinae (also Scarabaeidae) and the tribe Stigmoderini of the family Buprestidae. Marked adaptations to the reception of chemical signals (chemoreception) in beetles are seen in the presence of receptive organs (sensilla and associated fovae) on the antennae (Lawrence and Ślipiński 2013). The sensilla of Coleoptera are differentiated into specialised receptors of organic substances and pheromones.

The physical environment can profoundly influence movements and activity of pollinators and anthophilous insects including those that are not necessarily pollinators). Storms, cloudiness, wind, rain, temperature and humidity all make an impact on the foraging patterns of the flower-associated fauna, and phytophagous invertebrates generally (Ramírez *et al.* 2016; Stangler *et al.* 2015), with precipitation gradients known to influence high spatial turnover of plant communities and their associated pollinators (Devoto *et al.* 2009); the replacement of pollinators (by pollinators of lesser effectiveness) participating in specific plant–pollinator mutualisms along environmental and ecological gradients of decreasing precipitation, potentially resulting in eventual plant reproductive failure (e.g. Chalcoff *et al.* 2012). In such cases of 'pollination mutualism erosion' pollinators can serve as important biotic filters influencing plant distribution, and warn against prediction of geographical range shifts based only on climatic variables (Chalcoff *et al.* 2012). In addition, plants within the forest understorey occupy unique spatial and temporal environments. During the day they are subject to ever-changing sunlit and shaded mosaics and these are modified by seasonality (Herrera 1997). Thus pollinator visitation, and the variety of pollinators, is influenced by the location of the plant. Sun flecks may encourage greater rates of visitation to otherwise shaded understorey flowers or extend the foraging periodicity of crepuscular species (Chazdon 1988). Some pollinators, such as bees, preferentially visit sunlit or shaded conditions, though various groups (e.g. Hymenoptera generally) may forage over a greater range of the irradiance gradient; the foraging range of pollinators within the light gradient of the forest understorey being in part related to body size and length, and body temperature (Herrera 1997; see also Turton 1988). Cloudiness, by restricting movements of specialised or traplining pollinators (e.g. some bees), may allow an increase in the contribution of geitonogamy (pollination through transfer of pollen between flowers of different plants). Temperature can have a considerable influence on pollinator systems, both by affecting the activity of insects (and other pollinators) and by altering the volatilisation of attractants and nectar flow and viscosity (Abrahamczyk *et al.* 2011; Willmer 2011).

The energy balance of pollination systems has both plant and animal components. Plant and pollinator communities are interacting systems in which plants compete for pollinators and pollinators compete for floral resources. Plant–pollinator mutualisms may represent an optimisation of fecundity for both partners – plant and animal. And while pollination success entails an energy cost to the plant (in terms of nectar and or pollen) there is an inherent energy cost to the pollinator during foraging (Kearns and Inouye 1993; Willmer 2011). Pollination may also entail a reproductive (offspring) cost to the pollinator (Zurbuchen *et al.* 2010), and pollinator populations may suffer losses through predation (e.g. Heiling *et al.* 2004). This is a direct loss of individuals during foraging movements, for many predators (such as assassin bugs, wasps, spiders and insectivorous birds), also frequent flowers and wait patiently in ambush (Plate 111). Pollinators may make trade-offs between foraging strategies that optimise floral resource acquisition and those that reduce predator threats, but at the cost of lesser resource returns. The Australian crab spider *Thomisus spectabilis* (Thomisidae) waits in ambush upon inflorescences, the spider discriminating (selecting) in favour of the most 'productive' flowers; the selection for these being a response to floral odours, the recruitment cue that also solicits visits from the pollinators, that advertise those flowers with high floral rewards (Heiling *et al.* 2004). Crab spiders are known to have a negative effect on the rate at which flowers are visited by honey bees (*Apis mellifera*), however, this can be mediated by changes in foraging strategies (Llandres *et al.* 2012). But predators of pollinators may themselves perform pollination.

For both obligate and facultative pollinators, nectar is the most commonly recorded reward. There is general correlation between nectar production and composition, and the energy requirements and foraging periodicity of pollinators (Willmer 2011). The daily pattern of nectar production in rainforest trees can vary. It correlates generally with visitations from pollinators, which may feed upon floral resources in various forms (nectar, stigmatic exudate, pollen, resin, amino acids, staminodes) without any identifiable reproductive benefit to the plant (Carpenter 1976). These species are referred to as 'thieves', where feeding is non-destructive, and 'robbers', where damage to plant structures accompanies feeding. Ants are classically considered to be nectar thieves (Plates 93, 94). Although their feeding activities do not damage floral structures, the antibiotic secretions of numerous ants are likely to disrupt or inhibit pollen germination and growth of pollen tubes and so limit their effectiveness as pollinators (Harriss and Beattie 1991; Hölldobler and Wilson 1990). The presence of large numbers of nectar-feeding ants at flowers or the adjacent location of aggressive tree ant colonies (e.g. *Oecophylla smaragdina* in tropical regions, *Polyrhachis pilosa* in subtropical coastal rainforests) may preclude flower visitation by other insects.

Not all true pollinators are equally effective. Some, such as ants, may primarily transport pollen within the confines of single tree crowns or even single inflorescences. Others may transport pollen to conspecific plants over moderate or greater distances. Pollinator efficacy is also influenced by its population frequency, specific foraging behaviour and the structural modifications, characteristics or adaptive specialisations that physically permit pollen collection and carriage. The introduction of alien flower visitors (e.g. *Apis mellifera, A. cerana, Bombus terrestris*), or removal of plants to areas of cultivation or remnant vegetation within cultivation zones, gardens or regions removed from their natural range may obscure pollination modes and pollinator relationships, though such instances are examples of plasticity in plant pollination strategies. Flowering plants that disperse over long distances to islands are likely to leave their usual pollinators behind (Simberloff 1978). The lack of pollinator diversity on remote islands may act as a biotic mechanism that sifts and restricts recruitment of plant immigrants (Elmqvist *et al.* 1992).

Pollen represents a food source to some flower visitors, but to others it may be an encumbrance (e.g. impeding flight) or irrelevant. This introduces the caveat that, while visitors may function as pollinating agents, in many cases they are probably acting unwittingly and pollination is incidental.

POLLINATION OF SPARSELY FLOWERING SPECIES

Pollen flow between individuals of Neotropical rainforest plants with low population densities may be facilitated by large bees (Euglossini), leading to the suggestion first put by tropical ecologist Dan Janzen that this might permit the continued existence of tree species at low frequencies within the community (Janzen 1971). Janzen termed long-distance foraging activity 'trap-lining' and postulated that this activity was important to plants because it could permit effective out-crossing between plants at low densities, only small amounts of reproductive energy need be committed (because large numbers of blossoms were not required to attract pollinators), and plants could occupy shaded habitats and be relatively assured of pollination.

Low-frequency inter-tree movements by solitary bees, once placed in the family Anthophoridae (but now Apidae), have been found to be sufficient to account for observed fruit-set success in widely spaced, self-incompatible *Andira inermis* (Fabaceae) trees in disturbed lowland dry rainforest in Costa Rica (Frankie *et al.* 1976). Similar trap-lining patterns of host fidelity and repetitive foraging in the understorey of Australian subtropical rainforest and wet sclerophyll forest are exhibited by large solitary bees (e.g. *Xylocopa* and *Amegilla*) (G. Williams 2020). *Xylocopa* bees have been recorded foraging in a trap-lining manner on flowers of the rainforest and wet sclerophyll forest pioneer shrub *Melastoma affine* (Melastomataceae) in tropical Queensland (Gross 1993). The phenomenon of trap-lining behaviour by anthophorid/apid and halictid bees is also recorded in South-east Asia. In western Sumatra, long-tongued species of *Amegilla* and *Elaphropoda* (Anthophorini-Apidae) and *Nomia* (Halictidae) have been recorded trap-lining at the flowers of understorey Zingiberaceae and Gesneriaceae

(Kato *et al.* 1993). This form of foraging appears geographically widespread, at least with regard to tropical and subtropical rainforest environments.

How do trap-lining insects locate flowering plants? Attraction to specific floral fragrances (mobilised in air currents), rather than floral phenology or morphology, could recruit pollinators and facilitate pollen transfer. Whether this is the entrainment mechanism for all trap-lining insects is unknown, but there is ample evidence that phytochemical stimuli direct and modify insect pollinator responses to floral resources (Willmer 2011).

POLLINATION OF MASS-FLOWERING SPECIES

Mass-flowering rainforest trees that are in their peak phase of flowering attract large numbers of pollinators. These may quickly deplete the energy resources offered and, as a consequence of depletion, move to other trees of the same species that are in flower. Inter-plant movement and pollen transfer occurs, with out-crossing generally postulated as taking place primarily during the peak flowering phase (Heinrich 1975). This has been termed the 'conditioning-depletion' hypothesis. However, out-crossing may occur outside peak-phase flowering. In red-flowered forms of *Metrosideros collina* (Myrtaceae) in Hawaii, the principal out-crossing events appear to take place in early and late phase, with self-pollination predominating in the peak phase of flower availability (Carpenter 1976). In mass-flowering *Catalpa speciosa* (Bignoniaceae) from North America, out-crossing predominates in the late-phase, resource-depleted stage of flowering. A high proportion of receptive flowers (more than 50%) in peak phase remain unpollinated, which suggests that the flowers retained their nectar reward (Stephenson 1982). With *Catalpa speciosa*, early and peak flowering events appear to act as recruitment and conditioning mechanisms, with inter-tree movements increasing as conditioned pollinating insects feed on the resource-diminishing late-phase trees.

In her study of North Queensland populations of *Litsea leefeana, Neolitsea dealbata* and *Diospyros pentamera* trees House (1985), found that the majority of insect movements initially occurred within single tree canopies during the early flowering phase. Since male trees begin to flower two to three days earlier than females, inter-tree transport of pollen commences before female trees becoming receptive. This could result in initial waste of reproductive effort (because flowering female trees are not then available) in the early (male) flowering phase, but a later reproductive benefit of having pollen already mobilised at opening of female flowers. In all three tree species insect numbers increased and greatest quantities of pollen were mobilised at peak periods of flower availability. However, the greatest transfer of pollen to male trees took place towards the end of the flowering period, suggesting resource depletion may induce inter-plant movement of the accumulated anthophilous insect population.

SHARING OF POLLINATORS: THE 'GUILD' CONCEPT

Floral rewards, ecological gradients and variation in geography are considered primary determinants of pollinator guild structure. The composition of plant–pollinator guilds and the mosaic-like nature of their spatial, temporal and taxonomic composition might be primary factors contributing to speciation.

Guilds can consist of plants or animals (e.g. see Hawkins and MacMahon 1989; Kitching *et al.* 2005, 2007; Sakai *et al.* 1999a). In the former, guilds are composed of an assemblage of plants (often taxonomically closely related or sharing morphological attributes of similar gross floral form and floral resources attuned, in nectar volume and concentration, to a particular suite of visitors) that draw upon the same shared group of pollinators to achieve pollination within that group of plants (e.g. Kato and Miura 1996; Pauw 2006), and that may themselves serve as keystone species for the survival of particular pollinators (e.g. trap-lining bees [Sakai 2000]). In the latter, the guild comprises the actual pollinators, which may have diverse taxonomic affinities and morphological attributes (e.g. Kato 1996). These characterise the pollinator fauna (guild) found on particular plants, or groups of plants (which might be phylogenetically unrelated but share similar floral form, as in simple, open actinomorphic flowers). The term 'assemblage' may better describe the concept of participating members of plant–pollinator formations, as it carries no inference of cognisant awareness of the function the pollinators are jointly undertaking. It is important to remember that there can be considerable temporal variation in the visitation of pollinators within any loosely defined forest layer, and no uniform linear characterisation of the pollinator assemblage may be strictly applied.

With regard to the second guild category, taxonomically unrelated pollinators (though these might be related at higher taxonomic ranks, such as that of family or order) may form guilds which collectively undertake pollination of many individual plant species. It is common to see references to such functionally interactive entities as the 'small beetle–fly pollinator guild', 'the 'bee' or 'bird–bee guild' (or detailed refinements thereof [e.g. Sakai 2000; Sakai *et al.* 1999a]) – cohorts of particular insects or other animals jointly undertaking the role of pollinators. No one species is necessarily pre-eminent or solely responsible for pollination. Ideally, all participants in the guild should occur within a discrete region or site, and be co-active within a definable time period.

It follows that ecosystems contain seemingly functionally redundant animals which can all perform the same ecological roles (pollination, seed dispersal, decomposition of organic matter), even if they vary in quality of service. Less kindly, some players may be expendable, though *keystone* species by definition are generally not. These cohort guilds are moulded by adaptation to shared resources, and the relative dominance by subgroups and species within each guild is reallocated in response to differences in the availability of the shared resource. Rarely is a single plant, within a plant guild, pollinated by a single species of pollinator. Its pollinators may comprise one, or a small number, of pollinator orders or lesser taxa. For example, a plant may be pollinated only by various bee families or by a single family of bees (e.g. Halictidae) in consort with a single family of flies (e.g. Syrphidae); as in *Pollia crispata* (Williams and Walker 2003) and *Drymophila moorei* (Conran 1988) (Plates 5, 21).

A strong influencing factor in the establishment of guilds is floral morphology, the same factor that underpins the expression of pollination syndromes. Floral morphology generally determines what kinds of pollinators are able to access the floral rewards offered by the plant. Flower depth is often cited as a principal determinant of pollinator restriction, but other factors such as length of stamens and style, and structures that facilitate landing by a potential pollinator, can also contribute to the selection of guild members (e.g. Boberg and Ågren 2009). The variable expression of these characters, and the consequent interactive niches they establish, can result in overlap within guild membership as well as more tightly constrained plant–pollinator relationships.

Insect and insect–bird–bat dominated pollinator guilds are documented from tropical Queensland rainforest plants (aroids, palms, trees etc.), with beetle–fly guilds predominant in many of the trees studied (Boulter *et al.* 2005; Crome and Irvine 1986). But the composition of the visiting fauna at a flowering plant may, and often does, change over the length of the flowering event. Beetle–fly guilds also tend to dominate subtropical rainforest systems (G. Williams 1995). Flexibility, or conversely lack of specificity, in the plant–pollinator relationship exhibited by several Queensland tropical rainforest trees is related to the pollinator's physical capacity to achieve pollen transfer. Pollinator taxonomy is less important, with losses of pollinating species alleviated by recruiting from a pool of available pollinator guild members. Inherently, broadly defined plant–pollinator relationships are predicted for all Australian rainforests. Pollinator replacement and broader oligotropic (oligolectic, in which pollinators visit related plant taxa only) and polytropic (polylectic, visiting many plant taxa) relationships by pollinators seems to be the general rule at least for subtropical rainforest, where insects are responsible for vast amount of visitations and few trees are adapted for pollination by vertebrates. However, pollinator guilds in the rainforest understorey seem more taxonomically circumscribed, as exemplified by visitors to South-east Asian Zingiberaceae (e.g. Sakai 2000; Sakai *et al.* 1999a) and Australian Eupomatiaceae (G. Williams 2020).

The scenario of a broad Australian rainforest pollinator guild and flexibility in the pollination system mirrors findings in the fruit/seed disperser relationships of African, Oriental, Central and South American forests. In addition to pollination systems, few rainforest tree species appear to have tightly delimited (1:1) dispersal links with the dispersal vectors, and a broadly defined disperser guild allows flexibility in the reproductive strategy should individual dispersers be removed from the community (Terborgh 1990; Williams and Adam 2019).

However, it should be stated that tolerance (the point beyond which breeding mechanisms will begin to fail following critical fragmentation of plant and or pollinator populations) in the flexibility of the breeding systems of Australian rainforest trees is largely untested, since only a small number of species has been investigated. The apparent flexibility within the pollinator guild or breeding system may be a disturbance-induced disharmonic state (where the perceived relationships are artefacts of past disturbance or fragmentation) or part of an evolutionary shift in pollination strategy (e.g. from insect to vertebrate, animal to wind dispersion of pollen) – the perceived flexibility may be a transitional state, not a facultative adaptation.

The general conclusion from both the literature and field observations is that the majority of out-crossing tree species in Australian subtropical rainforests (and perhaps in tropical and temperate rainforests) are insect-pollinated, though some species may be cryptically wind-pollinated, or facultatively so (see Bullock 1994; Plate 59). The pollinator guilds from which mass-flowering canopy trees recruit have not generally been defined beyond simple recognition that small Coleoptera, Hymenoptera and Diptera are overwhelmingly dominant, and that individual trees recruit pollinator guilds or subguilds (subsets) that are reformulated from within spatial and temporal mosaics of available taxa (whatever is available at the time). Emergent, or occasionally emergent, rainforest trees (e.g. *Syncarpia hillii* [Myrtaceae], *Flindersia australis* [Rutaceae]), which project individual canopies above the uppermost canopy stratum, have received little attention as their sheer size generally precludes access. Do such trees attract a disproportionate amount of pollinators or pollinator classes that are better at transferring pollen between conspecific (of the same species) flowering trees and over greater distances? Do emergents maximise their pollination outcomes at the expense of their non-emergent neighbours?

Sands and House (1990) recognised two broad reproductive strategies in tropical entomophilous woody plants that can be applied to rainforests in general. The first is the production of a large conspicuous flower crop which attracts relatively unspecialised and often taxonomically divergent insects (the 'whatever is available' guild). In this system, distance between conspecific plants, and thus movement of pollen, may be small and seed (fruit) is characteristically produced from a small percentage of the flowers produced. The reproductive effort is channelled into attracting opportunistic pollinators (a high excess in flower production, often in combination with high abortion rates of developing fruit, is widespread among flowering plants). The second reproductive strategy is the production of small inconspicuous flowers over longer time periods which may be pollinated by vectors (e.g. bees) with memories enabling repeated foraging patterns, often in spatially disjunct plant populations. In this scenario, the pollinator guild would be constituted by members with a strong learning and discrimination ability, i.e. high-quality pollinators. This type of flowering has been recorded from Malaysian, Costa Rican and North Queensland rainforests (Bawa and Webb 1983, Bullock and Bawa 1981). The fruit number set often matches the numbers of flowers produced (Sands and House 1990).

Nest site availability, the variety of nest sites available and nest site resources are also thought to play a significant determining factor in how individual pollinator taxa might be available for guild membership. Such resources are known to be important for bees (Houston 2018) and more broadly have a significant role in the structuring of the entire guild community. For bees, the nesting resources may relate to the availability of bare ground (for many Halictidae and Colletidae) and the occurrence of suitable nesting cavities (Prendergast 2018a, 2018b). Eusocial bees (e.g. *Tetragonula*) are limited by their requirement for suitable cavities in which to found and establish hives; the number of available cavities will dictate the number of colonies that can occupy an area of rainforest. Pithy and hollow plant stems and pre-existing cavities are important for megachilids and some solitary apids (M. Batley pers. comm.). Large carpenter bees, such as *Xylocopa*, may require suitable woody substrate into which to excavate a nest (Michener 2007).

We are entirely ignorant of the life histories and immature stages of most invertebrate members of pollinator guilds, so no well founded statements can be made on how nest site and larval host availability might influence guild structure and membership. In addition, the effects of future climate change may alter the composition of pollinator guilds, or change the foraging and dispersal behaviour of individual pollinators (e.g. Phillips *et al.* 2009). How this will impact on plant reproduction and plant guild structure is uncertain.

7 Pollination and the Australian flora

In Australia, at least 28 families of beetles and 44 families of flies contain species that are attracted to or are encountered on flowers (e.g. Bernhardt 2000; Lawrence and Ślipiński 2013). The bee fauna appears to be exclusively anthophilous (Houston 2018); among the other Hymenoptera (of which bees are only one taxon) several families of 'wasps' (e.g. Crabronidae, Sphecidae, Thynnidae, Vespidae, Plates 88, 89, 90, 91, 92) contain species regularly observed on flowers (Allsopp 1992; Williams and Adam 1995). Dyer *et al.* (2012) consider that flowering plants in Australia are likely to have independently evolved spectral signals (owing to flowering plants in Australia being geographically isolated for some 34 million years) that maximise colour discrimination by Hymenoptera.

In arid regions the incidence of bird pollination has been reported as greatly reduced, pollination contributions by terrestrial mammals appear to be absent or uncommon and, among entomophilous plants, pollination by flies of the family Bombyliidae (and by bees) becomes more important (Barlow 1981). Studies of Australian tropical rainforest ecosystems indicate that many groups of insects contribute substantially to pollen transfer and reproductive success, and that within Australian tropical and subtropical rainforest at higher altitudes bee pollination may be less frequent or entirely absent (see House 1985, 1989; Irvine and Armstrong 1990). Birds and mammals, both terrestrial and flying, seem to be more important as agents of pollination in Australian landscapes than in northern temperate forests (Bawa 1990; Sussman and Raven 1978). Although bird and mammal pollination is relatively well documented for Australian sclerophyllous communities, in other plant communities it may be highly seasonal, incidental to other feeding activities, localised, unreliable or absent.

Myrtaceae (considered in the following section), Proteaceae (e.g. *Banksia, Grevillea*) and Fabaceae/Mimosaceae (*Acacia*) are conspicuous and iconic Australian plants in open sclerophyllous environments; many of which have flowers that attract a diversity of birds that forage, and potentially serve as agents of pollen transfer. But at the landscape level, with its profusion of different plants, pollinator assemblages of small insects remain the dominating theme throughout, the significance and number of individual higher taxa as pollen vectors varying between different plants (and likely also spatially and temporally). For example Hawkeswood (1990), in a study of the pollination ecology of *Bursaria spinosa* (Pittosporaceae), a white-flowered shrub – small tree distributed from northern coastal Queensland to southern South Australia, noted that a taxonomically diverse assemblage of insects (i.e. Lepidoptera, Coleoptera, Diptera, Hymenoptera) comprised the dominant flower visitors and that the beetle families Buprestidae, Cerambycidae, Scarabaeidae and Mordellidae, in concert with *Scolia* (Scoliidae–Hymenoptera), were considered the most important pollination agents of geitonogamous pollination, though more active insects such as various Diptera were likely to facilitate xenogamous pollen transfer. Although the introduced honey bee *Apis mellifera* was present native bees were generally absent. In addition to Hawkeswood's example, within the genera of Rutaceae encountered in open sclerophyll vegetation in eastern Australia *Boronia* is pollinated mostly by bees (e.g. *Exoneura, Homalictus, Lasioglossum* [Plate 86]) but also by occasional beetles, flies and lepidopterans, *Philotheca* mostly by butterflies but also beetles, flies and bees, and *Eriostemon* mostly by beetles but occasionally by bees and butterflies. But in *Correa* birds (mainly honeyeaters, parrots) are the major pollinators (Clifford and Drake 1981). In northern and north-western Australia hawkmoths are recorded as pollinators of the distinctive boab tree *Adansonia gregorii* (Plate 140) (Groffen *et al.* 2016). These included *Agrius convolvuli, Theretra oldenlandiae* and *Eudesmeola lawsoni*, as well as two unidentified species. However, bees and other insects, birds, sugar gliders (*Petaurus breviceps*) and Black flying foxes (*Pteropus alecto*) also foraged at flowers.

Although numerous birds have been recorded either as residents or transients in Australian 'closed' mangrove communities (which can be considered distinctive salt-tolerant maritime rainforest formations; see Plate 132) the pollination ecology of Australian mangrove communities is relatively unknown with most records based on little critical assessment of the function of flower visitors as actual agents of pollen carriage. Worldwide, pollination by birds and by insects (and entomophilous-adapted floral structures) is commonly recorded in mangroves. Although numerous bird species are recorded inhabiting or visiting mangroves in Australia (Schodde *et al.* 1982), there are few records of birds as pollinators. Studies by Hermansen *et al.* (2014b) and Homer (2009) indicate that Australian temperate and subtropical populations of *Avicennia marina* (Acanthaceae/Avicenniaceae) (Plates 105, 132) are wholly or primarily pollinated by insects, with Hermansen *et al.* (2014b) suggesting that *Apis mellifera* may have supplanted native pollinators at some sites. Contrary to this, studies undertaken in the Manning River estuary (northern New South Wales) indicate the native fauna visiting flowering *Avicennia marina* is very diverse (>170 spp.), and that most species carry pollen loads (G. Williams pers. obs.). However, small mangrove stand size may have a negative impact on the abundance of individual flower-visiting species and subsequent pollen deposition (Hermansen *et al.* 2014a), and daily cycles of tidal influence and inundation may also obstruct visitation by terrestrial species such as ants. With regard to other Australian mangrove-associated plants, in northern Queensland the tree *Sonneratia alba* (Sonneratiaceae) and sprawling shrub *Clerodendrum inerme* (Lamiaceae/Verbenaceae) appear to be pollinated primarily by hawkmoths, sunbirds are the likely pollinators of the herb – sprawling shrub *Acanthus ilicifolius* (Acanthaceae), birds are recorded as visitors and potential pollinators of the parasitic epiphyte *Lusiana subfalcata* (Loranthaceae), and the bee *Tetragonula hockingsi* is a possible pollinator of the prostrate herb *Sesuvium portulacastrum* (Aizoaceae) (Primack *et al.* 1981).

POLLINATION IN AUSTRALIAN MYRTACEAE

The Australian Myrtaceae are an important element in the Australian flora (Keith 2017), and include the world's tallest flowering plant *Eucalyptus regnans* (occurring in Tasmania and Victoria), exceeded in height only by the redwood conifers of North America (Boland *et al.* 1984). The iconic diversity and dominance of Myrtaceae in many Australian communities are features which distinguish Australian vegetation, however, the entry into Australia of the fungal pathogen Myrtle rust currently has impacted more than 360 myrtaceous species, posing a potentially cataclysmic threat to plant communities and their associated biota at a landscape scale.

In north-western Australia, among the nectar sources seasonally available, Myrtaceae are the dominant family in both the number of species of nectarivores that visit and their frequency of use (Franklin and Noske 2000); a theme likely to be expressed in many Australian plant communities. The dry-fruited Leptospermoideae (including *Eucalyptus, Melaleuca* and *Leptospermum*) are often major canopy dominants in sclerophyll forest, woodland and shrublands. Some elements of the Leptospermoideae are also important in rainforests (e.g. *Backhousia* and *Tristaniopsis*) and wet sclerophyll forest/rainforest transitions (*Lophostemon* and *Syncarpia*). It is believed that the sclerophyll members of the Leptospermoideae have evolved from rainforest ancestors, but the details of the separation of rainforest and sclerophyll elements in the flora remain obscure (Adam 1992; Briggs and Johnson 1979).

However, the evolution of fleshy fruits is suggested to be a key innovation for rainforest Myrtaceae

(Biffin *et al.* 2010). The typically hermaphrodite fleshy-fruited Myrtoideae (Lughadha and Proença 1996) (e.g. *Archirhodomyrtus, Rhodomyrtus, Syzygium* and *Rhodamnia*) (Plates 54, 57, 58, 59, 60), include many species that occur in Australian tropical and subtropical rainforests; although the Myrtoideae have been relatively unsuccessful in colonising fire-prone habitats.

Although the reproductive biology of the Australian Myrtaceae is generally understood (Beardsell *et al.* 1993), their pollination has been poorly studied despite the family's great economic importance. *Eucalyptus* species are known to be pollinated by large fruit-bats or flying-foxes (Pteropodidae) (Eby 1991), birds (Sampson 1998) and insects (e.g. Griffin *et al.* 2009), or by a combination of all. Pollination by birds (ornithophily) has arisen in several genera and is found throughout the ecological range of the family (Briggs and Johnson 1979). Nectivorous birds have been shown to be important pollen vectors in isolated populations of *Eucalyptus caesia,* this resulting in multiple paternity pollination events (Bezemer *et al.* 2016). Phillips *et al.* (2009) consider that bird pollination may confer an advantage of greater pollen movement and may represent a mechanism for minimising inbreeding in naturally fragmented populations. But the importance of individual pollinator taxa can differ, potentially in response to the height at which the flowers are presented in the crown, as in *Eucalyptus globulus* where anthophilous birds appear to forage mostly higher in the trees and to which they provide more efficient pollination services than co-foraging insects (Hingston *et al.* 2004a, 2004b). By comparison, large numbers of inefficient pollinators foraged at lower *Eucalyptus globulus* flowers (Hingston *et al.* 2004a). Six genera of Brazilian Myrtaceae (*Eugenia, Siphoneugenia, Blepharocalyx, Campomanesia, Myrica* and *Psidium*) are known to be pollinated by bees (Proença and Gibbs 1994), and native bees figure as significant pollinators or visitors to Australian native rainforest trees such as *Tristaniopsis* and *Syzygium* (G. Williams 2020). Adaptation to vertebrate and invertebrate pollinators is strongly expressed in tropical members of the genus *Syzygium* (e.g. Boulter *et al.* 2005; Lughadha and Proença 1996).

In rainforests, the inflorescence of myrtaceous genera may be massed (e.g. *Backhousia, Xanthostemon,* some *Syzygium* species) or less conspicuous, often with large unmassed flowers (e.g. *Gossia*, some *Syzygium* species). Ramiflorous and cauliflorous positioning of flowers (e.g. *Syzygium cormiflorum*) (Plate 59) and associated large, fleshy, animal-dispersed (zoochorous) fruit is exhibited by numerous species, thus reflecting selection influences from both pollinators and dispersers.

Forty-five species of nectar-feeding (nectarivorous) animals have been recorded visiting the flowers of tropical Australian *Syzygium tierneyanum* (Hopper 1980). Day-active foragers included seven species of birds (predominantly Meliphagidae), four moths (Sphingidae, Uraniidae, Erebidae), seven butterflies (Nymphalidae, Papilionidae, Pieridae), and the introduced honey bee *Apis mellifera*. Nocturnal foragers comprised the Queensland blossom bat *Syconycteris australis* and five species of moth (Sphingidae, Hypsidae, Noctuidae). There were also lesser numbers of miscellaneous visitors comprising beetles, ants, a wasp, and flies. On the basis of abundance and foraging behaviour, Hopper (1980) considered honeyeaters (Meliphagidae) and hawkmoths the most important pollinators, though honey bees were also prolific visitors. However, Brooks (1948) also records numerous scarabaeid beetle species, especially large Cetoniinae (e.g. *Lomapteroides duboulayi, Lomaptera wallacei yorkiana*), as well as curculionid and dermestid beetles, collected from *Syzygium tierneyanum* blossoms in North Queensland in December. Cetoniines frequently make long interplant flights so their contribution to out-crossing is likely seasonally significant. Pollinators and general flower visitors recorded from the inflorescences of North Queensland *Syzygium gustavoides* included beetles, Lepidoptera and flies, the abundance and species diversity of various higher order taxa varying between samples (Kitching *et al.* 2007). Volkovitsh *et al.* (2003) consider large adult *Metaxymorpha gloriosa* (Buprestidae) beetles in tropical Queensland important pollinators of *Syzygium tierneyanum*, *Syzygium kuranda*, *Eucalyptus* spp. and *Leptospermum* spp., as well as the additional rainforest taxa *Melicope elleryana* (Rutaceae), *Elaeocarpus angustifolius* (Elaeocarpaceae), *Psydrax odorata* (Rubiaceae), *Buckinghamia celcissima*, *Grevillea baileyana* and *Macadamia whelani* (all Proteaceae). This beetle is capable of relatively long distance flights between co-flowering plants.

The largely out-crossing related species *Syzygium cormiflorum* (see Appendix 1) exhibits varying degrees of self-compatibility, ranging from 10% to 32% (Crome and Irvine 1986). This degree of self-compatibility confers resilience in the absence of major pollinators. In *Syzygium cormiflorum* birds, insects and particularly blossom bats (which include the genera *Macroglossus* and *Syconycteris*) are effective pollinators (Crome and Irvine 1986). Crome and Irvine considered that blossom bats in combination with hawkmoths, were responsible for the greatest percentage of successful pollinations. Law and Lean (1999) found that in a fragmented North Queensland landscape *Syconycteris australis* visiting *Syzygium cormiflorum* carried about six times more pollen than birds but were less frequent visitors suggesting that the quantity of pollen distributed by bats and birds may not greatly differ. Blossom bats and honeyeaters contribute about half the pollination rate of North Queensland *Syzygium sayeri* (a largely self-incompatible species), the rest effected mainly by butterflies, flies, thrips and wasps (Boulter *et al.* 2005, 2008). Although the greatest number of visits by potential pollinators to *Syzygium sayeri* occurs during the day there is also a significant level of visitation by nocturnal flower visitors, both vertebrates (bats) and invertebrates (moths). These suites of diurnal and nocturnal flower visitors appear to contribute about equally to observed rates of pollination. The episodes of pollination mirror and respond to the large quantities of nectar that *Syzygium sayeri* produces both day and night, quantities which ensure continual rates of flower visitation by large numbers of pollinators.

The pollination ecology of tropical Queensland species of *Syzygium* can be compared with that of *Syzygium syzygiodes* from the lowland rainforests of Sulawesi (Lack and Kevan 1984). These are forests of lower density than the dipterocarp-dominated rainforests of the Sunda Shelf and the Malayan Peninsula. Trees are synchronous in flowering and populations flower over a period of four to five weeks. *Syzygium syzygiodes* is strongly self-incompatible and its pollination is dependent upon insects. Pollination success is directly related to the number of insect visits. Pollinators are recruited from a guild of short-tongued unspecialised insects with observed behaviour patterns that indicate little promotion of outcrossing. Although Sulawesi has a rich bat fauna, these and other vertebrates do not appear to visit the flowers of *Syzygium syzygiodes*. This suggests that the reproductive biology of the species is adapted to a species-poor community where specialisation would be disadvantageous; a situation markedly different to that recorded for the three North Queensland species discussed above, in which birds and or bats play key roles. Rather, the agents of pollen transport in *Syzygium syzygiodes* have broad taxonomic similarities with the putative pollinator assemblage recorded from North Queensland *Syzygium gustavoides*; a tree seemingly reliant on small insects rather than vertebrates for pollen transfer (Kitching *et al.* 2007). The invertebrate-driven pollination strategy exhibited by both *Syzygium syzygiodes* and *Syzygium gustavoides* is also similar to that of the fleshy-fruited Australian subtropical trees *Syzygium smithii, Syzygium floribundum* and *Rhodomyrtus psidioides*, and dry-fruited *Tristaniopsis laurina* (G. Williams 2018; Williams and Adam 2019).

Syzygium floribundum (Plate 60) is a bisexual, largely self-incompatible (Table 4.1), mass-flowering tree found in lowland rainforest and wet sclerophyll forest, usually restricted to stream and riverbanks, from northern New South Wales to central coastal Queensland (Floyd 2007) where it can be the dominant canopy species. Its flowers are coloured creamish-white and the floral structure and presentation is similar to that of *Eucalyptus* species, the stigma is receptive at anthesis and the stamens readily absciss following pollination. Though large eucalypt-visiting fruit-bats (*Pteropus poliocephalus, Pteropus scapulatus;* and less commonly *Pteropus alecto*) are locally abundant in the range where *Syzygium floribundum* occurs, unlike *Eucalyptus*, upon which they are nocturnal visitors, they generally do not visit its massed canopy flowers. Its pollinators are essentially insects for even few birds visit its prodigious blossom displays (Williams and Adam 2019). Insects can be profuse at peak flowering. Most insect visitors, on flowering trees studied in lowland rainforest in northern New South Wales, are less than 6 mm in size (e.g. 74–98% of total at Wingham), and largely comprise Coleoptera, brachyceran flies, and to a lesser degree Hymenoptera; these three groups can range between ~81–99% and 84–94% of individuals and species, respectively, of the total abundance (G. Williams 1995). However, the abundance of individuals, and the richness of species, varies within and between seasons, and time of day. The suite of anthophilous insects collected from *Syzygium floribundum* includes ~30 families of Coleoptera, 24 families of Diptera and 21 families of Hymenoptera (Williams and Adam 2019). Of the Hymenoptera small to minute Euryglossinae bees can be particularly common. Although numerous visitors are specialist, fully adapted anthophiles, visitors also include many that are poorly adapted and likely as individuals contribute little to pollination; for example the beetle families Anthicidae, Chrysomelidae, Coccinelldae, Latridiidae and Phalacridae, the fly families Chloropidae and Phoridae, and Hemiptera in general.

The related *Syzygium smithii* is a small tree found in rainforest, and sometimes wet sclerophyll forest, distributed from Wilsons Promontory in southern Victoria to southern Queensland (Floyd 2007) and includes several forms (see Harden *et al.* 2006). These can exhibit markedly different floral displays ranging from relatively sparse blossom clusters, somewhat obscured by foliage, to profuse unobscured inflorescence masses. The flowers of *Syzygium smithii* are white to creamish-white, bisexual and strongly fragrant and can produce large volumes of nectar (though this is variable) (Williams and Adam 2019). Insects can be abundant during peak flowering but abundance and richness of the fauna can vary dramatically dependent on the extent of mass flowering. Insect visitors, at higher taxonomic ranks, exhibit similar patterns of size, abundance, and seasonal and daily variation in visitation to that found in *Syzygium floribundum* (Williams and Adam 2019). The number of families, and the composition of visiting species, varies between sites but again these are mixed assemblages of adapted and poorly adapted anthophiles. The related *Rhodomyrtus psidioides* (Plate 58) is predominantly visited by small bees, beetles and flies, but may also be wind-pollinated (discussed in Chapter 8) (Williams 2018); a facultative adaptation to wind pollination is also possibly present in *Archirhodomyrtus beckleri* (Plate 54). In comparison to *Syzygium floribundum* and *Syzygium smithii* the abundance of insects visiting flowering *Rhodomyrtus psidiodes* is relatively low, this despite seasonally conspicuous and fragrant floral displays (G. Williams 2018). However, *Rhodomyrtus psidioides* has been heavily impacted by the fungal pathogen Myrtle rust with mature trees killed throughout its former range, such that the species may not survive in nature (B. Makinson pers. comm.).

The dry-fruited tree *Tristaniopsis laurina* is distributed from eastern Victoria to southern Queensland (Floyd 2007). It is often restricted to stream banks where it can occur as the dominant canopy species, or as a co-dominant with other dry-fruited Leptospermoideae such as *Lophostemon confertus*, or with *Syzygium floribundum* (Adam 1992). The flowers are bisexual, self-incompatible (Table 4.1), bright yellow and strongly honey-scented, the stamens fused into five antepetalous clusters, the petals abscissing in the senescent stage. The species is pollinated principally by Coleoptera, Diptera and to a lesser degree Hymenoptera; in lowland rainforest these comprising 53–100% of observed flower visitors, the proportion of higher taxa and overall abundance of individuals varying with respect to season and time of day. The majority of visitors (91–99%) are less than 6 mm in size (Williams and Adam 2019).

However, it is also useful to compare visitor records, from wet forest flowering Myrtaceae, with myrtaceous species occurring in open sclerophyll forest, woodlands, heathland, and more arid-adapted vegetation communities. Many of the available studies and field observations focus on beetles (e.g. Brooks 1948), in particular Buprestidae associated with Myrtaceae (e.g. *Leptospermum, Angophora, Eucalyptus*) (Bellamy *et al.* 2013); though these may have only been the more conspicuous species rather than the sole or primary visitors.

Hawkeswood (1980) records buprestid beetles (*Temognatha heros, Castiarina mustelamajor*) as pollinators of *Melaleuca pauperiflora* in semi-arid habitats of South Australia and Western Australia and considers that buprestids may play important pollinator roles in arid and semi-arid zones, and especially of Myrtaceae. Baskorowati *et al.* (2010) considered thrips as likely to be important pollinators of *Melaleuca alternifolia* (a species grown commercially for its oil content), but noted that other frequent visitors to flowers were bees, flies and wasps. *Melaleuca lanceolata* and *Verticordia picta* in Western Australia are pollinated by small melolonthine scarab beetles *Phyllotocus ustulatus* (Hawkeswood 1993), whilst members of the tribe Chamelaucieae include taxa that are adapted for bird pollination (*Balaustion, Cheyniana microphylla*) and insect pollination (*Cheyniana rhodella*) (Rye 2009). Rye (2013) found that various species of *Astartea* (a genus endemic to south-western Australia) are predominantly pollinated by native bees but also by beetles and flies. Brooks (1948, 1965) lists numerous Buprestidae, Scarabaeidae, Cerambycidae, Dermestidae and Cleridae (Coleoptera) variously collected from flowering North Queensland *Eucalyptus gummifera, Eucalyptus torrelliana, Eucalyptus hemiphloia* and *Eucalyptus ochrophloia*. Insects are pollinators of *Eucalyptus regnans* but 92% of the insects collected from flowers were Diptera (Griffin *et al.* 2009), although most of these were small in size, uncommon or carried little pollen, and are thought to contribute insignificantly to pollination. Mean pollen loads ranged from 20 pollen grains for sepsid flies to 84 000 for large tachinid flies. Based on population size and pollen loads the most important insect pollinators are likely calliphorid, tachinid and syrphid flies, and sphecid wasps (Griffin *et al.* 2009). Armstrong (1979) records bombyliid flies competing with day-flying moths, and butterflies, for the nectar of *Calytrix longiflora* in central Queensland, and thynnine wasps are recorded visiting the blossoms of Western Australian *Darwinia* species (Keighery 1975). Hawkeswood (1987) records beetles of the families Buprestidae, Cantharidae, Mordellidae, and Scarabaeidae foraging at the flowers of *Baeckea stenophylla* in heathland in northern New South Wales. The variety and abundance of insects to *Baeckea stenophylla* were mostly low, and this Hawkeswood suggested might be due to a poor emergence season for beetles or a reduction in available pollinators due to man-induced disturbance.

Birds can be conspicuous visitors to myrtaceous flowering plants in sclerophyllous environments, for example the Scarlet honeyeater (*Myzomela sanguinolenta*) which seasonally forages in large numbers on flowering *Melaleuca saligna* (*Callistemon salignus*) adjacent lowland moist sclerophyll forests in New South Wales (G. Williams pers. obs.). Sampson (1998) notes the specialised floral structure for bird pollination in Western Australian *Eucalyptus rameliana*, and Collins *et al.* (1984) record the meliphagid honeyeaters *Lichmera indistincta* and *Phylidonyris nigra* as pollinators of *Calothamnus quadrifidus*, these in competition with alien honey bees *Apis mellifera*. The endangered swift parrot *Lathamus discolor* is a pollinator of *Eucalyptus globulus* ssp. *globulus* (Hingston *et al.* 2004b). Though *Eucalyptus globulus* is also visited by native insects and exotic bees (*Bombus terrestris*, *Apis mellifera*), Hingston *et al.* (2004a, 2004b) found that the swift parrot facilitated on average 76% of the maximum possible seed set of open-pollinated flowers in single visits to a flower, but that single visits by native insects did not result in any seed production. The glider *Petaurus gracilis*, an endangered marsupial with a highly localised distribution, visits flowers, and consumes nectar and pollen almost exclusively, from Myrtaceae species of *Eucalyptus*, *Corymbia* and *Melaleuca* (Jackson 2001).

Commonly, however, visitors to flowering Myrtaceae in Australian sclerophyllous plant communities comprise suites of invertebrates and vertebrates, as in *Melaleuca quinquinervia* which is visited by insects and birds during the day, and at night by the blossom bat *Syconycteris australis* (Breeden and Breeden 1970).

Indigenous Australians and pollination ecology – untapped insight

But what of Indigenous Australians' interplay with the reproductive ecology of native plants and pollinators? Although the lives of contemporary Indigenous Australians are richly entwined with the ecology of Australia's flora and fauna, there seem to be few records that date from European occupation and settlement in the late 18th century. Nevertheless, there are early recordings of tribal wisdom and legend that hint at a vast body of knowledge. The Australian writer and social activist Mary Gilmore wrote in the 1930s (Gilmore 1934) that Aboriginal peoples of the Wiradjuri in New South Wales purposely cross-pollinated 'quandongs', *Santalum acuminatum* (Santalaceae [see Rugkhla *et al.* 1997]). This is a plant of arid sclerophyllous environments, its pulp and oily kernels a nutritious food source. K. Langloh Parker (1905), writing about the Euahlayi from north-west New South Wales, refers to 'sugar-bag' (honey) trees, which were hollow and held the hives of native stingless bees. Of the collection of honey from these trees, she related:

> *I deplored the sacrifice of the bees one day, but was told it was all right. Whoever had dropped the nest out would take home a waxy stick they had used to help get the honey out; they would throw the stick in the fires, then all the dead bees would go to a paradise in the skies, when next season they would send Yarragerh Mayrah, the Spring Wind, to blow the flowers open, and then down they would come to earth again.*

Later, referring to the prediction of seasonal climate impacts, Parker stated:

> *The old man said, 'It is drought now, but it will be worse. Bya mee has sent the manna by the little Dulloorah birds and the black ants, because there will be no flowers for the bees to get honey from, so he sent this manna.'*

Similar knowledge of the ecology of native bees (and impacts upon them) in their uncertain environment permeates Indigenous culture, to the point where names are attributed to particular species of bee and to individual components of nest structure. The Yolŋu people of northern Australia (Rudder 1977) recognise different species of bees by characteristics of the way the hive is built, where the nest is located and the nature of its honey. Collectively all bees are termed 'dawurr' or 'worrutj'; 'bayarrak' is a word used to describe a working bee that is collecting pollen from flowers and that can be followed back to the hive. The Yolŋu give names to individual species. For example, 'yarrpany' denotes a very small species of bee that is not savage, makes a large hive in the top half of a tree and produces yellow sweet-tasting pollen balls and very good honey. 'Niwnda' (also called 'nyiwuda' or 'gämu') is a large, yellowish coloured savage bee that produces orange-coloured pollen balls and good tasting honey. The Yolŋu also apply a specific nomenclature to the architecture and contents of bee hives. 'Nurru' means the hive entrance, 'gurak' the entrance passageway, 'mapu' the pollen balls stored by the bees, 'djukurr/djana/gutitj' are terms for the top of the honeycomb within the hive, 'djukurr/djana/gutitj' refers to the bottom of the honeycomb and 'gula/dhudi' means the debris from the hive, or bee faeces. Though it was common to track the hive's location by following the movements of foraging bees, during periods of relative inactivity (as in winter) hives could be found by searching for pellets of bee dung lying around the base of nest trees (Roth 1897).

8 Pollination syndromes: who brings the 'flower children' in rainforest?

'The current debate on pollination syndromes focuses on two major issues: the role of pollination specialisation and the predictive power of the syndrome.'

(Martén-Rodríguez et al. *2009)*

Pollination syndromes (inferred from floral features) have been considered unreliable predictors of the effective pollinators (e.g. Hingston and McQuillan 2000). While there are observations of flower morphology resulting in incorrect inferences of pollination mode, and some perceived syndromes seem profoundly procrustean in construction (Procrustes, a brigand of Greek mythology who tied his captives to a bed, stretching or amputating their legs so that they would fit neatly upon it), generalised correlations between flower structure and pollination vectors have permitted the recognition of serviceable pollination syndromes. This allows broad groups of putative pollinators to be inferred from flower structure and other attributes such as colour (Faegri and van der Pijl 1979; de Souza *et al.* 2016). In addition, morphologically based pollination syndromes have been correlated with phytochemical syndromes (Baker and Baker 1990; Cronk and Ojeda 2008), differences in nectar volume and the sucrose proportion of nectar (Perret *et al.* 2001; Vandelook *et al.* 2019), and in some cases daily patterns of nectar production and sugar composition; for example, discriminating patterns of 'bee' and 'bird' nectar production and sugar concentration. But there can be quite divergent pollination vector relationships *within* families. Barfod *et al.* (2011), in a review of pollination mechanisms in palms (Arecaceae), noted that individual species were pollinated by beetles, bees, flies, and occasionally by bats and marsupials; and that comparative studies of palm-pollinator interactions in closely related palm species documented transitions in floral morphology, phenology and anatomy correlated with shifts in pollination vectors.

Although there are instances of apparent proboscis – corolla length mismatch between long-tubed flowers and their short-tongued pollinators (e.g. Huang and Fenster 2007; Xic *et al.* 2012), nevertheless, there is a growing body of evidence that demonstrates from floral traits the accurate prediction of the types of pollinators (e.g. Kuriakose *et al.* 2018a; Liu and Sun 2019; Wilson *et al.* 2017). For example, Neotropical *Costus* (Costaceae) flowers that are bee-pollinated possess distinctive colour and morphological characteristics that distinguish them from species that are pollinated by hummingbirds (Kay and Schemske 2003); the hummingbird syndrome being considered a derived state estimated to have evolved independently seven or more times in the Neotropics (Kay *et al.* 2005). In the florally diverse Antillean Gesnerieae distinct hummingbird and bat pollination syndromes have been recognised, in addition to ecological generalist syndromes (Martén-Rodríguez *et al.* 2009). This study also demonstrated that the occurrence of ecological generalisation, where some plants are visited by many species, does not necessarily prevent the evolution of sets of floral traits that coincide with established pollination syndromes (Martén-Rodríguez *et al.* 2009).

Pollination syndromes. Real or envisaged?

Pollination syndromes are a suite of phenotypic flower traits that are hypothesised to reflect convergent adaptations of flowers for being pollinated by specific kinds of animals (e.g. Faegri and van der Pijl 1979); such that the syndromes evolved under the reciprocal selection of pollinators and plants (Paulus 2019). However, worldwide, the recognition and application of pollination syndromes, as suggested by floral traits, has been the subject of considerable contention and debate (e.g. Aguilar *et al.* 2015; Higham and McQuillan 2000; Hingston and McQuillan 2000; Martén-Rodríguez *et al.* 2009; Ollerton *et al.* 2009, 2015; Paudel *et al.* 2019; Perret *et al.* 2001; Rosas-Guerrero *et al.* 2014; Vandelook *et al.* 2019).

Ollerton *et al.* (2009) used flowers from six communities from three continents to score for the expression of floral characters used in published descriptions of floral pollination syndromes, and simultaneously characterising the pollinators of as many species as possible. Flowers were ordinated in multivariate 'phenotype' space but almost none fell within the discrete syndrome clusters generated. And in ~2/3 of plant species the most common pollinator could not be successfully predicted by assuming that each plant species belonged to the syndrome closest to it in the ordination outcome; thus Ollerton and his colleagues (Ollerton *et al.* 2009, and subsequent discussion of data flaws in Ollerton *et al.* 2015) considered that pollination syndromes, as usually articulated, may not successfully describe the diversity of observable floral structural or functional properties, or serve to reliably predict the pollinators of most plant species. Consequently, they cautioned the use of pollination syndromes for organising floral diversity or for inferring agents of flower adaptation.

Conversely, from studies of the floral traits of 417 plant species Rosas-Guerrero *et al.* (2014) found support for the syndrome concept, their results indicating that convergent floral evolution is driven by adaptation to the most effective pollinator group and that the predictability of pollination syndromes is greater in pollinator-dependent species and in plants from tropical regions. Ollerton *et al.* (2015) questioned this study. In response to the ensuing controversy regarding the recognition of the pollination syndrome concept Aguilar *et al.* (2015) mirrored the findings of Rosas-Guerrero *et al.* (2014), adding that many plant species also have secondary pollinators that often correspond to the ancestral pollinators documented in evolutionary studies.

Abrahamczyk *et al.* (2017) utilised quantitative datasets of flower morphometry, signal and reward traits for 86 species of Balsaminaceae to explore the expression of pollination syndromes. Cluster analyses of the floral traits, used in combination with independent pollinator observations, yielded seven clusters, six of which corresponded to plausible pollination syndromes, and one well supported cluster comprising highly divergent floral architecture. They considered, with the caveat of the need for data quality, it is possible to identify pollination syndromes from large datasets and to reliably extrapolate them from taxa for which observations are unavailable. Vandelook *et al.* (2019) tested 57 species of Balsaminaceae, performed using cluster analysis and ordination, to derive clusters of species resembling each other in nectar composition. Results showed that Balsaminaceae species (i.e. *Impatiens* spp.) with the same pollination syndrome tend to have similar nectar characteristics, whereas nectar traits between syndromes differ significantly. Vandelook *et al.* (2019) considered that there was strong evidence that the evolution of certain nectar characters has occurred as an adaptation to pollinator preferences. Analysis indicated that especially nectar volume and sucrose proportion of Balsaminaceae nectar has evolved in close association with pollination syndromes, while there were indications that the same is true for sugar and amino acid concentration. However, no evidence was found that amino acid composition in nectar evolved as an adaptation to pollination syndromes, this possibly due to its sensitivity to environmental factors such as nutrient availability.

In contrast Paudel *et al.* (2019), in a study of three Himalayan alpine *Roscoea* spp. (Zingiberaceae), found that the suggested pollination syndrome (ie. reliance on long-tongued pollinators) was a weak predictor of an efficient contemporary pollinator community. *Roscoea* floral traits suggested that long-tongued insects acted as the major pollinators. However >90% of pollination services, in all three of their study species, were contributed by morphologically mismatched short-tongued bumblebees (*Bombus*), while long-tongued butterflies and moths acted as robbers. Paudel *et al.* (2019) cautioned that the accuracy of the syndrome, among closely related plant species, may vary temporally and spatially – this suggesting the need for standardised, comprehensive evaluation of factors influencing the matching of a pollination syndrome of a plant species to the predominant pollinator community. Their cautionary note of potential temporal and spatial variation, in being able to accurately predict the participants suggested by an apparent of pollination syndrome, deserves greater exploration.

But characterisation of a particular plant species as representative of a single pollination syndrome can obscure, at least occasionally, wider plant–pollinator relationships and 'the notion that syndromes reflect natural selection to enhance pollen transfer by principle pollinators assumes that specialisation into functional groups of pollinators has been important in shaping evolution' (Martén-Rodríguez *et al.* 2009). It needs to be stressed that members of several taxonomically divergent animal groups may pollinate a particular plant species (Kirk 1988b), as is often the situation in mass-flowering rainforest trees and other plants with open floral structures that allow unimpeded access. The red-flowered South American tree, *Embothrium coccineum*, a supposedly bird-pollinated member of the Proteaceae, exemplifies the divergent nature of pollinators and the risks inherent in oversimplifying the assignation of pollinator syndromes, even where floral structure suggests a restricted pollinator guild. In Chile *Embothrium coccineum* attracts hummingbirds and passerines, and in Argentina it attracts hummingbirds (Smith-Ramírez and Armesto 2003). However, at several sites in Argentina *Embothrium* flowers are also pollinated by flies of the nemestrinid genus *Trichophthalma* (Plate 99), which also occurs in Australia (G. Williams 2020). Because hummingbirds are a relatively modern group and *Trichophthalma* is considered a Gondwanan taxon, pollination by Nemestrinidae has been interpreted as the ancestral condition, with ornithophily a derived later-evolving character; and indeed, based on Mesozoic fossil evidence (*Protonemestrius*) Nemestrinidae may be the earliest candidate for obligate pollinators of angiosperms (Grimaldi and Engel 2005). Variation in ecological and geographical gradients is thought responsible for changes in the associated pollinator community. In New Zealand, flowers of the bird-adapted mistletoe genus *Peraxilla* are also pollinated by short-tongued bees (Robertson *et al.* 2005). As with *Embothrium*, the apparent ornithophilous adaptation of *Peraxilla* is misleading, for although the bees provide a pollination service of lower quality (i.e. less efficient transfer) than birds, with pollen transport taking place over shorter distances, they can occupy the role of true pollinators. However, transfer of pollen over shorter distances, compared to the greater distances moved by pollen transported by birds, may have significant genetic consequences for plant populations.

Neither the most conspicuous flower visitors nor those which appear morphologically best adapted to the flowers they visit are necessarily the most efficient pollinators (Kirk 1988b). Pollination relationships need not be symmetrical. For example, while certain beetles may only visit flowers with the beetle-pollinated (cantharophilous) morphological syndrome, a flower seemingly agreeing with that syndrome might be visited and pollinated by organisms from a range of taxonomic groups.

The differences between some described pollination syndromes are slight and thus, in the course of evolution, only small changes to floral structure might have been required to result in a change of pollination vector. As an example of biotic vector variation between closely related plants, within the Zingiberales, members of the Musaceae are pollinated by bats, birds and tree shrews, Heliconiaceae by birds, bees and bats, Lowiaceae by dung beetles, Strelitziaceae by birds, bats and lemurs, Marantaceae, Costaceae and Zingiberaceae by bees and birds, and Cannaceae by birds (e.g. Cronk and Ojeda 2008; Duchen and Renner 2010). In the genus *Heliconia* (Heliconiaceae), species from the Solomon Islands are pollinated by bats but species in South America are primarily pollinated by hummingbirds (Kress 1985). In *Cayaponia* (Cucurbitaceae) there are examples of repeated evolutionary shifts from bats to bees raising the question of direction and frequency of pollinator transitions (Duchen and Renner 2010). Ancestral-state reconstructions suggest that the genus originated in tropical forests, from where open savannas were reached several times and that bee pollination arose from bat pollination, roughly concomitant with the shifts from forests to savanna habitats.

WIND POLLINATION IN FLOWERING PLANTS AND THE BALLISTIC RELEASE OF POLLEN

However, pollination may be achieved by abiotic and biotic means, as in species of *Plantago* (Plantaginaceae) and Indian *Syzygium alternifolium* (Myrtaceae) from which wind and insects are known or are likely to be agents of pollination (Raju *et al.* 2014; Sharma *et al.* 1993).

Pollination by wind (anemophily) is the most common form of abiotic (non-animal) pollination, has evolved at least 65 times in flowering plants from biotically pollinated ancestors (Linder 1998), and is found in at least 18% of angiosperm families (Ackerman 2000). Wind pollination occurs in families in which members have otherwise evolved highly specialised biotic pollination relationships (e.g. Moraceae). Successful as some abiotically pollinated groups (e.g. grasses) are, interestingly, the diversification rates of abiotically pollinated angiosperms is only about half that of biotically pollinated lineages (Eriksson and Bremer 1992). Historically, many botanists considered that the most primitive of the angiosperms were those that were pollinated by wind, and that all other types of angiosperm flowers had been derived from them. Their development was consequently linked to evolution from the wind-pollinated conifers, however, wind pollination in angiosperms is now thought to have evolved via various ancestral pathways, each possibly independently derived. The faculty of wind pollination in at least some angiosperms was achieved by selection away from animal means of pollination (suggested for the primitive *Tasmannia* [Winteraceae] [Gottsberger 1977]). Regardless of their derivation, both wind- and insect-pollinated flowers are among the earliest known fossils of flowering plants.

The selective shift to wind pollination from an ancestral biotic pollination state is thought to have been a response to changing ecological conditions, for example a decline in pollinator abundance; such changes bringing about a reduction in the advantages to be gained by relying on animal vectors for pollination (Friedman and Barrett 2008). However, Friedman and Barrett emphasise that the specific ecological and evolutionary mechanisms that might have driven the transition from biotic to abiotic (wind) pollination strategies are not well understood in comparison with other major reproductive transitions that have taken place during the evolution of the angiosperms, these including the evolution of self compatibility from an out-crossing state, and a move to dioecy from hermaphroditism. Whilst there appears to be no correlated evolution between the number of flowers per inflorescence and wind pollination, wind pollination does appear to have evolved more frequently in angiosperm lineages that already possessed either monoecious or dioecious states (Yamasaki and Sakai 2013), with wind pollination possibly evolving in dioecious lineages as a response to pollinator scarcity (Weller *et al.* 1998). Dioecy is usually correlated with small flowers, without showy petals or sepals, which are usually pollinated by ecologically generalist pollinators (usually small insects). The floral morphology of these plants may facilitate the evolutionary shift to wind pollination because pollen transfer is not obstructed by large and complex perianths.

A greater representation of wind-pollinated angiosperms is recorded from higher latitudes, arid temperate environments, open vegetation communities and island floras (Regal 1982; Whitehead 1969), and several studies have shown that altitudinal gradients can select for wind pollination (e.g. Berry and Calvo 1989) because changes in altitude impact on both the pollinator community and the habitat type. The aerodynamic requirements that underpin pollen transfer by wind possibly explain these ecological and geographical correlations (Friedman and Barrett 2008).

The occurrence of wind-pollinated plants is greater in temperate regions and less in the tropics and subtropics (Ollerton *et al.* 2011; Rech *et al.* 2016). This heightened frequency of wind pollination is particularly marked among canopy trees in deciduous forests, some using the strategy of flowering and release of pollen before the onset of leaf growth. In contrast, wind pollination has been rarely recorded from tropical and subtropical evergreen forests, though its true extent has been poorly investigated. Reasons for this putative rarity include the reduced impact and reliability of wind in the understorey, inefficiencies in the transfer of pollen between widely spaced plants of the same species in structurally complex and species-rich forest communities, and adverse impacts of rain removing pollen from the air (a disadvantage faced generally by plants in high-altitude cloud forests where moisture-laden air regularly shrouds the forests in a veil of mist) (e.g. Corlett 2004). An example of the rarity of wind pollination in tropical forests was found in a study plot in Brunei rainforest where only one out of 760 species was wind-pollinated, a species restricted to high ridges (Linskens 1996). However, Aizen and Ezcurra (1998) point out that in comparison to the temperate forests of the Northern Hemisphere, in which there is a high incidence of wind-pollination, the temperate forests of southern South America depend heavily on animals for pollination and seed dispersal. Though the fauna of this region is depauperate, plant-animal mutualisms rely on animals that are abundant and that show a high degree of residency. Thus human-induced rates of forest clearance (deforestation), fragmentation and habitat degradation (as is increasingly occurring) may be especially threatening for plants with dependent pollinator mutualisms due to the low redundancy in alternative partners.

The spatial and structural complexities of tropical and subtropical rainforests, as well as climatic conditions, may preclude the wide expression of anemophilous pollination strategies. However, it might be difficult to determine whether this is because of physical barriers to dispersal imposed by dense vegetation or because air humidity decreases the dispersal of pollen. Whilst there may be constraints on the origin of wind pollination in closed habitats, nevertheless a diversity of wind pollination strategies, in concert with insect pollination strategies, occurs in the closed forests of tropical regions (e.g. Berry and Gorchov 2004; Raju *et al.* 2014).

Forests with lower diversity of woody taxa are likely to favour wind pollination (Corlett 2004), thus a greater prevalence of anemophily in Australian rainforests may be anticipated in floristically and structurally simple hinterland and maritime vine thickets, coastal rainforests, *Backhousia sciadophora* (Myrtaceae) dominated dry rainforests (which structurally can resemble *Eucalyptus* dominated dry sclerophyll forests) and *Nothofagus* (Nothofagaceae) dominated cool temperate rainforest, as well as gap phase communities (Plates 118, 122). There may be a heightened incidence of wind pollination in gap phase rainforests because the initial open structure favours less impeded pollen transfer to conspecific plants. *Backhousia sciadophora* and *Nothofagus*-dominated rainforests are examples of species gregariousness, with the close spacing of individuals of the same species favouring pollen transport between plants (in optimum conditions *Nothofagus cunninghamii* and *Nothofagus moorei* can occur in almost pure stands). In coastal littoral rainforests (Plate 133), the frequent and dependable occurrence of off-shore winds might also favour anemophily as a pollination strategy as well as favouring a greater proportion of dioecy among wind-pollinated angiosperms.

Although there is a higher incidence of wind pollination among temperate angiosperms, most flowering plants are not wind-pollinated. This does not hold true for all families. A notable exception is the grasses, which characteristically possess large feather-like stigmas that trap the pollen dispersed from the anthers (although several grass species, including some bamboos, are not wholly wind-pollinated). Wind pollination is also dominant in the Cyperaceae (sedges) and Juncaceae (rushes), but even in plant families that are predominantly pollinated by insects a level of wind pollination may occur, with some species within a particular family being highly adapted for insect pollination and others equally specialised towards anemophily (e.g. *Ficus* versus *Streblus* in the Moraceae [Williams and Adam 1993], Plate 51). Many dioecious rainforest plants thought to be pollinated by small insects may also be facultatively wind-pollinated (Bullock 1994). Recorded instances of wind pollination in dioecious trees include species of Achatocarpaceae, Amaranthaceae, Anarcardiaceae, Capparaceae, Euphorbiaceae, Flacourtiaceae, Julianaceae, Moraceae, Nyctaginaceae, Oleaceae, Opiliaceae, Rutaceae and Urticaceae. Albuquerque *et al.* (2013) record wind pollination in *Myrsine parvifolia* (Myrsinaceae/Primulaceae) in Brazilian Atlantic rainforest; though this strategy represents a marked divergence from that of obligate thrips pollination in Australian *Myrsine variabilis* and *M. howittiana* (Williams *et al.* 2001). However, the flower structure of *Myrsine parvifolia* is open and dish-shaped rather than the enclosed urceolate forms of *M. variabilis* and *M. howittiana*. Arecaceae were considered to be wind-pollinated (e.g. as in *Wendlandiella* sp. [Listabarth 1993]) but pollination by insects has been widely reported (e.g. Schatz 1990; Kitching *et al.* 2007), the inflorescences of *Archontophoenix cunninghamiana* in Australian subtropical rainforest attracting a cohort of numerous small nematoceran and brachyceran flies (Scatopsidae, Drosophilidae, Syrphidae, Sarcophagidae, Tephritidae), small-sized beetles (Nitidulidae, Anthicidae, Chrysomelidae, Curculionidae, Scirtidae), wasps (Sphecidae, Vespidae), and solitary apid (Halictidae, Colletidae) and social nesting *Tetragonula* bees (G. Williams pers. obs.). Similar taxonomically diverse cohorts of flies, beetles and '*Trigona*' (the putative pollinator) have been recorded from the threatened Asian palm genus *Johannesteijsmannia* (Chan and Saw 2011). But dual insect-wind pollination strategies amongst the Arecaceae are not uncommon. In the Brazilian forest-dominant palm *Orbignya phalerata* the only likely biotic pollinator is a small nitidulid beetle *Mystrops mexicana*, but pollination by wind also occurs (Anderson *et al.* 1988). In the dioecious Neotropical understorey palm *Chamaedorea pinnatifrons* movements by numerous thrips and small Ptiliidae beetles in the pendulous male inflorescences cause the dry powdery pollen to be released from the anthers, the pollen then being carried in air currents to female flowers; this being termed 'insect-induced wind pollination' (Listabarth 1993). Yamasaki and Sakai (2013) noted that in dioecious *Mallotus japonicus* (Euphorbiaceae) occurring in Japanese temperate forests, and *M. wrayi* from tropical Borneo, pollination was by wind but likely also by insects; the two species being ambophilous. Such combinations of biotic and abiotic pollination strategies enhance a species adaptability to a wide range of habitats.

Flowers adapted for wind pollination usually have stamens well exposed and sometimes are structurally modified (e.g. with forked, filamentous or elongate stigmas) to increase the surface area available to snare quantities of dry air-borne pollen. Using filamentous and enlarged stigma morphologies recorded from anemophilous flowers as a guide (see Bullock 1994), wind pollination might be predicted in numerous species of Australian rainforest Euphorbiaceae (e.g. *Homalanthus populifolius* and *Croton verreauxii*). The stigmas of wind-pollinated flowers may also be sticky, to further increase the flowers' ability to snare pollen grains, although the size of pollen grains influences both pollen dispersal and the ability of the stigma to ensnare them (Faegri *et al.* 1992; Kearns and Inouye 1993; see Mariani *et al.* 2016 for a discussion of pollen dispersal modelling). However, all wind-pollinated plants are vulnerable to pollen being washed out by rain, not just those living in the rainforest environment.

Many of the species investigated possess anthers or flower buds that do not open unless the weather conditions are favourable. In addition to the adversity of conditions brought on by rain, air humidity can decrease pollen buoyancy and therefore reduce the distance over which grains might be transported. Ideally, pollen is best retained until weather conditions are warm and dry. Convection currents potentially facilitate pollen dispersal, and species which flower well above ground have particularly favourable prospects of grains becoming air-borne – a possibility that requires investigation. Although rainforest is partly defined by the wealth of trees that crowd the upper limits of the canopy, many vines and shrubs (gap phase species) that occupy openings formed by disturbance to the canopy might also take advantage of an upwelling air column to enhance pollen dispersal. Given that the pollen of anemophilous plants studied to date is usually released during the day, there is potential for thermal currents to lift pollen higher in the heated air column and so effect greater distances of dispersal.

In northern temperate forests, tree species such as the various birches (*Betula*, family Betulaceae) possess long male-flowered (staminate) catkins which droop downwards and release large quantities of dry pollen into the air when agitated by wind (Aira *et al.* 1998; Hjelmroos 1991). The quantity of pollen released can be staggering. Ten-year-old branch stems of the European beech, *Fagus sylvatica* (Fagaceae), may hold more than 28 million grains of pollen, and a similar aged stem of *Betula* more than 100 million grains. But the male flowers of such plants characteristically do not shed their pollen until moved by wind, the grains being held within scale-like structures until agitated. The distance over which the pollen of anemophilous species can be dispersed has been the subject of much study in the Northern Hemisphere (e.g. Leonardi *et al.* 2012; Poska and Pidek 2010). Large quantities of *Betula* pollen have been recorded far out to sea, many kilometres from the nearest forest. The pollen of *Betula* is considered a good flier, recorded being transported in quantity over distances of 2000 km (Friis *et al.* 2011), but even anemophilous pollen thought to be less buoyant and thus less adapted to long-distance dispersal has been found far removed from its source.

In Australian cool temperate rainforests, *Nothofagus* (closely related to *Betula*) is often an abundant angiosperm tree within the rainforest canopy, commonly dominating tree diversity. Individuals also occur as emergents. *Nothofagus sensu lato* has a distribution that comprises south-east Australia, New Caledonia, New Guinea, New Zealand and South America (Harden 1990), however, based upon fossil pollen evidence, the Betulaceae are of earlier derivation than *Nothofagus*. *Nothofagus moorei* is a large evergreen tree growing to 50 m, restricted to mountainous areas of northern New South Wales and southern Queensland (Boland *et al.* 1984). Individual trees possess both male and female flowers (monoecious) but the male flowers, which are held below the leaves, mature before the females, a reproductive strategy that promotes outcrossing via wind-dispersed pollen. Because of the abundance with which *Nothofagus moorei* occurs in cool temperate rainforest (Adam 1992), pollen is readily transported with a high probability of pollination occurring even though trees do not flower each year. The frequency with which strong turbulence occurs at high altitudes would be further expected to promote pollen transport, cloud cover permitting. Only where individuals of *Nothofagus moorei* are outliers or relics from past wide-scale land clearing (as at the Comboyne Plateau of northern New South Wales [Bale and Williams 1994]), might pollen transport be problematic, and sustainment of populations by sexual reproduction alone fail. In the related monoecious *Nothofagus cunninghamii*, which is restricted to Tasmania and Victoria, the separate male and female flowers are borne near the ends of the branches to facilitate pollen movement away from the tree. The height to which *Nothofagus cunninghamii* can grow (~40 m) also promotes dispersal of pollen into air currents, potentially over large distances and between stands. The species exhibits low range-wide genetic structure, typical of wind-pollinated trees, across most of the species' range, however, the Victorian populations form a distinct genetic group from that of Tasmania (Duncan *et al.* 2016).

Among Australian rainforest plants, an example of the secondary development of wind pollination from an original insect dependency is possibly provided by the endangered *Rhodomyrtus psidioides* (Myrtaceae) (Plate 58). *Rhodomyrtus psidioides* is a small tree related to the eucalypts, which occurs from southern Queensland to the central coast of New South Wales (Floyd 2007). Plant populations were often gregarious within rainforest stands (G. Williams pers. obs.). Its flowers are white in colour, bisexual and sweetly aromatic in fragrance like that of cloves. The stigma is flat and broad and held above the anthers, thus providing a wide surface unimpeded by obstacles for the receipt of pollen. Insects, predominantly native bees (e.g. *Leioproctus* [Plate 83], *Amphylaeus* [Plate 84], *Heterapoides, Hylaeus* and *Homalictus*), but at times also including small-sized flies and beetles, visit the flowers of *Rhodomyrtus psidioides* (G. Williams 2018). Most insect visitors to *Rhodomyrtus psidioides* plants growing in rainforest and wet sclerophyll forest in New South Wales are less than 6 mm in size and, with the exception of some of the visiting bee taxa, are generally poorly adapted, opportunist species. In addition, flower visitors are often infrequent. Long hours of field observation commonly show few species and low levels of visitation, making the role of visitors in the greater facilitation of pollination uncertain. The flowers of *Rhodomyrtus psidioides* appear to produce little nectar. This might not be unexpected, for plants that are primarily pollinated by wind normally do not produce nectar. In those that appear to have a dual wind and biotic pollination faculty, food rewards are in the form of the pollen itself or the secondary provision of nectar or other flower exudate.

A dual wind–insect pollination mechanism is known for the Australian rainforest vine *Trimenia moorei* (Bernhardt *et al.* 2003) (Plate 73). The species is heavily scented and flowers attract syrphid flies (*Melangyna* and *Triglyphus*), sawflies, native colletid and halictid bees, and honey bees. Pollen is also mobilised directly into the air, circumventing a total dependency on biotic means of pollen transport. However, the spathulate tepals of *Trimenia* fall off when the flower opens, such that they take no part in optical attraction of the open flowers (Endress 2001); though their loss reduces obstacles to the effectiveness of pollen transfer by wind.

As mentioned earlier, carriage of pollen by wind is unpredictable and a compensatory development by anemophilous species is the production of large quantities of dry (or relatively dry) smooth-surfaced pollen, typically of a small size. Size, in addition to shape, sculpture and ratio of external area to internal mass, matters when it comes to buoyancy in air currents (see Tanaka 2000). Anemophilous pollen types do not appear to occupy the smallest size classes, as minute pollen grains tend to adhere to one another – which is detrimental to buoyancy. Occasionally, masses of pollen can be found on the lower leaves of anemophilous plants. This pollen was released but not transported; it was simply deposited below the flowers without appreciable movement away from the parent plant. Though lost to the process of pollination, such 'unused' pollen can attract pollen feeders such as syrphid flies (Syrphidae), thrips and beetles, which consume a quantity of the pollen but without transfer to female flowers.

Some authors have attempted a definition of anemophily as it is encountered in angiosperms (see Culley *et al.* 2002; Faegri and van der Pijl 1979). The constructed syndrome is one in which there is a moderate to high density of plants, high pollen to ovule ratio, the flowers are male or female only, are held above the leaf clusters or crown profile to avoid impedance in the flow and settlement of pollen, the flower petals and sepals are reduced in size and thus individually insignificant (or absent entirely, or shed), anthers and stigmas are exposed, there is absence of attractants (e.g. fragrances absent, no nectar, absence of pronounced flower colour or floral guides), pollen is dry and relatively small in size, the exine is smooth and pollen is generally not sticky (free or nearly free of oil). Nectaries were typically lost after the evolution of wind pollination (Friedman and Barrett 2008) but where nectaries are present they are usually non-functional (Faegri and van der Pijl 1979; see Willmer 2011). Clearly 'no size fits all', as many of these attributes are also found among biotically dispersed pollen types, or contradict individual characteristics of species known to be wind-pollinated. As with all attempts to characterise the diversity of the Earth's many types of flowers, exceptions, if not inclined to rule, are not lacking.

Anemophily offers release from the requirement to flower during periods of insect abundance, so anemophilous species need not conform to phenological patterns of entomophilous or otherwise biotically pollinated plants in the rainforest. However, reliance on wind dispersal by passive means of pollen release, even if a vigorous shaking is frequently central to the phenomenon, is not always the case among flowering plants. Some disperse their pollen episodically, literally catapulting the dry pollen at a particular moment from dehiscent anthers that are held under tension until the flower bud opens. The dramatic expulsion of pollen visually parallels the clouds of spores released by some groups of fungi. The release of filaments (the long structures that support the anthers) and the opening of buds and anthers can occur simultaneously, the expulsion of pollen happening almost instantaneously. This occurs in several members of the Urticaceae, the family which includes the stinging trees of the genus *Dendrocnide*, and herbaceous *Elatostema* (Plate 74) found along damp rainforest stream banks. The mechanisms by which pollen is dispersed from Australian rainforest Urticaceae remain unstudied.

Among the Australian rainforest plants are species that ballistically expel their pollen. Two of these are *Trophis scandens* (Plate 52), a vine found growing within gaps formed by tree fall or wind disturbance to the canopy or along track margins, and *Streblus brunonianus* (Plate 51), a small tree, normally of the understorey (Williams and Adam 1993). Both are members of the family Moraceae, the group which includes the figs. Plants of *Trophis scandens* can be dioecious or monoecious, and produce small, inconspicuous creamish-white flowers. *Streblus brunonianus* is dioecious and possesses greenish-white flowers borne in dense spikes. Both possess inflexed stamens held in tension before the opening of flowers. As their flowers open the anther-carrying filaments fling out, releasing small clouds of pollen that are readily highlighted within sunlit forest gaps. Pollen release has only been observed from mid-morning to noon and is apparently restricted to flowers in direct sunlight. Although seemingly cued by sunlight and attendant temperature increase, pollen release is not especially synchronised; it appears to be randomly expelled from individual flowers.

Ramírez (1980) hypothesised that among the Moraceae wind pollination rather than insect mediated pollination is the ancestral condition. Wind pollination in the Moraceae has been recorded for the Costa Rican rainforest understorey tree *Trophis involucrata* (Bawa and Crisp 1980) and has been suggested for *Artocarpus rigidus* from Malaysia, for *Sorocea sprucei* from Venezuela, and for African lowland rainforest Moraceae more generally (Corner 1952; Zapata and Arroyo 1978). The method of pollen release reported for *Trophis involucrata* appears identical to that of *Streblus brunonianus* and *Trophis scandens*. Although ballistic pollen release by itself is not proof of wind pollination, more exhaustive studies of *Trophis involucrata* support wind as the mode of pollen transfer.

Let us not leave the intricacy of pollen dispersal by wind too soon. Upon the sticky glass slides set high in the atmosphere by ecologists and other scientists, to capture the air-borne pollen of anemophilous plants, is frequently found pollen grains from plants which are believed to normally use animal vectors. How does such pollen get dislodged from anthers and elevated into the upper air? Disturbance by wind may be the reason, as might other methods of physical agitation such as the beating of the wings of Lepidoptera, bees, beetles and birds – a veritable 'butterfly effect', if it can be so called.

POLLEN SCULPTURE IN SUBTROPICAL RAINFOREST PLANTS: IS WIND POLLINATION MORE COMMON THAN SUSPECTED?

The incidence of wind pollination may prove to be much more common in rainforest than was previously anticipated (Bullock 1994; Williams and Adam 1999). Though wind pollination in general could be truly rare in rainforests comprised principally or wholly by angiosperms (as opposed to boreal and temperate coniferous rainforests), it might prove otherwise for particular stands and individual species within particular families, as appears to be the case for members of the Moraceae (a family exhibiting diverse pollen dispersal strategies). Or it might be masked by characters of flower shape and form that fool observers into believing the plant relies only on biotic means of pollination.

While direct observations of the method of pollen release can provide a strong indication of a plant's pollen dispersal strategy, further clues reside in the sculpture of the exine, the outer coat of the pollen grain. There is considerable variation in the nature and pattern of pollen exine sculpture, and it has been suggested that possession of particular patterns of sculpture may be related to the mode of pollen dispersal and to specialised adaptation for carriage by pollinators (Faegri *et al.* 1992; Labandeira 2010). Pollen grains adapted for dispersal by wind generally possess relatively smooth surfaces, but insect-dispersed ('insect-pollinated') types are not necessarily sticky. However, individual families (with both wind- and insect-pollinated species, as in Primulaceae/Myristicaceae) may include a diversity of different pollen shapes, aperture forms and microsculpture across the family (Sauquet and le Thomas 2003), and superficially similar pollen types from the same family may represent genera or species with very different habits and or habitat/environmental conditions (Kodela 2006). Strongly sculptured exine surfaces (as in *Hibiscus*) are thought to characterise insect-transported pollen types, but individual insects may collect a variety of pollen – most insects that visit flowers are generalists and collect from any plants that happen to be in flower. For example, Neotropical bees have been noted to collect pollen that includes smooth and moderately sculptured pollen grains as well as kinds whose surfaces possess distinct spines (Roubik 1989). Collection of pollen by an insect is not in itself proof that the species is functioning as an agent of pollination.

To gain insight into dispersal strategies by examining the external microscopic structure of pollen, Paul Adam and I collected pollen from a range of plant species (Williams and Adam 1999). These were used to document the distribution of pollen types within four arbitrarily chosen classes (Appendix 3). Samples were collected from 130 native species (in 58 families) of trees, shrubs, vines and herbs from subtropical rainforest, and adjoining mangroves and wet sclerophyll forests in coastal New South Wales. We also investigated pollen from *Aleurites moluccana* (Plate 32), a tree native to North Queensland rainforests. The majority of woody plants in Australian subtropical rainforest communities are pollinated by small insects and possess flowers conforming to a generalised entomophilous syndrome. The majority of species we investigated conformed to that broad characterisation, however, exceptions to this dominant theme among the plants we studied were *Castanospermum australe* and *Banksia integrifolia* (which are primarily pollinated by vertebrates) (Plate 67) and *Podocarpus elatus*, *Trophis scandens* and *Streblus brunonianus*, which are wind-pollinated, or putatively so. Only *Alocasia brisbanensis* (Araceae), *Senecio minimus* (Asteraceae), *Stephania japonica* (Convolvulaceae), *Aleurites moluccana* (Euphorbiaceae), *Geranium ?homeanum* (Geraniaceae), *Hibiscus splendens* and *Hibiscus heterophyllus* (Malvaceae), *Jasminium volubile* (Oleaceae), *Pterostylis* spp. (Orchidaceae) and *Tasmannia insipida* (Winteraceae) were found to possess highly sculptured (echinate) pollen (Appendix 3). Some authors suggest this indicates specialised adaptation for carriage and or collection by insects. With the exception of *Aleurites moluccana* (Plate 32), all are understorey, forest edge or gap phase species. In *Tasmannia insipida* (Plate 78) calyptra droplets occur in both female and male flowers (males have reduced female structures), but it is unlikely that a 'pollination drop' mechanism, analogous to that found in gymnosperms is at work in *T. insipida* (Frame 2003). Frame proposed that the droplets function as a reward to potential pollinators in advance of flowers opening. *Hibiscus splendens*, *Hibiscus heterophyllus*, *Alocasia brisbanensis* and *Pterostylis* spp. are pollinated by related (oligophilic) flower visitors or by single or very closely related (monophilic) species. The pollinators of the remaining six species are unknown. Several other species with known specialised pollinator relationships (*Linospadix monostachyos* [Arecaceae], *Commelina cyanea* [Commelinaceae], *Eupomatia laurina* [Eupomatiaceae], *Wilkiea huegeliana* [Monimiaceae] and *Myrsine howittiana* [Myrsinaceae/Primulaceae]) were found to possess relatively smooth exine surfaces. *Trophis scandens* and *Streblus brunonianus* possess smooth-surfaced pollen but the gymnosperm *Podocarpus elatus* possesses relatively ornate sculpture.

The form of the pollen grains is partly correlated with taxonomy but there can be variation within families (see Appendix 3). For example, while echinate sculpturing is a characteristic feature of the Malvaceae other families contain species with very dissimilar sculpturing (e.g. Euphorbiaceae and Lauraceae). Further, it appears that pollen types might not be randomly distributed among growth forms. The majority of pollen is smooth, and trees dominate the growth forms that possess smooth pollen.

Although many trees with smooth pollen may be visited by a diverse range of insects, which can transport pollen between flowers, many of these smooth-pollen species may also be facultatively wind-pollinated (ambophilous/ambophily). The examination of pollen sculpture in New South Wales rainforest plants suggests that a greater incidence of wind pollination may occur in tropical and subtropical forests. Plants participating in plant–pollinator mutualisms have reduced pollen: ovule ratios (owing to the efficiency of pollen transfer between flowering conspecifics) (Pellmyr *et al.* 2020), however, wind-pollinated plants tend to have very high pollen: ovule ratios, thus data on those ratios could provide a means of examining the hypothesis that many apparently insect-pollinated rainforest flowers may also be wind-pollinated (Williams and Adam 1999).

GENERAL ENTOMOPHILY: POLLINATION BY THE SMALL AND THE MANY

Generalised (ecologically unspecialised) entomophily is essentially an 'open to all' adaptation – any potential pollinating insect is welcome and flowers have no specialised structures that obstruct pollinator visitation; an extreme example of this being the minute 'open' flowers of *Cissus* species (Vitaceae) (Plates 76, 77), but equally so in such mass-flowering trees as *Olea paniculosa* (Oleaceae) (Plate 62), *Argyrodendron actinophyllum* (Malvaceae/ Sterculiaceae) (Plate 41) and *Tristaniopsis laurina* (Myrtaceae) (Plate 100). The price paid for such a strategy is that a great number of insects that provide poor or no pollination service may be attracted, as exemplified by numerous native hylaeine and euryglossine bees, and many small Coleoptera (e.g. Coccinellidae, Dermestidae, Phalacridae). Although a variety of small insects can serve as agents of pollen transfer these can differ substantially in the distances over which they may carry pollen. Indeed the multiple fates of the greater proportion of pollen is a perilous one for most grains fail to reach conspecific stigmas 'with each avenue of pollen loss representing an opportunity for selection to act on traits that reduce the amount of wasted reproductive potentional' (Minnaar and Anderson [2019] give a value of ~98% loss).

In rainforests worldwide, many groups of insects, rather than vertebrates, are primarily responsible for pollination of sexually reproducing trees. Historically, bees and syrphid or hover flies have been considered more efficient and adapted collectors of pollen and, consequently, major pollinators of flowering plants. The foraging behaviour of eusocial Apidae, particularly honey bees, is well documented but the pollination contributions of other bee families are varied.

House (1985) found that pollination in three North Queensland dioecious rainforest tree species was achieved by guilds of insects (principally flies, wasps and beetles), generally less than 6 mm in size. Worboys and Jackes (2005) investigated the pollination ecology of *Idiospermum australiense* and found that numerous thrips and a variety of small beetles and other invertebrates were present on its flowers. Similar guilds of small insects also dominate the flower visitors frequenting subtropical rainforest trees and North Queensland dry rainforests and, although their pollen-carrying capacity is individually limited due to body size, this may be compensated for by their collective abundance.

In comparison to the numbers of small insects visiting the flowers of rainforest trees, larger insects (>9 mm) are relatively infrequent visitors, particularly in Australian subtropical and temperate rainforests (e.g. Ettershank and Ettershank 1990; Williams and Adam 1998; see Appendix 2). These larger insects characteristically include butterflies, large predatory wasps, several groups of flies (e.g. nemestrinids and tabanids) and beetles (e.g. buprestids, cetoniines and cerambycids). Butterflies are classically considered specialised and efficient pollinators but large flies, wasps and beetles also possess structures that permit carriage of large quantities of pollen. Individuals of many large insects undertake frequent and long distance inter-plant flights which, with their capacity for pollen carriage, potentially allows involvement in cross-pollination of widely dispersed plant populations in disproportion to their lower abundances (see Appendix 2). The flight movements of male thynnid wasps, for example, could result in individual pollen movements exceeding 130 m (Peakall 1990), with long distance pollen flow in the Australian ground orchid *Caladenia tentaculata* promoted by its wasp pollinator's avoidance of visits to more than one flower in a patch. Commulatively, and employing a bigger spatial and temporal brush, such patterns of pollen transport and dispersal influence neighbourhood size and inbreeding rates in plant populations and are critical for understanding important evolutionary processes such as population differentiation and speciation.

Many insects have been dismissed or considered inefficient pollinators, unable to transport pollen, or meaningful pollen loads, upon their body surfaces. However, scanning electron microscopy of large flower-frequenting insects has revealed heavy pollen loads ensnared by surface sculpture and structures (Williams and Adam 1998). Even small anthophilous insects manage to trap grains of pollen on their bodies.

Pollen-collecting hair-like structures (setae) and ventral punctation and depressions that retain pollen have been postulated to function as food-collecting surfaces and facilities, explaining their development in flower-frequenting insects (Williams and Adam 1999). Although the development and function of many structures on the body surface of flower-frequenting insects remains unclear, rainforest plants achieve varying levels of pollination via insect presence and movements.

In entomophilous blossoms such as *Cuttsia viburnea*, *Tristaniopsis laurina* and *Guoia semiglauca* (Plate 71) there is little depth effect in the perianth, the flowers are actinomorphic and open and the sepals are often reduced. Stamens are usually free and reduced in size and daily nectar volumes may be small. General entomophilous pollination has been recorded in a variety of plant taxa from a wide range of Australian rainforest types. These include *Eucryphia lucida* and *Eucryphia milliganii* in

Tasmanian cool temperate rainforest, *Acradenia euodiiformis, Cuttsia viburnea, Syzygium smithii* in New South Wales subtropical rainforest and *Alphitonia petrei, Diospyros pentamera, Flindersia brayleyana, Litsea leefeana* and *Neolitsea dealbata* in Queensland tropical rainforest. Unspecialised entomophilous floral syndromes also occur in many New Zealand rainforest species, which have been considered an adaptation to highly variable, ecologically unspecialised insect pollinators that dominate the available pollinator assemblage (Primack 1978). Such a hypothesis applies to species worldwide that exhibit a general entomophilous flower syndrome.

Generalised entomophily might be an adaptive reproductive strategy in mass-flowering trees in rainforest with depauperate pollinator guilds (where taxa are absent or occur irregularly) (Lack and Kevan 1984), however, the wide occurrence of pollination by small unspecialised insects recorded from many rainforest trees (not just from Australian rainforests) does not accord with such a prediction (e.g. Boulter *et al.* 2011). Nor does general entomophily appear to be an adaptation favoured by lower temperatures (Bawa 1990). The unspecialised entomophilous syndrome should not necessarily be construed as a primitive or conservative feature. By recruiting pollinators from a wide range of small insects, plants may be buffered from environmental fluctuations and the spatial and temporal fluctuations of pollinators which might affect the abundance of individual species but not the total insect population. After all, adaptive success is measured by persistence through time, not necessarily by ecological specialisation. While ecological specialisation can maximise opportunities at a particular moment, it may lead to loss of adaptive flexibility in the face of environmental change or loss of a host (González-Varo *et al.* 2013).

POLLINATION BY BEETLES (CANTHAROPHILY)

Beetles (Coleoptera) have a long evolutionary history with the oldest beetle-like fossils dating from the Lower Permian (fossil records reviewed in Lawrence and Ślipiński 2013), and have been suggested as the earliest pollinators of flowering plants (Gottsberger 2015). Though beetle pollination is common in relict angiosperms cantharophily has been considered too specialised to be ascribed solely to the earliest angiosperms and their forebears (Bernhardt and Thien 1987). Nevertheless, cantharophilous angiosperms must have evolved by the Mid to Late Cretaceous, joining with pre-existing beetle pollinated gymnosperms (Bernhardt 2000).

Beetles are pollinators of an ecologically diverse, phylogenetically heterogeneous assemblage of plants, in which cantharophily has apparently evolved independently in at least 14 families of eudicots and six families of petaloid monocots (Bernhardt 2000). In Australia cantharophily has been recorded in more than 30 families of flowering plants with specialist and generalist flower structures. In rainforest, flowering plants pollinated by beetles, either solely or as part of a larger guild of pollinators, occupy canopy, subcanopy and ground strata. Beetle pollination may be more important in Australian tropical rainforests than in tropical America, where there are relatively few records of beetle-dominated pollination relationships (Irvine and Armstrong 1990). Differences in pollinator prominence may reflect differences in dominance of flower types in North Queensland and the Neotropics. In tropical Queensland rainforest, populations of the dioecious tree *Diospyros pentamera* beetles carry the largest pollen loads and move more frequently between male and female trees, and are thought to be the main pollinators (House 1985, 1989). Small weevils and nitidulid beetles are pollinators of the dioecious Queensland subcanopy tree *Myristica insipida* (Myristicaceae) (Armstrong and Irvine 1989). *Myristica insipida* is related to the nutmeg tree, which in southern India is pollinated by small anthicid beetles (Armstrong and Drummond 1986). The monoecious understorey tree *Ryparosa kurrangii* (Achariaceae) is pollinated by small insects (Webber *et al.* 2008). Flower visitors include ants and thrips, but chrysomelid beetles (*Monolepta* sp.) and mirid bugs (*Coridzolon australiense*), which occur in large numbers on both male and female flowers, are considered the likely pollinators. *Ryparosa kurrangii* is restricted to three small areas of lowland rainforest in North Queensland. Flowers are borne as long drooping racemes on the main trunk, and at night these produce heavily scented nectar, upon which the insects feed. The nocturnal production of nectar coincides with the peak period of visitation by invertebrates, with estimates of over 300 individual *Monolepta* beetles on individual trees on some nights. In tropical Queensland anthophilous beetles form diverse guild assemblages (variably of Curculionionidae, Mordellidae, Chrysomelidae, Nitidulidae etc.), in concert with other insects, on the palms *Calamus radicalis* and *Normanbya normanbyi* (Kitching *et al.* 2007). Large cetoniine beetles (*Dilochrosis balteata, Lomaptera cinnamomea, L. wallacei yorkiana*

Lomapteroides duboulayi,) are recorded from flowering *Melicope elleryana* (Rutaceae) in North Queensland (Brooks 1948), and lycid beetles (*Metriorrhynchus lateralis*) are pollinators of the vine *Marsdenia fraseri* (Forster 1989) and are associated with the flowers of mass-flowering trees in subtropical rainforest (G. Williams 1995).

The cantharophilous syndrome encompasses a broad range of floral characteristics which overlap with those of several flower syndromes, however, some beetle-pollinated flowers lack characters historically associated with cantharophily (e.g. Goldblatt *et al.*1998). Classical cantharophily includes plant species whose flowers emit sweet fragrances or strong fruity or aminoid odours. Other characteristics include nocturnal opening of flowers, endothermic heat production by the inflorescence, and floral structures (including brush, chamber and bowl types) that provide a degree of closed or filamentous protection. Individual beetle syndrome characters are variously beneficial depending on the size of the beetles. They can be exclusive if the participating beetles are large (as in the Cetoniinae), and inclusive where dense, long stamen filaments allow small beetles to secrete themselves deep within (as in various Myrtaceae). In some instances the odours emitted by 'beetle-flowers' are dung-like, attracting their primary (scarab beetle) pollinators by deceit, as in the geoflorous *Orchidantha inouei* (Lowiaceae) from Borneo (Sakai and Inoue 1999) and the Australian *Typhonium brownii* (Araceae) (Monteith 1986). Many Oriental species of *Amorphophallus* (Araceae), known to be pollinated by beetles, emit carrion-like smells that mainly consist of dimethyl oligosulphides (Kite and Hetterschield 1997). Floral fragrance compounds that have been demonstrated to be especially attractive to anthophilous Oriental beetles include eugenol, benzyl acetate, methyl benzoate and linalool (Corlett 2004). Experiments with the fragrance compounds p-anisaldehyde and geraniol in New South Wales subtropical rainforest remnants elicited no response from any beetle family (G. Williams 1995).

The emission of strong odour, sweet or foetid, appears to be the unifying theme linking the diverse range of plants from which beetles have been reported as pollinators. This correlates with the marked development of olfactory senses in Coleoptera (Crowson 1981; Lawrence and Ślipiński 2013). Several beetle-pollinated plants, such as *Typhonium*, volatise odours by raising the temperature of their flowers (e.g. Gottsberger 1999; Monteith 1986). Certain Coleoptera show pronounced responses to heat (e.g. *Merimna* and *Diadoxus* [Buprestidae]), but the possible direct function of increased temperature in plants as an attractant to pollinators has not been exhaustively investigated (Seymour and Matthews 2006 suggest thermogenesis may also function as an 'energy reward' to large beetles that require high body temperatures to be active).

Beetles are thought generally to lack colour vision (Crowson 1981) but, while individual flowers may lack strong visual attractions, cantharophilous plants may produce large massed inflorescences that cumulatively establish a strong visual stimulus.

Present-day Coleoptera, adapted to feed from floral resources or strongly associated with flowers in other ways (floricolous), can be divided into two groups – a larger group of families that visit flowers solely for adult feeding and a second group whose larvae develop in flowers and fruits (Crowson 1981). Some families (Nitidulidae, which are pollinators of Australian subtropical rainforest *Hibiscus splendens, Hibiscus heterophyllus* [Malvaceae], *Alocasia brisbanensis* [Araceae] and *Pittosporum revolutum* [Pittosporaceae] [G. Williams 1995]) are represented in both groups. The first group, however, possess taxa with marked morphological adaptations to the floricolous habit, including specialised mouthparts for nectar-feeding, prolongation of the front of the head (frontoclypeus) and long and dorsally inserted antennae (Lawrence and Ślipiński 2013).

In flower-frequenting beetles the perpendicular (orthognathus) positioning of the mouthparts is considered to be the original or primitive state. It is characteristic of the many phyllophagous forms, in which the mandibles are well developed for chewing. The orthognathic condition of the mouthparts is indicative of unadapted (dystropic) and poorly adapted (allotropic) beetles. Prognathus mouthparts characterise specialised floricolous Coleoptera. In these, the mouthparts are acuminate and produced parallel to the body-axis. Particular segments and organs (e.g. the maxillary palps, the galae and lacinae) are specialised for nectar-feeding (Lawrence and Ślipiński 2013). There is no clear phylogenetic delineation of orthognathous and prognathous forms; the determining influence is largely ecological. The prognathic condition, with prolonged mouthparts and acumination of the frontoclypeus, reaches extreme development in nectar-feeding Buprestidae (of the tribe Stigmoderini [Bellamy 2003]) and the Cerambycidae (e.g. tribes Calliprasonini and Psilomorphini [Lawrence and Ślipiński 2013; Ślipiński and Escalona 2016). Prolongation of mouthparts is also characteristic of flower-visiting Ripiphoridae,

Tenebrionidae (subfamily Alleculinae), Meloidae, Oedemeridae, Cantharidae, Mordellidae, Lycidae and Scarabaeidae (Cetoniinae and Sericini) (Lawrence and Ślipiński 2013). These appear to be eutropic, fully adapted, obligate blossom visitors. Other adaptive attributes considered to characterise anthophilous beetles include compound eyes usually relatively large and finely facetted, antennae long and tending to be inserted towards the top of the head rather than laterally, adaptations for licking nectaries (i.e. maxillary lobes and ligula usually elongate and pubescent), well developed often pigmented wings, and a tendency towards an elongated body form.

In Australia the anthophilous Coleoptera are characteristically associated with sclerophyllous and mesic-adapted Myrtaceae. However, caution is needed in the interpretation of pollen removal by beetles and other vectors. In the Australian sclerophyllous *Banksia menziesii* (Proteaceae), large amounts of pollen are removed by staphylinid beetles but the beetles deposit no pollen on the stigmatic surfaces of the flowers (Ramsey 1988).

Beetle pollination is often associated with primitive families of angiosperms (e.g. Chloranthaceae, Magnoliaceae, Annonaceae, Eupomatiaceae and 'Idiospermaceae' [e.g. Gottsberger 1974, 1977, 1999; Irvine and Armstrong 1988; Tosaki *et al.* 2001]) but, as discussed earlier, the evolutionary development of beetle pollination in living relictual plants might be explained by several developmental pathways that represent conditions that are ancestral or derived (see Bernhardt and Thien 1987; Grimaldi and Engel 2005). Protogyny is often associated with cantharophily (as in the spadices of beetle-pollinated aroids, and some palms) and many 'primitive' or evolutionarily conservative angiosperms are protogynous (e.g. Eupomatiaceae). This may have obviated the need to evolve self-incompatibility systems (Schatz 1990), but protogyny within individual trees does not preclude self-pollination via geitonogamy (transfer of pollen from different flowers on the same plant). Beetles certainly appear to be among the first pollinators of flowering plants and, considering their ancient lineage among terrestrial organisms, were likely part of a proto-pollinating, pollen-feeding insect guild that began visiting early angiosperms. Feeding on pollen is considered central in the evolution of early Chrysomeloidea (Cerambycidae and Chrysomelidae [Lawrence and Ślipiński 2013]), this association with pollen food sources presumed to have occurred before the end of the Jurassic and before the emergence of angiosperms (Samuelson 1994). There is a close relationship between Neotropical Arecaceae (which are now considered doubtfully primitive) and the beetle families Nitidulidae and Curculionidae (Acalyptini/Derelomini) (Anderson *et al.* 1988; Uhl and Moore 1977). Weevils, possibly *Amorphoidea linospadicis* (Acalyptini/Derelomini), are known from flowers of the Australian subtropical understorey palm *Linospadix monostachyos* (Plate 13) (G. Williams 2020), and the genera *Eudela* and *Nodocnemus* are recorded from the northern Queensland palm *Normanbya normanbyi* (Plate 14) (Kolesik *et al.* 2006). Species of *Meriphus*, *Cyttalia* and *Storeus* occurred commonly in flowers of many tree species sampled in subtropical lowland rainforest remnants of the New South Wales north coast (G. Williams 1995). Curculionidae were dominant on flowering *Normanbya normanbyi* during all sampling occasions undertaken in North Queensland by Kitching *et al.* (2007). Both Nitidulidae and Curculionidae are represented in Late Jurassic fossil deposits (Crowson 1981; Gratshev and Zherikhin 2003; Ponomarenko 2004) and, if the geologically early derivation of Arecaceae is correct (Uhl and Moore 1977), the evolutionary relationship between Nitidulidae, Curculionidae and palms might be a long one.

There is historical controversy over the precursory polarity of the beetle feeding state (herbivory *v.* pollen-feeding), with implications as to the mode of pollination and evolution in primitive angiosperms (Irvine and Armstrong 1990). However, the pollen-feeding association of curculionid weevils (there are several weevil-like families, of which true weevils Curculionidae are but one [Gratshev and Zherikhin 2003; Lawrence and Ślipiński 2013; Ślipiński and Lawrence 2019]) with cycads, and their apparent role in pollen transfer (e.g. Chadwick 1993; Ornduff 1989, 1990), lend support to the pre-existing selection for pollen feeding and facultative pollination for beetles before the Late Cretaceous radiation of the Magnoliophyta (Anderson 1995). Beetles can be floral predators but often overlooked by pollination ecologists is acknowledgment of the evolutionary transition of predacious Coleoptera (e.g. Cantharidae, Melyridae, Staphylinidae and Cleridae) to pollen-feeding (Crowson 1981), providing evidence of the acquisition of a pollen-feeding (pollinivorous) habit independent of a pathway predicated upon transitional herbivory.

The diversity of beetle families and species found on the flowers of tropical rainforest angiosperms is massive in number of individuals and great in diversity of taxa. Overwhelmingly they are members of the suborder

Polyphaga (e.g. Buprestidae, Cerambycidae, Elateridae, Mordellidae, Nitidulidae, Ripiphoridae and Scarabaeidae) (see Lawrence and Ślipiński 2013). Their cousins within the remaining suborders Adephaga (e.g. Carabidae), Archostemata (e.g. Cupedidae, Ommatidae) and Myxophaga (e.g. Sphaeriusidae) (Lawrence and Ślipiński 2013; Ślipiński and Lawrence 2019) are relatively or completely absent. The situation is the same for Australian subtropical rainforest flower fauna (G. Williams 2020). Only arboreal members of the adephagous Carabidae are sometimes found on blossoms, their predacious habit giving them occasional reason to move from tree trunks and branches onto flowers in search of prey. Of the smaller-sized subtropical rainforest beetle fauna the Mordellidae (e.g. *Mordella*) frequent flowers as floristically and structurally diverse as *Abrophyllum ornans, Acradenia euodiiformis, Ackama paniculosa, Alphitonia excelsa, Cryptocarya microneura, Diospyros australis, Drypetes deplanchei, Elaeocarpus obovatus, Euroschinus falcatus Rhodomyrtus psidioides, Scolopia braunii, Schizomeria ovata, Syzygium floribundum,* and *Tristaniopsis laurina.* These are predominantly canopy trees that exhibit general entomophilous floral syndromes. Smaller-sized beetles that characterise the anthophilous fauna of such trees in Australian subtropical rainforests include Anthicidae (e.g. *Egestria*), Cantharidae (*Heteromastix*), Chrysomelidae (*Ditropidus, Monolepta*), Cleridae (*Elaele* and *Lemidia*), Coccinellidae (*Orcus*), Corylophidae (*Sericoderus*), Curculionidae (*Eristinus, Haplonyx* and various members of the tribe Tychini), Dermestidae (*Trogoderma*), Elateridae (*Microdesmes*), Latridiidae (*Corticara*), Melyridae (*Helcogaster*), Phalacridae (*Litochrus*), Scarabaeidae (*Cheiragra, Phyllotocus*), Scraptiidae (*Scraptia*) and Staphylinidae (*Anotylus*) (G. Williams 1995, 2018, 2020; Williams and Adam 2019). Canopy flowers are frequented by families that include larger-sized species, such as the Scarabaeidae (*Storeyus*), Cerambycidae (*Distichocera, Tropocalymma* [Plates 103, 104]), Buprestidae (*Calodema, Metaxymorpha* [Plates 100, 102]) and Elateridae (*Ophidius*). These lists are not exhaustive, and the genera cited in parentheses are only indicative. Families associated with the flowers of understorey angiosperms are more restrictive, most typically Nitidulidae (e.g. *Aethina, Epuraea*), Curculionidae and Staphylinidae.

Elleschodes *and* Eupomatia

The classical beetle pollination studies in Australian rainforest ecology are on the protogynous understorey shrub or small tree *Eupomatia laurina* (Plates 35, 36), commonly known in New South Wales as 'Bolwarra' (e.g. Hamilton 1897; Hotchkiss 1958; Irvine and Armstrong 1990; G. Williams 2020). The species provides one of the few described examples of close linkages between plant and pollinator among Australian rainforest plants. The original detailed study of pollination and breeding system in *Eupomatia laurina* was undertaken by Alexander Hamilton in the late 19th century in the subtropical rainforests of southern coastal New South Wales, although brief observations had been recorded in the early and mid 19th century. The first observations were made by Robert Brown, a botanist who visited Australia as naturalist on board the *Investigator*, commanded by Matthew Flinders (Brown 1814). Hamilton (1897) described the pollen loads observed on the feet, antennae and setae of the plant's pollinators (weevils), the shedding of the fused petal and sepal cap (forming a eucalypt-like operculum or calyptra) on anthesis (Plate 36), and the abscission of the uniquely united flower body.

Eupomatia laurina is distributed from eastern Victoria to North Queensland and altitudinally from approximately sea level to 1200 m (Harden *et al.* 2006). It is not restricted to rainforest and can occur as a common understorey small tree in wet sclerophyll forest. *Eupomatia laurina* can reproduce vegetatively from the base of the stem and surface roots. It is one of the relatively few rainforest species in Australia which can recover from severe bushfire damage, though this depends on the frequency of disturbance and damage – frequent fires ultimately result in death of plants.

In *Eupomatia laurina* large, white waxy flowers with strong fruity odours are borne singly or sometimes in small groups. Flowers are formed of basally united stamens and large fleshy staminodes that collectively form a unit called a synandrium. Synandria fall off ~14–30 h after opening (Irvine and Armstrong 1990; G. Williams 1995). Individual subtropical *Eupomatia laurina* populations show between-site asynchronous flowering cycles but within individual populations flowering is broadly synchronous. Intra-population flowering of individual trees extends over approximately two to three weeks. Flowers of single plants normally open synchronously between ~3.30 a.m. and 6.30 a.m., at which time they are in female phase. The male phase occurs in the evening. In this *Eupomatia laurina* differs from the related *Eupomatia bennettii* – while the latter is also protogynous its female phase occurs on day 1 and its male phase on day 2, before shedding of the synandrium (Endress 2003). Although small numbers of flowers may open several days before and following the main flowering episode, more than 500 flowers on a single tree may open synchronously. The pollinators are quickly attracted to these.

Throughout its range, *Eupomatia laurina* is pollinated by the weevil genus *Elleschodes* (Plate 35). The related *Eupomatia bennettii* and *Eupomatia barbata* are pollinated by the same genus, though a different species to that which pollinates southern populations of *Eupomatia laurina*. The southern pollinating species of *Eupomatia laurina* is *Elleschodes hamiltoni* but the weevils that pollinate North Queensland populations are undescribed and are apparently a new species to science. Adult weevils leave the flower at dusk through the dehiscing protogynous anthers and move to flowers opening at dawn the next day. They fertilise stigmas as they move around on the floor of the floral chamber without damaging the stigmatic surface. The synchronisation of flower anthesis and abscission of the synandrium facilitates out-crossing.

In *Eupomatia* the enveloping staminodia protect the adult weevils against predators, a putative predator being the northern Queensland green tree ant *Oecophylla smaragdina* (Armstrong and Irvine 1990). The apical tips of the staminodia function as food sources, and staminodia enhance visual display, carry osmophores for the production of attractant odours and produce sticky pollen secretions that facilitate carriage of heavy pollen loads (300+ grains) on individual weevils. These features have been proposed as characteristics of the floral morphology of evolutionary relict Magnoliales.

In terms of beetle larval development, the abscissing floral synandrium provides food substrate removed from the developing gynoecium. Little is known of the life history of *Elleschodes*; however, large fat bodies are found in individual weevils. This suggests that they may be able to remain inactive within the soil, awaiting suitable environmental conditions or emergence cues. In New South Wales *Elleschodes hamiltoni* populations pupate in the soil for periods of eight to 18 days and under laboratory conditions emerge over a 10-day period, with >98% of adults emerging within a 4-day period (G. Williams 1995). Such generally synchronous emergence, if coinciding with *Eupomatia* flowering, would be advantageous in maximising utilisation and consequent reproductive success of the equally synchronised anthesis of the host plant population.

An interesting phenomenon is displayed by *Elleschodes* weevils when *Eupomatia* flowers are investigated: weevils, when disturbed within the synandrium, will frequently make a rapid exit and drop from the flowers. This is a predator avoidance mechanism that is independent of the protection function of the staminodia postulated by some ecologists. Rather than hiding within the protective wall of enveloping flower structures, the beetles choose the opposite: escape, seeking refuge in the leaf matter on the forest floor. Similar 'drop–escape' responses are found in several beetle families. They are characteristic of many phyllophagous Buprestidae, Chrysomelidae and Curculionidae, each beetle simply tucking its legs close to its body and tumbling away from the perceived threat.

Typhonium *and dung beetles*

The herbaceous understorey arum lily *Typhonium brownii* (Araceae) is an example of a specialised plant–pollinator adaptation and mutualism (Monteith 1986). The extraordinary production of heat that the spadix appendage of *Typhonium brownii* is capable of when emitting odours attractive to its beetle pollinators has been early highlighted. However, the adaptive relationship between protogynous *Typhonium* and its scarabaeine pollinators is more complex than a bad smell and beasts that find the promise of pleasure from it. Flowers of *Typhonium brownii* have a one-day cycle. They open in the afternoon and their peak temperature level is achieved about three hours after anthesis. Commensurate with this is the presence of a strong dung-like odour. Many small flies, in addition to the true beetle pollinators, are drawn to this odour. The flies are little more than opportunistic itinerants loitering in hope of a meal, and their presence is not constant. The structure of the *Typhonium* flower, with its separated male and female flowers along the central spadix, denies flies any role as pollinators. Similarly, *Alocasia brisbanensis* sifts out its visitors. The beetles, carrying pollen on their bodies from other plants (the male flowers of *Typhonium* are not yet presenting pollen) fall into the lower chamber of the flower, formed by the bulbous collared base of the spathe. The bristles that encircle the lower spadix, and the smooth inner walls of the spathe, temporarily deny escape. The following day the flower repositions itself somewhat horizontally, allowing the beetles to exit the imprisoning chamber. But in escaping they inadvertently collect on their bodies pollen that has been shed from the male flowers, and so disperse it to other plants. Out-crossing is achieved.

POLLINATION BY DIPTERA (MYOPHILY AND SAPROMYOPHILY)

There are ~125 000 described species, in 110 families, of Diptera (Raguso 2020). Flies are most appreciated as generalised pollinators, or as specialised pollinators in brood-site deceptive plants that mimic fungi, faeces or carrion (Woodcock *et al.* 2014).

Based upon morphological characters the earliest evidence for an anthophilous adaptation is found in flies. These include fossil Nemestrinidae from the Early Cretaceous Yixian formation of China and the long-tongued brachyceran *Cratomyia macrorrhyncha* from the Early Cretaceous Santana formation of Brazil (Grimaldi and Engel 2005). However, whilst some groups of Diptera, such as syrphids, exhibit temporal flower constancy in their foraging behaviour fly responses to flowers do not appear to be necessarily cued by colour (e.g. Goulson and Wright 1998). Consequently, they have not been considered major drivers in the evolution of flowers (Dyer *et al.* 2012; Zhang and Wang 2017).

The fly blossom syndrome is often characterised as one of unspecialised actinomorphic flowers with no depth effect, in which colours are generally light or dull and nectar guides are frequently present; though this could equally exemplify a floral expression that targets a diversity of non-Dipteran flower visitors. However, the syndrome of myophily, as classically characterised (Faegri and van der Pijl 1979), is also well developed in more morphologically complex flowers such as those of the Asclepiadaceae (Ollerton and Liede 1997), that possess varying degrees of depth effect in their blossoms, and in the zygomorphic Orchidaceae (e.g. Sumatran *Dendrochilum longibracteatum* [Pederson 1995], 'Tuncurry Midge orchid' *Genoplesium littorale* [Bower *et al.* 2015]). Interestingly, both Asclepiadaceae and Orchidaceae exhibit complex mechanisms of pollen presentation and pollen mass formation (pollinia); pollinia greatly increasing the probability that pollen removed by a pollinator will reach a specific stigma (Harder and Johnson 2008). Myophily may have arisen independently several times, not only because of the morphologically divergent flowers that attract flies (e.g. Plates 75, 95, 96, 97, 99) but because of the diverse phylogenetic relationships of the participating plant taxa.

The greatest diversity of anthophilous Diptera occurs on accessible unspecialised blossoms. There is a correlation between short proboscis length and unspecialised shallow blossoms, and increasingly longer proboscis length and specialised tubular blossoms (Armstrong 1979). Species with long probosces (e.g. as in Bombyliidae, Nemestrinidae, Tabanidae, flower visiting Empididae), however, can also take exposed nectar in shallow blossoms. In unspecialised myophilous flowers, odours are generally absent or imperceptible (at least to humans) and nectar is easily obtainable. Sexual organs are well exposed, though some fly-pollinated blossoms (though these may also be pollinated by other insects) can be short and tubular (e.g. *Leucopogon* and *Trochocarpa* [Ericaceae/Epacridaceae], see Plate 108) with relatively inaccessible nectaries and concealed stamens.

Visitation and pollination by flies, often in combination with pollination by beetles and Hymenoptera, is a dominating trait among unspecialised entomophilous rainforest angiosperms (e.g. House 1985; Lloyd and Wells 1992; G. Williams 1995). Flies of the families Sciaridae (*Sciara* sp.) and Chloropidae (*Apotropina bispinosa)* are considered pollinators of Australian *Marsdenia cymulosa* (Asclepiadaceae) (Forster 1992b). Calliphoridae, Tabanidae and Sarcophagidae are visitors to the flowers of Tasmanian Cunoniaceae/Eucryphiaceae and chironomid flies are common visitors to female flowers of New Zealand *Pseudowintera colorata* (Winteraceae) (Lloyd and Wells 1992). Myophilous pollination is recorded in tropical *Aristolochia, Rafflesia* and *Drimys* (Australasian species of the latter now in the closely related *Tasmannia* [Doust and Drinnan 2004]), and is also reported in basal angiosperms such as *Cabomba* (Cabombaceae) (Endress 2001). Pollination by syrphid flies (Syrphidae, hover-flies) is recorded from several Asian and Neotropical orchid genera (e.g. *Paphiopedilum* and *Phragmipedium*) (e.g. Bänziger 1996, 2002; Bänziger and Disney 2006).

Fly pollination is well known from tropical herbs, but has generally been considered rare in tropical trees (Ashton 1969). Pollination of 'cocoa' (*Theobroma cacao* – Malvaceae), the source of chocolate, is primarily achieved by small ceratopogonid flies of the genus *Forcipomyia* (Toledo-Hernández *et al.* 2017). However, in *Theobroma cacao* bees and ants also facilitate pollination and increased crop yields; the ants disturbing pollinators and this resulting in increased levels of out-crossing and heightened levels of fruit production owing to pollinators moving to other flowers. Small Diptera (in addition to co-visiting Hymenoptera and Coleoptera) are pollinators of dioecious *Neolitsea dealbata, Litsea leefeana* and *Diospyros pentamera* trees in Queensland tropical rainforest (these species also occur in subtropical rainforest), as well as being visitors (and potentially pollinators) to the flowers of such Australian tropical species as *Syzygium cormiflorum* (Crome and Irvine 1986; House 1985, 1989).

Flies of the suborder Brachycera are the most important diurnal Dipteran pollinators of rainforest angiosperms. Syrphidae, and particularly the Bombyliidae or 'bee-flies' (both of which are generally adaptively specialised flower-feeders [e.g. Neff *et al.* 2003]), are not as common on Australian mass-flowering rainforest trees as they are on sclerophyll species (e.g. *Leptospermum* [Myrtaceae]). In comparison with syrphids and bombyliids, Phoridae, Lauxaniidae, Tephritidae, Chloropidae, Drosophilidae and other small unspecialised Brachycera are well represented on the flowers of Australian subtropical rainforest trees (e.g. G. Williams 2020). Chloropid flies (*Conioscinella, Cadrema*) are the exclusive pollinators of the critically endangered *Genoplesium littorale* (Orchidaceae), a species restricted to dune-associated habitats in the Forster–Tuncurry region of northern New South Wales, and which offers nectar as reward (Bower *et al.* 2015).

The second Diptera suborder, the paraphyletic 'Nematocera' (which includes mosquitoes, biting midges and gnats) in Australia includes at least 12 families including Ceratopogonidae, Culicidae, Chironomidae, Mycetophilidae, Bibionidae and Cecidomyiidae. These contain anthophilous species but most nematocerans possess short mouthparts and are generally restricted to shallow flowers of open and readily accessible structure or small tubular blossoms. However, some Culicidae (mosquitoes), a family with notably elongate mouthparts, are pollinators of orchids (e.g. Lahondère *et al.* 2020).

Specialised pollination relationships exist between Nematocera and terrestrial Australian orchids (e.g. *Pterostylis, Acianthus* and *Corybas*), and although these are principally non-rainforest genera several species occur in subtropical and cool temperate rainforests or associated ecotones. Nematocera are implicated in the pollination of *Daphnandra micrantha* (Monimiaceae/Atherospermataceae) (Williams *et al.* 2001), a small understorey tree restricted to coastal rainforests in New South Wales. Male fungus gnats (Mycetophilidae) are pollinators of New Zealand *Pterostylis oliveri* (by *Mycetophila latifascia*), *P. irsoniana* (*Morganiella fusca*) and *P. venosa* (*Tetragoneura* sp.), each apparent species-specific mutualism possibly achieved by the deceptive emission of floral volatile chemicals that resemble the sex pheromones of respective females, so as to lure their male pollinators (Thalwitzer *et al.* 2018). Midges of the family Cecidomyiidae are pollinators of Ecuadorian *Siparuna* and *Artocarpus integer* (Moraceae) in Sarawak lowland rainforest (Feil 1992; Sakai *et al.* 2000). Cecidomyiids (*Forcipomyia*) are also recorded as pollinators of Neotropical *Trichosalpinx* orchids (Bogarín *et al.* 2018). In the dioecious Chinese *Schisandra henryi* (Illiciaceae/Schisandraceae) female cecidomyiids of the genus *Megommata* are attracted to female flowers, deceived by their resemblance to more abundant pollen-producing male flowers; pollination is thus achieved by deceit (Yuan *et al.* 2007). In Australian lowland subtropical rainforest, large numbers of diurnally active *Bibio imitator* (Bibionidae) flies, especially females, can sometimes be found on the flowers of dioecious *Claoxylon australe* (Euphorbiaceae) (G. Williams 1995). However, their role as pollinators is arguable owing to their preference for visiting male flowers only. Female *Claoxylon australe* flowers possess a prominent three-branched stigma exerted above the gynoecium, which suggests at least a partial adaptation to wind pollination. *Bibio imitator* is also recorded on the flowers of *Scolopia braunii* (Flacourtiaceae/Salicaceae) and *Syzygium floribundum* (Myrtaceae), and in addition on flowering *Avicennia marina* (Acanthaceae/Avicenniaceae) (G. Williams pers. obs.) where individuals spend long feeding bouts in flower clusters, and in so doing acquire dense pollen loads (see Plate 96).

Small Diptera, vulnerable to desiccation, are better represented in mesic habitats or more moist microhabitats than in sclerophyllous communities and so Nematocera, in terms of diversity, are probably better represented among the anthophilous fauna of rainforest understoreys. However, numerous Nematocera can sometimes be found at night on the flowers of rainforest trees; the true pollination contribution made by nematocerans to mass-flowering rainforest trees may be understated as their presence is generally low during daylight hours when many observations are undertaken. Nocturnally active nematocerans recorded from flowering trees in lowland rainforest remnants in New South Wales include Cecidoymiidae (on *Acradenia euodiiformis, Alphitonia excelsa, Cuttsia viburnea, Euroschinus falcatus, Guioa semiglauca* and *Syzygium floribundum*), Ceratopogonidae (*Alphitonia excelsa*), Chironomidae (*Acradenia euodiiformis, Cuttsia viburnea, Guioa semiglauca, Syzygium floribundum* and *Syzygium smithii*), Culicidae (*Alphitonia excelsa, Cuttsia viburnea, Guioa semiglauca* and *Syzygium floribundum*), Mycetophilidae

(*Acradenia euodiiformis* and *Alphitonia excelsa*), Psychodidae (*Alphitonia excelsa* and *Cuttsia viburnea*), Scatopsidae (*Alphitonia excelsa, Cuttsia viburnea, Rhodomyrtus psidioides* and *Syzygium smithii*), Tipulidae (*Acradenia euodiiformis, Alphitonia excelsa, Guioa semiglauca* and *Syzygium smithii*), as well as miscellaneous Nematoceran records (*Alectryon coriaceus, Acradenia euodiiformis, Alphitonia excelsa, Cuttsia viburnea, Guioa semiglauca, Euroschinus falcatus, Rhodomyrtus psidioides, Syzygium floribundum, Syzygium smithii* and *Tristaniopsis laurina*) (G. Williams 1995, 2020). These are all from flowers with open actinomorphic structures with little or no depth effect.

The pollen-carrying capacity of small flies is limited but this may be partially compensated for by their abundance. Their low energy demands are met by the minute quantities of nectar or pollen produced by small flowers.

In addition to the specialised nectar- and pollen-feeding Diptera there is a second, sapromyophilous, group of flies that are attracted to flowers by deceit (see Bänziger 2002; Bänziger and Pape 2004; Willmer 2011). These flies are attracted by foetid odours that stimulate feeding or oviposition instincts (as by *Austrobaileya scandens*). Such flies, typically Sepsidae, Calliphoridae, Sarcophagidae, Tachinidae and sometimes Drosophilidae, are poorly adapted allotropic species that derive part of a mixed diet from blossoms. In Australian subtropical rainforest, flies of these families are common on the foetid-sweet flowers of the tree *Alphitonia excelsa* (Appendix 1), while Sepsidae (e.g. *Australosepsis niveipennis, Sepsis dissimilis* and *Sepsis nitens*), commonly associated with vertebrate waste products, may be abundant on flowers of *Acradenia euodiiformis, Alectryon coriaceus, Elaeocarpus obovatus, Euroshinus falcatus, Guioa semiglauca, Syzygium floribundum* and *Tristaniopsis laurina* trees (G. Williams 1995).

A separate sapromyophilous flower syndrome has been distinguished from that of general myophily (Faegri and van der Pijl 1979; Willmer 2011). Sapromyophilous flowers are often greenish, purple or brown with no nectar guides but frequently with dark spots and motile hairs, capable of movement, or appendages. There is generally no nectar attractant but sapromyophilous flowers produce strong foetid odours. The sexual organs are generally concealed within the blossom. Sapromyophilous flowers, often with 'fly-trap' or 'kettle-trap' blossoms' (formed by a modified perianth or leaf spathe), occur in families such as Araceae, Aristolochiaceae, Asclepiadaceae, Burmanniaceae, Hydnoraceae, Orchidaceae, Rafflesiaceae, Sterculiaceae and Taccaceae; as well as in more primitive taxa such as Annonaceae. These taxa might represent a secondary development in plant pollination ecology, developing from an earlier established syndrome such as cantharophily.

POLLINATION BY HYMENOPTERA

Pollination contributions by Hymenoptera are both diverse and extensive, and occur over a wide range of plant taxa, plant communities and geographical and climatic regions. Several taxonomically and behaviourally diverse hymenopteran groups are found on flowers. Some are true nectivores (Apiformes [bees], Pergidae-Euryinae) that possess mouthparts adapted to reach hidden flower nectaries. Some are parasitic species (Evaniidae) that search on flowers for hosts. Others are primarily predators (e.g. Formicidae, Pompilidae, Crabronidae, Sphecidae, Scoliidae) which not only frequent flowers in search of prey but may directly imbibe nectar as a source of energy. Several of the last, for example large Thynnidae (species of *Anthobosca*, Plate 89), possess specialised combs on their forelegs that enable them to gather or remove from their bodies pollen that has been displaced from anthers (Plate 112). Regardless of the primary ecological adaptation, all these groups have the potential to function as pollinators. The large gregarious wasps and bees may be important agents of out-crossing, as they often undertake long-distance movements between flowering plants.

POLLINATION BY WASPS (SPHECOPHILY)

Evidence for early pollination by wasps is recorded from the Mid Cretaceous. Grimaldi *et al.* (2019) record the fossil basal aculeate wasp *Prosphex anthophilos* (from Mid Cretaceous Burmese amber) feeding on the eudicot genus *Tricolporoites*. Pollen grains were associated with the mouthparts, and a pollen mass was identified within the buccal cavity. *Prosophex* feeding on *Tricolporoites* supports the hypothesis that relatively small generalised anthophilous insects were important pollinators of early angiosperms.

However, there appears to be no specific syndrome of sphecophily, as wasps and sawflies (e.g. Symphyta-Pergidae) are associated with numerous blossom types. These include highly modified flowers, such as those in Orchidaceae and Moraceae, and relatively unspecialised, readily accessible actinomorphic flowers with highly fragrant sweet or foetid odours (e.g. respectively *Tristaniopsis laurina, Alphitonia excelsa*). Wasps have been recorded taking nectar from bat-pollinated plants, with comparative studies of wasp diversity in Neotropical and northern temperate habitats indicating greater diversity in the former and overlap with bees in resource utilisation (Heithaus 1979).

The term 'wasp' encompasses a heterogeneous assemblage of 'primitive' unspecialised and highly evolved families. One superfamily, the Sphecoidea, has recently been subsumed within the superfamily Apoidea, once the sole preserve of true bees (Melo 1999). Worldwide, the enlarged membership of Apoidea now comprises Spheciformes or sphecoid wasps (families Ampulicidae, Sphecidae and Crabronidae) and Apiformes or true bees (families Stenotritidae, Colletidae, Andrenidae, Halicitdae, Melittidae, Megachilidae and Apidae, including Anthophoridae [Michener 2007]). An important phylogenetic distinction between Spheciformes and Apiformes is that, although they are kin groups (Apoidea), there is strong morphological and molecular evidence that bees are derived directly from ancestral Spheciformes (Branstetter *et al.* 2017; Debevec *et al.* 2012; Peters *et al.* 2017).

Wasp mouthparts may be adapted to capture prey, and excision from the pupal cell or host. In higher aculeate groups (e.g. Vespoidea and Spheciformes), thought to have originated ~150 mya (their origin coinciding with the appearance and radiation of the angiosperms [Grimaldi and Engel 2005]), there are tube-like modifications to the mouthparts (particularly prolongation and development of the glossae, paraglossae and prementum) that allow nectar to be imbibed (see Naumann 1991). Species of *Eurys* and *Neoeurys,* genera of the sawfly subfamily Euryinae, though small in size, have elongate mouthparts which enable them to reach otherwise inaccessible nectaries (e.g. of Orchidaceae). In the tribe Bembicini (family Crabronidae) the morphological distinction between wasps and true bees (Apiformes) becomes fine and somewhat arbitrary. Although some sphecoid wasps are regular and frequent flower visitors, they retain predacious habits.

There are highly specialised instances of plant–wasp pollination documented for *Ficus* (Moraceae)-Agaonidae, Orchidaceae-Ichneumonidae, Orchidaceae-Scoliidae and Orchidaceae-Thynnidae. Pollination by wasps in Australian rainforest plant taxa, other than for several species of *Ficus* (figs), is poorly known although records exist for higher wasp families (e.g. Thynnidae, Vespidae and Scoliidae) visiting Mimosaceae, Myrtaceae, Loranthaceae, Orchidaceae, Proteaceae, Sapindaceae, Goodeniaceae, Gyrostemonaceae and Lamiaceae in Australian non-rainforest communities (e.g. Hawkeswood 1981; Inouye and Pyke 1988; G. Webb 1984, 1989). Pergidae are also associated with Campanulaceae, Trimeniaceae, Orchidaceae, Myrtaceae and Rutaceae blossoms, and pollination of *Macadamia* (Proteaceae) by *Radumeris tasmaniensis* (Scoliidae) has been recorded (Bernhardt *et al.* 2003; Vithanage and Ironside 1986). '*Campsomeris*', in association with other unidentified wasps, has also been recorded visiting the flowers of *Syzygium sayeri* in tropical Queensland rainforest (Boulter *et al.* 2005). Wasps of the superfamilies Vespoidea and Apoidea-Spheciformes visit flowers of tropical Queensland populations of Ebenaceae and Lauraceae (House 1985). They also visit Anacardiaceae, Apocynaceae, Arecaceae, Cunoniaceae, Ebenaceae, Escalloniaceae/Rousseaceae, Myrtaceae, Rhamnaceae, Sapindaceae and Vitaceae species in lowland subtropical rainforest remnants of New South Wales (G. Williams 2020; Williams and Adam 1995).

Little is known about the pollination role and contributions of flower-frequenting wasps within rainforest communities. Butterflies, birds and bees seem to have captured the spotlight yet there are data on the putative pollinating function of wasps in non-rainforest ecosystems. The data suggest that they play a significant role (or at least are an important anthophilous insect component) in plant communities as diverse as Greek phrygana shrubland, Japanese primary beech and deciduous forest, Costa Rican savannah, and tropical deciduous and oak forests (e.g. Heithaus 1979; Inoue *et al.* 1990; Kato *et al.* 1990, 1993; Petanidou and Ellis 1993). The last two are similar to Australian communities classified as 'dry rainforest'. Nayak and Davidar (2010) record wasps, within a larger taxonomic suite of pollinators, as pollinators in tropical dry evergreen forests of eastern India. In Japanese beech forests, wasps are reported as presumably understorey specialists, since the canopies of primary beech forest are wind-pollinated. In South-east Asia aculeate wasps are associated with several angiosperms, though not always clearly as pollinators. Possible examples include pollination of *Casearia grewiaefolius* (Flacourtiaceae) by a vespid, and *Allophyllus serratus* (Sapindaceae) by sphecids (Aluri *et al.* 1998).

Sphecidae (inclusive of Crabronidae in earlier works), Vespidae and Pompilidae are widely recorded from the flowers of Neotropical rainforest trees and are well represented in a variety of Northern Hemisphere plant communities. Nakase and Kato (2012) record the night-visiting *Provespa* species (Vespidae) as the probable pollinator of the epiphytic orchid *Coelogyne fimbricata* in montane rainforest of peninsular Malaysia. Equally, Crabronidae (e.g. *Bembix*), Sphecidae (e.g. *Sphex*), Vespidae (e.g. *Polistes, Abispa, Deuterodiscoelius* – Plate 92) and Pompilidae (e.g. *Calopompilus* and *Heterodontonyx*) are common visitors to the flowers of Australian tropical and subtropical rainforest trees with general entomophilous flower syndromes. Thynnidae are well represented on Australian subtropical mass-flowering rainforest trees (~41% of wasp diversity), greater than that recorded from Neotropical plant communities, where they may comprise less than 10% of the anthophilous wasp fauna (Heithaus 1979; G. Williams 1995). This may be partly due to the greater representation of thynnids in Australia. Thynnids are commonly encountered on the flowers of Australian sclerophyllous plants, notably terrestrial orchids. But the effectiveness of thynnids as pollinators of orchids may depend on the number of pollinia carried and deposited, and whether they mediate cross- or self-pollination. In subtropical rainforests (including mangrove communities) in coastal New South Wales members of such thynnid genera as *Acanthothynnus, Agriomyia, Dimorphothynnus, Hemithynnus, Rhagigaster* and *Zeleboria* can be numerous on flowers of *Alphitonia excelsa, Avicennia marina, Elaeocarpus obovatus, Guioa semiglauca* and various Myrtaceae (G. Williams pers. obs.; Williams and Adam 1995). The large males are often encumbered by the attachment of wingless females.

Fig–fig wasp mutualism

The fig–fig wasp mutualism is a classic example of coevolution between plants and insects (Janzen 1979; Thompson 2014; Wang *et al.* 2010), and the pollination of figs (*Ficus* spp.) is the subject of a substantial body of scientific literature. Worldwide, the genus *Ficus* (Moraceae) (e.g. *Ficus coronata* [Plate 49]) contains ~750 species (Rønsted *et al.* 2008), is mostly distributed in the tropics, and is generally dependent on wasps of the family Agaonidae (superfamily Chalcidoidea), subfamily Agaoninae (fig wasps), for pollination, though under certain environmental conditions some *Ficus* can self-pollinate (Bronstein 1989). Pollinating wasps 'show remarkable conservatism in host *Ficus* preference having concordant and ancient associations in the vicinity of 60 million years' (McLeish and van Noort 2012), however, Milton (1991) hypothesises that patterns of flower and fruit fruit production in Moraceae and the pre-*Ficus* ancestor reflect ancient traits that have little or nothing to do with wasp pollinators.

The Agaonidae are the only group of non-aculeate (see Naumann 1991) wasps known to be important in the pollination of plants, but fig wasps exhibit different behaviours that variously facilitate pollen transfer, for while some are active pollinators others are passive. Active pollinators use coxal corbiculae ('pollen baskets') or specialised folds to move pollen to and from mesosternal pollen pockets. Passive pollinators do not perform this behaviour (Lopez-Vaamonde *et al.* 2002; see Naumann 1991).

Ficus species are dioecious (or functionally dioecious) or monoecious (Milton 1991). The monoecious state is thought to be an ancestral one, with dioecy in figs originating several times, and with reversion to monoecy from a dioecious condition occurring probably at least twice. Such changes in the evolution of breeding systems would have required a selective shift in fig–fig wasp partnerships, involving speciation and instances of figs sharing more than one wasp pollinator, and more than one species of wasp sharing a particular fig species (e.g. see McLeish and van Noort 2012; Molbo *et al.* 2003; Wiebes 1979). Shared expressions of a mutualism might represent intermediate stages in the speciation process.

Volatile chemicals in figs attract fig wasps (Grison-Pigé *et al.* 2002; Willmer 2011). On the basis of morphological characters, true fig wasps appear to have possibly developed from primitive prototorymid (Torymidae) seed-eaters (Boucek 1988). These pre-agaonid wasps eventually co-evolved with different species of figs (Bronstein 1992; Weiblen 2002; Wiebes 1979), developing different morphologies in relationship to differences in syconia morphology (e.g. thickness of syconia walls or size of fig).

The *Ficus*–agaonid association is thought to date from some time in the Cretaceous, greater than 100 mya; though fossil fig wasps date only from ~34 mya (Compton *et al.* 2010). Fossil fig wasps, belonging to the modern genera *Pegoscapus* and *Tetrapus*, are known from Miocene Dominican amber (c. 23–16 mya) (Compton *et al.* 2010). Fig–fig wasp relationships are generally considered host-specific, with both fig and fig-wasp being mutually dependent. However, breakdown (or evolutionary realignment or intermediate expression of speciation) in pollinator specificity has been recorded for several species (Ware and Compton 1992; Dixon *et al.* 2001; see also McLeish and van Noort 2012). The Australian fig wasps still are poorly known, the pollination ecology of few Australian fig species has been investigated (Boucek 1988; Naumann 1991), and we remain overly reliant for ecological insight on mechanisms exposed in the figs of foreign forests. Current Australian fig–agaonid records include *Pleistodontes froggattii* (on *Ficus macrophylla* and *Ficus rubiginosa*), *Pleistodontes greenwoodi* and *P. xanthocephalus* (on *Ficus obliqua*), *Pleistodontes imperialis* (on *Ficus macrophylla* and *Ficus rubiginosa*), *P. nigriventris* (on *Ficus watkinsiana*) and an undescribed species of *Kradibiella* (on *Ficus coronata*) (Dixon *et al.* 2001; Lopez-Vaamonde *et al.* 2002). Shared agaonid records for *Ficus macrophylla* and *Ficus rubiginosa* indicate lack of complete host specificity.

The tiny flowers of fig trees are enclosed in an urn-shaped inflorescence (the 'fruit' of the fig) known as a syconium; a distinctive hollow structure that sets the genus *Ficus* apart from other members of the Moraceae, and that will later develop to protect the maturing seeds. Individual flowers consist of either a single ovary and style or a small group of anthers, with male flowers near the top of the syconium and female flowers near the bottom. However, some syconia may only possess female flowers. Wasps can only gain entry to the syconium via a small opening (ostiole) at the apex. The female wasp, carrying pollen upon its back, enters the syconium via the ostiole and lays eggs in the ovaries of sterile (neuter) female flowers. This is usually followed by the death of the female wasp, whose body remains within the fig. A single male or female wasp develops within each ovary. These emerge within the syconium and mate, often with the offspring of other females who have also oviposited in the fig. Males emerge first and seek out galled ovules containing females. They chew through the ovules and mate with the females using a telescopic organ known as a gaster. The winged females then fly to other receptive figs, pollen from male flowers being passively dusted onto their bodies as they leave, however, some species actively collect pollen in thoracic pollen pouches, this facilitated by specialised leg combs. There is potential for males to mate with females from the same founder. Because males have low mobility, the choice of ovules into which the founding female has laid eggs may have dramatic outcomes for the mating structure of wasp populations inside each fig; for example, there may be localised dominance of full siblings in adjacent galled ovules, or clustering of male and female galled ovules.

Fig wasps may disperse over distances greater than 160 km (Ahmed *et al.* 2009). This dispersal capability potentially enables the wasps to reach and pollinate hyper-dispersed fig populations, as well as recolonise sites previously devastated by climatic or volcanic events (Shanahan *et al.* 2001). This long-distance dispersal attribute is significant in terms of the geographical isolation of many fig species in tropical and subtropical rainforests, where trees might be reproductively isolated as well. In Australian subtropical rainforests, for example, individuals of the rainforest emergent *Ficus macrophylla* are commonly separated by many kilometres (G. Williams 1993), so that the ability of its dependent pollinators to move between isolated trees is important for out-crossing and for the maintenance of the mutualism when fig trees are at low densities. However, there are recorded instances of localised extinction of agaonid wasps owing to severe drought, and subsequent delayed recolonisation, suggesting that limitations to wasp dispersal can occur (Harrison 2001).

Although the relationship between fig and wasp is beneficial, and thus a true expression of plant–pollinator mutualism, the pollination system in figs has a reproductive cost. In Costa Rican species of *Ficus*, seed loss (due to development of fig wasp larvae within the syconia) as high as 75% is common (at least 25% loss occurs) and a general loss of ~55% is the average reproductive (offspring) price for successful out-crossing (Janzen 1979). This degree of loss is probably experienced by most, if not all, rainforest figs. Consequently, the fig–fig wasp mutualism involves a high level of symbiotic parasitism by the agaonid wasp, each fig being forced to sacrifice several of its nascent children in return for the development of the remainder. But there is a grim twist. Unlike newborn winged female wasps, male agaonid wasps are wingless (and bizarrely ant-like in appearance) (Plate 87) and are unable to escape their dark home. In death they join the remains of their mother within the fig, the fig reaping a gruesome harvest in return for the prior loss of its own young. A charnel house and ossuary by any account, of which frugivores (humans included), in their devouring of the fig's succulent flesh, are completely ignorant.

Some researchers have called the fig–fig wasp mutualism a form of co-slavery in which wasp and fig participants are tied inextricably in an obligatory reproductive interdependency; the fig needs the wasp and the wasp is equally bound to the fig. But the relationship is also one of conflict, a conflict of selection. Each fig tree is selected to produce seeds and pollinators (the latter from the portion of its ovules expended within the constraints of the mutualism). In contrast, wasps concern themselves only with maximal production of a new generation of wasps. The ghoulish interdependency is enacted over the proportion of female flowers that remain to produce a seed – veritably, children wagered for children.

Less well known, in the relationship of wasps and figs, is the association of non-agaonid Chalcidoidea with figs (Boucek 1988). These equally minute wasps are not pollinators and may adversely affect the fig–fig wasp pollination mutualism. A number use fig ovules already occupied by agaonid larvae, and many are gall-makers. Most known non-pollinator chalcidoid wasps seem to be specific to certain fig species, although numerous species may exhibit obligate relationships with a single fig species (asymmetrical mutualisms). Regardless of whether the Chalcidoidea wasps associated with *Ficus* species are pollinators or non-pollinators, the chalcidoid fig fauna is a specialist one whose members appear to co-exist in stable fig relationships that likely reflect long-term speciation.

However, fig wasps and the fig-fig wasp pollination mutualism are not immune from the impacts of climate change. Jevanandam *et al.* (2013) have shown that an increase of 3°C or more above current temperature experienced across much of the Equatorial tropics would markedly reduce the lifespan of the four agaonid wasps they studied, and suggested that this would affect multiple trophic levels within ecosystems.

POLLINATION BY ANTS (MYRMECOPHILY)

Ants (Formicidae) are placed in the Aculeata – the clade of Hymenoptera that also contains the stinging wasps and bees (Naumann 1991). Ants appeared from a single aculeate wasp ancestor during the Lower Cretaceous (~100 mya), with fossil evidence indicating most of the subfamilies representing extant taxa diversified during the Late Cretaceous-Early Eocene; this is the period that also witnessed the rise of angiosperms and most herbaceous insects (Moreau *et al.* 2006). All known living ants are eusocial, however, the emergence of eusociality in ants is relatively late in geological time (Wilson and Nowak 2014), and though ants today are among the dominant animals of the world's terrestrial biosphere, based on fossil evidence in the Mesozoic they may have been a comparatively minor element of the fauna (LaPolla *et al.* 2013).

Although ants are now numerically abundant in many communities, their role and function in effecting pollination has been discounted or questioned (Hölldobler and Wilson 1990; Peakall *et al.* 1987; Willmer 2011). Possible limitations to ant pollination include the small size of many flower-visiting species (which purportedly precludes contact with stigma or stamens), the smoothness of the body integument, spatially or behaviourally restricted foraging patterns which restrict pollen transfer, and the pollenicidal nature of metapleural glandular secretions that may interfere with pollen germination or stigma function (Dutton and Fredrickson 2012). Nevertheless, there are records of ants pollinating plants (e.g. Gómez 2000; Hölldobler and Wilson 1990; Peakall *et al.* 1991; Ramsey 1995) although some plants may possess morphological or chemical defences that deter ant visits (Ghazoul 2001). Such defences include small hair-like structures on the perianth segments, and enclosed flowers that restrict access. Ant-repellent chemicals have been recorded from floral tissues (and floral volatiles can mediate ant attraction), however, there is a general palatability of floral nectar to ants (Feinsinger and Swarm 1978; Rico-Gray *et al.* 2004). This would appear to discount suggestions that chemically unpalatable nectar substances militate against ant visitation to flowers, as has been suggested for plants in some tropical habitats (Janzen 1977). Flowering plants may also seek to deter ant damage and nectar theft by the provision of extrafloral nectaries (Hölldobler and Wilson 1990). Differences in the structure and placement of these, particularly in co-occurring species, can result in the attraction of different numbers and species of ants (Apple and Faener 2001).

Tentative ant pollination syndromes have been postulated (Hölldobler and Wilson 1990). Features of plants possibly pollinated by ants are suggested to include high plant density, small inconspicuous flowers borne near the ground, sessile flowers with low nectar volumes and few concurrently open flowers, low ovule number (thus requiring smaller levels of pollen transfer), small amounts of dry sticky pollen, inter-digitation of branches from adjacent plants and a warm dry habitat (abundance of ant workers during peak flowering and relative general abundance with respect to the numbers of other visitors may also be key factors in successful ant pollination outcomes [Gómez and Zamora 1992]). However, ant pollination is not confined to plants with these traits, so plant structure may not be the determining factor in the evolution of ant pollination. Rather, pollination by ants may require the absence of pollenicidal secretions, or the evolution of plant mechanisms that allow the avoidance of these secretions (Peakall *et al.* 1987). Gómez and Zamora (1992) noted that *Proformica longiseta* ants made up more than 80% of the total number of insects found on flowers of the mass-flowering woody plant *Hormathophylla spinosa* (Cruciferae) in the Sierra Nevada mountains, Spain. These proved to be the true pollinators even though pollen coming into contact with their bodies exhibited lower germination rates. Whilst the viability of pollen may be reduced through ant contact, nevertheless experimental outcomes have also shown that viable pollen can remain following ant contact (Schneemilch *et al.* 2011).

The secretions from the metapleural glands of ants act as powerful antiseptics against body surface and nest microorganisms and as a defence against predators. Possession of metapleural glands is widespread in the family Formicidae and characterises the family throughout its phylogenetic history. However, in some species of the genera *Camponotus, Dendromyrmex, Oecophylla* and *Polyrhachis* the metapleural glands are secondarily atrophied or lost (Hölldobler and Wilson 1990). These genera include many arboreal species (*Polyrhachis* in particular is well represented in Australian subtropical littoral rainforest), but there is little direct evidence in the anthophilous or nectarivorous ant fauna that demonstrates correlation of absence (or modification) of the metapleural gland with an ability to act as pollinators.

A seemingly unusual behaviour of several otherwise highly aggressive and predatory ants, is their predilection for imbibing the nectar of rainforest flowers. The large and gregarious Australian bulldog ant *Myrmecia nigrocincta* (Plate 93) is a common visitor to flowers of subtropical rainforest trees, shrubs and aroids, while *Myrmecia urens* is recorded as pollinating the orchid *Leporella fimbricata* (Peakall *et al.* 1987).

Regardless of the conjecture surrounding the role of ants as pollinators, many Australian rainforest trees are frequently visited by ants, and ants are the most abundant and diverse flower visitors at some tree species. Ants of the genus *Crematogaster* appear to be pollinators of the small subtropical rainforest tree *Alectryon coriaceus* and can be common on the flowers of *Mischocarpus pyriformis* (Sapindaceae). *Polyrhachis* ants are likely pollinators of *Alectryon coriaceous* in subtropical littoral rainforests (Plate 94) G. Williams 1995, 2020).

Although ants frequently function to deter the attacks of herbivores on plants (though the quality of protection may vary owing to different visitation rates [Apple and Faener 2001]), equally, ants may interfere with the function of pollinators. In the threatened endemic Mauritian plant *Roussea simplex* (Rousseaceae) the invasive ant *Technomyrmex albipes* disrupts the gecko *Phelsuma cepediana,* the sole agent of pollination and seed dispersal (Hansen and Müller 2009). The ants visit to forage on nectar and fruit pulp and also tend honeydew-producing mealybugs on the fruits, and their aggressive interference prevents access by pollinating and seed dispersal vectors. Seed set in plants infested by *Technomyrmex albipes* is greatly reduced and geckos ingested and dispersed fewer seeds from fruits where ants were present. Thus ant presence and activity resulted in a double disruption to plant reproductive success. However, Kuriakose *et al.* (2018b) recorded *Technomyrmex albipes* as the dominant (and only nocturnal) visitor to the flowers of *Syzygium occidentale* (Myrtaceae) in the forests of the Western Ghats, India. The species is partially self-compatible and exhibits considerable levels of autogamy. Although there was a range of other insect (bees) and bird visitors fruit set in flowers exclusively visited by *Technomyrmex albipes* was significantly higher than those visited by any other animal.

Currently only conjecture surrounds the impact of the native ant *Oecophylla smaragdina* on the reproductive ecology of rainforest plants in tropical Australia. Commonly known as the 'green tree ant' or 'weaver ant', *Oecophyyla smaragdina* forms large colonies with numerous nests located throughout the canopies of individual trees and shrubs and has been recorded visiting the monoecious inflorescences of *Ryparosa kurrangii* in tropical North Queensland (Webber *et al.* 2008). The species is extremely aggressive, is highly defensive, vigorously seeks out prey, and recruits in large numbers to food resources; such that it is able to repell pollinators and in effect can serve as a biocontrol agent against herbivores (Tsugi *et al.* 2004).

POLLINATION BY BEES (MELITTOPHILY)

Bees are considered specialised crabronid wasps (Branstetter *et al.* 2017) and though fossil evidence dates the emergence of Spheciformes to the Early Cretaceous (Michener 2007; Peters *et al.* 2017), with 'sphecoid' wasps having existed in the Late Jurassic or earlier (Grimaldi and Engel 2005), the first appearance of bees (Apiformes) in the development of terrestrial invertebrates is uncertain. Although a fossil of the *Tetragonula/Trigona*-like social apid *Cretotrigona prisca* dates from the Late Cretaceous no solitary bees (the great majority of taxa) are known from fossil evidence that dates conclusively to an earlier time (Grimaldi and Engel 2005; Michener 2007).

However, bees may have originated earlier, perhaps in the Early-Mid Cretaceous (~125–120 mya) (Poinar 2020), a time roughly synchronous with the appearance of the angiosperms. In addition to the enigmatic and contentious 'bee' *Melittosphex burmensis* (Melittosphecidae), recorded from Early Cretaceous Burmese amber (Danforth and Poinar 2011; Patiny 2012), and the Mid Cretaceous *Discoscapa apicula* (Discoscapidae) (Poinar 2020) – see Box for a discussion of both – the postulated origin for bees is otherwise based, for example, on suggested fossil halictine nests and a fossil nest trace of the ichnospecies *Cellicalichnus dakotensis*. *Cellicalichnus dakotensis* is assigned to the Cenomanian (~94 mya) of Arizona, with the fossil trace evidence comprised of irregular cells distributed around a vertical shaft where short horizontal tunnels arise bearing opposite pairs of cells (Grimaldi and Engel 2005; Michez *et al.* 2012). There also are mid-Eocene, and younger, leaves with damage typical of leaf-cutter bees (Megachilidae); a derived family, again indicating a much earlier evolution of the group (Wappler and Engel 2003).

Regardless of their origin, it appears that the earliest bee lineages included taxa that were host-plant specialists. With a present known fauna of more than 16 000 species, bees represent one of the most important pollinating insect taxa (e.g. Michener 2007; Dainese *et al.* 2017 and references therein). This massive fauna feeds almost exclusively on the floral resources offered by flowering plants.

Historically, the short-tongued bee family Colletidae (e.g. Plates 83, 84) has been considered the basal family in the phylogeny (the history of the genealogical development of a group) of bees, because its members possess several morphological attributes linking them to crabronid and sphecid wasps (e.g. *Bembix, Sceliphron* respectively) (Melo 1999; Michener 2007; Pulawski 2018). However, the Melittidae (also short-tongued, though believed to be the progenitors of long-tongued bees) have more recently found strong support as a family that deserves a place somewhere near the base of the bees' phylogenetic tree, more than the Colletidae (Danforth *et al.* 2006). Such matters might seem trivial in the broader scheme of pollinator–angiosperm history, but upholding the basal placement of Colletidae suggests that bees originate from Australia or South America. The first bees may have been an assemblage of angiosperm generalists, and possibly specialists; colletids include some specialists but in Australian subtropical rainforests, as elsewhere, they are mainly generalists, minor pollinators and thieves. Passing the basal title to the Melittidae suddenly involves a probable African origin for bees and implies that from their very beginnings bees were floral specialists, not generalists, since melittids are more specialised in their host plant relationships. If Melittidae are indeed the basal group, as recent critical analysis of molecular and morphological characters appears to support, this has important implications for the antiquity of bees, their radiation and the broader coevolution (or co-radiation) of bees and flowering plants. However, Melittidae do not occur in Australia or South America. Thus, the coevolutionary interaction of bees and angiosperms in Australia may have been different, particularly with regard to the Myrtaceae and especially eucalypts. Placing the Melittidae as a basal group may help explain difficulties in interpreting the fossil record. While Melittidae (and long-tongued Apidae and Megachilidae) are present in Eocene (c. 55–34 mya) and older Cretaceous amber (Michener 2007), the Colletidae are absent from the fossil record until the later Miocene (Engel 1999). The relatively good fossil representation of melittids and their long-tongued associates has been argued to be a reflection of a bias in favour of resin-collecting bees (Michener 2007).

Pollination by bees is known as melittophily and flowers adapted to be pollinated primarily by bees are termed melittophilous. Many bee species are highly efficient foragers. They possess a wide range of foraging behaviours and varying degrees of specialisation as flower visitors and potential pollinators (Batley and Hogendoorn 2009; Cunningham *et al.* 2013; Gross 1993), and can exert a precise flower selection choice mainly driven by their pollen requirements (Aronne *et al.* 2012). Nevertheless, a low pollen and nectar supply can impact adversely on bee reproduction (Peterson and Roitberg 2006; Zurbuchen *et al.* 2010).

Bees normally restrict their foraging activity to daylight hours, however, instances of nocturnal visitation are known; for example from *Xylocopa tenuiscapa* in India and *Xylocopa tranquebarica* from Thailand (Burgett and Sukumalanand 2000; Somanathan and Borges 2001) and probably from the genus *Reepenia*, which occurs in northern Queensland and the New Guinea region (Michener 2007). However, nocturnal foraging by bees in Australian subtropical rainforests has yet to be recorded.

The foraging behaviour and specialisation in the diversity of plant taxa visited varies between opportunistic polytropic (polylectic) bees that visit many species of plants, to oligotropic (oligolectic) or monotropic (monolectic) bees that visit a restricted range of flowers.

Bees in amber

Amber results from the fossilisation of various tree resins (Langenheim 2003). In Australia deposits are known from periods dating from the Late Triassic (Tasmania) to the late middle Eocene (Victoria), this spanning some 230–40 million years (Stilwell *et al.* 2020). The insect inclusions trapped within amber provide a unique visual opportunity for engaging with early invertebrate organisms, not least bees and wasps (e.g. Antropov 2016; Engel 2001; Melo and Rosa 2018; Michener 2007). Michener (2007) highlights that fossil bees are rare, and that only those preserved in amber allow a degree of insight into their phylogeny; though these may be biased to those species that collect resin for the construction of nests. Bees recorded from amber include the apid *Cretotrigona prisca* from North America (Late Cretaceous, ~80 mya), the halictid *Electrolictus antiquus,* the apids *Boreallodape mollyae* and *Melissites trigona,* and the megachilids *Ctenoplectrella grimaldi* and *Glyptapis mirabilis,* from the Baltic region (Eocene; 70–40 mya), the apid *Apis henshawi* from Germany (Oligocene; 40–25 mya), and the halictids *Halictus petrefactus* from Spain and *Oligochlora eikworti* from the Dominican Republic (Miocene; 25–10 mya).

Michener (2007) cites 36 bee species known from the Late Eocene (35–40 mya), 34 of which were Apidae and Megachilidae. By ~20–30 million years ago the bee fauna exhibits modern features, however a post-Late Eocene extinction event resulted in the loss of many taxa, this event associated with global cooling that characterised the Eocene-Oligocene boundary (Michener 2007). This was followed by a second radiation producing the Recent/Holocene bee fauna.

The primitive 'bee' *Melittosphex burmensis* (Danforth and Poinar 2011; Poinar 2009) is also recorded from Burmese amber (Early Cretaceous). This was placed in the family Melittosphecidae, this erected to emphasise that *Melittosphex* is not within the monophyletic group that includes all living bees, but possesses a combination of wasp and bee features making it an important transitional form that bridges the gap between crabronid wasps and extant bees. Danforth and Poinar (2011) argued that the presence of branched (plumose) hairs in *Melittosphex burmensis* suggested it was a pollen collector, and that the possession of branched hairs was evidence that pollen-collecting Apoidea (bees) were present ~20 million years after the emergence of eudicots (~120 mya), the major angiosperm lineage with extensive reliance on bee pollination. However, the assertion that branched hairs automatically indicates pollen collection has been considered erroneous as some predatory wasps (e.g. Sapygidae; absent from Australia) have branched hairs while masarine wasps and a crabronid wasp (*Krombeinictus*) lack them, yet are able to successfully collect pollen (see Patiny 2012). In addition bees of the subfamilies Hylaeinae and Euryglossinae (Colletidae) effectively lack plumose hairs, but transport pollen in the crop (Houston 2018; Michener 2007; Patiny 2012).

A rare find of an early bee in association with attached pollen (and carrying beetle triungulins) is that of *Discoscapa apicula* (Poinar 2020). This fossil species was found in Mid-Cretaceous amber, and 'shares with modern bees plumose hairs, a rounded pronotal lobe, middle and hind leg scopae containing pollen grains and a pair of spurs on the hind tibia. But its narrow hind basitarsi, extremely low placed antennal sockets and some wing vein features are those of apoid wasps' (Poinar 2020). A new family, Discoscapidae, was established by Poinar to receive *Discoscapa apicula,* owing to this taxon having a bifurcated scape. He considered this a unique diagnostic character of the family, not otherwise found in living or extinct families of apoid wasps or bees.

Discoscapa apicula represents an example of an early bee in an ecological context (that of pollen carriage and role of host). But while amber allows us to glimpse fossil bees in almost perfect form, it provides less evidence regarding co-occurring angiosperms, for only occasionally are fossil flowers found in amber (Friis *et al.* 2011), and none in which bee and flower have been snared in company.

Consequently, there is no single bee-specific flower syndrome. Specialised, highly adapted, trap-lining and eusocial bees visit zygomorphic robust flowers that have hidden nectaries and whose pollen is often presented as discrete pollinia units (as in orchids, where the pollen grains are held together as a single unit) rather than individual sticky grains. Such flowers are rarely massed in large inflorescences and often have deep tube-like corollas. Flowers are commonly bright yellow or blue, usually with prominent or distinct nectar guides which may be ultraviolet-positive. The colours are obvious and fragrant. 'Bee flowers' produce relatively concentrated nectar in moderate quantities and characteristically have concealed sexual organs with few stamens and many ovules per ovary (Willmer 2011). In flowers that are buzz-pollinated, however, nectar production is not obvious and pollen appears to be the only primary energy reward of significance. The depth effect of the perianth partitions the type of bees that can forage successfully, so long- and short-tongued species tend to utilise flowers of different depths (Real 1981; Kato *et al.* 1993). Robust bees can, however, break through the base of the corolla to reach otherwise inaccessible nectaries. Opportunistic polytropic bees (and some more highly adapted species) commonly visit flowers with general entomophilous traits of open dish-like corollas (mostly coloured white or dull), often in massed branched (paniculate) inflorescences with small to moderate quantities of nectar and sweet odours.

Bees collect nectar, pollen and occasionally resin, but they rarely collect nectar and pollen during the same foraging event because efficient collection of either requires different behaviour. However, these are not the only floral rewards offered by angiosperm flowers to anthophilous insects, and in the case of the Australian orchid *Thelymitra nuda*, a species that lacks floral nectar, bees harvest pseudopollen from staminodes (Bernhardt and Burns-Balogh 1986). Not all bees are exclusively anthophilous; instances of necrophagy (attraction to the products of animal decay) have been observed (Michener 2007; Roubik 1989), including *Amegilla* attracted to the rotting bodies of dead cane toads in the Australian Northern Territory (G. Williams pers. obs.) and some bees are attracted to sweat, presumably for minerals. Bees will also collect from plants without nectar rewards (e.g. Gross 1993; Petanidou and Ellis 1993). While bees are popularly considered classical agents of plant pollination, all is not as it may seem. The pollen collected in the specialised pollen baskets (corbiculae) of social Apidae (honey bees, native *Tetragonula* species) is not available for pollination and some reduction in pollen viability may occur, although this is not necessarily significant, when pollen comes into contact with the body integument of individual bees (Harriss and Beattie 1991). Corbiculae are located only on the hindlegs (see Plate 82). Each corbicula has a smooth, wide concave face and can carry waxes and resin as well as pollen. Around the rim of each corbicula are hair-like setae, the length of which effectively increases the area of corbicula available for pollen storage. Many species of solitary bees pack and carry large amounts of pollen in specialised arrays of hairs (scopae) on the undersurface of their bodies, usually on the abdomen. These can be found on many Megachilidae (e.g. *Megachile pictiventris*, Plate 85). The scopae are formed by stiff, backward-facing setae across the width of each abdominal segment, though usually the scopae on abdominal segments closest to the thorax are better developed (Michener 2007; see Naumann 1991). Often the whole ventral surface of the abdomen has tightly packed masses of bright yellow pollen.

Pollen collected within the scopae is often mixed with floral oils, which may also be harvested and separately ingested (Houston *et al.* 1993; Michener 2007). These are collected by using modified brush-like tarsal segments (feet-pads, or tarsomeres), hind-leg combs and toothed leg spurs. In some instances, the spurs are used to penetrate lipid bodies formed by oil-producing elaiophores (as found in Maliphagidae). Oil is scraped from the flower and transferred by the legs to the scopae. Such oil-gleaning behaviour is exhibited by tropical, subtropical and temperate bees. In the females of the southern African bee *Rediviva emdeorum* (Melittidae) the forelegs are exceptionally long, equal to or longer than the bee's body (Michener 2007). The forelegs of different *Rediviva* species vary in length and are inserted into the long floral spurs of their flower hosts (principally *Diascia* [Scrophulariaceae]) to collect the rich oil resources held within; this relationship representing an exceptional example of insect–flower co-adaptation. How was such a profound relationship engendered? Were there long oil-rich floral spurs that patiently awaited the arrival of a malformed bee, or was there a subtle dance of co-responding partners?

Depending on the bee taxa, collection of pollen can range from combing movements of fore- or hindlegs to the use of long bristles on the mouthparts, which fit into narrow flowers. Not all bees collect and carry pollen among scopal hairs or specially modified pollen baskets. Australian bees of the subfamilies Hylaeinae and Euryglossinae, both members of the family Colletidae (Plates 83, 84), carry pollen internally in the crop, and may possess specialised brush-like structures (e.g. on the forelegs) with which pollen is swept into the mouth (Houston 2018). In so doing, they harvest quantities of pollen for their own nutritional benefit. This is a loss to the plants, as only the lesser numbers of grains incidentally caught on their bodies are potentially available for pollination. This does not fully preclude colletid bees from the role of pollinator. For example, in Brazilian savannah woodland a species of the colletid genus *Ptiloglossa* pollinates the tree *Siphoneugenia densiflora* (Myrtaceae) (Proença and Gibbs 1994), and Colletidae are implicated in the promiscuous pollination strategies of numerous Australian plants of the rainforest canopy and understorey.

The role of bees in several tropical ecosystems, and their response to fragmentation, has been extensively studied (e.g. Ferreira *et al.* 2015; Ramírez *et al.* 2016; Smith and Mayfield 2018). Many tropical bees are important pollinators but the broader role, function and behaviour of bees vary in response to climate, latitude, vegetation structure and aggregation, vegetation stratification, size of individual plants and seasonal abundance (Heithaus 1979; Smith-Ramírez *et al.* 2014; Williams and Adam 1997). Foraging patterns are influenced by temperature, light intensity and humidity, though some species forage during cool periods (Burgett and Sukumalanand 2000; Somanathan and Borges 2001), for example *Xylocopa tenuiscapa* that forages in temperatures as low as 2°C in the cloud forest of the Western Ghats, India. Interactions between bees and flowers may also be strongly influenced by the actual temperature of the flower, ambient temperature, and the temperature of nectar, and bees may prefer to land on flowers that are warmer rather than similarly coloured alternatives which might contain an equal nutritional reward (Dyer *et al.* 2006). However, Norgate *et al.* (2010) found that while raising the temperature of a flower above ambient levels can result in an advantageous attraction of pollinators a preference for warmer flowers is context-dependent. For example, above a certain temperature *Tetragonula carbonaria* (Plate 82) could exhibit a preference for nectar at ambient temperature over warmer nectar, the behavioural switch being associated with the body temperature rising above that maintained during flight. Thus flower–pollinator interactions are dependent upon ambient temperature and might alter with respect to different thermal environments (Norgate *et al.* 2010).

Bee pollination dominates in biotically pollinated temperate plants, however, bee diversity can be markedly reduced in cooler months; as has been recorded in the northern Ryukyu Archipelago where only a single species of *Colletes,* in concert with *Apis cerana*, foraged during winter (Kato 2000). In lowland wet sclerophyll forest and subtropical rainforest on the mid north coast of New South Wales bee diversity during cooler months is largely restricted to that of *Apis mellifera* and *Tetragonula carbonaria* (G. Williams pers. obs.). In Neotropical lowland rainforest, the majority of trees in many common families (e.g. Burseraceae, Euphorbiaceae, Clusiaceae, Fabaceae, Flacourtiaceae, Lecythidaceae, Melastomataceae, Orchidaceae and Sapotaceae), particularly canopy trees, are pollinated by bees (Bawa 1990). Highly eusocial bee species comprise a larger proportion of Neotropical faunas than do temperate ones, though the species richness of bees in the tropics is apparently less than that of warm temperate regions. Bee pollination is widely reported in numerous plant families in the Asian region, including Acanthaceae, Balsaminaceae, Lamiaceae, Fabaceae, Sapindaceae, Verbenaceae, and Zingiberaceae (Corlett 2004). The ecological diversity of tropical bees, however, is much greater than that recorded from elsewhere (Roubik 1989).

In Australia, the primitive bee family Colletidae is well represented and its members occur on a heterogeneous range of plant taxa throughout much of the continent (Houston 1975, 2018). Most Colletidae are oligolectic on Myrtaceae (Cardale 1993) and thus constitute, with the beetle family Buprestidae (particularly the Stigmoderini – see Bellamy 2003; Lawrence and Ślipiński 2013), an anthophilous, taxonomically distinct insect group characteristic of this distinctive family of Australian plants. Several Halictidae (which are also short-tongued) also exhibit a preference for Myrtaceae (K. Walker 1986), however many are generalist feeders. Oligolectic bees show little foraging constancy within Myrtaceae and forage on whatever plants in that family are in blossom (Williams and Adam 1997). Oligolectic bees may not be true pollinators but rather depleters of nectar – robbers or thieves. Colletidae appear to be implicated in the pollination of few primitive angiosperms, and most bee pollination in primitive flowering plants appears to depend on the family Apidae (Bernhardt and Thien 1987). However, Bernhardt *et al.* (2003) record the colletid genera *Amphylaeus*, *Leioproctus* and *Meroglossa*, in addition to apids and halictids, visiting flowering *Trimenia moorei*; though overall syrphid flies were the most common visitors. Bee pollination in primitive plant taxa probably evolved from more generally entomophilous systems and an evolution to bee specificity has been demonstrated in divergent, allied entomophilous plant taxa (Ashton 1969; Bernhardt and Thien 1987).

The majority of Australian bees are solitary (Houston 2018). Social members of the Apidae are largely represented by the native genera *Tetragonula* and *Austroplebeia*, whose distributions are essentially tropical–subtropical, and the introduced honey bee *Apis mellifera*, represented by several races. *Apis mellifera* competes with Australian native bees and other native insects for floral resources (Paton 1993; Sugden and Pyke 1991), though its impact on native plants is complex and may be reliant upon many factors, including the interaction with co-flowering species and the mating system of the plant (Gilpin 2017). However, *Apis mellifera* is known to preferentially forage for pollen and only secondarily for nectar (Aronne *et al.* 2012), this being a temporal phenomenon related to brood provision. Although *Apis mellifera* has been recorded commonly from Australian tropical and subtropical rainforest trees and shrubs (e.g. Williams and Adam 1997), the function and impact of introduced honey bees in Australian rainforests has not been well studied. Native Apidae (e.g. *Tetragonula* species) are efficient pollinators of the macadamia nut, *Macadamia integrifolia* (Proteaceae) (Heard 1993), in plantations and frequently visit the flowers of the related subtropical rainforest trees *Macadamia tetraphylla* and *Hicksbeachia pinnatifolia*. But, worldwide, not all Apidae are effective pollinators. *Euglossa* bees have been recorded as nectar thieves on Panamanian *Drymonia serrulata* (Gesneriaceae) flowers, various abundant species of Neotropical *Trigona* visit *Prockia crucis* (Flacourtiaceae/Salicaceae) but probably pollinate only rarely, and a high rate of resource depletion of Neotropical *Heliconia psittacorum* flowers by *Trigona trinidensis* has been observed (Bullock *et al.* 1989). The Australian *Tetragonula carbonaria* is known to have a foraging range of over 700 m (Smith *et al.* 2017). By comparison, the foraging range of the introduced *Apis mellifera* is recorded up to a potential range of eight kilometres (Visscher and Seeley 1982), thus giving it a capacity to outcompete smaller native species, especially when populations of flowering plants are highly dispersed. Nevertheless, *Tetragonula carbonaria* is frequently encountered in the lowland coastal rainforests of New South Wales (Williams and Adam 1997). Although an effective pollinator of small-flowered plants growing at subcanopy (e.g. *Acradenia euodiiformis*, Plate 80) and ground levels (*Pollia crispata*, Plate 22) it has been considered a pollen robber of the pioneer shrub *Melastoma affine* (Melastomataceae) in tropical Queensland – bees may bite through the plant's poricidal anthers to steal the pollen within (Gross 1993). However, various '*Trigona*' species are pollinators of Malaysian and Thailand rainforest trees (e.g. of *Shorea siamensis* in dipterocarp forest [Ghazoul and McLeish 2001]) and automimetic attraction is recorded for trigonid bees by synchronously flowering, protogynous *Piper* (Piperaceae) and *Pothomorphe* (Piperaceae) (Bernhardt and Thien 1987; Momose *et al.* 1998). *Tetragonula* bees commonly harvest pollen from the spadices of herbaceous *Alocasia brisbanensis* (Araceae) in Queensland and New South Wales subtropical rainforests (G. Williams 1993). They are pollinators of Australian *Dendrobium*, *Cymbidium* and *Caladenia* orchids and are probable pollinators of the orchid *Sarcochilus moorei* (Adams *et al.* 1992).

There are differences of opinion as to the spatial distribution of bee pollination systems in tropical rainforests with contrasting views as to the characterisation of bee responses to shaded zones, understorey strata and canopy-flowering plants (e.g. Bawa 1990).

Foraging behaviour and abundance can be expressed laterally through horizontally layered forest strata, or vertically in response to resource availability in canopy or subcanopy strata (e.g. Stangler *et al.* 2016). However, highly adaptive or specialised foraging behaviour does not automatically confer high levels of pollination and many bees, including highly developed eusocial species, may contribute little to reproductive success (Bullock 1994). In the Neotropical tree *Prockia crucis*, for example, visitation is by a taxonomically and behaviourally diverse guild of bees but most of these are minor resource thieves and not pollinators (Bullock *et al.* 1989).

Nectar depletion in bee-visited plants also serves to diurnally partition the types of bees that will forage. In the Costa Rican rainforest vine *Ipomoea batatas*, bees show staggered visiting times with larger-sized species visiting early in the morning and late in the afternoon, during periods of nectar maxima. Smaller bees visit during the interim period when nectar availability is at a minimum (Real 1981). Nectar concentration can be a powerful determining influence in the establishment of such patterns (Willmer 2011). Bees also respond to the overall display size of an inflorescence. The age of flowers also appears to modify foraging patterns – bees quickly learn to discriminate between young flowers that offer higher levels of nectar rewards and senescent flowers that produce less nectar or have ceased production altogether.

Buzz-collection of pollen

Buzz-collection, buzz-pollination and sonication are terms coined to characterise the active release of pollen from flower anthers, and subsequent transport, by certain bee species (Willmer 2011). Approximately 15 000–20 000 species of plants possess flowers that release pollen only through small openings located at the tip of the anther. Insects capable of buzz-collecting pollen resort to vibrating the anthers to release the pollen. Worldwide, about 10% of the extant angiosperm flora are postulated as buzz-pollinated (Dukas and Dafni 1990) with this strategy considered to have evolved (across disparate families) in response to increased competition between plants and pollen-consuming floral visitors (including pollen thieves and true pollinators) to control the rate of pollen removal (gamete loss) from flowers (De Luca and Vallejo-Marín 2013).

Bees capable of performing buzz-pollination range in size from small Colletidae, less than 6 mm, to large Apidae (*Xylocopa*) whose body length exceeds 3 cm, and include both oligolectic and polylectic species (Michener 2007). Bees normally restrict this activity to flowers that withhold their pollen until the anthers are physically vibrated (sonicated, by wing muscles). Pollination by sonication has evolved in several plant families (e.g. Boraginaceae, Fabaceae, Solanaceae, Melastomataceae, Lamiaceae) and has apparently arisen several times within the Ericaceae/Epacridaceae (De Luca and Vallejo-Marín 2013; Dukas and Dafni 1990; Houston and Ladd 2002; Johnson and McQuillan 2011). In plants that conform to the buzz-pollination syndrome the anthers are usually exposed, readily accessible to foraging bees. However, in the Western Australian buzz-pollinated *Conostephium pendulum* (Ericaceae/Epacridaceae) the corolla tube encloses, thus hiding, the anthers (Houston and Ladd 2002). Flowers that require buzz-pollination usually offer only pollen as a reward, as such angiosperms seem to possess neither obvious nectar rewards nor nectariferous glands (Buchmann 1983). In such plants (various species of *Senna* [Caesalpinoideae]) the anthers have pores (poricidal) or small slits through which the pollen is released. Many species of Ericaceae (Epacridaceae) have porate anthers but in the Western Australian *Conostephium pendulum* a slit elongates from the apex as the anthers rupture, and this may function to limit self-fertilisation (Houston and Ladd 2002). Some buzz-pollinated plants do produce nectar (e.g. in the Boraginaceae) and these, or their forebears, may have provided an evolutionary pathway to the non-nectariferous state typical of buzz-pollinated species (Dukas and Dafni 1990).

When bees actively buzz anthers, pollen is mainly deposited over certain parts of their bodies, although the often frenetic movements result in individual grains being collected wherever there is hair or space to snare them. Commonly, pollen gathers most profusely on the forehead, face and exposed mouthparts, the front of the thorax and ventral surface of the thorax generally, and the legs (especially between individual segments).

Though small buzz-collecting bees are often inconspicuous at their task, the noise they emit is quite audible to the human ear (often a tell-tale 'zztt, zztt'), and since their arrival and departure at flowers can be swift their sound frequently provides the first indication of their presence. Buzz-collection of pollen has been widely recorded from many bee species, including Neotropical and Australian taxa (Duncan *et al.* 2004; Roubik 1989; G. Williams 1998). In Australia, these include members of the families Halictidae, Apidae and Colletidae. Native bees that are able to buzz-collect may use the same strategy irrespective of whether they are foraging on an exotic or native species of the same genus (recorded from *Lasioglossum musicum* foraging on exotic *Senna* [G. Williams 1998]).

Recourse to the strategy of buzz-collection is not obligatory in those bees capable of it. Caroline Gross, then at James Cook University, Townsville, observed solitary apid and halictid bees in northern Queensland collecting pollen from the pioneer shrub *Melastoma affine* (Melastomataceae), though only *Xylocopa bombylans*, *X.* sp. near *gressitti*, *Amegilla anomala* and *Nomia* sp. consistently contacted the stigma while foraging (Gross 1993). Highest bee visitation rates occurred when conditions were sunny, with consequent higher pollination rates (>90%) and levels of fruit set. *Melastoma affine* possesses poricidal anthers and thus requires buzz-pollination before pollen in substantial amounts can be removed. Species of *Xylocopa* are able to 'buzz' the anthers of *Melastoma affine*, and were among a larger pollinating bee guild able to effect buzz-release of pollen. However, the bees she observed did not always vibrate the flowers they landed upon. Similar alternate buzz/non-buzzing collection strategies have been noted from other Australian bees. Likewise, the bees that have been observed visiting the threatened Australian subtropical rainforest shrub *Senna acclinis* (Plate 81) do not always vibrate its anthers, even though such an action is required for pollen to be released in any quantity (G. Williams 1998). It depends on the kind of flowers bees are collecting pollen from, as to an appropriate means of accessing the pollen and whether a particular kind of bee is adaptively tuned to undertake buzz-pollination. Oddly, or perhaps by force of habit, bees that utilise buzz-collecting strategies to release pollen from poricidal anthers sometimes attempt bouts of buzzing anthers of flowers that readily expose their pollen. *Lasioglossum musicum* has been recorded attempting to buzz-harvest pollen from the flowers of the subtropical rainforest herb *Pollia crispata* (Commelinaceae, Plate 21) (Williams and Walker 2003).

Many families of bees (e.g. Apidae, Colletidae, Oxaeidae, Halictidae and less commonly Andrenidae, Strenotritidae and Mellittidae) include species that are capable of vibrating anthers, but bees that can buzz-collect are not necessarily specialist foragers. They are usually promiscuous with respect to the types of flowering plants that they are attracted to and, like most bees and other flower visitors that utilise pollen as a source of sustenance, tend to collect pollen from whatever is in flower and whatever they can gain entry to – even if they have to forcibly break in (Willmer 2011). Northern Hemisphere bumblebees (*Bombus*) commonly access hidden floral structures by tearing and bursting through the protective walls of the flower's petals although species of *Bombus* can buzz-collect from flowers with poricidal anthers if they choose. The collection of floral rewards is pillage and plunder, even if it is done politely.

The collection of pollen by the strategy of buzz-collection is a behavioural mechanism used by bees to access a protected valuable food resource. The targeted pollen is typically dry and small-grained, as it could not otherwise fall through the anther perforations unless the anthers were physically dismembered by the animal seeking out the pollen, as do some tropical species of Trigonini bees. The anthers holding the pollen are most commonly tube-shaped. Bees cling to them and vibrate their flight muscles; pollen is released from the tips of the anthers and collected on the hairs that cover the bee's body surface. Some pollen may miss being deposited on the bee and may land on the flower's petals. It would thus be available for later collection by the bee or other anthophilous insects. Individual bees may buzz flowers for up to 10 s, though much shorter vibration bursts are more normal. The length of time spent vibrating flowers has been found to be related to the quantity of pollen available and the ease with which it can be extracted during individual foraging bouts. More pollen leads to shorter visits. Known vibration frequencies expressed by bees are ~4–5 kHz (Roubik 1989); field ecologists have cheated flowers and artificially induced pollen release from the anther by using an appropriate tuning fork (Kearns and Inouye 1993).

Roubik (1989) defined three basic behavioural strategies used by bees to collect pollen from angiosperms with poricidal anthers. The first involved buzzing while actually clinging to the anthers, the second (a variation on the first) entailed 'milking' softer anthers with the bee's mandibles, and the third was a gleaning strategy that collected pollen that had been dislodged by buzzing and gathered on other parts of the flower. Roubik pointed out, however, as did Buchmann (1983) before him, that such collection strategies are not restricted to flowers with poricidal anthers. They are useful behavioural responses to many floral barriers that bees confront daily. Pollen robbers may choose to break apart the enveloping anthers, while long-tongued bees may insert their proboscises into the anthers to retrieve the small pollen grains secreted within. It is of interest to note that the most abundant Neotropical bees (Apinae and Trigonini) apparently do not possess appropriate behavioural adaptations for removing pollen by buzzing.

It is arguable whether a bee's ability to buzz-collect pollen from poricidal anthers is learned through the various experiences entailed in daily foraging, or is inherent. Clever experiments and qualified supposition based on field observations show that the behaviour can be at least partly learned, for example, by fortuitously clinging to anthers in the normal process of foraging. Equally, variation in vibration frequency and length of individual vibration bouts may be a response to foraging from different flower species. Each flower species may provide its own suite of experiences (e.g. difference in pollen quantity and anther size, gross flower structure, presence of confounding non-rewarding structures such as pseudanthers), thus necessitating an appropriate vibration frequency, length of collection bout and physical grasp attitude (orientation) in relation to floral morphology. Trial and error are likely precursors to a finely honed craft. Michener (2007) notes that vibrating behaviour is widely known among bees and wasps, and is used to navigate tight spaces and remove small obstacles such as pebbles during the construction of nests; his hypothesis being that such an ability, though not exhibited by all modern bees, is probably ancestral, transferable to the task of releasing pollen from anthers reluctant to offer the resource they hold.

Fidelity of bee taxa to subtropical plant species

Although bees are sometimes the primary observed visitors to flowers, the frequency with which they appear can be low; though 'visitor frequency' alone may be a poor proxy for pollen transfer efficacy. However, less than one visit an hour, by any species (as can be the situation on flowers of the mass-flowering Australian subtropical rainforest tree *Abrophyllum ornans*) (Plate 30), over the duration of flowering may still result in levels of pollination above those from selfing alone (*Abrophyllum ornans* exhibits high levels of selfing) (G. Williams 1995). In addition, such apparently insignificant visitation events contribute to heterozygosity through out-crossing.

Distinct species assemblages of bees can exist between adjacent sites even when encountered on the same plant species (Williams and Adam 1997). These differences may occur where trees grow in floristically and topographically different rainforests, as well as floristically related rainforest subformations. The reason is not understood, largely because so little is known of the life history of individual bee species. Also, the interactive influences of site microhabitat availability, foraging strategies and founding events remain elusive variables, thus defying meaningful measurement. The presence or abundance of a particular bee taxon might mean nothing more than '*it got there first*' and consequently came to dominate the best nest sites. General similarities in the bee fauna can also be encountered within a limited geographic area, suggesting rainforest stand structure and floristic composition, by themselves, can have a relatively minor influence on the bee assemblage – the 'ponderables' are both bewildering and exciting.

The recruitment and visitation of bees to most trees appear to be greatly influenced by the density of available blossoms. For example, small trees of *Alectryon coriaceus*, common in littoral rainforests, that possess few flowers or have many senescent flowers and developing fruit, are visited only occasionally by bees. These are usually species of small *Exoneura*, or Hylaeinae and Colletinae (i.e, *Amphylaeus, Hemirhiza, Heterapoides, Hylaeus, Leioproctus*), and sometimes *Homalictus* (Halictidae) (G. Williams 1995, pers. obs.), all of which are generalists. Bees, in company with species of *Geron* (Bombyliidae), are often prominent visitors to the related *Alectryon subcinereus* (Plate 97) in lowland wet sclerophyll forest (G. Williams pers. obs.). Species include *Tetragonula carbonaria, Palaeorhiza viridifrons*?, *Hemirhiza melliceps, Leioproctus* sp.?, and *Homalictus megastigmus*. Isolated, often shaded, sparsely flowering (e.g. *Acradenia euodiiformis* [Rutaceae]) trees generally do not recruit native bees even though densely flowering conspecific trees nearby might be visited by large numbers of bees. But the size of individual blossoms has also been experimentally demonstrated to act as a powerful recruitment cue for bees to floral resources (Bateman 1951).

Although abundance of individual bees and generally of species increases with an increase in availability of blossoms, floral morphology does not appear to strongly influence recruitment. For example, in lowland New South Wales rainforest, high numbers of visiting bee taxa occur on *Euroschinus falcatus, Acradenia euodiiformis, Tristaniopsis laurina* and *Syzygium floribundum* (Williams and Adam 1997). These plants include species with open dish-like corollas with ovaries that are partly exposed (half-inferior; e.g. *Tristaniopsis laurina* and *Syzygium floribundum*) and those with fully exposed (superior) ovaries (*Euroschinus falcatus* and *Acradenia euodiiformis*). The latter group is often visited by eusocial Apidae; unadapted or semi-specialised Colletidae characterise bee visitors to the former. High numbers of bee species are especially associated with *Tristaniopsis laurina* and *Syzygium floribundum*. The former is yellow-flowered and the latter possesses creamish-white blossoms, both flower profusely and their individual blossoms exude large amounts of nectar.

However, bee recruitment to some flowering rainforest plants can be poor. At flowers of the subtropical rainforest vine *Aphenopetalum resinosum* (Aphanopetalaceae) only single bee species, such as *Hylaeus trilobatus*, may seasonally forage; though in this instance the visitor is at best a poor pollinator and is not a rainforest specialist being found in other habitats such as coastal grassy woodland, where it forages at the flowers of *Jacksonia scoparia* (Fabaceae) (G. Williams pers. obs.).

Surprisingly few bee species may be shared across the full suite of tree species in flower at any one rainforest site. For example, in a study of lowland subtropical rainforest in New South Wales, ~80% (78 species) of records for individual bee species were from three or fewer plant species or individual locations (Williams and Adam 1997). Only one species (a species of minute *Heterapoides* [Colletidae]) and the introduced honey bee *Apis mellifera* were recorded from 10 or more of the plants sampled. *Apis mellifera* is ubiquitous in Australian lowland subtropical rainforest, being found on virtually all mass-flowering trees, often when no native bees are present.

At both the family and genus level bees have been found to exhibit a broad polytropic response within the spectrum of available flowering subtropical rainforest trees, though much remains to be done to assess the plant fidelity of individual bee species. Where particular bees are recorded from single or few plant species this may be due to the rarity of that bee in an individual rainforest, rather than a close plant–pollinator mutualism.

In subtropical rainforest, bees of the family Apidae visit a diverse range of tree families. These include Ebenaceae, Euphorbiaceae, Myrtaceae, Rhamnaceae, Rutaceae and Sapindaceae (Williams and Adam 1997). Most apid diversity in subtropical rainforest is represented by the solitary nesting genus *Exoneura*, but at least two other solitary nesting genera, *Xylocopa* and *Amegilla*, are also commonly encountered. These are large, conspicuous and quick-flying bees that often forage in a trap-lining manner. Much rarer in subtropical rainforests is the relatively massive black and yellow *Xylocopa aruana*. Superficially resembling a bumblebee, *Xylocopa aruana* and its close relatives are often common in the tropical forests, savannahs and gardens of northern Australia. In subtropical rainforests, as well as other plant communities in subtropical south-eastern Australia, social Apidae are represented by native *Tetragonula* and the ever-present honey bee *Apis mellifera*. *Apis mellifera* can dominate the flower-frequenting insect fauna in fragmented Australian rainforests, it competes with native pollinators for floral resources and may modify the pollination ecology of native plants. It is common in nearly all rainforests, and in subtropical rainforest is found on flowers of numerous families including Anacardiaceae, Ebenaceae, Escalloniaceae/Rousseaceae, Flacourtiaceae/Salicaceae, Myrtaceae, Rhamnaceae, Rutaceae and Sapindaceae. These records are predominantly trees and large shrubs, but *Apis mellifera* will also forage at flowering vines in the rainforest understorey (e.g. *Ripogonum discolor* – Ripogonaceae) and rainforest margins (e.g. *Cissus hypoglauca, C. sterculifolia* – Vitaceae) (G. Williams pers. obs.). However, *Apis mellifera* often restricts its foraging activity to inflorescences in full sunlight, preferring to shun many shaded flowers. The occurrence of *Tetragonula carbonaria* is more mosaic-like, sometimes apparently completely absent from individual rainforest stands. *Tetragonula* seems more sensitive to the many impacts of fragmentation; its need for communal hive sites is more vulnerable to usurpation by introduced honey bees or the many kinds of human disturbance.

In Australia, Halictidae recruit to a diverse assemblage of mass-flowering subtropical rainforest trees (e.g. Anacardiaceae, Alstroemeriaceae, Commelinaceae, Cunoniaceae, Ebenaceae, Euphorbiaceae, Myrtaceae, Rhamnaceae, Rutaceae and Sapindaceae). Of the two numerically abundant halictid genera, *Homalictus* and *Lasioglossum*, individuals of *Homalictus* are more common in littoral rainforests and are frequent foragers on the flowers of many adjoining dune shrubs (e.g. *Scaevola* and *Hibbertia*). *Lasioglossum* (Plate 86) is a distinctive and characteristic genus of the bee fauna of mass-flowering Myrtaceae in lowland rainforests, and also visits sun-dappled flowering herbaceous plants of the forest floor. Leaf-cutter bees, Megachilidae, are more restricted in occurrence, both spatially and with respect to the diversity of trees at which they forage. They frequently visit the massed open nectar-rich blossoms of Myrtaceae but can also be found on less conspicuous flowers such as those of Ebenaceae. Often, megachilids are observed to be occasional, even rare, visitors to mass-flowering trees, or exhibit episodic flushes of larger numbers of taxa and individuals at single peak-flowering trees (Williams and Adam 1997). During such events megachilids may arrive and depart at breakneck speed. The length of time spent at individual blossoms is hurried, seemingly frenetic and driven by some unfathomable schedule.

In subtropical rainforests, the greatest diversity of bees is generally constituted by the family Colletidae. This is not an unexpected phenomenon given the speciose (species-rich, diverse) nature of the family across Australia, both in species and genera. Within the family, and within the context of Australian subtropical rainforests, there are exceptions to the rule, some genera being restricted in ecological and geographic range. *Paracolletes* (Colletinae), *Pachyprosopis* and *Sericogaster* (Euryglossinae) and *Palaeorhiza* and *Meroglossa* (Hylaeinae) have restricted distributions in lowland rainforests. Most members of the subfamily Colletinae, predominantly species of *Leioproctus*, are most predictably encountered on flowers of Myrtaceae, particularly those of *Syzygium*, although they are also recorded from Anacardiaceae, Escalloniaceae/Rousseaceae, Rhamnaceae, Sapindaceae and other mass-flowering rainforest taxa. *Euryglossa* (Euryglossinae) and *Leioproctus* are genera characteristic of the Colletidae that visit *Syzygium floribundum*. Hylaeinae, principally species of *Hylaeus*, contain the greatest number of species and characterise the native bee fauna of mass-flowering littoral rainforest trees. The subfamily is recorded from all plant families but can be particularly numerous on Myrtaceae. Most Hylaeinae genera appear to be polytropic, at best oligotropic, in their fidelity to mass-flowering actinomorphic rainforest plants. *Heterapoides* and *Hylaeus*, in particular, visit a broad spectrum of flowering subtropical rainforest trees. However, Hylaeinae rarely visit structurally specialised (e.g. Orchidaceae) or semi-specialised (e.g. Commelinaceae, Zingiberaceae and Fabaceae) zygomorphic rainforest plant flowers. Hylaeinae are often the only bees observed visiting isolated, partially shaded or understorey flowers of otherwise massed, crown-flowering rainforest trees. The hylaeine species *Hemirhiza melliceps*, for example, preferentially forages on shaded and often isolated flowers (e.g. of *Alectryon coriaceus*, *A. subcinereus* [Sapindaceae] and *Diospyros australis* [Ebenaceae]), and at least one species of the hylaeine genus *Palaeorhiza* forages on shaded *Diospyros australis* flowers (Williams and Adam 1997, G. Williams pers. obs.). Consequently, hylaeines may contribute to pollination of spatially restricted, small blossom clusters.

Of interest is *Hyleoides concinna*, which commonly visits the flowers of the emergent tropical and subtropical rainforest tree *Brachychiton acerifolius* (Plate 42) and the exotic pomegranate *Punica granatum* (Punicaceae). Both trees possess vivid red flowers that, although broadly tubular in shape, possess no obvious nectar guides and are readily accessible to bees. The records are particularly interesting because they indicate foraging fidelity (in subtropical rainforest *Hyleoides concinna* is generally infrequently observed on other flowering plants) or at the very least temporal foraging constancy. They represent a departure from the bird-pollination syndrome suggested by the red-coloured flowers of *Brachychiton acerifolius* and *Punica granatum*. The records are of further interest as, reportedly and thus paradoxically, bees cannot see red (Chittka and Waser 1997; Dyer *et al.* 2012). Such records underline the risk of predicting pollination vectors based on flower colour.

Hylaeine visitation to a phylogenetically diverse mass-flowering subtropical rainforest flora follows the broad polytropic response by Hylaeinae, and Colletidae generally, to plant species in less mesic habitats (see Houston 1975, 1981). The unifying theme of plant visitation by Hylaeinae appears to be one of blossom morphology rather than shared plant phylogeny, in which visited plant species generally possess allophilic (with no structural characters for guiding visitors) or hemiphilic (intermediately adapted) blossoms capable of being pollinated by short-tongued and semi-specialised anthophilous insects. The ecological distribution of Hylaeinae appears to be diverse, and hylaeines represent a potential source of generalist pollination vectors upon which subtropical rainforest trees with unspecialised pollination requirements may be able to draw (Williams and Adam 1997). Though insects may ingest large quantities of the pollen encountered, some grains are deposited upon their body and are available for transport to other flowers.

Influence of flower availability on bee activity

Bees respond to increased availability of floral resources, and in Australian subtropical rainforest the greatest numbers of bee species and individuals are recorded in November and December (Williams and Adam 1997). In part this is explained by the greater number of trees that flower in late spring and early summer, and by bee life histories.

Daily abundance and diversity of bees on individual subtropical rainforest trees are influenced by changes in weather conditions (e.g. cloudiness, wind) but increased foraging frequency by individuals and increased number of bee species at flowers is also associated with increased numbers of blossoms. Decreased foraging frequency and reduction in taxonomic diversity of the bee fauna generally corresponds with reduction in available floral resources (depleted nectar in each flower, and gradual senescence and abscission of flowers).

Different species of bees demonstrate differential daily recruitment patterns to available flowers. Some species forage only during periods of maximum resource availability, while others (e.g. *Tetragonula carbonaria*) continue to forage during diminished and late resource phases (Williams and Adam 1997). Typically, the introduced *Apis mellifera* forages only at the daily peak phase of floral resource availability (e.g. on *Acradenia euodiiformis*, *Guioa semiglauca* and *Alphitonia excelsa*), potentially maximising the theft of resources available to native pollinators and flower-frequenting fauna more broadly. Trees that tend to flower over longer periods (more than three weeks), such as *Euroschinus falcatus*, *Guioa semiglauca*, *Syzygium floribundum*, *Tristaniopsis laurina* and *Alphitonia excelsa*, also demonstrate fluctuations in attendant bee frequency and diversity that generally correspond with increases and reductions in available floral resources.

Bees may preferentially visit particular trees within populations and, in dioecious species (e.g. *Euroschinus falcatus*), preferentially forage at female or male plants. Such patterns can have profound impacts on the successful pollination and fertilisation of individual plants, species and populations in individual rainforests.

Interaction between Apis mellifera *and* Tetragonula carbonaria

Although there is always a large array of anthophilous insects on subtropical rainforest trees, of which native bees are in general not a numerically dominant element (often fewer than 5% of individuals present) (G. Williams 1995), bees may nevertheless play an important role in the overall pollination ecology of rainforests. Introduced *Apis mellifera* has a large floral resource foraging overlap with native bees and is a potential threat to them (e.g. via a possible reduction in fecundity to native species through competition [Johanson *et al.* 2019]) and to the reproductive ecology of flowering rainforest plants and Australian biodiversity more generally. However, the full impact of their presence remains uncertain. Of interest is the study by Morillo *et al.* (2008) in which alien *Apis mellifera* bees visit flowers of *Hechtia schottii*, a dioecious bromeliad in Mexico. Though visited by several insect species, the introduced honey bee is considered the bromelliad's pollinator. It is unknown if *Apis mellifera* supplanted the original native pollinator, but irrespective, the plant maintains reproductive success. In addition, a study of the alien African honey bee, *Apis mellifera scutellata*, (Dick 2001) found that these were the predominant pollinators of the canopy-emergent tree *Dinizia excelsa* (Fabaceae) growing as isolated trees in Amazonian pasture. Trees in habitat fragments produced on average as much as three times as many seeds as trees in continuous forest.

In Australian subtropical rainforests there are few native social bee species, these being most diverse in northern Australia. They are especially rare in more southern latitudes, where *Tetragonula carbonaria* is usually the only native social bee. Interactions between the eusocial apids *Apis mellifera* and *Tetragonula carbonaria* are of interest because of possible resource competition and the potential for displacement of native stingless bees (*Tetragonula* and *Austroplebeia* spp.) by feral *Apis* populations more widely.

Apis mellifera is recorded as a common visitor to tropical, subtropical and cool temperate Australian rainforest plants. It is also commonly encountered at flowering plants in a variety of Australian non-rainforest vegetation communities, including sclerophyllous woodlands and forests and tidally influenced mangroves; as well as beach strandline environments where it can be a common visitor to *Cakile maritima* [Brassicaceae]) (G. Williams pers. obs.). *Tetragonula carbonaria*, though widespread in subtropical rainforests, has a more localised distribution. Nevertheless it is found on a floristically diverse range of species including the arum *Alocasia brisbanensis*, herbaceous *Pollia crispata* and *Rubus rosifolius*, the trees *Euroschinus falcatus*, *Rhodomyrtus psidioides*, *Abrophyllum ornans*, *Acradenia euodiiformis*, *Acronychia oblongifolia*, *Tristaniopsis laurina* and *Alphitonia excelsa*, a palm *Archontophoenix cunninghamiana* and the synchronous, mass-flowering vine *Austrosteenisia blackii* (Fabaceae). In wet sclerophyll forest on the north coast of New South Wales *Tetragonula carbonaria* visits the climbing orchid *Erythrorchis cassythoides*, but is too small to be a pollinator as it will not trip the pollinium deposition mechanism (though see Singer 2002), the true pollinators being *Amegilla* bees. It is a common visitor to many flowering garden herbs, annuals and orchard trees that are cultivated near lowland rainforest.

Apis mellifera is commonly observed in the proximity of other native bee genera and readily disturbs native Hylaeine foraging on the same flowers. The generally smaller Hylaeinae quickly fly away at the approach of their larger and more numerous competitor (Williams and Adam 1997). Such adverse impacts from honey bee foraging are not restricted to subtropical flowering plants, as similar impacts occur between honey bees and native flower-visiting fauna (not solely native bees) in Australian tropical and cool temperate rainforests.

The situation appears different with *Tetragonula carbonaria* and maybe other native social apids as well. These diminutive native bees (*Apis* >9 mm, *Tetragonula* <6 mm) frequently continue to forage, giving little indication of displacement or movement that suggests avoidance of the much larger and robust *Apis*. Admittedly such responses are likely to depend on the numbers of *Apis* present, but even when *Apis mellifera* occurs in moderate numbers *Tetragonula* may persist in foraging, seemingly undeterred. The subtropical rainforest plants on which *Tetragonula* and *Apis* co-forage include a diversity of floral morphologies and phylogenetic relationships (e.g. *Cordyline stricta*, *Alocasia brisbanensis*, *Archontophoenix cunninghamiana*, *Euroschinus falcatus* and *Rhodosphaera rhodanthema* [e.g. Plates 7, 10, 11, 16, 95]).

The apparent lack of disturbance or displacement of *Tetragonula* cannot be interpreted as evidence that the foraging of introduced *Apis mellifera* has not affected other native pollinators. Various anthophilous flies (e.g. Syrphidae, Lauxaniidae and Drosophilidae) and Coleoptera (e.g. Nitidulidae, Chrysomelidae and Scirtidae) are readily disturbed by the foraging activities of *Apis mellifera* (Williams and Adam 1997). When they visit mass-flowering rainforest trees in large numbers, honey bees often leave only a short period of uninterrupted flower visitation for native pollinators. The sheer magnitude of *Apis* activity overwhelms all attempts by native insects to find flowers unattended by honey bees. It comes down to an exercise in elbow room and pugnacity.

The presence and survival of *Tetragonula* colonies is heavily dependent on the availability of tree hollows (Eltz *et al.* 2003). *Apis* competes with native fauna (including vertebrates such as parrots, microbats, dasyurids and possums) for nest hollows (e.g. Oldroyd *et al.* 1994) though the degree to which this occurs in Australia, and the level of impact, remain largely unstudied. Introduced honey bees may be most affecting native social Apidae in rainforest at the nest sites rather than at the food source, even though limited studies in other environments have found that *Tetragonula* utilise different types of cavities from those occupied by *Apis*. Native social bees may be particularly vulnerable at nest establishment because their colonies are founded by young queens in concert with small numbers of workers, rather than by the massed swarms of honey bees. Such small founding colonies may be readily displaced. Additionally, *Tetragonula* bees may have low fecundity (as Eltz *et al.* [2003]) found for '*Trigona*' in lowland dipterocarp forests of Sabah, Malaysia), and old queens are unable to vacate nests, as can those of *Apis*, so that destruction or loss of a nest site may involve loss of the hive's means of re-establishment at an alternative location.

A significant battleground might be the location of the potential nest site, but the conflict might be the dynamics of colony foundation.

POLLINATION BY LEPIDOPTERA (BUTTERFLIES – PSYCHOPHILY, MOTHS – PHALAENOPHILY)

Lepidoptera are now the largest group of plant-feeding insects (Grimaldi and Engel 2005). The larvae of most Lepidoptera live and feed on angiosperms and the majority of adult butterflies feed on nectar (Braby 2000; Common 1990). Lepidoptera diversified on flowering plants during the Cretaceous and Tertiary (Grimaldi and Engel 2005), participating in a contemporaneous coevolutionary radiation (Grehan 1991; Wahlberg *et al.* 2013). The earliest fossil lepidopteran comes from the Jurassic, but with Lepidopteran diversity rates (through multiple lineages) considered to have evolved after the K/T–K-Pg mass extinction (Grimaldi and Engel 2005). Recent studies by Chazot *et al.* (2019) and Espeland *et al.* (2018), respectively, estimate the origin of butterflies to the Early (107.6 mya) – Late Cretaceous (119 mya). However, 'plots of Lepidoptera diversity and phylogeny through time suggest that a high proproportion of their evolutionary history remains undetected in the fossil record', for this is considered significantly incomplete (Sohn *et al.* 2015).

Although eagerly sought by collectors, little is known about the role of butterflies (the division or suborder of Lepidoptera formally termed the Rhopalocera) and moths (the second division of Lepidoptera, known as the Heterocera) as pollinators in Australian rainforests. While their small size, often drab colours, difficulties of identification and nocturnal habits may work against a broad interest in the plant–pollinator relationships of moths and flowering plants (Sphingidae being a notable exception [Moulds *et al.* 2020]), the apparent lack of knowledge on the pollination role of butterflies is surprising given the decades over which they, and rainforests in Australia, have been the subject of popular attention. Paradoxically, there is a relative wealth of data regarding their larval food plants; though again this is somewhat biased towards non-rainforest species.

Australian rainforest plant genera that exhibit exclusive floral characters suitable for moth pollination include *Psydrax*, *Cyclophyllum* and *Randia* (Rubiaceae), *Hymenosporum* (Pittosporaceae) and *Clerodendrum* (Lamiaceae/Verbenaceae), though there are many instances of moths visiting flowers with very open floral structures (as in *Syzygium sayeri*, *S. tierneyanum*) (e.g. Boulter *et al.* 2005; Hopper 1980). Moths (e.g. Noctuidae, Pyralidae) are important nocturnal pollinators of South-east Asian *Calamus* (Arecaceae) (Fah 1995; Lee *et al.* 1995), however, the role of moths as possible pollinators of Australian *Calamus* (e.g. *C. caryotoides*; Plate 12) remains to be investigated. Fragrance is a primary attractant cue in blossoms adapted to being pollinated by moths, but is less significant for the attraction of butterflies. The fragrances emitted by moth-blossoms are strong and odour production may occur or peak at specific times of the day, commonly at dusk and during the night. Moth-blossoms also tend to have nectar flows that peak near or at night.

In both butterfly- and moth-pollinated blossoms the floral tube is often very long and slender and the nectar produced is less viscous than in bee-flowers. The length of floral tube and the length of the lepidopteran proboscis (the long, curled feeding tongue) are inter-dependent limiting factors for both the plant and the types of lepidopterans that can effectively reach nectar resources. In the majority of Lepidoptera, food uptake can only occur via the proboscis – nectar viscosity (dilute solutions exhibit increased capillarity) is a limiting factor in the ability of moths and butterflies to feed from flowers. If the nectar is too viscous, even as a result of evaporation during the heat of the day, the nectar resource is unavailable. Many of the flowers to which lepidopterans are attracted contain nectar with significant levels of amino acids. These add to the nutritional value derived from the ingestion of nectar. However, not all Lepidoptera are nectar-feeders. Some species of Micropterigidae moths apparently feed directly on pollen, although pollen grains may not constitute their sole diet. Micropterigids have chewing mouthparts and superficially resemble caddis-flies (Trichoptera), a completely different order of insects, and are considered a primitive group that evolved very early from ancestral lepidopteran stock. They are known from specimens preserved in amber dating from the Lower Cretaceous. As discussed earlier Neotropical *Heliconius* butterflies also disseminate nutrients, derived from pollen grains, into their bodies.

Though fragrance is an important attractant in moth-blossoms, the role of colour in nocturnal moth-pollinated (phalaenophilous) flowers remains elusive. Plants that are pollinated, or at least visited, by day-active butterflies may have vivid flower colours but are also commonly white. There are marked divergences from the generalised Lepidoptera-pollinated syndrome – several moth-pollinated flowers, specifically those pollinated by hawkmoths, are without nectar, lack corolla tubes, do not produce scent and are not white (Johnson *et al.* 2016; Willmer 2011).

There is some similarity in syndrome between bird-pollinated (ornithophilous) flowers, and flowers visited by day-active lepidopteran pollinators. Direct competition is known between day-active hawkmoths and hummingbirds (Trochilidae) in the Americas. Pollination by both hawkmoths and birds occurs in several plant taxa (Ashton 1969; Hopper 1980), an example being *Clerodendrum tomentosum* (Lamiaceae/Verbenaceae, Plate 39) (Hamilton 1919), a small tree of Australian subtropical and dry rainforests. Hawkmoths are also recorded as primary pollinators of the mangrove-associated tree *Sonneratia alba* (Sonneratiaceae) and the shrub *Clerodendrum inerme* (Lamiaceae/Verbenaceae) in northern Queensland (Primack *et al.* 1981). Some rainforest trees whose flowers are considered to conform to the moth-pollinated syndrome may not be vector-specific at all. They are likely to be promiscuous, attracting moths, birds, bats and bees, and probably less-conspicuous pollinators as well. Hawkmoths are visitors to many species of Neotropical dry forest plants adapted for other pollination vectors such as bats, hummingbirds and bees, in addition to hawkmoth flowers in which the floral tube or perianth is shorter than the length of the moth proboscis. Perhaps we are looking for a level of correlation between plant form and pollinator adaptation that is rarely achieved, rather than a degree of plant–pollinator mutualism that is expressed far more loosely or subtly. However, Johnson *et al.* (2016) have found that hawkmoth proboscis length may be related to the length of floral tubes, with plants with longer floral tubes excluding hawkmoths with mismatched proboscises, and considered that proboscis length can account for the evolution of floral tube length (as well as nectar volume).

Significant behavioural differences distinguish the pollination contributions of moths and butterflies. Generally, the flight activity of butterflies is almost exclusively but not fully diurnal. Although the activity patterns of moths are characteristically crepuscular (during twilight) or nocturnal, there are several species that fly and visit flowers only during daylight (Common 1990). Hawkmoths (e.g. *Macroglossum micaceum* and *Macroglossum nubilum*) typify the former; among the day-flying species are the large, brightly coloured, tropical *Alcides zodiaca* (Uraniidae), *Amata* and *Nyctemera amica* (Erebidae). In lowland subtropical rainforests Erebidae (e.g. *Amata*, *Asura*; Plate 106) and various genera of nocturnal Noctuidae are commonly found on flowers of numerous mass-flowering trees. Although many nocturnal moths visit the flowers of subtropical rainforest trees, their contribution to plant pollination is unknown.

But for the great majority of moths that are nocturnal, their activity patterns, and consequent provision of pollinator services, may be adversely affected by artificial light sources (e.g. from sources adjacent forests) (MacGregor *et al.* 2015). However, behavioural divergence exists within the moths. Some groups, most markedly the Sphingidae, hover during feeding and others (e.g. Erebidae, Geometridae, Noctuidae) alight and clamber over individual blossoms and inflorescences. This difference in behaviour directs placement of pollen. In species that hover, pollen is deposited on the proboscis or head, and upper thorax. In species which alight, pollen deposition on the undersurface of the thorax is the norm.

Many of the 87 species and 31 genera of Australian hawkmoths recognised by Moulds *et al.* (2020) are found in tropical and subtropical rainforests, and wet sclerophyll forest. However, the association of this rich fauna with flowering plants, and thus as possible pollinators in mesic vegetation communities, remains poorly studied. Pollination by hawkmoths (sphingophily) has been more extensively investigated in tropical forests than has pollination by other moths (e.g. Corlett 2001; Schulze *et al.* 2001). Their hovering method of flight is energetically very expensive, so hawkmoths need to forage at blossoms that ensure a high return for the energy expended in seeking out and feeding from them (Willmer 2011). Flower presentation may also be a limiting factor to visitation with hawkmoths showing marked preferental foraging at upright flowers rather than pendant ones (Fulton and Hodges 1999). Food constancy is recorded in hawkmoths; that is, they repeatedly return to and concentrate on a given flowering species for a period of time, usually until that species has ceased to flower; and the length of visitation to flowers may be longer in instances where flowers provide greater nectar rewards (Johnson *et al.* 2016). Trap-lining foraging patterns are also known (Linhart and Mendenhall 1977), with pollen movement by hawkmoths under favourable foraging conditions (clear and warm nights) extending beyond 350 m. In such instances hawkmoths mirror the trap-lining behaviour recorded by some bees and hummingbirds. However, they also exhibit ecological release (showing flexibility away from native species when exotic plants are available, often preferentially [e.g. Moulds *et al.* 2020]) and commonly depart from native plants to forage at the flowers of introduced horticultural plants; for example, species of tropical Australian *Macroglossum* at dusk attend the flowers of paw paw, *Carica papaya* (G. Williams pers. obs.).

Hawkmoths do not hold a universal mandate as pollinators among the Heterocera. For example the diminutive moths of the genus *Epicephala* (Gracillariidae) are recorded as pollinators of *Breynia oblongifolia* (see Plate 65) (Finch *et al.* 2018). This is a commonly encountered shrub found in lowland wet sclerophyll forests and the margins of rainforest in south-eastern Australia. Larvae develop within the tiny enclosed floral structure. Pollen is deposited on the bodies of emerging adults. The moth families Noctuidae, Pyralidae and Geometridae include numerous species found on flowers at night, and to a lesser degree during the day. This holds equally true for the tropical rainforests of Africa, South and Central America, Australia and South-east Asia. Noctuids are visitors to the cauliflorous flowers of Queensland *Syzygium cormiflorum*, and are a rich and abundant component of the moth fauna that visits flowers in Oriental forests (Corlett 2004; Crome and Irvine 1986). In Thailand, Noctuidae are pollinators of *Dipterocarpus obtusifolius*, whose fragrant flowers open at dusk (Ghazoul 1997). Pyralidae are considered cross-pollinators of *Balanophora kuroiwai* in the Ryukyu archipelago of Japan (Kawakita and Kato 2002). The moths lay eggs on the inflorescences and the developing larvae do not cause damage, suggesting a mutualistic relationship between moths and plant. Noctuidae are also recorded as pollinators of *Gnetum* in Hong Kong and *Telosma* and *Pergularia* (Asclepiadaceae) in India (Bhatnagar 1986; Corlett 2001). Recorded instances of possible pollination relationships between Oriental geometrids and angiosperms include the Dipterocarpaceae, Moraceae and Orchidaceae, but are likely to be far more extensive. Notably, several of the plants mentioned here (e.g. *Balanophora kuroiwai, Breynia oblongifolia*) possess flowers that reflect little or no structures of form or colour characteristic of the classically defined moth pollination syndrome.

Butterflies are very visible components of the insect fauna of tropical ecosystems, with individual Amazonian forest sites likely to contain between 600 to 1600 species (Brown and Hutchings 1997). Although butterflies visit many brightly coloured and taxonomically unrelated flowers, butterfly pollination systems are among the least studied in tropical rainforests, including those of Australia. Many species of *Syzygium* (Myrtaceae) in peninsular Malaysia are pollinated by danaid butterflies (Appanah 1990), and are considered major pollinators of species of *Callicarpa, Clerodendrum* (Lamiaceae/Verbenaceae), *Trichilia* (Meliaceae) and *Vernonia* (Asteraceae) in the Western Ghats of India (Corlett 2004; Devy and Davidar 2003). Butterflies are also considered the main pollinators of Sri Lankan *Elaeocarpus amoemus* (Elaeocarpaceae) (Susanti and Sripanomyom 2006). The flowers of this species have distinctive red calyxes that contrast with the white petals, however, Susanti and Sripanomyom (2006) were unable to show experimentally what role calyx coloration may play in attracting pollinators. Several butterflies have been recorded from the related North Queensland *Syzygium tiernyanum* and *Syzygium sayeri*, as well as other tropical and subtropical Australian Myrtaceae (e.g. Boulter *et al.* 2005). The role of Australian rainforest butterflies as pollinators has been little investigated, but their potential for relatively long-distance flight (especially that of butterflies of the family Papilionidae [birdwings, swallowtails etc.]) may be important to out-crossing in individual plant populations. Hopper (1980), for example, records various papilionids (*Graphium macleayanum, Papilio ulysses, Ornithoptera priamus euphorion*) and nymphalid butterflies (*Vindula arsinoe ada, Hypolimnas bolina nerina, Mycalesis terminus terminus*), as well as a pierid (*Delias mysis mysis*), foraging at flowers of *Syzygium tierneyanum*. These species are capable of relatively long-distance movements between flowering plants.

Pollination by butterflies to Australian rainforest trees may also be seasonally episodic. That is, populations of a butterfly species may occur one year in large numbers at a particular plant then be almost totally absent from it for many following years. Such a scenario has been observed in *Delias nigrina* (Pieridae) visiting *Macadamia integrifolia* (Proteaceae) in northern New South Wales (G. Williams pers. obs.). Seasonal visitation episodes have also been recorded among vertebrates; certain Hawaiian honeycreepers (Drepanididae), for example, are common on flowering trees one year then defy all efforts to observe them at the same site in other years, though they occur at other locations (Evenhuis 2007).

Butterfly activity is directly affected by temperature (Xing *et al.* 2016) and rainfall patterns, as well as cloud cover, and at Australian subtropical latitudes the greatest species-richness of butterflies occurs in late spring-summer (C. Hill 1988), corresponding with the seasonal peak in the availability of flowering rainforest plants. Butterflies of any affinity, however, are rarely observed at many available flowering subtropical Australian rainforest plants. They are typically absent or rare visitors to species of such families as Ebenaceae and Lauraceae. Even the conspicuous massed inflorescences of *Syzygium floribundum* (Myrtaceae), *Alphitonia excelsa* (Rhamnaceae), *Melia azedarach* (Meliaceae) trees, in addition to numerous members of the Sapindaceae, in lowland subtropical rainforests attract little more than fleeting attention (G. Williams 1995). Nevertheless, butterflies do serve as pollinators among Australian rainforest plants; for example, the vines *Hoya australis* (Asclepiadaceae) and *Parsonsia straminea* (Apocynaceae), and the tree *Pararchidendron pruinosum* (Mimosaceae) (Forster 1992a; G. Williams 1993). Several species (e.g. *Danaus affinis, Hypochrysops cyane*) are visitors to flowering *Avicennia marina* mangroves (G. Williams pers. obs.), and the diminutive but seasonally abundant nymphalid *Hypocysta metirius* is a common associate of small flowering herbs, shrubs and epiphytic orchids growing in sunny forest gaps or on rainforest margins (see Plate 108) in northern New South Wales. But, like numerous day-active moths, many adult butterflies visit and feed at the flowers of various exotic plants that infest rainforest margins. For example, the pierid *Belenois java* (Plate 114) and the papilionids *Graphium macleayanum, Graphium agamemnon* and *Papilio ulysses*, and the various tropical and subtropical species of *Ornithoptera*, are frequent visitors to lantana (*Lantana camara* [Verbenaceae]) flowers. While the vast nectar resource represented by *Lantana camara* and various additional weeds supplements the food available to the butterflies and other native insects such as bees that attend their flowers, the impact of the diversion of native anthophilous insects (and vertebrates) on the reproductive ecology of indigenous rainforest angiosperms is unknown. It is tempting to arbitrarily dismiss such visits as negligible with respect to perceived adversity but, when rainforest ecosystems are highly fragmented, are heavily invaded by alien plants and have small populations of individual angiosperm tree and shrub species (e.g. Pisanu *et al.* 2009; Rossetto and Kooyman 2005), the diversion of potential pollinators (and native pollen) could be of greater consequence than casual observations suggest. Even where rainforest ecosystems are little fragmented, the diversion of butterflies from native plants may still affect pollen transfer and the reproductive success of particular plant species.

POLLINATION BY MISCELLANEOUS INSECTS AND OTHER INVERTEBRATE GROUPS, ESPECIALLY THRIPS

The pollination contributions of other invertebrates, including insects, are largely unknown. Crabs have been recorded as occasional pollinators of Arecaceae and of the Brazilian bromeliad *Aechmea pectinata* (Barfod *et al.* 2011; Canela and Sazima 2003). However, Canela and Sazima (2003) note, that whilst *Armases angustipes* (Grapsidae) crabs were recorded in more than 31% of the inflorescences of *Aechmea pectinata*, they consumed stamens and stigmas in newly opened flowers, consequently reducing plant reproductive success by destroying floral parts and also by detering visits by hummingbird pollinators.

Springtails (Collembola), which are not true insects but are closely related to them (Naumann 1991), have been recorded as pollen feeders from a variety of mainly northern temperate and polar plants (Kevan 1978), and may be pollinators of the subterranean, mushroom-like flowers of Asian *Aspidistra* species (Ruscaceae/Asparagaceae).

Terrestrial amphipods also are likely candidates as pollinators of various Asian species of *Aspidistra*, exploiting the protective floral structures as feeding chambers. *Aspidistra* exhibits an asymmetrical pollination system – although it has specialised floral structures (in most species the stigma's architecture prevents self-pollination and flowers are mostly located in the detritus of the forest floor) that restrict entry, its probable amphipod and possibly collembolan pollinators have a more generalised behaviour. Conran and Bradbury (2007) studied cultivated plants of *Aspidistra elatior* in the Adelaide Botanic Gardens (South Australia) and found that in the absence of its native pollinators *Aspidistra elatior* was able to exploit Australian soil invertebrates to achieve pollination and seed-set. But flies have also been recorded as pollinators of *Aspidistra* (e.g. *Aspidistra formosa, A. mararasmioides, A. multiflora* [Vislobokov 2017]); the genus thus exhibiting an exceptionally taxonomically divergent pollination agent strategy.

Among the lesser studied orders of anthophilous insects, Miridae and other Hemiptera are sometimes encountered on flowers of Australian mass-flowering rainforest trees (e.g. *Acradenia euodiiformis, Alectryon coriaceus, Alphitonia excelsa, Cuttsia viburnea, Euroschinus falcatus, Guioa semiglauca, Rhodomyrtus psidioides, Syzygium floribundum, Syzygium sayeri, Syzygium smithii* and *Ryparosa kurrangii*) (Webber *et al.* 2008, G. Williams 2020), as well as the mangrove *Avicennia marina*. In the latter, *Oncopeltus sordidus* (Lygaeidae) bugs feed on nectar and in so doing large amounts of pollen adhere to their bodies (Plate 110). Individual movements, however, are mainly to adjacent flowers in the same flower cluster, or to other clusters on the same tree such that pollination contributions are primarily geitonogamous (G. Williams unpubl. records). Mirids have been recorded as minor pollinators of some species of Asian Dipterocarpaceae (*Shorea*), however, Miridae (*Decomioides schneirlai*) and Anthocoridae (*Orius atratus*) are considered primary pollinators of dioecious *Macaranga tanarius* (Euphorbiaceae) growing in Okinawa and the Amami Islands, Japan (Ishida *et al.* 2009). *Decomioides schneirlai* and *Orius atratus* feed on nectar from extrafloral nectaries and are likely to breed in chambers formed by flower bracts. Vincent *et al.* (2015) proposed that the Western Australian *Beyeria cockertoni* (Euphorbiaceae), a non-rainforest plant, was apparently dependent on a species of Miridae for pollination. Hemiptera are also considered pollinators, within a larger pollinator suite of beetles and bees, of *Sarcandra glabra* (Chloranthaceae) in Japan (Tosaki *et al.* 2001). However, the presence of some hemipterans (e.g. Reduviidae) on influorescences during mass-flowering events may be, at least in part, an opportunistic response to increases in abundance of prey that rely on floral resources as well as other plant material.

Cockroaches (Blattodea) are postulated to be the main pollinators of *Uvaria elmeri* (Annonaceae) in a Sarawak lowland rainforest, a departure from the incidence of beetle pollination often found in the Annonacaeae (Nagamitsu and Inoue 1997). Flowers of dioecious *Clusia blattophila* (Clusiaceae) are visited by crickets, ants and moths but the principal pollinator is the cockroach *Amazonia platystylata* (Vlasáková *et al.* 2008); its body surface being rough enough to retain pollen grains. *Clusia blattophila* produces seed only sexually, and flowers at night. Cockroaches react to floral volatiles and acetoin suggesting that floral scent is a factor involved in attracting cockroaches to its flowers. Acetoin is a compound involved in chemical communication of several cockroach species and it is possible *Clusia blattophila* exploits the cockroach's sensitivity to this compound to attract them to its flowers as part of the pollination syndrome of this species. However, the density of male plants affects visitation rates because only males reward pollinators; with cockroaches spending less time on rewardless female flowers (Vlasáková 2015). Nocturnally active cockroaches visit the flowers of *Alphitonia excelsa* growing in New South Wales subtropical rainforests and those of *Syzygium cormiflorum* in tropical Queensland, but are likely to occur much more widely on the flowers of numerous rainforest plants. The nocturnal habit of most blattodeans means that their possible pollination contributions are overlooked by field naturalists. As with several other anthophilous insects, notably several families of Coleoptera, cockroaches commonly chew floral parts – any pollination is at a price to the plant. Where reproductive structures such as stigmas and anthers are devoured, the presence of such flower visitors may be negative or provide a reduced benefit. Also, blattodeans (like many other anthophilous insects) undertake relatively short-distance movements so that pollination contributions may result in little out-crossing. If movements are confined to a single flowering plant, no fertilisation will result if plants are dioecious or flowers are self-incompatible.

Distantly related to cockroaches are the herbivorous, sometimes predatory, Orthoptera; katydids, grasshoppers and crickets. Katydids are members of the family

Tettigoniidae, several Australian species (e.g. *Windbalea viride, Anthophiloptera dryas, Zaprochilus* spp.) of which are adapted to feed on nectar, pollen and whole flowers (Rentz 2010). Though destructive of floral parts their flower-visiting habits may result in instances of pollination, but the pollination contributions by Australian Orthoptera fauna are largely unstudied. However, on the Indian Ocean island of Réunion the orthopteran *Glomeremus orchidophilus* (Gryllacrididae) is the only pollinator of the orchid *Angraecum cadetii* (Hugel *et al.* 2010). *Glomeremus orchidophilus* is restricted to Réunion, though the orchid also occurs on Mauritius where nocturnally active orthopterans of the same genus also occur (Micheneau *et al.* 2010); yet their role as pollinators of *Angraecum cadetii* is uncertain (Hugel *et al.* 2010). In a review of the flower-visiting orthopterans from South-east Asia Tan *et al.* (2017) note that of 140 recorded incidences of orthopterans visiting flowers, 41 species have so far been recorded to visit 35 different plant species; these mainly in Singapore, Malaysia, part of Thailand and Brunei Darussalam. The authors concluded that Orthoptera are over-looked flower visitors in the region.

Small, readily overlooked and dismissed as pollinators, are thrips (Thysanoptera) (e.g. Mound and Marullo 1996, 1998]). Thrips have been recorded as pollinators of primitive angiosperms (e.g. magnoliids [Gottsberger 1977; Gottsberger *et al.* 1980]) but rarely as pollinators of gymnosperms; though putative pollen-collecting thrips are known from the Early Cretaceous (Peñalver *et al.* 2012). Thrips are thought to have radiated to flowering plants from an ancestral or detritivorous habit (Mound and Marullo 1998), and although poorly recognised, pollination by thrips appears to be widespread and across a diversity of plant taxa, including extant members of the basal ANA/ANITA grade (e.g. *Chloranthus* [Ma *et al.* 1997]) and species previously considered to be wind-pollinated (e.g. *Platycarya strobilacea* – Juglandaceae [Fukuhara and Tokumaru 2014]). The range of angiosperm families and genera occurring in rainforests in which thrips are implicated as pollinators or have close relationships, includes Sarawak and West Malaysian Euphorbiaceae (*Macaranga*), Annonaceae (*Popowia*) and Myristicaceae (*Horsfeldia*), India and Australian Myristicaceae (*Myristica*), Malaysian and Sri Lankan Dipterocarpaceae (*Shorea*), Australian Monimiaceae (*Wilkiea*), Myrsinaceae/Primulaceae (*Myrsine*), Smilacaceae (*Smilax*), Moraceae (*Antiaropsis, Maclura*), Arecaceae (*Linospadix*), Phyllanthaceae/Euphorbiaceae (*Breynia*), Calycanthaceae (*Idiospermum*), Amazonian Annonaceae (*Bocageopsis, Oxandra*), Brazilian Monimiaceae (*Mollinedia*), African Lecythidaceae (*Napoleonaea*) and Malipighiaceae (*Peixotoa*), and Panamanian Moraceae (*Castilla*) (e.g. Appanah and Chan 1981; Armstrong and Irvine 1989; Bernhardt and Thien 1987; Davies 1999; Del-Claro *et al.* 1997; Fiala *et al.* 2011; Gottsberger 1999; Momose *et al.* 1998; Mound and Marullo 1996; Sakai 2001; Sharma and Armstrong 2013; Williams *et al.* 2001; Worboys and Jackes 2005). This is only a cross-section of their presence as pollinators among modern flowering plants. Thrips are likely more significant in the pollination of Australian rainforest plants, and elsewhere, than has been considered (e.g. Williams *et al.* 2001; see Table 8.1). Their role as pollinators may have been overlooked because, on entomophilous flowers, they form part of a larger anthophilous guild dominated generally by Diptera, Coleoptera and Hymenoptera.

As already discussed, thrips are widely recorded as pollen-eaters, destroying at least part of the standing pollen crop otherwise available for pollination (Kirk 1985b). Feeding by thrips on pollen grains is influenced by pollen grain volume, pollen type and daily temperature patterns. Although feeding impacts may reduce plant fitness, thrips activities can still result in a reproductive benefit to plants if thrips pollinate as well as consume. Thrips are consumers of pollen and the individual pollen loads they carry are usually small, but this may be partially compensated for by their abundance and by the inability of other potential pollinators to enter specialised thrips-pollinated flowers (e.g. the monoecious *Breynia oblongifolia* [G. Williams pers. obs.]). Thrips may also be pollinators in Australian ecosystems other than rainforests. The common summer flowers of the Australian Alps, such as various *Celmisia* (Asteraceae) and *Ranunculus anemoneus* (Ranunculaceae), which do not have enclosed floral structures, nevertheless support enormous populations of *Thrips imaginis* and very few other insects. Unfortunately, the importance of this thrips species in the pollination biology of alpine meadow flowers has yet to be demonstrated (Williams *et al.* 2001).

Thrips are the most abundant visitors to dioecious *Myristica insipida* (Myristicaceae), an understorey tree in tropical Queensland rainforests (Armstrong and Irvine 1989), but their role as pollinators has been queried because some ecologists did not find pollen on thrips collected from female flowers and thrips behaviour suggested they did not make regular movements between

Table 8.1. Species of thrips (Thysanoptera associated with the flowers of Australian subtropical rainforest plants (after Williams *et al.* 2001, G. Williams pers. obs.).

Plant family and species	Thrips species
Anacardiaceae	
Euroschinus falcatus	*Haplothrips ?victoriensis*, *Haplothrips* sp., *Pezothrips kellyanus*, *Thrips imaginis*, *T. coloratus*, *T. australis*, *T. setipennis*, *Desmothrips bagnalli*, *Andrewarthaia kellyana*, *Erythridothrips cubilis*
Arecaceae	
Linospadix monostachyos	*Thrips setipennis*
Celastraceae	
Denhamia celastroides	Larvae indet.
Cunoniaceae	
Ackama paniculata	*Thrips* sp., *Haplothrips* sp., larvae indet.
Schizomeria ovata	*Thrips* sp., Phlaeothripidae indet.
Ebenaceae	
Diospyros australis	*Thrips setipennis*, *T. imaginis*, *Thrips* sp., *Helionothrips spinosus*, *Pezothrips kellyanus*, *Anaphothrips* sp., larvae indet.
Diospyros pentamera	*Thrips hawaiiensis*
Elaeocarpaceae	
Elaeocarpus obovatus	*Haplothrips* sp.
E. reticulatus	*Thrips imaginis*
Escalloniaceae/Grossulariaceae	
Anopterus macleayanus	*Thrips setipennis*
Escalloniaceae/Rousseaceae	
Cuttsia viburnea	*Pezothrips kellyanus*, *Haplothrips victoriensis*, ? *Haplothrips* sp., ? *Thrips imaginis*, *T. setipennis*, *Anisopilothrips venustulus*, *Parthenothrips dracaenae*
Flacourtiaceae/Salicaceae	
Scolopia braunii	*Thrips* sp., *Haplothrips* sp.
Lauraceae	
Cryptocarya microneura	*Thrips imaginis*
Monimiaceae	
Wilkiea huegeliana	*Thrips setipennis*, *Haplothrips* sp.
Moraceae	
Maclura cochinchinensis	*Thrips imaginis*, *T. hawaiiensis*, larvae indet.
Myrtaceae	
Rhodomyrtus psidioides	*Thrips* sp., *T. setipennis*, *Heliothrips haemorroidalis*, *Haplothrips* sp.
Syzygium floribundum	*Thrips australis*, *T. hawaiiensis*, *T. setipennis*, *Haplothrips* sp., *Haplothrips ? victoriensis*, *Andrewarthaia kellyana*, *Erythridothrips cubilis*
Syzygium smithii	*Thrips* sp., *T. australis*, *T. setipennis*, *T. imaginis*, *Haplothrips* sp., *Desmothrips* sp., *Anaphothrips* sp.
Tristaniopsis laurina	*Thrips* sp., *Haplothrips* sp.
Myrsinaceae/Primulaceae	
Myrsine howittiana	*Thrips setipennis*, *T. ? setipennis*
Myrsine variabilis	*Thrips setipennis*

Continued

Table 8.1. *(Continued)*

Plant family and species	Thrips species
Rhamnaceae	
Alphitonia excelsa	*Haplothrips* sp., *Thrips hawaiiensis, T. imaginis*
Ripogonaceae	
Ripogonum discolor	undetermined sp.
Rutaceae	
Acradenia euodiiformis	*Pezothrips kellyanus, Haplothrips ? victoriensis, Desmothrips bagnalli, Desmothrips* sp., *Erythridothrips cubilis, Thrips imaginis*
Sapindaceae	
Alectryon coriaceus	*Haplothrips* sp.
Guioa semiglauca	*Thrips* sp., *T. setipennis, T. australis, T. imaginis, Nesothrips* sp., *Haplothrips* sp., *Phlaeothripidae* indet.
Smilacaceae	
Smilax glyciphylla	*Thrips setipennis*

Myristica insipida trees. They are also the most abundant visitors to the flowers of *Idiospermum australiense* in North Queensland rainforest (Worboys and Jackes 2005). The floral morphology and phenological patterns in *Shorea* species (Dipterocarpaceae, dominant canopy trees in many Asian rainforests) possess adaptations to pollination by flower thrips (Appanah and Chan 1981; Kondo *et al.* 2011). Moreover, *Shorea* flower buds provide a thrips brood site similar to that found in species of Australian *Wilkiea, Breynia* and *Myrsine*. In *Shorea acuminata*, thrips are the principal flower visitors accounting for more than 60% of the collected insects during flowering. Visitation was synchronous with flowering, but they only made a small contribution to out-crossing; movements largely being to flowers on the same tree, with most seeds resulting from geitonogamous self-pollination (Kondo *et al.* 2011). Thrips have been found to represent over 97% of the total number of visitors to a single flowering *Shorea lepidota* tree in Malaysian rainforest, with a density at individual flowers ranging from two to 16 adults (Appanah and Chan 1981). Thrips transported small quantities of pollen grains (on average two grains) upon their bodies, which they preen when alighting on flowers. Similar small pollen loads have been reported from thrips collected on male flowers of Australian *Myristica insipida*. These individually tiny pollen loads, and the propensity for regular preening, may explain the absence of pollen from thrips collected at female *Myristica insipida* flowers during studies in northern Queensland rainforests. The habit of regular preening, which is typical of adult flower thrips when preparing their wings for flight, may lead to incorrect conclusions concerning the effectiveness of thrips as pollen transporters. An example of the variable nature of pollen loads carried by thrips is reported from *Neoheegeria* in West Malaya (Moog et *al.* 2002). *Neoheegeria* is a pollinator of *Macaranga hullettii* (Euphorbiaceae), the pollen carried found to range from 0–268 grains per adult.

Even in instances where thrips species are pollen robbers, consuming pollen with no apparent benefit to plant reproductive biology, such pollen sources may sustain thrips populations in the absence of flowering by species for which they are true pollinators. Thrips that pollinate and utilise *Shorea* flowers as brood sites have been found in the majority of other flowering Malaysian rainforest trees; the continuous availability of these alternative flowering species allows thrips to maintain populations (Appanah and Chan 1981). A similar scenario occurs among Australian Thysanoptera. *Thrips setipennis*, which pollinates *Wilkiea huegeliana*, also occurs at the flowers of many other subtropical rainforest plants, many of which produce open flowers in profusion. The ability to utilise pollen types from a wide range of flowers may provide buffering against fluctuations in the availability of food resources and brood sites. This wide ecological range of *Thrips setipennis*, contrasted with the apparently restricted pollinator mutualism of *Wilkiea huegeliana*, *Myrsine howittiana* and *Myrsine variabilis* (for which

Thrips setipennis is the only known pollinator), illustrates that specialised plant–pollinator mutualisms need not be strictly symmetrical (Williams *et al.* 2001).

How are thrips recruited to superficially unattractive rainforest flowers such as those of *Wilkiea huegeliana*? The flowers of *Wilkiea huegeliana* are dull coloured, appear to lack any conspicuous form and seem devoid of obvious recruitment signals. From a human perspective, its flowers are plain and unenticing. But for flowers in general, colour and the concentration and composition of floral scents are recruitment cues to thrips (Kirk 1985a; Teulon and Penman 1992). Visual cues such as flower shape, size, pattern and contrast also provide important recruitment stimuli, yet the flowers of *Wilkiea* and especially *Breynia* appear to lack obvious attraction stimuli such as nectar guides, conspicuous flower colour or size, and strong floral odour. However, at least in species of *Wilkiea*, flowers do 'yellow' before opening, which might provide an appropriate visual signal. They may also emit fragrance plumes, possibly highly periodically, which are less obvious to humans but are readily discerned by the sensory organs of thrips. In *Breynia oblongifolia* (Plate 65), a nondescript shrub of subtropical rainforest margins and ecotones (the floristically diffuse boundaries between two or more vegetation communities), years of casual observation found no pollinator (G. Williams pers. obs.). Then, on a cluster of plants, hundreds of thrips, a few at each minute blossom, were observed entering and leaving through the small ostiole at the apex of the tiny conical flowers (those examined carried single or several pollen grains). There was no obvious change in flower colour, just the same apple-green hue little distinguished from that of the leaves, and no obvious fragrance that might be the cause of attraction.

Tentatively, to encompass the variety of flowers upon which Thysanoptera are decided pollinators, a thrips pollination syndrome (thripophily) might include flowers which are white or yellow, sweet-scented flowers with compact floral structures, possible nocturnal presentation of pollen to distinguish thripophilous flowers from bee-flowers and globe to urn-shaped flowers (that provide enclosed brood sites for thrips eggs and larvae), but not necessarily flowers with an obvious fragrance (Kirk 1988a, 1988b).

The discussion of invertebrate pollination syndromes cannot be brought to conclusion without mention of malacophily – pollination by snails and slugs. Pollination by terrestrial snails has been recorded in several plants. Dayanandan *et al.* (1990) observed a 'shell-less' snail on the the inflorescence of *Shorea megistophylla* (Dipterocarpaceae), in lowland dipterocarp forest in Sri Lanka, and a helicarionid snail was observed at an exotic bromeliad inflorescence growing in a garden adjacent subtropical rainforest in New South Wales (G. Williams pers. obs.). Both were likely feeders upon the respective inflorescences. Snails (*Lamellaxis gracile*), in addition to bees (*Apis cerana*), are pollinators in India of the facultatively autogamous prostrate weed *Volvulopsis nummularium* (Convolvulaceae) (Sarma *et al.* 2007). Flowers open in the morning and last for only half the day. Snails are exclusive pollinators on rainy days when bees are not active. On sunny days snails arrive first, followed later by bees. Although *Lamellaxis gracile* consumes pollen grains and anthers some grains gather on the body and later come into contact with dehiscent anthers and stigma lobes effecting pollen transfer. Sarma *et al.* (2007) found that 74% of snail-visited flowers developed fruit. Gardère (2018) observed the hygromiid snail *Leptotaxis bollei* feeding on the anthers, and sheltering in the bottom of the corolla, of *Campanula jacobaea* (Convolvulaceae) in the Cabo Verde archipelago. Snails occasionally transported pollen on their shells, their movements favouring cross-pollination.

POLLINATION BY BIRDS (ORNITHOPHILY)

> *'Comparison with the behaviour of other pollinator groups suggests that birds promote pollen dispersal through a combination of high mobility, limited grooming, and intra- and inter-species aggression.'*
>
> *(Krauss* et al. *2017)*

Like bees, birds (Aves) originated in the Cretaceous, having likely descended from feathered theropod dinosaurs that were capable of flight within an arboreal environment; though other non-Aves groups such as troodontids (e.g. *Anchiornis*) and some dromaeosaurids (e.g. Microrapterinae) probably also were capable of some form of arboreal foraging assisted by 'feathered' flight or gliding within canopy or understorey habitats (Burnham 2007; Sullivan *et al.* 2017; Xu *et al.* 2014). Thus the development of the 'bird-pollinated' flower syndrome (proposed as inclusive of all these possible participants) has evolved over a timeframe that parallels the appearance and radiation of flowering plants. In some families (i.e. Costaceae, Gesneriaceae, Lamiaceae and Zingiberaceae) bird-pollination may be derived from an ancestral adaptation to being pollinated by bees (Kay *et al.* 2005; Wilson *et al.* 2017). Certainly, fossil evidence for Australian Proteaceae suggests that pollination by birds, and possibly small mammals, was well underway by the Early Cenozoic (e.g. R. Hill 1994). Evolutionary shifts to pollination by birds have occurred in numerous lineages of flowering plants, and these shifts have driven changes that affect bird attraction, deterrence of non-pollinating and destructive flower visitors, protection against vigorous foraging by birds (e.g. mechanical strengthening of floral parts, sheathing of the ovary by a staminal tube), and accurate placement of pollen upon birds' bodies. However, in the Australian genus *Hakea* (Proteaceae), a genus with a relatively young 'crown age' of 10 million years, bird pollination may be the primitive state, with multiple shifts to insect pollination representing later transitions (Mast *et al.* 2012).

Although individual Australian rainforest tree species (e.g. *Castanospermum australe* [Fabaceae]) show marked adaptation to pollination by vertebrates, these play only a minor pollination role in the overall subtropical rainforest plant community. But flower-visiting vertebrates are somewhat better represented in tropical Queensland rainforests, at least in terms of increased diversity. Within these northern mesic plant communities birds and mammals are recorded from a wider number of apparently vertebrate-adapted flowering plants, such as brush-flowered species of *Syzygium* (e.g. Boulter *et al.* 2005; Crome and Irvine 1986). When birds visit flowers pollen is normally deposited on the facial or throat feathers, or the bill. A novel departure from this is the carriage of asclepiad pollinia on the tongue by a South African sunbird (Pauw 1998).

Ornithophilous blossoms tend to produce larger volumes of nectar than flowers adapted to attract bees, but the sugar concentrations of bird-flowers are generally lower (Pyke *et al.* 2020). In the Australian tropical rainforest herb *Hornstedtia scottiana*, a species apparently pollinated by meliphagid honeyeaters (Ippolito and Armstrong 1993), nectar is reportedly to be largely sucrose, and this is atypical of ornithophilous angiosperms. Sucrose-dominant nectar is also recorded from humingbird-pollinated flowers, however, flowers pollinated by passerine perching birds tend to have hexose-dominated nectar (Cronk and Ojeda 2008). Bird-pollinated flowers generally contain low concentrations of amino acids. Specialised bird-flowers may have deep long-tubed corollas, are often brightly coloured (red or red-yellow) (K. Johnson 2013), lack odour and often possess perching or landing platforms (as in *Hornstedtia scottiana*). In *Hornstedtia scottiana* the scarlet flowers are positioned close to the ground, the inflorescences serving incidentally as perches whilst honeyeaters forage. However, in the imposing large herb *Doryanthes palmeri* (Doryanthaceae) (Plate 27), a species restricted to the steep mountains of the New South Wales – Queensland border, the distinctive reddish-brown flowers are displayed atop a tall stem; the flowers attracting honeyeaters (Meliphagidae).

The flowers of some Australian rainforest trees which do not possess bright flowers (e.g. *Syzygium sayeri*, in which the flowers are creamish [see Boulter *et al.* 2005, 2008]), however, are visited and pollinated by many species of birds. Regardless, recruitment to bird-flowers is primarily visual and red is classically considered a bird-attracting colour (Cronk and Ojeda 2008; Shrestha *et al.* 2013). Red coloration may also function to deter visits by bees whilst serving to enhance attraction to birds (Cronk and Ojeda 2008). In the Costa Rican *Combretum farinosum* (Combretaceae) red flowers, which produce very little nectar, apparently serve as flags to attract hummingbirds to nectariferous (nectar-producing) green-coloured blossoms (Schemske 1980). However, flowers generally assigned to the entomophilous syndrome attract bird species that exhibit specialised nectarivorous (nectar-feeding) adaptations. For example, the long and curved-billed Australian eastern spinebill *Acanthorhynchus tenuirostris* (Meliphagidae) will sometimes hang precariously to feed from the small flower clusters of the understorey shrub *Pittosporum revolutum* (Pittosporaceae), a plant with yellow-coloured tubular flowers usually visited and pollinated by insects including nitidulid beetles (e.g. *Cychramus picticollis*) and thrips (G. Williams pers. obs.).

Pollinators potentially play an important role in mediating selection of floral traits (Pauw 2006). In fully bird-adapted plants, visited by specialised (eutropic) birds, nectar is abundant, sometimes continuously flowing, and deeply held within the extended perianth. For such fully adapted plants, nectarivorous birds must be long-tongued and long-billed. In Hawaiian lobeliod (Campanulaceae) flowers, there is a correlation between length of the perianth tube and beak length of their honeycreeper (Drepanididae) pollinators (Pratt 2005). Such characteristics can serve as isolating mechanisms, affecting pollinator visitation (Mesquita-Neto *et al.* 2018) and promoting speciation among plant species.

Morphological traits of bird pollinators (e.g. bill length and or bill curvature) and plants (tubed-corolla length, width) potentially affect resource partitioning in mutualistic systems, with closely matched pollinators hypothesised to preferentially forage on morphologically closely matched flowers (Sonne *et al.* 2019). However, even in apparently highly co-adapted pollinators, such as the sickle-billed Hawaiian drepanidids (e.g. *Drepanis coccinea*), species that are presumed by their long and curved bills to be specifically co-adapted to flowers with deep corolla tubes (e.g. as in *Tremmattolobelia kauaiensis* [Campanulaceae]), may be more efficient foragers at short flowers than their short-billed drepanidid relatives (e.g. *Himatione sanguinea*), and in addition may readily forage on a wide variety of introduced plants and feed upon some insects (Carothers 1982; Fancy and Ralph 1998).

Pollination by birds is widely reported in tropical rainforests and southern temperate regions, however, pollination systems involving birds in the Northern Hemisphere are poorly developed (Sussman and Raven 1978; Proctor and Yeo 1975). In the Oriental Region the number of bird-dependent flowers and flower-dependent birds are reported as being much lower than such plant-pollinator dependencies recorded from the Neotropics and the Australian Region (Corlett 2004). Although birds may respond differentially to available floral resources in various forest strata (Chmel *et* al. 2016), there is little fidelity or constancy by pollinivorous or nectarivorous birds and most rainforest birds are polytropic or oligotropic in the range of plant taxa visited; that is, they visit a wide or narrow diversity of flowering plants, but their foraging behaviour is not restricted to a single or few species. This behavioural plasticity, or 'super generalist' feeding strategy, exhibited by many birds, may serve plant reproduction well; as with the rare Mauritian endemic plant *Nesocodon mauritianus* (Campanulaceae) which is now heavily reliant on the introduced bulbul *Pycnonotus jocosus* for pollination (Olesen *et al.* 2002). However, Crichton *et al.* (2019) caution that the broad flower visitation spectrum deployed by 'super-generalist' pollinators (not just birds) may facilitate weed invasion; as Goulson and Derwent (2004) note for the invasive woody weed *Lantana camara* (Verbenaceae), which throughout its range in Australia, is effectively pollinated by the introduced honey bee *Apis mellifera*.

Many rainforest birds show intermediate degrees of specialisation in their adaptation to flower visitation. In the Neotropical hummingbirds (Trochilidae) and sugarbirds (Coerebidae), Hawaiian honeycreepers (Drepanididae) and Old World/Australasian sunbirds (Nectariniidae) and honeyeaters (Meliphagidae) adaptations (such as bill and tongue structure and length) to utilising blossoms as the main source of nutrients are more fully developed, although many are partly insectivorous (insect-eating). Trochilidae, Coerebidae and Drepanididae are absent from Australia, Nectariniidae (in Australia) are restricted to tropical Queensland (though the single Australian species *Cinnyris jugularis* ranges into south-east Asia [Cheke *et al.* 2001]) but can be common in urban gardens, and the Meliphagidae (~175–180 spp. worldwide) have radiated widely in Australian ecosystems (Joseph *et al.* 2014) but are poorly represented in rainforests at higher latitudes (though somewhat more diverse if species that utilise flowering mangroves, arguably a rainforest community, are included). This rarity of bird pollination in Australian rainforests is difficult to explain, given its importance in other ecosystems.

Ornithophily in the Australian flora is a relatively recent phenomenon in a geological and evolutionary sense (Armstrong 1979). Many types of putative ornithophilous blossoms show differential adaptation to entomophily and chiropterophily (bat pollination), in which ornithophily may be transitional or be one component of a broader expression of pollinator adaptation within a particular family (e.g. bird-fly-bee systems within Ericaceae/Epacridaceae [see K. Johnson 2013]). In Australian plant communities generally, bright red colours in bird-pollinated flowers are relatively infrequent, though several red-coloured *Melaleuca* species found in sclerophyllous habitats are conspicuous exceptions. The many white and cream-coloured *Melaleuca* species (including those previously placed in *Callistemon*) to which numerous Meliphagidae (e.g. Scarlet honeyeater *Myzomela sanguinolenta*) are attracted further emphasise that bright colours do not serve as exclusive indicators of the ornithophilous pollination syndrome. Diversity in the colour range of flowers visited by specialised flower-adapted birds has been recorded elsewhere. Long-billed Oriental sunbirds, for example, are reported visiting plants whose flowers include red, orange, pink, and white (Corlett 2004).

However, there are several Australian tropical and subtropical rainforest trees with floral characters that meet the criteria for putative bird-pollination syndromes (e.g. *Alloxylon pinnatum*, *Grevillea robusta*, *Stenocarpus sinuatus* [Proteaceae], *Brachychiton acerifolius* [Malvaceae/Sterculiaceae], *Erythrina vespertilio* and *Castanospermum australe* [Fabaceae]). These include species with radially symmetrical, semi-adapted actinomorphic blossoms, and those with more specialised, bilaterally symmetrical zygomorphic blossoms. As noted, the bright red flowers of *Brachychiton acerifolious* attract particular species of native bees (rather than birds) that show high fidelity to that species when in flower. Several trees occurring in Australian rainforests (e.g. *Syzygium moorei* [Myrtaceae] and *Archidendron grandiflorum* [Mimosaceae]) possess pink brush-shaped flowers that suggest possible transition from insect- to bird-pollination (or adaptation to pollination by both) (Williams and Adam 2010). *Capparis arborea* (Capparaceae) also has prominent large brush-like blossoms. However, the flowers are coloured white. This is a small tree of dry rainforest and some upper estuary forests adjacent mangrove communities, north from the Hunter River, New South Wales. Although visited by cetoniine beetles (*Bisallardiana gymnopleura*), bees (*Apis mellifera, Amegilla bombyformis* [Plate 81], *A. ?pulchra*), and sometimes butterflies (*Elodina angulipennis*), the principal pollinators of plants growing adjacent estuary zones appear to be long-billed Brown honeyeaters (*Lichmera indistincta*) (G. Williams pers. obs.) (in southern India, the related *Capparis brevispina* and *C. zeylanica* are pollinated respectively mainly by butterflies and birds [Nayak and Davidar 2010]). Brown honeyeaters regularly contact the anthers (these being borne on long filaments) and the projecting stigma of *C. arborea*, and undertake long feeding bouts at individual flowers punctuated by movements to adjacent flowering plants. The cetoniine *Bisallardiana gymnopleura* also undertakes long feeding bouts at individual flowers, movements are usually restricted to flower clusters in the same tree, and individuals tend to forage deep within the stamen cluster; thus largely avoiding contact with both anthers and stigma. Although *Bisallardiana gymnopleura* likely contributes to geitonogamous pollination, the contribution to out-crossing in *Capparis arborea* populations is predicted to be less than that of the honeyeaters. However, *Lichmera indistincta*, as with other meliphagid honeyeaters, does not exhibit constancy to flowers of particular structure, for individuals also visit the flowers of Grey mangrove, *Avicennia marina*; these being open structures with short stamens partly fused to the dull orange coloured corolla.

Several tree species (e.g. *Brachychiton acerifolius*) are often characterised by highly dispersed populations where individuals may be separated from conspecifics by many kilometres. This in itself might suggest that pollination is achieved by vectors, such as birds, that range over long distances and allow tree populations to exist at very low densities. However, many Australian tropical and subtropical rainforest trees that are not pollinated by vertebrates (e.g. *Endiandra discolor* [Lauraceae]) also occur at very low densities over large ranges. Indeed, a great diversity of species, in which many are represented by few individuals in highly dispersed populations (sometimes referred to as metapopulations, although the definition includes interactive connectivity in addition to spatial pattern at the landscape scale), is a frequent condition of tropical rainforests worldwide (Thomas 1994). Therefore the spatial distribution and population density of rainforest plants may not provide clues to their pollination ecology.

The pollination ecology and breeding systems of Australian subtropical and warm temperate rainforest trees are largely unstudied, but birds play an important pollination role in Australian sclerophyllous heaths, woodlands and forests, particularly in the pollination of *Blandfordia* (Blandfordiaceae), *Banksia* (Proteaceae), *Eucalyptus* and *Melaleuca* (Myrtaceae) (Ford *et al.* 1979; Paton and Turner 1985; Pyke *et al.* 2020; Turner 1982). Paton (2000) showed that the bird-pollinated shrub *Astroloma conostephioides* (Ericaceae/Epacridaceae), a species native to southern Australia, was severely pollinator-limited in degraded and invaded habitats due to the absence of migratory nectarivorous birds. In the rainforests of New South Wales, birds rarely feed on nectar and pollen. The Meliphagidae, although an important element of Australia's northern tropical forests (e.g. Macleay's honeyeater *Xanthotis macleayana* and the Bridled honeyeater *Lichenostomus frenatus*) and open temperate vegetation, in subtropical rainforest formations are usually represented by a single species, Lewin's honeyeater *Meliphaga lewinii* (Adam 1987). However, secondary regrowth (seral) rainforests, rainforest trees along forest margins and ecotones, and species-poor rainforests may be able to recruit other meliphagid birds from adjoining sclerophyllous communities. Flower visitation by brush-tongued lorikeets, such as the Rainbow lorikeet *Trichoglossus moluccanus* and Scaly-breasted lorikeet *T. chlorolepidus* (Psittaculidae), is primarily restricted to eucalypts occurring in the rainforest–wet sclerophyll forest interface; pollination by parrots apparently does not contribute greatly to pollination in southern Australian rainforests. Only the Blue-browed fig-parrot, *Opopsitta diophthalma coxeni*, can be considered a subtropical rainforest specialist, but its role as a pollinator in subtropical rainforests is yet to be determined, and is currently so rare that for practical purposes is probably negligible.

This contrasts with the world's tropical rainforests where many birds such as hummingbirds, sunbirds and honeycreepers are conspicuous plant pollinators. In part, the relative paucity of bird pollination in Australian subtropical rainforests is because most flowers are not adapted for pollination by birds and the flowers of most trees and shrubs just don't produce sufficient nectar or pollen to meet birds' energy requirements. The birds that are observed on most flowering subtropical rainforest trees, even Lewin's honeyeater, are feeding on insects or fruit rather than nectar or pollen; some pollen may, however, adhere to their bodies and thus be transported between trees. It is puzzling that, given the presence and abundance of pollinating birds in the vicinity of subtropical rainforests (in adjacent sclerophyll forest and coastal heathland), appropriate floral structures have not evolved more widely.

POLLINATION BY FRUIT-BATS, FLYING-FOXES AND BLOSSOM-BATS (CHIROPTEROPHILY)

Large fruit-bats, also called flying-foxes, and the smaller blossom-bats (Pteropodidae [Megachiroptera]) are widely recorded as pollinators of tropical rainforest trees (Endress 1994), however, the role of bats as pollinators of flowering plants in the Australian Region is apparently more important than their recorded role as pollinators in the adjacent Oriental Regional (Corlett 2004). This is possibly due to the extent of eucalypt-dominated woodland and sclerophyll forest in the Australian landscape. Even the small microchiropteran bats, which are usually insectivorous, have evolved flower-frequenting and thus potentially pollinating species in the Neotropical Americas and New Zealand (Czenze and Thurley 2018; Saldaña-Vázquez *et al.* 2020). In Australia, only the Pteropodidae are recorded as pollinators, visiting blossoms of a variety of trees. The evidence for pollination by fruit-bats in subtropical rainforests is minimal, although several species roost there and fruit-bats are important fruit-dispersers.

In the islands of the South Pacific many fruit-bats are important keystone species, whose pollinator and seed dispersal activities play a critical role in the reproductive biology of individual plants and the wider ecology of the island ecosystem (Cox *et al.* 1991). They are important to the maintenance of overall island biodiversity. Island faunas tend to possess depauperate pollinator guilds (e.g. Samoa, which nevertheless has over 800 species of flowering plants). In the Pacific these decline further eastward from Australasia. Even large island complexes, like New Zealand and Hawaii, are not immune from relying on relatively impoverished pollinator assemblages, or possess plant species with specialised pollinator requirements that are vulnerable to extinction should their pollinator be lost. Fruit-bats in the Pacific are declining and several have recently become extinct (as in Guam) (Flannery 1995). Their further demise threatens the preservation of flowering plants and the maintenance of broader ecosystem function and mutualisms – a potential cascade scenario. Although continental fruit-bat faunas are less vulnerable to extinction than their island cousins, a reduction in fruit-bat abundance or diversity might have significant impacts on the ecology of Australian tropical rainforest, particularly on hyper-dispersed plant populations. Such a hypothesis needs testing.

Bat pollination is interpreted as specialised because it involves a distinct set of correlated adaptations (Duchen and Renner 2010). The floral syndromes characterising or linked to pollination by bats (such as ramiflorous and cauliflorous inflorescences) are well established and orientation by bats to blossoms is usually by sight or smell (Hall and Richards 2000; Kress 1985; but see von Helversen and von Helversen 1999). Specialised bat-flowers are relatively large, exposed as sturdy peduncles or suitably placed inflorescences. They open at night, normally for only a single night, are usually whitish in colour and are rarely brightly coloured. They emit strong, nocturnally produced, musky and sometimes sweet odours, although strong odours are absent in several Neotropical and Paleotropical *Heliconia* species that are pollinated by bats (Kress 1985). Classic bat-flowers produce large quantities of nectar nocturnally and large amounts of pollen, are large-mouthed and have strong single flowers, or are brush flowers (as in Myrtaceae, Capparaceae, Barringtoniaceae and some Proteaceae). In the Indonesian *Freycinetia insignis* (Pandanaceae), the food reward is in the form of large sugary bracts (Proctor and Yeo 1975). The positioning of bat-pollinated flowers may be flagelliflorous (the inflorescence is projected beyond the crown profile of the tree on a hanging pedicel), as in *Oroxylum indicum* (Bignoniaceae), cauliflorous, as in *Syzygium cormiflorum* (Myrtaceae) or more simply exerted from the foliage, as in *Passiflora mucronata*

(Passifloraceae) (Crome and Irvine 1986; Gould 1978; Sazima and Sazima 1978). An unusual departure from the 'bat-pollination' syndrome is the New Zealand geoflorous root parasite *Dactylanthus taylori* (Balanophoraceae) which is pollinated by the Lesser short-tailed bat *Mystacina tuberculata* (Czenze and Thurley 2018).

Within the group of plants described as bat-pollinated, floral morphology and phenology may further restrict their utility to particular bats. In *Pseudobombax tomentosum* (Bombacaceae) in Brazil, the erect position and morphology of the flowers facilitate visits by large bats of the family Phyllostomidae that land to feed, but restrict visits by small, hover-feeding glossophagine bats (Gribel 1988). Small New World glossophagine bats are known to locate flowers by echolocation, and acoustic nectar guides have been described from the flowers of *Mucuna holtoni* (Faboideae) (von Helversen and von Helversen 1999). The nectar guide contains a small concave 'mirror' that functions by redirecting energy from the bat's echolocation calls back to the direction of the incidence. Visits by bats to receptive flowers cause *Mucuna holtoni* to release pollen explosively. Echo responses enable the bats to discriminate between flowers that have been visited (which contain larger nectar rewards; 100 μL) and those that have released pollen (and contain less nectar; 10–20 μL). An exceptional form of an acoustic bat pollination adaptation is found in the Cuban vine *Margravia evenia* (Margraviaceae). *Margravia evenia* displays dish-shaped foliage leaves. The stalks of the leaves are twisted, bringing the leaf blade into an upright position, with the upper concave side facing the approaching pollinators. The 'dish' leaves give an attracting echo that in effect serves as a location beacon for the plant's pollinators (Simon *et al.* 2011).

Although many nectarivorous birds and bats visit the same flowers, their diurnal and nocturnal feeding patterns mitigate against direct temporal foraging competition (Sussman and Raven 1978). Differential plant adaptation to bat visitation can be reflected in the timing of nectar production. *Melonycteris fardoulis* (as *Melonycteris woodfordi*) is endemic to the Solomon Islands (Flannery 1995), and is a pollinator of *Heliconia solomonensis* (Kress 1985). Nectar production in *Heliconia solomonensis* commences at ~1900 h and ceases at ~2100 h, a nocturnal timespan unlikely to attract the attention of birds.

The furred surface of floricolous bats has great pollen-carrying potential and bat pollination is recorded in many plant families (Law 1992; Willmer 2011). Large fruit-bats are damaging (dystropic) as the bats are sometimes destructive to blossoms, and can also be destructive of the rainforest tree canopies in which they rest during the day (Corlett 2004; Elmqvist *et al.* 1992). This results in the loss of rainforest flower crops, and an adverse impact on plant reproductive success, in later seasons. However, where fruit-bats are dystropic they may be able to transfer pollen at some reproductive cost to the plant (destruction of a proportion of available blossoms). The pollen of chiropterophilous blossoms contains higher levels of protein than that of entomophilous blossoms, however, pollen is a complex structure and vertebrates demonstrate differential physiological abilities to utilise it as a protein source (see Willmer 2011). In the specialised nectar- and pollen-feeding Queensland blossom-bat *Syconycteris australis*, pollen digestion rates and alimentary retention time (retention in the intestine), in addition to pollen type and size, are primary determinants of the species' capacity to use pollen efficiently (Howell 1974; Law 1992). Physiological constraints, in combination with pollen size and structure (and floral morphology, anthesis and nectar production and composition), are probably strong selective determinants of bat visitation to, and thus pollination of, rainforest plants.

Small blossom-bats are significant nocturnal and crepuscular 'high energy demanding' pollinators in tropical forests generally (Crome and Irvine 1986; Kress 1985; Sussman and Raven 1978) but the low nectar levels of blossoms of Australian rainforest trees at higher latitudes may not fulfil the energy requirements of larger pteropodid bats (*Pteropus* species). Though considered partially dystropic, large Pteropodidae contribute to the pollination of eucalypts (Armstrong 1979), so the role of fruit-bats in the pollination of some mass-flowering rainforest Myrtaceae (e.g. *Syzygium sayeri*, known to be visited occasionally by *Pteropus conspicullatus*), and possibly Capparaceae, cannot be discounted. Pteropodidae have been considered pollinators of *Hicksbeachia pinnatifolia*, a cauliflorous tree restricted to the subtropical rainforest of northern New South Wales and south-east Queensland. Pollen (of *Banksia serrata, Banksia integrifolia, Grevillea ?robusta, ?Stenocarpus sinuatus* [Proteaceae], *Erythrina* sp. [Fabaceae], Myrtaceae and Lauraceae) has been recorded from the droppings of the Grey-headed flying-fox *Pteropus poliocephalus* in coastal New South Wales (Parry-Jones and Augee 1991a, 1991b). However, large pteropodid fruit-bats seem to generally avoid many of the mass-flowering trees (e.g. *Syzygium floribundum*) seasonally available in Australian subtropical rainforest.

Several species of small blossom-feeding megachiropteran bats (e.g. Queensland tube-nosed bat *Nyctimene robinsoni*, northern blossom-bat *Macroglossus lagochilus* and Queensland blossom-bat *Syconycteris australis*) occur in the tropical rainforests of North Queensland, and feed from such genera as *Syzygium* and *Banksia*. The single blossom-bat occurring in New South Wales rainforests, *Syconycteris australis*, primarily visits trees with chiropterophilous blossoms but such flowers are absent or infrequent in Australian subtropical rainforests. However, *Syconycteris* may commonly feed at banana flowers where plantations are near rainforest roost sites. *Banksia integrifolia* (Proteaceae), comprising several subspecies found growing in Australian coastal woodland and rainforest communities, possesses an inflorescence that partially agrees with the chiropterophilous syndrome (Plate 67). The flowers of coastal *Banksia integrifolia* populations are visited by *Syconycteris australis*, although their attraction to plants in flower may vary between localities (Law 1992). *Syconycteris australis* is known to move between isolated tropical rainforest remnants across cleared agricultural landscapes, flying distances of up to nearly 6 km (mean 1 km). Consequently, this small blossom-bat, able to transport significant pollen loads, is capable of effecting cross-pollination in geographically disjunct rainforest communities (Law and Lean 1999).

POLLINATION BY NON-FLYING MAMMALS

Although many arboreal herbivorous mammals visit flowers and include them in their diet, the evidence supporting the transfer of pollen by non-flying mammals in tropical forests has been considered to be largely indirect and weak. Flower visitation by such animals has been viewed as opportunistic and partly destructive, a response to the copious availability of food resources. However, the very ready availability of floral food rewards to non-flying mammals might provide resources during periods of low fruit availability (Bawa 1990). Thus, the reproductive benefits may be measured as dispersal of genetic material as seeds.

There is increasing evidence for pollination by mammals other than bats (e.g. Gribel 1988; Willmer 2011; Zoeller *et al.* 2016). Interestingly, the regions in which pollination by non-flying mammals is more commonly recorded are those with a depauperate flower-visiting bat fauna (e.g. Madagascar and temperate Australia). An early hypothesis of non-flying mammal pollination systems is that these may be relictual, resulting from more recent competitive displacement by flower-visiting bats (Sussman and Raven 1978). Such a proposition may or may not be the correct one, but how can it be tested?

Examples of non-flying mammal pollinators include nocturnally active lemurs (Lemuridae and Daubentoniidae) in Madagascar, tree shrews (Tupaiidae) in West Sumatra, bush-babies (Lorisidae) from West Africa and the honey possum *Tarsipes rostratus* (Tarsipedidae) from south-west Australia (e.g. Baum 1995; Itino *et al.* 1991; Sussman and Raven 1978). In tropical forests there is also a diverse range of incidental flower visitors that, in frequenting canopy and understorey trees, may serve as occasional pollinators: these include monkeys, squirrels, civets, rodents and martens (Corlett 2004; Joshi *et al.* 1995; Yumoto *et al.* 2000). Some, such as rodents, flying squirrels and monkeys, may consume or damage many of the flowers they visit while feeding upon floral resources.

General characteristics of the non-flying mammal pollination syndrome include dull-coloured flowers in tight inflorescences strongly attached to stems, nocturnal anthesis and nectar production, copious nectar production, channelling of nectar to the ground, flowers positioned close to the ground (geoflorous) and hidden in foliage, and a strong musky or yeasty odour (Turner 1982; Wiens and Rourke 1978). These characteristics reflect nocturnal behaviour and poor colour reception but well developed olfactory responses in many mammals, but the diversity of floral traits found in flowers visited by non-flying mammals (and frequently by other pollination vectors) does not necessarily support the idea of a specialised pollination syndrome (e.g. Saffer 2004).

As well as *Tarsipes rostratus*, there are several additional instances of flower visitation and pollination by non-flying mammals in the Australian flora (*Antechinus stuartii, Sminthopsis murina, Sminthopsis griseoventer, Phascogale tapoatafa* [Dasyuridae], *Cercartetus nanus, Cercartetus caudatus, Cercartetus concinnus* [Burramyidae], *Trichosurus caninus, Trichosurus vulpecula, Petaurus norfolcensis* [Phalangeridae], *Rattus tunneyi* and *Rattus fuscipes*, as well as the exotic *Rattus rattus* [Muridae]) (e.g. Collins and Rebelo 1987; Crome and Irvine 1986; Goldingay 2000; Hackett and Goldingay 2001; Johnson *et al.* 2011; Turner 1982, 1984). Endress (1994) notes 'marsupials' (in combination with bats) as probable pollinators of the tropical Queensland and New Guinean tree *Cynometra lenticellata* (Fabaceae).

The examples above are primarily non-rainforest examples, although some animals include rainforest in their distribution (e.g. *Antechinus stuartii* and *Trichosurus caninus*). Rats are potential pollinators within the putative pollinator guild of *Balanophora fungosa* (Balanophoraceae) (Irvine and Armstrong 1990), a low-growing geoflorous herbaceous root parasite of Queensland rainforest trees (Plate 18); though ants might be facultative pollinators of the related Japanese species *Balanophora kuroiwa* (Kawakita and Kato 2002), and as stated earlier, the related New Zealand *Dactylanthus taylori* is pollinated by the Lesser short-tailed bat *Mystacina tuberculata* (Czenze and Thurley 2018). Using motion cameras, in North Queensland forests Pierce and Ogle (2017) recorded a previously unknown broad guild of vertebrate visitors to the flowers of *Balanophora fungosa*; bush rats (*Rattus* spp.), Musky rat kangaroo (*Hypsiprymnodon moschatus*), Giant White-tailed rat (*Uromys caudimaculatus*), Long-nosed bandicoot (*Parameles nasuta*) and four honeyeater species (i.e. *Meliphaga notata, M. gracilis, Xanthotis macleayana, Myzomela obscura*). Of these bush rats and Musky rat kangaroos were the most frequent visitors. Some 'repeat visiting' individual visited the same plants where they fed from the same flowers, however, Pierce and Ogle (2017) considered that the role of all recorded visitors in the reproductive ecology was not clear.

Pollination by small rodents is recorded for Central American Melastomataceae and South African geoflorous species of *Protea* (Lumer and Schoer 1986; Wiens and Rourke 1978; Zoeller *et al.* 2016); the latter possess cryptic blossoms obscured from external view by overhanging foliage. Introduced rats are apparent pollinators of the Hawaiian, bird-adapted *Freycinetia arborea* and rodents visit the flowers of the small understorey tree *Helicia niligirica* (Proteaceae) in the Western Ghats of India (Devy and Davidar 2003; Proctor and Yeo 1975). Pollen loads on the fur of non-flying mammals (e.g. *Rattus tunneyi, Petaurus norflcensis*) caught on *Banksia* species in northern New South Wales had variable pollen loads on their fur or in their faeces (Griffin *et al.* 2009). The pollen loads on their fur were comparable to nectarivorous birds sampled from the same plant species; the inflorescences of *Banksia ericifolia* exposed to nocturnal pollinators had consistently high fruit set (Hackett and Goldingay 2001).

In Australian rainforests, there is a relatively large specialist arboreal mammal assemblage in tropical north-east Queensland (e.g. *Dendrolagus bennettianus, Pseudocheirus archeri, Phalanger orientalis* [Strahan 1983; Winter 1988]). Although the role of these species in pollination is unknown, a level of incidental pollination (as suggested for arboreal mammals in Oriental rainforests) would not be unlikely. In Australian subtropical and temperate rainforests the variety of arboreal mammals is small (e.g. *Trichosurus caninus, T. vulpecula, Pseudocheirus peregrinus* [Strahan 1983]), and none is a rainforest specialist (Adam 1987, 1992). It is unlikely that non-flying mammals are significant pollinators in southern rainforests.

POLLINATION BY REPTILES (SAUROPHILY)

Reptiles are a group of organisms that may represent some of the oldest agents of biotic pollen transfer for they, and their dinosaur kin, were present during the dawning of the angiosperms.

There are more known instances of pollination and flower visiting by present-day reptiles than might be surmised (Valenciá-Aguilar *et al.* 2013). The nectar-feeding lizard *Podarcis lilfordi*, a member of the family Lacertidae found on the Spanish Balearic Islands where it is the true pollinator of *Euphorbia dendroides*, achieves higher rates of fruit- and seed-set than do insect visitors (Perez-Mellado *et al.* 2000; Traveset and Sáez 1997). *Podarcis lilfordi* is only one member of a diverse assemblage of lizards that visit flowers and include nectar, pollen, individual floral parts, whole flowers and fruit in their diet. As consumers of fruit, reptiles influence seed dispersal and subsequent seedling establishment (Godínez-Álvarez 2004; Olesen and Valido 2003). There are strong arguments that plant–reptile interactions have affected the evolutionary development of flowers and fruit.

For several plants, lizards are important keystone dispersers of seeds and play a central role in their reproductive ecology, particularly on some islands; which tend to have highly generalist pollinators and plant–pollinator complexes (see Olesen *et al.* 2002). In insular environments, reptiles seem to interact with a greater number of plant species. Lizards may also be keystone pollinators. On the Ile aux Aigrettes, Mauritius, the gecko *Phelsuma ornata* has been suggested as capable of carrying pollen of nearly 80% of the island's plant species (Nyhagen *et al.* 2001). On Mauritius *Phelsuma ornata* is a possible pollinator of the rare *Nesocodon mauritianus* (Campanulaceae) and, with the related *Phelsuma cepediana*, is a pollinator of *Trochetia boutonia* and *Trochetia blackburniana* (Hansen *et al.* 2006). *Phelsuma cepediana* is the sole pollinator of the endangered *Roussea simplex* (Rousseaceae) on Mauritius (Hansen and Müller 2009). However, reptiles may also eat pollen, as for example, the

Seychelles Islands species *Phelsuma sunbergi*, which consumes that of the palm *Lodoicea maldivica* (Noble *et al.* 2011). Like the scenario for thrips pollination, in which pollen is consumed in return for pollen that is carried, this is a price paid for pollen transport and potential pollination by flower-visiting reptiles.

Further to the herbivory records cited above there are additional examples of lizards known to eat fruit, flowers and or nectar. In Cordylidae, *Platysaurus capensis* eats fruit of *Ficus cordata*; in Gekkonidae *Hoplodactylus* consumes nectar, flowers and fruit, *Naultinus* consumes nectar and fruit, and *Phelsuma* consumes nectar. In Igaunidae, *Iguana iguana* eats fruit, flowers and other plant parts, and *Liolaemus pictus* and *Tropidurus torquatus* consume fruit. In Lacertidae, *Gallotia* and *Podarcis* eat mainly flowers and fruit. As well as nectar from *Euphorbia dendroides*, *Podarcis liliford*i also consumes nectar of *Crithmum maritimum* and *Rosmarinus officinalis*. *Lacerta lepida* eats mainly beetles but when insect populations are low it may switch to *Capparis spinosa* and *Ephedra dystachia* fruit: the latter species is a member of the relictual Gnetophyta, not an angiosperm (Frohlich and Chase 2007). In Teiidae, *Aspidoscelis murinus* is a generalist feeder that includes flowers and fruit plus other plant parts in its diet, and consumes nectar of *Aloe vera*. *Teius teyou* consumes various fruit. In Scincidae, *Cyclodina* and *Leiolopisma* eat fruit and in Varanidae, *Varanus olivaceus* is a fruit-eater.

Australian records are few, almost non-existent, though the paucity of published observations does not reflect the actual extent of plant–reptile mutualisms. The Tasmanian skink *Niveoscincus microlepidotus* consumes nectar and petals of *Richea scoparia* (Ericaceae/Epacridaceae), the lizards gaining access to nectar by tearing apart the fused petals (calyptra). Olsson *et al.* (2000) found that *Niveoscincus microlepidotus* selected for *Richea* flowers with above-average nectar content, which suggested skink behaviour may have evolved in response to plant characteristics. *Richea scoparia* benefits from its mutualism with *Niveoscincus microlepidotus*, as the fused petal structure remains intact and attached unless torn apart. Experiments demonstrated that lizard behaviour resulted in a significantly greater release of seeds – 60 experimental plants with intact calyptra failed to release any seed. Thus the interacting behaviour of nectivorous *Niveoscincus microlepidotus* with the reproductive morphology of *Richea scoparia* seems to represent a mutualism that benefits both participants. In this case the skink does not serve as a pollinator. Rather, it facilitates pollination by removing the fused calyptra at a time suitable for pollination (since *Richea* grows in an exposed mountainous, cold and snow-prone environment), the subsequent exposure of reproductive organs allowing access by pollinators.

Scincidae, Lacertidae and Iguanidae are overwhelmingly fruit-eaters, and there is a high reported incidence of nectar-feeding in the family Gekkonidae (Perez-Mellado *et al.* 2000; Willson *et al.* 1996). With regard to pollination by lizards, there are few reliable observations that leave little room for untested speculation: *Hoplodactylus duvaceli, Hoplodactylus maculatus* and *Hoplodactylus pacificus* on *Metrosideros excelsa*, *Hoplodactylus maculatus* on *Phormium tenax* and *Rhopalostylis sapida*, *Phelsuma ornata* on *Gastonia mauritiana* and *Lomatophyllum tormentorii*, and *Podarcis lilfordi* from *Crithmum maritimum* and *Euphorbia dendroides*. All but *Podarcis* are members of the Gekkonidae. Though these lizards carry pollen upon their bodies they spend different amounts of time visiting flowers (as do all flower-visiting taxa), occur in significantly variable abundances on flowering plants and express different behaviours and movement patterns that influence the potential quality of their pollinator role. For example, *Podarcis lilfordi* may spend between one and five minutes in individual plants of the shrub *Euphorbia dendroides*, generally remaining less than a minute on each flower (Traveset and Sáez 1997). The geckos *Hoplodactylus duvaceli* and *Hoplodactylus maculatus* have been found in densities of seven or eight individuals per square metre of flower cover.

Pollen seems to adhere easily to the lizards' bodies, as their scales offer numerous opportunities for deposition and carriage. However, pollen may be more restricted in its location, for example on the ventral surface and in the snout and throat of *Podarcis lilfordi*, or only within the throat of *Hoplodactylus duvaceli*.

A last example of pollination by reptiles is appropriate as it draws our attention back to a peculiar instance of plant ecology – the flowering of a deciduous host plant before renewed onset of foliage growth. It comes from an article titled 'Little dragons prefer flowers to maidens: a lizard that laps nectar and pollinates trees' (Sazima *et al.* 2005) that appeared in *Biota Neotropica*. It is a joyous title that sits comfortably with Janzen's 'How many babies do figs pay for babies?' (Janzen 1979, in *Biotropica*) and 'Putting your sons in the right place: the spatial distribution of fig wasp offspring inside figs' (Zavodna *et al.* 2005, in

Ecological Entomology). The 'little dragon' is not a dragon at all, and is certainly not the agamid most would associate with the popular name. It is a skink, *Euprepis atlanticus*, found only on the Fernando de Noronha Archipelago off north-east Brazil. *Euprepis* does not frequent flowers of plants close to the ground (the stratum of all other records). It is an adept climber and will ascend up to 12 m into the trees of yellow-orange flowered *Erythrina velutina* (Fabaceae) to imbibe the large volumes of dilute nectar in the base of its corollas. Between feeding bouts *Euprepis* sometimes basks upon inflorescences – a possible thermoregulation strategy – and will descend to the ground to forage for insects. It is a general scavenger and forages mostly at ground level on food items ranging from the remains of its own eggs to animal corpses it can tear apart. *Erythrina velutina* flowers during the dry season when it is without foliage; the interest of *Euprepis* in nectar, in addition to that of supplying energy needs, may be related to the poor availability of fresh water during drought conditions on the islands. A single *Erythrina* flower may accumulate up to 106 µL of nectar in less than an hour. In seeking out nectar *Euprepis* clambers from flower to flower and tree to tree, collecting pollen on its scales as it moves about. Thus, through its behaviour and its ability to transport pollen, *Euprepis* is a potential pollinator. Experimental observations bear this out. Skinks have also been observed aggressively interacting with one another. This forces individuals to move between trees, which may be a mechanism for out-crossing.

Pollination of a deciduous tree by a skink endemic to off-shore islands in Brazil might at first be dismissed as having little to do with the pollination ecology of Australian rainforest plants. But the angiosperm genus *Erythrina* also occurs in Australian rainforests, and in subtropical and tropical Australia there are several deciduous trees that set flowers before the onset of foliage growth (see Chapter 3). Within the rainforest environment live several arboreal and semi-arboreal lizards, including skinks, agamids and geckos. A long and disparate bow? Perhaps we have overlooked a mechanism of pollination, ancient among vertebrates, that survives more widely than we know. As with flowers, 'conspicuousness' tends to be noticed, and ecologists are as likely to be beguiled by events as any creature given so fundamentally to observation.

This chapter on pollination syndromes is concluded with a small personal contribution. After days of observation in New South Wales subtropical rainforest, a small skink, possibly *Saproscincus galli*, was seen to climb with considerable agility over and between flowers of the rainforest herb *Pollia crispata* (Commelinaceae). It used its tail to grasp *Pollia* stems, an act well practised if measured by its agile hold, and on one occasion leapt more than 25 cm from leaf to branch. No small feat. In these movements, the skink made frequent contact with flowers (Williams and Walker 2003).

Such events and participants may contribute in only a minor way to the pollination of rainforests plants, but from grains of sand, beaches are made.

9 Pollination ecology of Australian subtropical rainforests: implications for the conservation of remnant communities

The area of Australian rainforest has been reduced from ~8 million hectares, before European settlement, to ~2 million hectares (Adam 1992). Most of the surviving rainforest is situated along Australia's eastern coast, and over 54% occurs in Queensland. Rainforest communities on the eastern coastal fringe frequently exist as disjunct remnants threatened by invasion of alien plants and animals, fire and drought, and by human impacts.

Although threats to rainforests are a major focus of Australian conservation concerns, remarkably little is known about many aspects of their ecology. Even though much of the Australian rainforest is now conserved in formal conservation reserves or through the operation of planning regulations, we lack much of the biological knowledge on which to base management programs.

It is frequently argued that conservation is better served by conserving communities rather than species. This is intuitively the case, but for rainforest we are not able to define 'community', or the interactions within a community, beyond a coarse scale. Efforts are centred on plants (e.g. Harden *et al.* 2004; Parkes *et al.* 2012), with the implicit assumption that if we get the plants right the rest of the community will sort itself out. But we have scarcely begun to test this.

Our knowledge of the reproductive ecology of Australian rainforest plants remains poor. Only a small fraction of the rich diversity of trees, shrubs, vines and herbs have been addressed, yet such knowledge is crucial to the preservation of fragmented communities and populations. The majority of Australian studies on these topics have concentrated on tropical Queensland forests, although more recently studies have investigated the ecological dynamics of species in southern latitudes (e.g. Boulter *et al.* 2011).

BACKGROUND

Pollination of Australian subtropical rainforest trees, particularly canopy species, is principally achieved by a taxonomically and ecologically diverse guild of insects comprising mainly flies (Diptera), beetles (Coleoptera) and bees and wasps (Hymenoptera). At the community level, and the level of individual tree species, generalist pollination systems appear to prevail. Trees with generalist pollination systems include dry-fruited species (e.g. *Tristaniopsis laurina, Alphitonia excelsa*) and those with fleshy or partially fleshy fruit (*Guioa semiglauca,* and *Syzygium smithii*). The abundance of individual pollinators varies and the membership of pollinator guilds, occurring on flowering trees at any one site, may fluctuate over space and time. Membership of pollinator guilds can vary dramatically over short distances and between conspecific flowering plants in different rainforest subformations. While guild variation may reflect some disruption to communities (or disruption of specific pollinator life histories), it is equally likely to be influenced by local environmental conditions, pollinator plant host and nest site availability and the mosaic-like distribution of rainforest biotas.

Worldwide, generalist insect pollination systems at the plant community level are probably universal. Their prevalence at the community level occurs in alpine floras, in deciduous, evergreen and cloud forests, in shrubland and in Northern Hemisphere chaparral and matorral communities. In Australia, generalist pollination by diverse insect guilds occurs in all rainforest formations irrespective of latitude and altitude, and is prevalent in sclerophyllous plants and the Australian Myrtaceae generally (G. Williams 2020; Williams and Adam 2010). Thus, the dominant pollination system in Australian subtropical rainforest subformations does not appear to represent a substantial departure, other than the obvious one of participating taxa, from the generalist theme recorded from a diversity of ecosystems elsewhere.

Dependence on pollinators is greatly influenced by the nature of plant breeding systems. Pollinator dependence is particularly high in biotically pollinated dioecious and self-incompatible species, or where ecosystem survival may be linked to keystone pollinators (e.g. Cox *et al.* 1991). Australian subtropical rainforests include several tree species which are principally vertebrate-pollinated, and an unknown proportion of species that may be at least facultatively wind-pollinated. Though such systems appear to be relatively uncommon, some rainforest trees that are visited by small insects may also be cryptically wind-pollinated. The contribution of vertebrates to pollination in tropical regions is well recorded, and taxonomically narrower pollination mutualisms by insects have been extensively documented in tropical rainforest communities. Whether the occurrence of these narrower plant–pollinator mutualisms represents a truly greater proportion of specialist pollination systems in tropical rainforests, as opposed to subtropical rainforests, or is simply an artefact of the choice of the rainforest plants studied, remains to be investigated.

Pollination in the canopy layer of subtropical rainforest formations is predominantly generalist but there is a higher incidence of specialist, or interactively narrower, plant–pollinator relationships in the understorey (G. Williams 1995, 2020). Plants participating in these pollination mutualisms are small trees, shrubs, herbs and some epiphytes, rarely vines. Vines, which are generally gap phase and ecotonal species not confined to individual forest strata, include species with specialist (e.g. *Maclura cochinchinensis* [Moraceae] and *Smilax glyciphylla* [Smilacaceae]) and generalist (e.g. *Cissus antarctica, Cayratia clematidea* [Vitaceae] and *Parsonsia purpurascens, P. straminea* [Apocynaceae]) pollinator relationships. With regard to the influence of population density on pollen flow and plant fecundity, understorey species tend to have much larger populations than those of canopy trees but may still be vulnerable to reproductive failure if populations are fragmented (House 1993) or occupy disadvantageous microlocations (Englund 1993; Handel and Peakall 1993). An increased incidence of specialist plant–pollinator mutualisms in the understorey has been reported from tropical rainforests worldwide (e.g. Armstrong and Irvine 1989; Janzen 1971; Roubik 1989). Vertical stratification of the rainforest plant community, and plant membership within individual strata, may have a strong influence on the nature of the pollination strategy employed by individual plant species. However, many small subtropical rainforest communities, particularly in coastal New South Wales, have been subject to structural modification or total removal of the understorey vegetation, to provide recreation facilities or parkland settings (e.g. Khappinghat National Park) (G. Williams 1993). In addition to destroying whole components of the rainforest ecosystem and causing the piecemeal extinction of ecological interactions and microhabitats, such practices probably impact more widely on the pollination ecology of remnants.

Several subtropical rainforest plant species with specialised plant–pollinator mutualisms (e.g. *Myrsine variabilis, Myrsine howittiana, Wilkiea huegeliana, Breynia oblongifolia, Eupomatia laurina, Smilax glyciphylla* and *Maclura cochinchinensis*) are not restricted to primary rainforest; they also occur in dynamic rainforest–wet sclerophyll forest ecotones (e.g. *Eupomatia laurina*). Several are known to coppice (depending on fire intensity and plant size), and thus re-establish by vegetative means following fire damage (Table 9.1). Whether species gain an ecological advantage by participating in a specialist pollination mutualism remains to be investigated, but the incidence of close mutualisms in various ecotonal plants suggests that pollinator-specific relationships are successful in diverse environments.

Unspecialised floral syndromes have been viewed as an adaptation to highly variable pollinator assemblages (Primack 1978), but the low populations of many Australian subtropical trees in relatively intact rainforest stands, the high species-richness of complex subtropical

Table 9.1. Examples of subtropical rainforest trees, shrubs, herbs and vines that regenerate from coppice growth following bushfire (Adam and Williams 2001; G. Williams pers. obs. 2019–2020).

For a discussion of the plant communities below see G. Williams (1993).

Family	Species	Family	Species
Aphanopetalaceae/ Cunoniaceae	*Aphanopetalum resinosum*[3]	**Monimiaceae**	*Wilkiea huegeliana*[2]
Araliaceae	*Polyscias elegans*[2]	**Moraceae**	*Ficus ?macrophylla*[3] *Maclura cochinchinensis*[2] *Streblus brunonianus*[3] *Trophis scandens*[2]
Asparagaceae/ Laxmanniaceae/Dracaenaceae	*Cordyline stricta*[3]	**Myrsinaceae/Primulaceae**	*Myrsine variabilis*[1]
Bignoniaceae	*Pandorea pandorana*[2]	**Myrtaceae**	*Backhousia sciadophora*[3] *Lophostemon confertus*[1] *Rhodamnia rubescens*[1] *Rhodomyrtus psidioides*[2] *Syzygium oleosum*[2] *Syzygium smithii*[1,2]
Cannabaceae/Ulmaceae	*Trema tomentosa*[3]	**Oleaceae**	*Notelaea longifolia*[1,2,3]
Capparaceae	*Capparis arborea*[3]	**Phyllanthaceae/Euphorbiaceae**	*Breynia oblongifolia*[2] *Bridelia exaltata*[2] *Glochidion ferdinandi*[1]
Colchiaceae/Liliaceae	*Tripladenia multiflora*[2]	**Pittosporaceae**	*Hymenosporum flavum*[1] *Pittosporum multiflorum*[2,3]
Cunoniaceae	*Schizomeria ovata*[1]	**Proteaceae**	*Banksia integrifolia*[2]
Ebenaceae	*Diospyros australis*[1]	**Putranjivaceae/Euphorbiace**	*Drypetes deplanchei*[2,3]
Euphorbiaceae	*Alchornea ilicifolia*[3] *Baloghia inophylla*[2] *Croton verreauxii*[1] *Mallotus philippensis*[1,3]	**Rhamnaceae**	*Alphitonia excelsa*[2]
Eupomatiaceae	*Eupomatia laurina*[1]	**Rubiaceae**	*Cyclophyllum longipetalum*[2]
Fabaceae	*Austrosteenesia blackii*[3]	**Rutaceae**	*Acronychia imperforata*[2] *Acronychia oblongifolia*[1] *Melicope micrococca*[1]
Flacourtiaceae/Salicaceae	*Scolopia braunii*[2]	**Sapindaceae**	*Arytera divaricata*[2] *Cupaniopsis anacardioides*[2] *Elattistachys nervosa*[3] *Guioa semiglauca*[1,2] *Jagera pseudorhus*[2] *Mischocarpus pyriformis*[2]
Lamiaceae/Verbenaceae	*Clerodendrum ?tomentosum*[2,3]	**Sapotaceae**	*Pouteria australis*[2,3]
Lauraceae	*Cryptocarya microneura*[1]	**Smilacaceae**	*Smilax australis*[2,3]
Malvaceae	*Hibiscus heterophyllus*[3]	**Solanaceae**	*Duboisia myoporoides*[1]
Meliaceae	*Dysoxylum fraserianum*[2] *Melia azedarach* var. *australasica*[1] *Synoum glandulosum*[1]	**Vitaceae**	*Cayratia clematidea*[2] *Cissus antarctic*[2] *Cissus sterculifolia*[2] *Tetrastigma nitens*[3]
Menispermaceae	*Sarcopetalum harveyanum*[2] *Stephania japonica*[2]		

Fire event: 1. Lansdowne State Forest (31°44′54′S, 152 ° 32′56′E), 1994, (lowland subtropical rainforest-wet sclerophyll forest); 2. Harrington (31 °52′06′S, 152 °42′03′E), Nov-Dec, 2019, (littoral rainforest on sand); 3. Breakneck Lookout/Kiwarrak State Forest (32 ° 00′17′S, 152 ° 25′02′E). Nov-Dec 2019, (dry rainforest on steep rocky scree slope).

rainforests (e.g. Sea Acres Nature Reserve, northern New South Wales) and the spatial and temporal variation in the incidence of flowering may have acted as a barrier to the evolution of specialist mutualisms, especially if populations and individual rainforest stands are prone to high extinction or recruitment rates over geological time-spans. Factors that have historically driven such episodes of rainforest expansion, contraction or total extinction include variation in temperature and rainfall patterns, sea level rise and retreat, and the localised influence of fire and seasonal water availability. These can be linear and progressive (at least for protracted periods), broadly curved oscillations of recruitment and loss that are gently extended in time, or cataclysmic events that come and go with varying degrees of amplitude (punctuated by geologically measured time-spans of relaxation and predictability) and frequency.

'Community' has an abstract meaning and a concrete one: rainforest communities in the abstract sense may be long-standing, but in the concrete sense may be relatively young. When we look at rainforest we gaze upon an assemblage of species that includes those which are ecologically non-functional (failing to reproduce and for all practical purposes already extinct, though still standing), others that are in the throes of successful colonisation, those that are tinkering at the extent of their distribution owing to a variety of environmental factors (e.g. soil moisture and temperature gradients), and those which are core species capable of recruiting to and surviving in a broad floristic or geographical range of rainforests. Many areas of rainforest in tropical Queensland and maritime New South Wales, rather than representing long-standing communities, are less than 10 000 years old. The full establishment of rainforest on a new site may take up to 1500 years, although it can be much quicker. Given the longevity of numerous trees, relatively few generations may have cycled between the time of their establishment and our encountering them in a given rainforest community. We have a fundamentally different understanding of time, when we measure our fleeting experience of the world against that of the forest in which we stand. And while rainforest communities (if left undisturbed) may move towards achieving a sense of structural climax, for most individual plant species, and therefore the floristic diversity of the community, *change* seems to be the usual circumstance. What, therefore, is the fate of pollinators in a fluctuating plant–pollinator environment? If links are broken, how are new ones founded in periods of rainforest and plant re-establishment, particularly when rainforest is renewed from scratch? (Williams and Adam 2010). A prospect many of the 2019–2020 fire- and drought-impacted stands in south-eastern Australia will now have to face.

In the generalist-dominated subtropical rainforest canopy there are no obvious ecological barriers to the movement of pollination vectors. Potentially, individual members of the pollinator guild can move within and across the tree community, their visits to flowers largely constrained only by local climatic conditions, phenological patterns, floral resource availability and vector life history. Individual trees within populations, and individual tree species within communities, may flower intermittently (flowering events sometimes being separated by many years) and asynchronously within seasons, but whenever flowering occurs a pool of potential pollinators will be present.

Plant species without obligate requirements for specific pollinators may be able to track spatial shifts in the availability of microhabitats (Thomas 1994) independent of pollinator responses. Benefits of generalist pollination systems are that some pollinator activity, and concomitant pollen flow, can still occur when weather conditions are not optimal for all components of the pollinator guild. Native bees and large beetles, for example, generally cease to forage during cold periods. While the influences of diel climate patterns may be less significant in lowland tropical rainforests, subtropical rainforests are subject to more variable weather patterns. Plants visited by a diverse pollinator guild may occupy a wide geographical and ecological range (the case for many trees in subtropical rainforest) but the selection pressure resulting from visitation by a taxonomically and ecologically diverse native pollinator guild may preclude the establishment of specialist plant–pollinator mutualisms. There will be limited opportunity for a single taxon to drive evolution of a specialist relationship. Where plants are served by guilds of pollinators that vary in a spatial and temporal mosaic, the unpredictability of the variation may constrain the establishment of specialist mutualisms. Even in specialist relationships (e.g. *Wilkiea huegeliana* – *Thrips setipennis* [Thysanoptera]) the relationship may be asymmetrical, the plant being pollinated by a single vector but the pollinator visiting numerous kinds of flowering plants (Williams *et al.* 2001).

Taxonomically diverse and fluctuating pollinator faunas have been hypothesised to act as an obstacle to the further evolution of floral characters (Petanidou and

Ellis 1993). This may or may not be proven. Natural experiments are long-running and require a measure of hindsight spanning unnumbered human generations. Whether a dynamic pollinator guild may select for the dominance of generalist entomophily in subtropical rainforest communities, or whether the prevalence of entomophilous flowers selects for a diverse pollinator guild is a 'chicken and egg' problem. However, the abundance of individuals and species in the pollinator guild on a single flowering subtropical rainforest tree, and their variation in pollination effectiveness, may defy investigation of co-evolution and quantification of selection pressure. Plant reproduction strategies may have already resulted in, or reflect, the loss of high-risk pollinators from the rainforest pollinator guild. Equally, it has been proposed that self-pollination could have evolved rapidly in response to pollinator failure and that selection may have altered floral structure and mechanisms to allow visits by alternative pollinators (Bond 1994). The fossil archive, no matter how perfect, can never adequately speak of such events. Nor can it do more than hint at discarded interdependencies or partners whose moment has gone – few discovered fossils suspend flowers and their attendants in a moment of embrace.

In predominantly generalist pollination systems, polylectic pollinators (pollinators which visit many kinds of plants) may not reliably recruit to individual flowering trees due to lack of a dominant entrainment cue. Trees must compete for pollinators in an environment in which pollen flow, owing to vector behaviour, may be largely non-directional. Pollinators potentially fly off in all directions, stumbling from tree to tree and from flower to flower. These chance patterns of pollinator visitation usually result in considerable variation in plant reproductive success. Consequently, rare or uncommon tree species, and widely separated and thus isolated individual trees, may be unable to compete against the attraction stimuli of more numerous or beguiling mass-flowering species, and may suffer reduced fecundity. This potential for lowered reproductive success, due to competition for polylectic pollinators in an uncertain pollination environment, may be an underlying causal mechanism contributing to the high extinction rates recorded among tropical rainforest trees generally (Primack and Hall 1992), although reduced fecundity in species may not make the system unviable as a whole (Bullock 1994). Rather, there may be a lot of turnover or ecological sifting of species, the rainforest community remaining intact. However, if flowering (phenological) patterns in tree populations and individuals are typically heterogeneous (see Newstrom *et al.* 1994), such that entrainment or recruitment cues are not temporally or spatially predictable, then pollinator/pollen movement and subsequent reproductive success may not be precarious. The ecosystem will function, over geological time, to accommodate the potential for random pollen movement. Efficient pollination, resulting in a high level of plant fecundity, may be uncommon in rainforest due to the spatial and temporal complexity of plant populations and flowering episodes.

IMPACTS OF FRAGMENTATION AND CONSERVATION OF REMNANTS

Human occupation of Australia has modified the nature, extent and connectivity of rainforests in the surrounding landscape matrix (Adam 1987; Laurance and Bierregaard 1997; Pavlacky *et al.* 2015). Even in rainforests that are structurally intact, high levels of alien plant invasion may have occurred within the canopy, along exposed margins and throughout the understorey (e.g. Fensham *et al.* 1994). Historically, natural rainforest remnants have acted as biotic refugia capable of expansion upon the return of conducive climatic conditions (Schneider and Moritz 1999; Truswell 1990). Two hundred years ago the distribution of rainforest in New South Wales, and in many other areas of eastern and northern Australia, was largely a reflection of topography (that affected rainfall and fire), climate and soil. Rainforest has become much more fragmented since European settlement, even though some formations (littoral or maritime rainforests, monsoonal rainforest and vine thickets) were already highly disjunctive and spatially heterogeneous (e.g. Floyd 1990; Russell-Smith and Lee 1992). Although we now have a broad understanding of the history of Australian rainforest, the detailed histories of individual sites and individual taxa are likely to remain elusive.

Many of the present stands, particularly monsoonal thickets in northern Australia and floodplain and littoral rainforests in New South Wales, are typically no larger than several hectares. These frequently survive within an increasingly altered (often cleared, heavily grazed or frequently burnt) vegetation matrix. Effectively, they are islands in a sea of agriculture, coastal urban development or manipulated savannah; the continuum of rainforest, wet sclerophyll and dry sclerophyll forest has been completely disrupted.

The major loss in area must have resulted in extinction of discrete populations at the local scale, in altered flow of genetic material and in reduction of the genetic component of biodiversity. However, to the best of our knowledge it was not accompanied by total broad-scale extinction of vascular plant species or vertebrates. Although the size of some rainforest fruit and the relictual nature of their present dispersal mechanism (e.g. *Idiospermum australiense, Endiandra globosa, E. floydii* and *Pouteria australis*) suggests loss of an original (maybe megafauna) seed disperser (e.g. G. Williams 1993; Rossetto *et al.* 2015; also Rule *et al.* 2012), we have insufficient knowledge of the present biota, let alone that of the past, to draw conclusions about loss of invertebrates or non-vascular plants.

The rainforests studied in subtropical coastal Australia are all remnants of post-European settlement, forest-clearing, sand-mining (for rutile, zircon etc.) and logging (e.g. Adam 1992; Bradshaw 2012; Chisholm 1925; see White and Vesk 2019 for a study of fire and logging impacts on the 'rainforest' understorey assemblages in wet sclerophyll forests, and Hickey 1994 for a study of post-logging and post-fire regeneration in Tasmanian old growth wet sclerophyll forest). It is important to note that use of fire by Aboriginal and Torres Strait Islander peoples also altered the extent and nature of rainforest within a matrix of sclerophyll vegetation (e.g. see Plate 117) (Benson 1991; Rule *et al.* 2012). Many remnants are, or at least appear, capable of regeneration in a broad sense via advance of boundaries or seedling recruitment of some canopy species on the forest floor (Kanowski *et al.* 2009; Tng *et al.* 2012). In broad community terms the regeneration process (a function of pollination and seed recruitment) is being sustained, although individual species may locally fail to recruit or regenerate; or may re-establish only over long time spans (White and Vesk 2019). This suggests that, though individual plants and animals may be vulnerable due to habitat and population fragmentation (e.g. Laurance and Bierregaard 1997; Thomas 1994), or disruption of biological processes such as pollination and seed dispersal (Rossetto *et al.* 2015), fragmented and spatially heterogeneous subtropical rainforest communities are potentially viable over long timespans (climatic conditions allowing). However, the remnants and mutualisms may be affected by exotic animals and plants, pathogens (e.g. Myrtle rust *Austropuccinia psidii,* root-rot pathogen *Phytophthora cinnamoni*), pesticide and herbicide drift from adjacent lands, edge effects, isolation of microhabitats, and mismanagement by landholders (Bond 1994). How individual rainforest stands, and individual plant and animal species within stands (especially endemic 'short-range' taxa and ecological specialists) respond to extreme events of fire and drought remains uncertain. Although drought-induced impacts on plant survival can be dramatic (e.g. crown collapse and death of *Archontophoenix cunnighamiana* palms, mortality of *Cyathea* tree ferns [G. Williams pers. obs.]), nevertheless rainforest trees can be so long-lived that the impacts of land use, loss of vegetation connectivity and modification of microclimate on rainforest communities, plants and associated fauna may be masked for very long periods. As Stride *et al.* (2018) note, the longevity of trees in 'non-equilibrium state' fragments and populations may mask unfolding extinction events (e.g. loss of seedling recruitment) for centuries. Some plants with specialist pollination mutualisms (e.g. *Hibiscus splendens* and *Hibiscus heterophyllus*) can exhibit high recruitment levels, but the point at which fragmentation of specific populations will result in the breakdown of reproductive strategies, and ultimately population extinctions, is unknown.

Significantly, several subtropical rainforest remnants (those remaining from clearing for farming or sandmining) that have been subject to studies of stand dynamics generally have a higher proportion of animal-dispersed species (with fleshy fruit) (Evans 1993). Dry-fruited species are more vulnerable to localised extinction. This indicates that, although impacts on the pollination ecology of remnant mass-flowering canopy tree populations are not apparent (due to the great diversity of taxa, the subtlety of interactions, and the absence of base data and the longevity of individual trees), there is considerable disruption to diaspore dispersal ecology and patterns of within-stand species recruitment and extinction (Evans 1993). Extinction of individual plant species (as appears likely for *Rhodomyrtus psidioides* [Makinson 2018; G. Williams 2018]) may have much broader impacts on pollen movement and subsequent plant fecundity, and reduce the diversity and composition of the pollination guild (if larval or adult pollinator foodplants are lost). Where species are part of a sequentially available flowering suite and provide temporally complementary floral resources, their localised extinction may have substantially broader community impacts on the sustainment of pollinator (and frugivore) populations.

Being able to identify plants and mutualisms at risk has advantages to conservation managers, though the failure of pollinator and seed-disperser relationships will only affect extinctions when plant populations are constrained by the plant's dependency on adequate levels of seed dispersal, seed stores and germination. Plants with a high risk in one aspect of their reproductive ecology, such as a requirement for a single pollinator, may compensate by a low risk in other aspects (Bond 1994). For example, they may be able to reproduce vegetatively or utilise members of a broad seed-dispersal guild, or achieve dispersal through a range of biotic (e.g. birds and ants) and abiotic (wind, water and gravity) means. Factors that provide a conceptual assessment index for identifying a plant's vulnerability to extinction include the probability of the plant–pollinator or disperser mutualism failing, the degree of reproductive dependence on the mutualism and the importance of seeds in the plant's demography (Bond 1994).

The impact of rainforest fragmentation on plant breeding systems and biotic communities generally is important in the design of conservation reserves, although reserves are normally based on what is available rather than on theoretical design considerations. Maintenance of biodiversity and the preservation of ecosystem processes are also vital. Is the artificial fragmentation of rainforest populations (through logging and clearing, imposition of roads etc.), and subsequent impacts on plant fecundity, a significant disturbance given the natural spatial disjunction of rainforest tree populations, the rarity of many tree species in primary (undisturbed) complex rainforest and the non-equilibrium (unstable) dynamics of rainforest ecosystems generally? Smith and Mayfield (2015) noted that small tropical forest fragments can act as reservoirs of dipteran diversity (as one example/surrogate indication of taxon response), but for some fly functional groups, such as pollinators and predators, small forest fragments and their edges have lower functional redundancy than large forest patches and their edges; though dipteran diversity was more positively associated with local forest cover than patch size alone. In a later paper Smith and Mayfield (2018) observed that in small rainforests fragments, compared with larger ones, there was lower taxonomic diversity of bees visiting flowers of trees, and that bees visiting trees in small fragments also had higher mean body sizes than those in larger fragments. They considered that this suggested small-sized bees may be less likely to persist in small rainforest fragments. Smith and Mayfield (2018) also found reductions in the abundance of eusocial stingless bees visiting flowers in small fragments compared to large ones, their results suggesting that pollinator visits to native trees living in small tropical forest remnants may be reduced (with consequent reductions in plant reproductive success).

Highly dynamic plant recruitment patterns are not necessarily restricted to small rainforest stands or to artificial remnants. For example, high mortality rates for uncommon and rare trees have been recorded in Sarawak lowland rainforests, where ~35–50% of total species recorded at any location were represented by a single individual (Primack and Hall 1992). In the same forests, a very high proportion (approaching 50% at one location) of tree species were recorded only as adults, not as saplings, which indicates lack of recruitment and regeneration. For ~12% of species in the same Sarawak study plots became extinct over a 20-year period. Such instability in the floristic composition of individual rainforest stands must influence the establishment and retention of specialised plant–pollinator mutualisms.

The capacity of plants to be pollinated by a generalist fauna (particularly if pollinators can be recruited from a broader regional forest pool), to self-pollinate or to reproduce vegetatively may permit species to persist in forest remnants where pollination is unreliable or not constant (Bond 1994; Williams and Adam 2019). Several Australian subtropical rainforest species (e.g. *Alchornea ilicifolia* [Euphorbiaceae], *Melia azedarach* var. *australasica* [Meliaceae] and *Eupomatia laurina* [Eupomatiaceae]) are able to regenerate vegetatively or reproduce by agamospermy. These faculties are useful in sustaining populations following disturbance (e.g. from fire, storm damage or damage from adjacent tree collapse) but may play no role in founding events and do not allow new genetic combinations.

Reproduction by sexual means is the norm and, for most species, pollination will be required to maintain populations. However, while a high proportion of rainforest trees may be facultative or obligate outbreeders and pollen movement between individuals and populations is required to maintain genetic variation, many rainforest trees show potential for self-pollination. This may reduce the requirement for pollinators that forage over long distances, as well as reducing the risk of extinction (Bond 1994). Where species have the potential to self-pollinate the need for pollinators diminishes, and the reduced distance over which pollen needs to travel can be met by the multitude of generalist insect pollinators.

Managing rainforest reserves to preserve or sustain a diverse pollinator guild can be critical to the long-term viability of individual plant populations and the rainforest ecosystem in general. At a fundamental level, the promotion of pollinator diversity can be enhanced by retaining floristic and structural diversity in the remnant, by re-establishing connectivity between isolated fragments and by the encouragement or retention of vegetated buffers on adjacent lands. Plant reproductive success can be enhanced by supplementary plantings of the species (particularly obligate out-crossing species) surviving at low population levels within remnants, and by removing exotic plant species that may compete for native pollinators (e.g. the introduced *Senna pendula* which potentially competes with the threatened native *Senna acclinis*). A diverse pollinator guild may be important in that it provides flexibility and resilience in breeding systems in the face of environmental change. However, a reduction in pollinator guild diversity may result in the loss of pollinator subsets required for pollination of particular plants. A potential threat to pollinator guild diversity and flexibility in plant reproductive ecology is posed by introduced *Apis mellifera*, a species widespread in Australia and capable of dominating flower-frequenting native insect faunas in fragmented forests. Yet feral honey bee populations are present in all Australian rainforest types, and are likely to remain entrenched members of the insect fauna that visits the flowers of the greater majority of rainforest angiosperms. Frequently, they are the only readily observed insect visitors.

Plants with obligate single pollinator mutualisms will be particularly vulnerable to extinction if the pollinator is lost. Species in this category may include *Wilkiea huegeliana* and *Eupomatia laurina*, and though individual *Eupomatia laurina* plants can reproduce vegetatively this does not permit establishment of additional plants from adventitious root stock. Such extinctions may lead to the loss of other members of the rainforest biota. For example, *Wilkiea huegeliana* is the only foodplant of the rare and endemic regent skipper butterfly (*Euschemon rafflesia rafflesia* Hesperiidae) (Braby 2000) in the southern part of its range.

The conservation of rainforest remnants requires some exchange of genetic material, normally in the form of seeds, between spatially isolated remnants (Evans 1993; Russell-Smith and Lee 1992). Ecological functions in individual rainforest isolates do not occur in a vacuum, so conservation efforts and management practices need to address regional remnants as a collective interactive unit. Management strategies need to consider the influence of the surrounding landscape matrix and the re-establishment of connectivity between fragments and contiguous vegetation and look beyond the simple restoration of canopy tree cover. Knowledge of breeding systems and population structures, such as the need to retain sex balance in dioecious species, is also important.

The general tendency has been for conservation programs to be directed at single reserves, to the neglect of additional regional sites (Evans 1993). Such strategies reflect a poor understanding of plant reproductive ecologies and the dynamic, often transitory nature of floristic composition. They assume that biodiversity is replicated uniformly in each remnant, which it is not. Plant species in individual remnants frequently possess very low populations, often represented by few or single trees. Such species are acutely prone to local extinction due to windthrow, flood, fire or storm impacts. Species may also survive in rainforest remnants as spatially discrete (gregarious) clumps which are equally vulnerable to chance disturbance events. In these circumstances, managers need to consider whether enhancement by artificial plantings and manipulation of microhabitats is appropriate. Although small and isolated remnants are potentially viable over the long term, passive management is unlikely to achieve the long-term preservation of individual rainforest remnants and individual plant and animal species, or to retain existing interactive reproductive ecologies and pollination mutualisms; as was demonstrated in regenerating littoral rainforest remnants at Harrington, northern New South Wales, in late 2019, where dense, highly flammable, intervening *Leptospermum* (Myrtaceae) growth permitted inroads and destruction by an intense fire event. Re-establishment of connectivity and an increase in conservation reserve size is important to the preservation of small and fragmented communities because of the amelioration of edge effects and potential increase in species populations, and retention of the interactions between flora and fauna within individual reserves, are fundamental to the preservation of biodiversity.

FURTHER CONTRIBUTIONS TO THE DARK SIDE: FRAGMENTATION AND RISKS TO PLANT BREEDING SYSTEMS

How might a plant's breeding system, and the degree of specialisation it relies upon for pollination, be affected by rainforest fragmentation, and can breeding system and specialisation in plant–pollinator mutualisms be useful predictors of threat in a fragmented environment? Being able to predict a plant's risk of extinction is likely to assume increased significance in a world of unfolding climate change.

Aizen *et al.* (2002) investigated whether self-incompatible plants in fragmented habitats were more vulnerable to decline in pollination and reproduction success than self-compatible species, whether specialist plants were more susceptible than generalist-pollinated ones and whether self-incompatible species offset their expected greater vulnerability to fragmentation by tending towards more generalist pollination strategies. Their rigorous review of the literature suggested that no generalisations could be made about susceptibility. Though most species responded negatively to fragmentation, the pollination and reproductive success of self-compatible and self-incompatible plants seemed equally at risk of decline, the hypothesis that specialised species were more vulnerable than generalist ones was not upheld and there was no clear trend to offset differences in the strategies of self-incompatible and self-compatible species.

Pertinent also is the investigation of Honnay and Jacquemyn (2007). They discussed whether so-called common species were as vulnerable to habitat fragmentation as rare species: rare species have historically been the focus of attention. Common species have been assumed to be safe, or safer, from the threats of random genetic drift, increased levels of selfing (inbreeding) and cross-fertilisation of siblings (and sibling crosses with parents). They tested whether plant rarity and mating system, as well as plant longevity, might be potential moderators of the relationship between population size and genetic diversity. The most alarming conclusion was that common plant species were not only as likely as rare species to be at risk of genetic erosion from habitat fragmentation, they were potentially even more susceptible to loss of genetic diversity. In other words, there was no safety in numbers. This might greatly increase the number of plant species at risk of extinction in remnants that face ongoing fragmentation processes. Many fragments are unable to support plant populations that are too small to maintain a mutation-drift balance, and population isolation might be such that gene flow is insufficient to replenish lost alleles.

At this stage in our understanding, it cannot be assumed that most rainforest plants with generalist pollination and reproductive strategies, or those that appear common within a rainforest remnant, have an automatic level of safety.

In Australia numerous isolated rainforest remnants are now subject to artificial regeneration and corridor restoration projects, and these have made substantial contributions to re-establishing habitat connectivity and ecosystem integrity, extending boundaries of remnants and removing invasive plant species. However, a study by Kaiser *et al.* (2008) based on the island of Mauritius gives cause for concern. Less than 2% of the native forest of Mauritius remains, and this figure includes degraded forest invaded by alien plants. They studied the reproductive success of the threatened tree *Syzygium mamillatum* (Myrtaceae) in both restored and unrestored sites. *Syzygium mamillatum* is a cauliflorous species once pollinated by specialist nectar-feeding birds, however, these are locally extinct, *Syzygium mamillatum* now largely pollinated by the generalist endemic Gray white-eye *Zosterops mauritianus*. Contrary to their expectation trees in unrestored forest experienced greater visitation rates and increased levels of flower probing, than those in restored forest, and trees in unrestored forest experienced greater reproductive success. Though the initial number of buds per tree in the restored area was somewhat higher than that for trees in the unrestored forest, during the budding stage most trees in the restored forest suffered severely from the attack of lepidopteran herbivores, and fruit set and number of seeds per fruit were also lower than that in the unrestored area. The case is one of disruption to native plant–animal interactions caused by competition for pollinators between alien and native plants (Crichton *et al.* 2019; Moragues and Traveset 2005), in which the restoration strategy failed to take into account the wider ecologically impacts of the removal of invasive species. The findings demonstrate the importance in restoration projects of mimicking the original structure of habitats and the interactive structure (web) of interspecific relationships (Kaiser *et al.* 2008; Menz *et al.* 2011). This is not an easy task given our imperfect knowledge and the possibility that individual native species may have already become locally extinct and may have been replaced by exotic ones (as Cause *et al.* 2012 note for the pollinator role played by the invasive wasp *Vespula pensylvanica* on flowering Hawaiian *Metrosideros polymorpha* [Myrtaceae]). Nevertheless the study by Kaiser *et al.* (2008) illustrates the potential futility in restoring habitat conditions and floristic composition without incorporating information on ecosystem function, not least the dynamics of plant–pollinator relationships.

Appendix 1. Case studies of pollination in the Australian rainforest flora

CASE 1. THE FOREST FLOOR: MIXED HOVER-FLY (SYRPHIDAE) AND BEE POLLINATION IN *POLLIA CRISPATA* (adapted from Williams and Walker 2003)

In Australia, herbs in the genus *Pollia* (Commelinaceae) comprise two endemic species, *Pollia crispata* and *Pollia macrophylla*. *Pollia crispata* occurs in New South Wales and Queensland and *Pollia macrophylla* is restricted to Queensland (Hnatiuk 1990). *Pollia crispata* is found growing along sunlit stream banks in mountainous (montane) and subcoastal rainforest and wet sclerophyll forest. It can form dense spreading mats wherever soil moisture levels are dependably high throughout the year.

Pollia crispata is a perennial, annually flowering self-compatible herb. Once seed dispersal has occurred and population-founding events have taken place, individual plants are able to grow vegetatively from adventitious stem roots. Vegetative colonisation can be aggressive in moist sunny open locations, and plants can invade areas otherwise denuded of forest cover and achieve dense growth up to waist height. In shaded situations, plants progressively die back and they are generally absent from dimly lit areas below regenerated canopy cover.

The flowers of *Pollia crispata* are actinomorphic and bisexual, ~8–9 mm wide, in terminal inflorescences with petals (three) coloured white (Harden 1993). The sepals (also three) are greenish-white and the gynoecium is superior, with a long and narrow style. Each flower has three stamens and two or three long staminodes or non-functional stamens. Flowers appear not to produce nectar and there are no obvious nectar guides. The flowers open sequentially each morning and appear slightly protogynous. The perianth is open and shallow and does not restrict access. However, each petal can act as a landing platform if oriented downwards, in which case the lower petal is pseudolobellate. Pollen is bright yellow, dry and easily displaced from anthers and the pollen exine pattern is obviously sculptured with net-like (reticulate) patterns. The bright-coloured pollen appears to convey the main visual recruitment signal, or at least enhance the visual signal projected by the flower.

The flower structure of *Pollia crispata* allows access by a taxonomically diverse assemblage of pollinating insects. Patterns of flower opening and pollen removal suggest that the species is adapted to diurnal flower visitors. Flowers are pollinated by insects, but being self-compatible these can produce viable seeds independent of insect pollination. The composition of the associated pollinator fauna in New South Wales subtropical rainforest is dominated by syrphid flies, halictid bees and the apid bee *Tetragonula carbonaria*, but most of the insects that visit its flowers have a broader flower host range. In general, the insects visiting *Pollia crispata* flowers are restricted to the understorey and ground stratum of subtropical rainforest. Skinks (e.g. *Saproscincus*) may facilitate levels of pollination.

Most insects seem to visit flowers between early daylight and midday, and generally avoid visiting flowers during the night or during rain. The activities of flower visitors usually result in the removal of available pollen from flowers by early afternoon. The abundance of insects normally decreases when flowers are in full sun and insects tend to be absent from late afternoon onwards. Most insects visit shaded flowers and flowers in dappled sunlight, even when flowers are hidden by low overhanging foliage of other plants. However, there are exceptions, these being species of sun-loving hover-flies (e.g. *Melangyna damastor*, *Betasyrphus* spp. and members of the syrphid tribe Syrphini generally). In contrast to these sun-preferring syrphids, at least one hover-fly, a species of *Episyrphus*, generally visits only shaded and dapple-lit flowers. Numerous small and secretive bees (especially Halictidae) forage timidly at *Pollia crispata* flowers, and trap-lining bees are occasionally observed. Many flowering rainforest plants, irrespective of the strata within which they occur, have annual and daily variation in the number of visiting insect species. On some days flowers appear to be totally devoid of visitors, although in most years the fly genus *Episyrphus* and the bee genus *Lasioglossum* are the most commonly observed visitors. Visits by *Tetragonula carbonaria* tend to coincide with peak periods of flowering in individual *Pollia crispata* populations; this ubiquitous little bee is often totally absent when blossom numbers diminish later in the flowering season.

Pollen loads of *Pollia crispata* visitors (see Table A1.1) are partly related to the amount of time spent foraging, and visitors may carry only scattered grains upon their body. Some species may really be pollen thieves that actively clean away and consume pollen that has gathered on their bodies (e.g. the hylaeine bees *Amphylaeus nubilosellus* and *Hemirhiza melliceps*), or feed more directly on pollen grains still held upon the anthers (e.g. Syrphidae). The introduced honey bee *Apis mellifera* frequents the open flowers of *Pollia crispata* and in attempts to remove pollen occasionally agitates flowers by shaking, but it does not 'buzz' flowers. At least one native bee, the halictid *Lasioglossum musicum*, sometimes attempts buzz-collection of *Pollia*'s pollen, even though its anthers require no such manipulation to release the pollen. Many of the bees that frequent *Pollia crispata* flowers exhibit foraging constancy when the species is in flower. They show no interest in adjacent flowering plants even when these occur in the same forest stratum. Many foraging *Lasioglossum* bees remain upon individual flowers for some minutes, using their fore- and mid-legs to pack pollen into hind-leg scopae and setae on the undersurface of the abdomen (abdominal ventrites).

The syrphid–bee pollinator guild found on *Pollia crispata* can be compared with that of the Commelinaceae and ground stratum flora elsewhere. In Japan, small *Lasioglossum* bees and the beetle genera *Protaetia* (Scarabaeidae) and *Chlorophorus* (Cerambycidae) are recorded from the related *Pollia japonica* (Kato 2000). Halictid and syrphid-dominated plant–pollinator relationships are recorded for Australian ground orchids in the genus *Thelymitra* and the ground stratum more generally (Dafni and Calder 1987), and are probably pollinators of the rainforest understorey herb *Drymophila moorei* (Conran 1988). *Tetragonula carbonaria* is recorded from numerous flowering rainforest plants. *Commelina cyanea*, similarly a member of the Commelinaceae, is visited and pollinated by bees and syrphid flies. Its pollinators include small halictids and colletids, larger trap-lining solitary apid bees (e.g. *Amegilla pulchra*) and syrphid flies (e.g. *Syritta* spp.) Contrary to the actinomorphic open and white flowers of *Pollia crispata*, flowers of *Commelina cyanea* are bright blue and slightly zygomorphic, suggesting that flower shape and colour may be poor indicators of pollinator taxa. This is supported by the observations of Kato and Miura (1996), who recorded syrphid flies as the pollinators of blue-flowered actinomorphic *Commelina communis* and pink-flowered zygomorphic *Murdannia keisak* (also of the Commelinaceae) in a Japanese lowland marsh community. Equally, many syrphid species and genera are known to participate in adaptively specialised pollinator relationships with zygomorphic-flowered slipper orchids in tropical Asian rainforests. The participating syrphid genera include those, such as *Episyrphus* and *Betasyrphus*, involved in the pollination of the simple actinomorphic flowers of *Pollia crispata*, though obviously involving different species.

The known visitors to *Pollia crispata* (see Table A1.1) have broader flower associations in Australian rainforest and wet sclerophyll forest communities. *Episyrphus* occurs on the flowers of *Aneilema acuminatum* (Commelinaceae) and various Syrphini are attracted to the flowers of *Clematis glycinoides* (Ranunculaceae). *Melanogyna damastor* is encountered in sunlight wet forest margins on the New South Wales north coast on numerous low-flowering plants. In addition to the association with *Commelina cyanea* already noted, *Nomia aurantifer* is recorded from the flowers of the small mangrove tree *Aegiceras corniculatum* (Myrsinaceae/Primulaceae) and the mass-flowering canopy vine *Austrosteenisia blackii* (Fabaceae). *Lasioglossum musicum* visits the bramble *Rubus rosifolius* (Rosaceae) that grows on rainforest margins, *Lasioglossum atronitens* is attracted to the herbaceous *Dianella caerulea* (Asphodelaceae/Liliaceae) and the scrambling vine *Hibbertia scandens* (Dilleniaceae), and *Lasioglossum tooloomense* is a visitor to *Dianella caerulea*. *Lasioglossum polygoni* is recorded from the shrub *Senna acclinis* (Caesalpinioideae) and the subcanopy tree *Diospyros australis* (Ebenaceae) (G. Williams 1998; Williams and Adam 1997). *Amphylaeus nubilosellus* and *Hemirhiza melliceps* have been regionally recorded from several rainforest canopy and understorey trees and shrubs. The native bees *Amegilla pulchra, Amphylaeus nubilosellus, Hemirhiza melliceps, Lasioglossum atronitens, Lasioglossum convexum, Lasioglossum leichardti, Lasioglossum musicum, Lasioglossum polygoni* and *Tetragonula carbonaria* have ecological and floral host ranges not limited to wet forest types (Cardale 1993; K. Walker 1995).

Pollination of *Pollia crispata* (and potentially that of the additional native Commelinaceae mentioned here, *Commelina cyanea* and *Aneilema acuminatum*) differs

Table A1.1. Pollen loads on vectors visiting *Pollia crispata* flowers (after Williams and Walker 2003).

Taxa	Main sites of pollen lodgement	% home pollen
DIPTERA		
Bombyliidae		0
Geron nigrocciput		
Syrphidae: Syrphini		
Betasyrphus sp.	Face, pro-, meso- and hind femora, ventrites generally, lateral margins of basal metasomal segments, particularly sternopleura and mesopleura	100
Episyrphus sp./spp.	Ventral areas, thorax, fore, mid and hindlegs, inner ocular margins (comb-like long setae on mid femora)	>99–100
Melanogyna damastor	Thorax and forelegs generally	100
miscellaneous Syrphini spp.	Scattered grains on whole body surface and face, but particularly ventrites, anepisternites, sternopleural and pteropleural regions	~50–100 (0–2)
HYMENOPTERA		
Apidae		
Amegilla pulchra	Hind tibiae and femora	>50
Apis mellifera	Corbiculae – hindleg, scattered grains elsewhere	100
Tetragonula carbonaria	Corbiculae – hindleg, scattered grains elsewhere	100
Colletidae-Hylaeinae		
Amphylaeus nubilosellus	Scattered few grains on body and wings generally	100
Hemirhiza melliceps	Few (or no) grains scattered on body surface generally, or few grains restricted to meso- or metasoma	100
Halictidae		
Lasioglossum (*Austrevylaeus*) spp.	Ventral surface and legs generally, particularly hind femora and basal ventrites, propodeum and mesasomal metapleuron, ventral scopae of metasoma	~60–100 (0–2)
Lasioglossum (Callalictus) tooloomense	Ventral surface and hindlegs generally, especially hind femora	100
Lasioglossum (Chilalictus) convexum	Base of propodeum, metapleura, base of metasoma and mid legs, but particularly hindlegs	0 (2)
Lasioglossum (Chilalictus) polygoni	Ventral surface generally, but particularly propodeum, hindlegs and ventrites 1–2	<40–100 (2,?3)
Lasioglossum (Parasphecodes) atronitens	Hind femora, tibiae and tarsomeres, ventral surface of metasoma, base of propodeum, scattered grains on body surface generally	>99–100 (1,?2)
Lasioglossum (Parasphecodes) leichhardti	Base of propodeum, ventral surface generally, but particularly hindlegs and basal half of metasoma	>99–>90 (2)
Lasioglossum (Parasphecodes) musicum	Petiole, propodeum and mesosomal metapleura, basal segment of metasoma, ventral surface generally, hind trochanter, femora, tibiae and tarsomeres	>95–100 (2)
Lasioglossum sp.	Hind tibiae and tarsomeres, ventral scopae of metasoma	100
Nomia aurantifer	Scattered grains over whole body, but especially ventral surface generally, apex of metasoma, hind tibiae and hind femora	100
Formicidae		
Myrmecia nigrocincta		0
Myrmecia ?tarsata		0
REPTILIA		
Scincidae		
Saproscincus ?galli	Not determined	

Home pollen = *Pollia* pollen. Numbers in brackets indicate the range of other pollen species carried.

from the entomophilous pollination strategy of mass-flowering subtropical rainforest canopy trees. In the latter, pollination is primarily achieved by a diverse assemblage of ecologically generalist Diptera, Coleoptera and Hymenoptera, of which Syrphidae (especially Syrphini) and native Apiformes form only a small or inconsequential proportion (Williams and Adam 1997). The pollinator relationships found in *Pollia crispata* are more typical of those documented in the rainforest understorey, where plant–pollinator mutualisms are generally more specialised and pollinator assemblages on individual plants are less diverse, and groups such as bees are more prominent.

CASE 2. THE FOREST SUBCANOPY: BEE POLLINATION AND BUZZ-COLLECTION OF POLLEN IN THE THREATENED AUSTRALIAN SHRUB *SENNA ACCLINIS* (adapted from G. Williams 1998)

The genus *Senna* (family Fabaceae, subfamily Caesalpinioideae) worldwide comprises ~350 species of shrubs and herbs (Harden 1991). Fifteen species are recorded from New South Wales but six of these are exotic, introduced from the Americas and Africa, and have become naturalised to varying degrees.

Senna acclinis is a threatened native shrub listed as endangered on New South Wales *Biodiversity Conservation Act 2016*. It is restricted to the central and northern coastal forests of New South Wales, and Queensland. In New South Wales it is chiefly recorded from subtropical rainforest remnants and is now rare, owing to clearing of habitats. It is considered a gap phase species.

Senna acclinis (Plate 81) flowers are bisexual but at least partly protogynous, yellow in colour and ~30 mm wide at full anthesis (in female-phase flowers the corolla is only partly open and the stigma abuts the anthers; in male-phase flowers the corolla is fully open). Flowers are lightly but sweetly fragrant. The style is long, curved and narrowly tapered to the apex. The stigma is a light creamish green that turns brownish in colour and recurves away from anthers when no longer receptive. The stamens usually number 10 and immature anthers are approximately the same colour as the corolla. Anthers dehisce longitudinally along a central suture but pollen does not readily dehisce unless stamens are vibrated ('buzz' or vibratile dehiscence). Flowers do not possess obvious nectar rewards or nectariferous glands and there is no depth effect in the corolla. Consequently, floral morphology does not limit visitor access but the poricidal anthers restrict access to pollen.

The species is pollinated by bees that are able to buzz-collect pollen, though some bee taxa that have been observed to visit its flowers are likely robbers as they have not been observed to exhibit this behaviour. Visiting bee taxa include *Hylaeus turgicollaris*, *Amphylaeus nubilosellus* (Colletidae-Hylaeinae), *Lasioglossum polygoni*, *Lasioglossum victoriellum*, *Lasioglossum* (*Parasphecodes*) sp. (Halictidae) and *Exoneura* sp. (Apidae). All species have been recorded carrying *Senna acclinis* pollen, with *Lasioglossum polygoni*, *Lasioglossum victoriellum*, *Lasioglossum* (*Parasphecodes*) sp. and *Exoneura* sp. at times carrying large quantities of pollen principally massed on the sternum (ventral metasoma) and hind femora and tibiae (see Table A1.2). Other insects known to visit the flowers of *Senna acclinis* include the large myrmecine ant *Myrmecia nigrocincta* and the pompilid wasp *Phanagenia ?fasciata*. *Myrmecia nigrocincta* commonly forages for nectar from numerous flowering trees and shrubs in subtropical rainforest (Plate 93). Although *Senna acclinis* is apparently self-compatible, the inability of *Myrmecia nigrocincta* to appropriately release pollen would restrict its capacity to effect pollination.

Visitation rates of individual insects vary, but shaded flowers generally seem to attract only occasional hylaeines (e.g. *Amphylaeus nubilosellus*) and many shaded flowers are never visited. Only flowers in full sunlight are visited frequently. Individual flowers throughout the day are sometimes revisited. Time between bee visits to flowers ranges from less than one to 48 min, and time spent foraging at individual flowers may range from less than one minute to more than 12 min.

Amphylaeus nubilosellus, *Hylaeus turgicollaris* and *Lasioglossum polygoni* frequently (but not always) remove pollen from dehiscent anthers by landing on stamens then vibrating anthers by 'buzzing' their flight muscles. The duration of individual buzzing events is generally less than two seconds but the sound is quite audible over a distance of at least 1 m and is often the only indication that bees are present at flowers. In addition to bees being potentially dusted with pollen, buzz-removal of pollen results in pollen deposition on the inner surfaces of petals directly below the anthers. This pollen may become available for transport by other flower visitors. All three bees collect the pollen from the petal surfaces, so that in effect

Table A1.2. Pollen loads on insects visiting *Senna acclinis* flowers (after G. Williams 1998).

Taxa	Sites of pollen lodgement	% home pollen
HYMENOPTERA		
Apidae		
Exoneura sp.	Hind legs, ventrite 1, ventral surface generally	100
Colletidae		
Amphylaeus nubilosellus	Vertex, hind femora, abdominal ventrites, scattered grains on body surface generally	71–100 (1–2)
Hylaeus turgicollaris	Fore tibia, thorax	0–100
Halictidae		
Lasioglossum polygoni	Ventrites 1–2, hind femora and tibiae, thorax generally, base of wings, fore, mid and hindlegs	~60–100 (1–2)
Lasioglossum victoriellum	Frons, ventrites 1–2, hindlegs	>95 (1)
Lasioglossum sp.	Ventrites 1–6, hindlegs	>95–100 (1–2)
Formicidae		
Myrmecia nigrocincta	Mesotibia, mid trochanter, mesepisternum	100
Pompilidae		
Phanagenia ?fasciata		0

Home pollen = *Senna* pollen. Numbers in brackets indicate the range of other pollen species carried.

the petals function as secondary pollen presenters (see Willmer 2011). After collecting pollen, *Amphylaeus nubilosellus* and *Hylaeus turgicollaris* have been observed to rest on sunlit leaves and groom their bodies. Such behaviour is characteristic of many bee species, though it does not result in the removal of all pollen that has landed upon the body. Though Hylaeinae carry pollen in their crop, small amounts of pollen remain isolated on their bodies. That bees would groom following visits to previously buzzed flowers indicates that single bee visits result in only partial removal of pollen. This leads some researchers to suggest that restricted removal of pollen from flowers with poricidal anthers encourages pollen transport by greater numbers of pollinators and, in turn, acts to promote successful pollen dispersal.

Although *Senna acclinis* appears to be self-compatible it is dependent on a pool of pollinators capable of removing the generally indehiscent pollen. The bees recorded visiting *Senna acclinis* are generalists (though they may exhibit temporal or spatial foraging constancy to individual flowering species) and have been recorded from other plant taxa. The ability of *Senna acclinis* to recruit pollinators from a pool of bees that are generalist foragers, rather than depending upon a single pollinator, may reduce the vulnerability of *Senna acclinis* populations to local extinction.

CASE 3. THE FOREST SUBCANOPY: VERTEBRATE–INVERTEBRATE POLLINATOR PLASTICITY IN THE AUSTRALIAN TROPICAL RAINFOREST TREE *SYZYGIUM CORMIFLORUM*

Syzygium cormiflorum (Plate 59) is a primarily outcrossing, small tree of the rainforest understorey, though the genus as a whole is a rich and diverse one with over 50 species known to occur in Australia, most of which are endemic (Crome and Irvine 1986).

Many rainforest species in the genus *Syzygium*, and the family Myrtaceae more generally, expose their flowers as dense and conspicuous blossom masses at the forest canopy level. However, in *Syzygium cormiflorum* two growth forms present flowers in different manners (Crome and Irvine 1986). Larger-sized trees possess ramiflorous blossoms, that is, flowers are located on lateral branches; a second smaller form occurs in which blossoms are produced directly on the trunk. This latter condition is called cauliflory and is found occasionally in Australian rainforest angiosperms, for example the understorey shrubs and small trees *Ficus coronata* (Moraceae), *Phaleria clerodendron* (Thymelaeaceae) and *Davidsonia jerseyana* (Cunoniaceae). Individual cauliflorous species can also exhibit ramiflory. Crome and Irvine studied the cauliflorous form of *Syzygium cormiflorum*.

The cauliflorous inflorescences of *Syzygium cormiflorum* are small cymes. Each flower is ~9–16 mm in diameter, is coloured white, emits a slightly unpleasant odour and produces large quantities of nectar of low sugar concentration (varying each day between 2.8–9.8% sucrose equivalents) whose flow pattern is attuned to both night and day flower visitors. In general appearance the flowers are what Crome and Irvine aptly termed the 'shaving brush' type – blossoms possesses numerous filaments that are up to 30 mm long. Flower presentation (generally pointed downwards, but may be oriented in numerous directions), architecture and odour agree with the bat-pollinated (chiropterophilous) floral syndrome. The species has a dual pollinator strategy that involves vertebrates and invertebrates. It can be hypothesised that the brush-like floral structure was first an adaptation to vertebrate pollinators, but its architecture allowed subsequent visitation by many invertebrates. The vertebrates comprise a mixed guild of mammals and birds. The mammals are nocturnal or crepuscular and include the pygmy possum *Ceratetus caudatus*, as well as small nectarivorous blossom bats (*Macroglossus lagochilus* and or *Syconycteris australis*, which can be hard to distinguish at night). By day the suite of vertebrate flower visitors switches to a cohort of melaphagid honeyeaters (*Meliphaga lewinii, Acanthorhynchus tenuirostris, Xanthotis macleayana* and *Lichenostomus frenatus*). Although the bats, and the eastern spinebill *Acanthorhynchus tenuirostris*, are capable of hovering while feeding, all bat and bird species land upon the flowers to feed. Overall, birds are the most numerous of the vertebrate visitors. Experimental manipulation (e.g. caging flowers to exclude potential pollinators) demonstrated that bats were far more effective pollinators than birds, even though birds were the most numerous visitors to flowers.

Insects and other invertebrates that are associated with the flowers of *Syzygium cormiflorum* include mites, cockroaches (Blatellidae), small brachyceran flies (Drosophilidae and Muscidae), ants, several families of beetles (Curculionidae, Staphylinidae and Chrysomelidae) and at least two families of moths (Sphingidae and, particularly, Noctuidae). Not all are necessarily pollinators, though the noctuid moths were readily observed to contact numerous stamens and, like the flies and cockroaches, moved between individual trees.

Pollination in cauliflorous *Syzygium cormiflorum* is mainly achieved by vertebrates, particularly by nectarivorous bats. Invertebrates also play a role as pollinators, though admittedly a reduced one. Regardless of the effectiveness of individual invertebrate taxa, nocturnal and crepuscular moth visitation probably accounts for a significant level of pollination achieved by invertebrates, especially that of invertebrate-induced out-crossing.

The lesson is that the conspicuousness, abundance and heightened diversity of related groups of flower visitors do not necessarily reflect status as the most important pollinator or pollinators. Nor are observations that are limited to a particular time of day a wise strategy for gauging the suite of pollinators. No group of flower visitors, no matter how seemingly inconsequential should be dismissed as pollinators; in the reproductive strategy of *Syzygium cormiflorum* there is contingency for diverse episodes of pollinator presence or absence.

CASE 4. THE FOREST CANOPY: POLLINATION OF THE RAINFOREST PIONEER TREE *ALPHITONIA EXCELSA* (adapted from Williams and Adam 2001)

The genus *Alphitonia* consists of ~20 species distributed in Australia, Malesia and Polynesia. Four of the six species occurring in Australia are endemic. *Alphitonia excelsa* (Rhamnaceae) occurs in the Northern Territory, Western Australia and Queensland, extending to the far south coast of New South Wales (Harden 1990). Isolated populations occur inland in the Pilliga Scrub and the Nandewar Ranges (Mt Kaputar National Park) on the New South Wales northern slopes. In New South Wales *Alphitonia excelsa* occurs commonly on the margins of littoral rainforest, in floodplain rainforest remnants, submontane subtropical rainforest and wet sclerophyll forest. It is also found as occasional trees in dry sclerophyll forest. This distribution indicates that the species is a successful coloniser (pioneer) of a broad range of forest subformations and soil types.

The small white flowers of *Alphitonia excelsa* are massed in terminal or axillary cymes and conform to the general entomophilous floral syndrome. The flowers are not obscured by foliage, are protandrous and strongly scented with an odour reminiscent of urine. They do not possess nectar guides. Flowers consist of five sepals and four or five petals (rarely six), which are reduced and spoon-shaped (spathulate). There are usually four or five stamens per flower and the style possesses two or three lobes, which develop after anther dehiscence. Petals and stamens are erect by the time the anthers dehisce, and recurved and held laterally as the ovary develops. Pollen is cream to greenish in colour, translucent and slightly sticky. The sticky pollen grains are exuded under pressure in an elongated mass, clear of the encapsulating petals, onto flower visitors. Following anther dehiscence, the stamens are carried 'free' or away from the developing ovaries as the encapsulating petals gradually recurve. There are no obvious nectariferous basal glands, but the receptacle is fleshy. Nectar is produced at full flower opening, which is diurnal and occurs between 1030–1300 h EST, and ceases upon apparent pollination/fertilisation. The stigmatic lobes become brown following pollination.

The insect fauna associated with flowering *Alphitonia excelsa* (see Table A1.3) include species filling a variety of roles. Not all are pollinators, nor even taxa normally associated with flowers (e.g. some genera and species of Hemiptera and Coleoptera). However, the flower-visiting assemblage does contain true pollinators; individual species vary in efficiency in effecting pollen transfer. There is variation in the diversity and abundance of anthophilous insects occurring between different trees, plant populations and seasons. This might be explained by temporal changes in the size and composition of the available pool of insects, varying patterns of daily foraging behaviour by individual insects, differential response to recruitment cues entrained by variation in the availability of floral resources, or environmental influences.

Alphitonia excelsa trees that exhibit seasonal peak-phase flowering generally attract the greatest number of insects. Nevertheless, there is considerable variation in abundance and diversity of visitors on adjacent trees with apparently similar stages of flowering and perceived amounts of resources. The variation can be at gross taxonomic ranks, for example some trees recruit large numbers of flies and others have a greater proportion of beetles. Flies may be dominant at the commencement of flowering cycles then steadily reduce in number as the flowering episodes progress. Trees that are undergoing peak flowering may attract large numbers of wasps, in particular conspicuously large Thynnidae and Scoliidae.

Mid- and late-phase flowering tends to be accompanied by a high number of flower visitors. The late-phase peak has been interpreted as a concentration of insects to diminishing floral resources at the conclusion of flowering, a lack of alternative mass-flowering trees late in the season, or changes in the regional pool of available insects. In many cases there is no obvious pattern of recruitment, rather, many seasons hold their own peculiar display of visitors and spectacle.

At night the suite of visitors to flowers changes dramatically. Small Diptera (e.g. Nematocera) and Coleoptera (e.g. Anthicidae, Scirtidae and Scarabaeidae – Melolonthinae) emerge to replace the frenetic activity of diurnal pollinators. Such changes in the composition of the anthophilous fauna, between day and night shifts, on mass-flowering rainforest trees is the norm.

Vertebrates, including nocturnal mammals, do not appear to be significant visitors to the flowers of *Alphitonia excelsa*, however, several species of passerine birds (e.g. *Zosterops lateralis* and *Meliphaga lewinii*) occasionally forage for insects among foliage and seed clusters. Thus, as with the vast number of Australian rainforest trees, pollination is achieved principally or solely by insects. Daytime flower visitors are largely Diptera, Hymenoptera and Coleoptera, which collectively constitute more than 90% of the fauna. Miscellaneous taxa principally include Hemiptera but also Blattodea, Thysanoptera, Lepidoptera, Psocoptera, Orthoptera and Collembola.

There can be considerable differences in the proportion of individual insect species and insect orders attracted to different flowering populations. Such variation can occur between seasons. Flower visitors recorded from *Alphitonia excelsa* at a littoral rainforest site in northern New South Wales were dominated by Hymenoptera (~37%) and Diptera (~49%), but at a nearby inland location Diptera were dominant and averaged ~72% of the diurnal anthophilous fauna. Regardless of the proportional representation by major groups of higher taxa (families and orders) and whether they are active by night or day, at least 80% (and sometimes over 90%) of visitors are within the <3–6 mm size range. The pollination efficiency of individual taxa is poorly known and, while very small insects can carry only minute pollen loads they may occur in such large numbers that they serve a significant role in pollination. Abundance matters, as do body size, body sculpture and individual behaviour.

Collectively 116 families, comprising ~200 genera, have been recorded from the flowers of *Alphitonia excelsa* growing in New South Wales subtropical rainforest. Many of these (and the many species therein) also visit other rainforest trees with entomophilous flowers. Some *Alphitonia excelsa* populations are rich in wasps of the families Crabronidae, Pompilidae, Scoliidae, Sphecidae, Vespidae and Thynnidae, while species of the fly families Platystomatidae, Stratiomyidae, Sepsidae and Tachinidae can also be prolific. Many of the beetle (e.g. Cetoniinae and Buprestidae) and bee groups (e.g. Euryglossinae) that can be common on rainforest Myrtaceae are frequently rare or absent, though honey bees are usually prolific and in some seasons represent a high proportion (>25%) of total insects foraging on flowers.

The pool of anthophilous taxa found on *Alphitonia excelsa* includes species (obligates) (e.g. within the families Buprestidae, Cantharidae, Mordellidae, Scarabaeidae, Thynnidae, Colletidae, Halictidae and Erebidae) that possess mouthparts adapted to feeding from flowers, and taxa with no obligate morphological adaptation to feeding from flowers (e.g. Chrysomelidae, Coccinellidae, Dermestidae and Melyridae). These families are typical of the beetle fauna to be expected on the blossoms of all rainforest trees with general entomophilous flower syndromes. The fauna also includes widely occurring Diptera from the families Calliphoridae, Lauxaniidae and Platystomatidae. These possess mouthparts adapted to feed from a variety of liquids, including the products of animal and fungal decay.

Butterflies are not a conspicuous element of the anthophilous fauna visiting subtropical populations of *Alphitonia excelsa*, but species occasionally encountered include *Junonia villibe callibe, Hypolimna bolina nerina, Tirumala hamata hamata* (all members of the Nymphalidae), *Delias nysa nysa* (Pieridae), *Candalides absimilis, Deudorix epijarbas diovis, Erysichton lineata lineata* (Lycaenidae) and *Graphium eurypylus lycaon* (Papilionidae). Butterflies are able to easily feed on nectar present in the small open-structured flowers. They undertake frequent inter-plant movements and are likely to make important contributions to out-crossing.

The slightly sticky pollen of *Alphitonia excelsa* is not easily displaced from the anthers, so wind pollination is unlikely to occur. The protandrous pattern of flower opening, and the recurving of petals which articulates the stamens away from the developing ovaries, may militate against self-fertilisation or interference with stigmatic function by self-pollen. Further, there is no obvious adaptation of flowers to visits by specialised insects (e.g. long-tongued bees) and the shallow, readily accessible perianth allows visitation by numerous insects, regardless of their size. Consequently, pollination in *Alphitonia excelsa* appears to be a flexible general entomophilous system in which the contributions made by individual insect orders and lower rank taxa vary spatially and temporally.

Many of the insects visiting *Alphitonia excelsa* also visit flowers of other species. Successful pollination in plants utilising a pool of generalist pollinators is likely to be enhanced if there is reduced competition between plant species for the same pool of insects. On the mid north coast of New South Wales, most lowland rainforest trees have ceased flowering by the beginning of summer, with the majority of species flowering from October to December (G. Williams 1995). Flowering of *Alphitonia excelsa* populations is relatively synchronous and occurs after this spring–summer flowering peak, with populations flowering for approximately six to eight weeks. *Alphitonia excelsa* is generally the last widespread lowland rainforest tree in coastal New South Wales to flower in abundance during summer. This may increase the chances of pollination, by flowering within a sequence of spring–summer flowering cohorts, and substantially extends the availability of floral resources to nectar- and pollen-dependent animals.

Irvine and Armstrong (1988) studied the pollination of the related *Alphitonia petriei* in tropical Queensland. Like *Alphitonia excelsa*, the species is a pioneer, mass-flowering, bisexual and protandrous rainforest tree. Unlike *Alphitonia excelsa*, *Alphitonia petriei* flowers from September to November, during the Australian tropical dry season. They found that *Alphitonia petriei* has a flexible entomophilous pollination system (similar to that suggested by the floral visitors to *Alphitonia excelsa*) dominated by Diptera and Coleoptera, and to a lesser degree by wasps. The ever-present honey bee, *Apis mellifera*, was also an active visitor to *Alphitonia petriei* flowers. They recorded 10 Coleoptera species from limited observations on *Alphitonia petriei*, and stated: 'In terms of cantharophily [beetle pollination syndrome], it is the specialised inflorescence type, attracting predominantly medium-sized, non-flower-damaging, nectar/pollen-feeding beetles'. Tropical Australian rainforest pollinator faunas generally dominated by combinations of Diptera, Hymenoptera and Coleoptera are documented from *Flindersia brayleyana* (Rutaceae), *Neolitsea dealbata* and *Litsea leefeana* (Lauraceae) and *Diospyros pentamera* (Ebenaceae). House (1985) found that Diptera were the most abundant visitors to flowers of dioecious *Neolitsea dealbata*, *Litsea leefeana* and *Diospyros pentamera* but that Coleoptera were the most important in carrying pollen to female *Diospyros pentamera* trees. Generalist insect pollination systems are widely recorded from several plant communities and from individual plant species, not just Australian ones. However, rainforest invertebrate assemblages dominated by Diptera, Hymenoptera and

Table A1.3. Invertebrate genera collected from flowering *Alphitonia excelsa* trees in New South Wales subtropical rainforests (not including all taxa).

Blank cells indicate genera/species not determined (after G. Williams 1995; Williams and Adam 2001, 2010).

Orders and families	Genera (where known)
BLATTODEA	
Blattellidae	*Ectoneura*
COLEOPTERA	
Anobiidae	
Anthicidae	*Anthicus*
Attelabidae	*Auletobius*
Buprestidae	*Castiarina, Diphucrania, Melobasis, Neocuris, Torresita*
Cantharidae	*Chauliognathus*
Carabidae	*Sarothrocrepis*
Cerambycidae	*Iphra*
Chrysomelidae	*Cryptocephalus, Ditropidus, Monolepta*
Cleridae	*Lemidia*
Coccinellidae	*Archegleis, Diomus, Cryptolaemus, Epilachna, Orcus, Rodatus, Rodolia*
Corylophidae	*Sericoderus*
Curculionidae	*Balanerhinus, Baris, Cyttalia, Meriphus, Neolaemaraccus*
Dermestidae	*Anthrenocerus, Thaumaglossa*
Elateridae	*Megapenthes, Ophidius*
Euglenidae	*Aderus*
Hydrophilidae	*Pseudohydrobius*
Latridiidae	*Cortinicara*
Lycidae	*Porrostoma*
Melyridae	*Carphurus, Dicranolaius, Helcogaster, Neocarphurus*
Mordellidae	*Hoshihananomia, Mordella, Mordellistena, Tomoxia*
Nitidulidae	*Epuraea*
Phalacridae	*Litochrus, Parasemus*
Pythidae	*Paromarteon*
Ripiphoridae	*Macrosiagon*
Scarabaeidae	*Cheiragra, Eupoecila, Glycyphana, Heteronyx, Neorrhina, Phyllotocus,*
Scirtidae	*Pseudomicrocara, Scirtes*
Staphylinidae	
Tenebrionidae	*Ecnolagria, Nocar, Ommatophorus*

Orders and families	Genera (where known)
COLLEMBOLA	
Entomobryidae	
DIPTERA	
Bibionidae	*Bibio*
Bombyliidae	*Geron, Ligyra, Pseudopenthes*
Calliphoridae	*Chrysomya, Paramenia, Stomorhina*
Cecidomyiidae	
Ceratopogonidae	
Chironomidae	
Chloropidae	*Apotropina*
Culcidae	
Dolichopodidae	*Krakatauia*
Drosophilidae	*Scaptodrosophila, ?Drosophila, Leucophenga, Nesiodrosophila*
Empididae	
Ephidridae	
Lauxaniidae	*Homoneura, Melanina, Sapromyza, Steganopsis, Trypetisoma*
Milichidae	
Muscidae	*Musca*
Mycetophilidae	
Nemestrinidae	*Cyclopsidea*
Phoridae	
Platystomatidae	*Duomyia, Euprosopia, Microepicausta, Pogonortalis, Rivellia*
Psychodidae	
Rhagionidae	*Chrysopilus*
Sarcophagidae	*Sarcophaga*
Scatopsidae	
Sepsidae	*Australosepsis, Sepsis*
Stratiomyidae	*Acanthasargus, Damaromyia, Hermetia, Odontomyia*
Syrphidae	*Dideopsis, Eristalinus, Mesembrius*
Tabanidae	*Scaptia*
Tachinidae	*Austrophorocera, Blepharella, Blepharipa, Prosena, Rutilia, Senostoma*

Orders and families	Genera (where known)
Tephritidae	*Bactrocera, Spathulina*
Therevidae	
Tipulidae	
HEMIPTERA	
Aphididae	
Cicadellidae	
Cicadidae	*Arunta*
Eurymelidae	
Flatidae	*Massila*
Lygaeidae	*Arocatus, Nysius*
Membracidae	
Miridae	
Psyllidae	
Tingidae	
HYMENOPTERA	
Agaonidae	
Bethylidae	
Braconidae	*Bracon, Iphaulax*
Ceraphronidae	
Chalicididae	
Chrysididae	*Stilbum*
Crabronidae	*Acanthostethus, Bembix, Bembicinus, Ectemnius, Larra, Pison, Sericophorus, Sphodrotes, Tachysphex, Williamsita*
Encyrtidae	
Eulophidae	
Eupelmidae	
Eurytomidae	
Evanidae	*Evania*
Figitidae	
Ichneumonidae	*Echthromorpha*
Mutillidae	*Euphotomorpha*
Mymaridae	
Pergidae	*Neoeurys*
Perilampidae	
Platygastridae	

Orders and families	Genera (where known)
Pompilidae	*Calopompilus, Heterodontonyx, Platyderes*
Pteromalidae	
Scelionidae	
Scoliidae	*Radumeris, Platyderes*
Sphecidae	*Sceliphron, Sphex,*
Thynnidae	*Acanthothynnus, Anthobosca, Diamma, Dimorphothynnus, Epactiothynnus, Rhagigaster, Tmesothynnus, Thynnoturneria, Zaspilothynnus, Zeleboria*
Torymidae	
Vespidae	*Abispa, Leptomenoides, Paralastor, Polistes, Pseudabispa*
Apiformes	
Apidae	*Apis, Exoneura, Tetragonula*
Colletidae	*Amphylaeus, Heterapoides, Hylaeus, Leioproctus,*
Halictidae	*Homalictus, Lasioglossum*
Formicoidea	
Formicidae	*Camponotus, Crematogaster, Iridomyrmex, Myrmecia, Polyrhachis*
LEPIDOPTERA	
Erebidae	*Amata*
Hesperiidae	*Trapezites*
Lycaenidae	*Candalides, Deuodorix, Eurysichton*
Nymphalidae	*Hypolimna, Junonia, Tirumala*
Papilionidae	*Graphium*
Pieridae	*Delias*
ORTHOPTERA	
Gyrillidae	*Cycloptilum*
NEUROPTERA	
Mantispidae	*Spaminta*
PSOCOPTERA	
Ectopsocidae	*Ectopsocus*
THYSANOPTERA	
Phlaeothripidae	*Haplothrips*
Thripidae	*Thrips*

Coleoptera are not restricted to anthophilous insects. For example, Basset and Kitching (1991) collected ~42 000 arboreal arthropods (which are likely to have included a proportion of anthophilous taxa) in tropical Queensland rainforest using specially designed intercept traps. Of their sample, 27% of individuals were Coleoptera, 41% were Diptera and over 11% were Hymenoptera.

CASE 5. A RAINFOREST TREE NEARLY TOO FAR AWAY: *GREVILLEA ROBUSTA*

A plant grown slightly removed from its natural geographic range stands a good chance of being found by one of its suite of co-adapted pollinators, especially if their flight patterns traverse great distances. If it is a generalist, it will probably recruit pollinators from the locally available guild.

But what of rainforest angiosperms that colonise newly formed volcanic islands and incipient oceanic reef cays, or species planted for purposes of forestry and agriculture far from home? An example of the latter is *Grevillea robusta* (Proteaceae), a semi-deciduous tree native to Australia's subtropical latitudes (Boland *et al.* 1984). It is normally associated with rainforest and, though naturally uncommon within its home range, is widely planted for ornamental purposes even in cold temperate regions of southern Australia and the arid interior. The species has been widely planted in Sri Lanka, India and Africa, as a shade tree and for honey production.

In Kenya *Grevillea robusta* is a popular species grown on farms. In Australia nectar-feeding birds (commonly lorikeets, wattlebirds and honeyeaters) and exotic honey bees are the most frequently observed daytime visitors to its flowers. In Africa birds and insects are the main visitors but the likely pollinators are sunbirds (e.g. *Nectarinia amethystina, Nectarinia cyanolaema, Nectarinia superba* and *Nectarinia venusta*) and white-eyes (*Zosterops kikuyuensis* and *Zosterops senegalensis*) (Kalinganire *et al.* 2001). No nocturnal flower visitors have been observed. Sunbirds are specialised nectar-feeders with long beaks that allow deep probing of flowers, white-eyes are short-beaked generalists that rely on nectar and insects.

Nectar is the principal reward offered by the zygomorphic flowers of *Grevillea robusta* but in Africa and Australia ants and honey bees are 'thieves' and rob this resource. In trees planted in Kenya, nectar concentration has been found to increase steadily throughout the day as temperature increases and humidity decreases. Nearly 90% of visits by birds occur in the morning when the volume of available nectar is at its highest. After depleting the available nectar most birds then move to other trees, thus maximising out-crossing. Opening of individual flowers is staggered over a two- to three-day period and nectar secretion by a whole inflorescence may last as long as seven days. This prolonged period of secretion increases the likelihood of visits by pollinators, and the chance of pollen transfer to other trees. Flowers that are covered by bags to exclude birds or that are hand-pollinated with their own pollen do not set fruit (are self-incompatible), and flowers that are left open but are not artificially pollinated set few fruit (0.9%). *Grevillea robusta* is heavily dependent on the availability of pollinators for pollen transfer and seed-set.

In some areas of Kenya, seed yield from *Grevillea robusta* has been low, possibly due to an absence of pollinators (Kalinganire *et al.* 2001). Though the quality of pollen transfer might be less than optimal, pollinators are available; the specialised floral structure, absence of natural pollinators and self-incompatible breeding system are apparently no obstacle to levels of seed-set, far from the rainforests of northern New South Wales and southern Queensland. Given that *Grevillea robusta* and many other rainforest trees can survive and flourish in areas and environments so different from their native localities, it can only be wondered why their present distributions are so restrictive. What hand holds them back?

CASE 6. LITTORAL RAINFOREST: BREEDING SYSTEMS AND FLOWERING PERIODS IN AN ENDANGERED MARITIME-ASSOCIATED ECOSYSTEM

Flowering plants and their pollinators do not interact independently of the ecosystems and plant complexes within which they live, and to which they are dependent. But many of these ecosystems are endangered.

Littoral rainforest, also known as 'beach scrub', is a distinct habitat-restricted rainforest formation found associated with beach dune zones, coastal headlands and islands (Plates 133, 134) (Miles and Kendall 2006; Young and McDonald 1987). Individual stands are spatially isolated (mosaic-like) within a coastal environment matrix otherwise comprising open forest, woodland, shrub communities or heath. Occasionally they abut locations of intense urban development and farmland. Many stands grow on sandy soils of low nutrient status, but a suite is also distributed on headlands of diverse geological ancestry.

Although proximity to the sea results in equitable temperatures and high humidity (Floyd 1990), the prevailing environmental conditions are marginal for rainforest. At lower latitudes they experience a seasonally dry climate, but throughout their range sites are subject to salt-laden winds. This results in wind-shearing and deformation of the frontal vegetation canopy, and levels of salt burn along the ocean face (Plate 133). Canopies may be lower than 1 m where plants are subject to frequent strong winds (particularly on headlands), but can extend to much greater heights in protected zones. For example, on the sandmasses of Fraser Island and Cooloola in south-east Queensland, canopy height can exceed 50 m (Adam 1992). Structurally well developed rainforest also occurs at Iluka and Sea Acres Nature Reserve, northern New South Wales, and though there is a trend in reduction in tree diversity with increases in latitude this is not consistent on headland stands because of the influence of more varied soil types, aspect and exposure.

Littoral rainforest extends in distribution from north-western Australia, Cape York Peninsula to eastern Victoria, stands are generally very small in size, and since European occupation have largely suffered from extensive clearing, a history of mining for mineral sands (e.g. rutile, zircon), high levels of ongoing alien plant invasion, and frequent impacts of fire of varying intensity. Fire is an important factor in determining the boundaries of stands, with catastrophic fire episodes resulting in multiple deaths of trees and loss of projective canopy cover, potentially leading to stand extinction. The small, often narrow and torturous boundary of remnants, with minimal if any 'core' area, leaves individual stands vulnerable to edge effects, fire and projected impacts of climate change (Suppiah *et al.* 2010). Sites are subject to heavy invasion by alien plants (e.g. Department of Environment and Energy 2019). Coconuts, *Cocos nucifera* (Arecaceae), are a significant invasive 'transformer' species in North Queensland (Spencer *et al.* 2010), and *Sphagneticola trilobata* (Asteraceae) infests the understorey of tropical remnants; this mirroring the understorey invasions by Asparagaceae typical of many northern New South South Wales stands. Major threatening weeds on the mid north coast of New South Wales include *Anredera cordifolia, Asparagus* spp. (Plate 147), *Bryophyllum delagoense, Chrysanthemoides monilifera, Delairea odorata, Gloriosa superba, Senna pendula, Solanum seaforthianum, Tradescantia fluminensis.* However, the suite of persistent and temporary invasive species can change dramatically from site to site, reflecting chance dispersal events and point-specific conducive circumstances for seed germination or vegetative growth following the discarding of plant material by humans. Although *Lantana camara* is widespread and invasive, and under conditions of drought can add to fuel loads, with respect to littoral rainforest it can also fulfill the interim role of 'fence', denying ready pedestrian access that results in the increase light penetration to the forest floor (with commensurate increases in alien plant invasion), trampling of native plant seedlings, and facilitating entry of alien animals.

As a consequence of their remnant status, high biodiversity attributes and nature conservation significance (see Adam 1987; G. Williams 2020), and the complexity of threatening processes, littoral rainforest has been listed as *critically endangered* under the Federal *Environment Protection and Biodiversity Conservation Act, 1999* (EPBC Act) (see Department of Environment and Energy 2019). One site (Iluka) is included within the *Central Eastern Rainforest Reserves of Australia (CERRA)* World Heritage Area (Adam 1987; Hunter 2003), and dune thickets are associated with the *Wet Tropics World Heritage Area* of north-east Queensland. Additional conservation listings for littoral rainforest include various state and local government environment protection ordinances and zonings (e.g. Queensland *Nature Conservation Act, 1992*; New South Wales *Biodiversity Conservation Act, 2016*). In practice, however, these listings do not serve as a bulwark against adverse environmental impacts, chemical inflow from adjacent residential areas, invasion by alien plant and animal species, fire ingress, ill-considered on-ground management activities or coastal development proposals, or point-specific clearing of vegetation by adjacent landholders.

Based on available mapping data the area of littoral rainforest before European occupation (1788) is estimated at 35 000–67 000 ha (Department of Environment and Energy 2019). This is now estimated to consist of ~16 000 ha in Queensland, ~1300–1600 ha in New South Wales and <300 ha in Victoria. Stands growing on sand dunes mostly exhibit 1–2 strata, with the canopy progressively rising in height from a low, wind-deformed frontal zone to trees in rear hind-dune zones in excess of 30 m and showing little crown deformation. Floristic composition varies with influences of latitude, soil nutrients and parent material, however, while soil nutrients are low there is heterogeneity in nutrient distribution. Canopy dominance, population abundance, and species composition changes at individual sites, nevertheless, commonly encountered tree species growing on sand-associated stands in northern New South Wales include *Elaeodendron australe* (Celastraceae); *Elaeocarpus obovatus* (Elaeocarpaceae) (Plate 90), *Baloghia inophylla* (Euphorbiaceae), *Syzygium hemilamprum, S. oleosum, S. smithii* (Myrtaceae), *Bridelia exaltata* (Phyllanthaceae/ Euphorbiaceae), *Drypetes deplanchei* (Putranjivaceae/ Euphorbiaceae), *Alphitonia excelsa,* (Rhamnaceae), *Acronychia imperforata* (Rutaceae), *Arytera divaricata, Cupaniopsis anacardioides, Guioa semiglauca* (Sapindaceae), and the gymnosperm *Podocarpus elatus* (Podocarpaceae) (Floyd 1990; see Table A1.4). In subtropical zones numerous fig trees (e.g. *Ficus coronata, F. fraseri, F. henneana, F. microcarpa F. obliqua, F. opposita, F. rubiginosus, F. watkinsiana*) occur (Harden *et al.* 2006), and large *Agathis robusta* and *Araucaria cunninghamii* (Araucariaceae) grow in deep sands of low nutrient status in south-eastern Queensland (e.g. Fraser Island, Cooloola) (Adam 1992). *Banksia integrifolia* is an associate, usually at hind-dune locations, and particularly on margins or as relics of earlier stages of stand development. The palm *Livistona australis* (Arecaceae) can be common in sclerophyll-rainforest ecotones (as near Forster and the vicinity of Crowdy Head, New South Wales).

Littoral rainforest on headlands and dunes have similarities of structure and floristics but may have different histories, and treated as a whole can be considered a maritime variant of subtropical rainforest (*sensu* Adam 1992). Dune rainforests have a strong overall similarity and floristically may represent a distinct lineage which has been present in Australia, as a separate entity, for many millions of years (Williams and Adam 1991). Nutrient levels in headland rainforests are likely higher than those on dunes, though owing to the variety of geologies the range of soil nutrient levels is probably considerable. Headland rainforests tend to be more floristically diverse (e.g. Sea Acres Nature Reserve, Port Macquarie). *Lophostemon confertus* (Myrtaceae) is a common dominating associate, the crowns of old trees gnarled and sheared by exposure to the pruning effects of wind. Other headland species reflect those found in adjacent littoral rainforest on dune systems, but the presence of species in headland stands is less predictable, and may include those characteristic of hinterland dry rainforest; for example, *Aphananthe philippinensis, Celtis paniculata* (Cannabaceae/Ulmaceae), *Capparis arborea* (Capparaceae), *Diospyros pentamera* (Ebenaceae), *Alchornea ilicifolia, Croton insularis, Mallotus philippensis* (Euphorbiaceae), *Backhousia sciadophora, Gossia bidwillii* (Myrtaceae), *Cleistanthus cunninghamii* (Phyllanthaceae/Euphorbiaceae), *Jagera pseudorhus, Rhysotoechia bifoliata* (Sapindaceae), *Dendrocnide photinophylla* (Urticaceae) – as at Redhead and Hallidays Point/Black Head near Taree (G. Williams 1993). Single trees, as of *Syzygium francisii* (Myrtaceae) at Crowdy Head (Williams and Adam 1991) and *Pisonia umbellifera* (Nyctaginaceae) at Camden Head (G. Williams 2000), can be interpreted as chance dispersal events, but such singular records may reflect solitary survivors of past land clearance of a once more extensive stand, with no record of prior composition or abundance.

At higher latitudes most canopy trees in littoral rainforest flower during mid spring to late summer (e.g. G. Williams 1995, 2000, Table A1.4), but not all species flower each year (G. Williams 1995). And as with Australian rainforests more generally, the dominating theme amongst mass-flowering trees and shrubs is one of small, predominantly white coloured flowers with an open actinomorphic flower form with little depth effect. These are predominantly visited and pollinated by small-sized insects, the majority of which are less than 5–6 mm in size. Lists of insect visitors to flowering littoral rainforest trees and shrubs are given in G. Williams (1993, 1995, 1998, 2018, 2020), Williams and Adam (1995, 1997, 2001, 2019) and Williams *et al.* (2001). Pollinator associations are also dicussed in G. Williams (2020).

In littoral rainforests there are relatively few instances of flower structure, colour, branch architecture and nectar volume suggesting primary adaptation to pollination by vertebrates. Thus the communities, overall, express a syndrome of 'general entomophily' indicating 'plasticity' in their requirements for pollinators. However, at the local landscape scale, subsets of available pollinator assemblages (e.g. see Williams and Adam 2019) respond to the availability of differing floral resources – in effect forming spatially and temporally dynamic mosaics of interactive species, but whose taxonomic and adaptive composition, fidelity and efficiency of pollination services cannot be predicted.

Overall, the majority of trees in littoral rainforest possess bisexual flower structures (e.g. Table A1.4) but instances of self-compatibility, or an obligate requirement by these for out-crossing, are known for very few. This is unfortunate because knowledge of the reproductive biology of a species is an important prerequisite that allows us to understand their resilience to environmental change (Adam and Williams 2001). Flower structure, by itself, does not provide a sure determinant of strategy.

Table A1.4. General flowering periods and breeding systems for trees and large shrubs in three littoral rainforest remnants, 1. Camden Head, 2. Harrington and 3. Hallidays Point/Black Head (flowering periods after Floyd 2007; Harden 1990, 1991, 1992, 1993; G. Williams 1995, 2000; G. Williams pers. obs.).

Breeding systems follow Harden (1990, 1991, 1992, 1993). Flowering may be irregular, differing between regions, sites and years, or not flowering at all in some years.

Family/Species	site	flowering	breeding system
Podocarpaceae			
Podocarpus elatus	2,3	Sept	dioecious
Amarylliidaceae/Liliaceae			
Crinum pedunculatum	1	'summer'	bisexual
Anacardiaceae			
Euroschinus falcatus	1,2,3	Oct–Jan	bisexual, or unisexual
Araliaceae			
Polyscias elegans	1,2	Feb–May	bisexual, occas. polygamous
Polyscias sambucifolius	3	Dec–Feb	bisexual, occas. polygamous
Arecaceae			
Archontophoenix cunninghamiana	1	Feb–Jun	unisexual
Livistona australis	2,3	Aug–Oct	bisexual
Asparagaceae/Laxmanniaceae/Dracaenaceae *Cordyline stricta*	1, 2		bisexual
Cannabaceae/Ulmaceae			
Aphananthe philippinensis	3	Sept–Nov	monoecious or dioecious
Celtis paniculata	1,2,3	Dec–Jan	monoecious, or polygamous
Trema tomentosa	2,3	Dec–Jan, or all year	monoecious or polygamous
Capparaceae			
Capparis arborea	3	Jan–Mar	bisexual
Celastraceae			
Elaeodendron australe	1,2,3	Aug–Nov	bisexual
Cornaceae/Alangiaceae			
Alangium villosum	3	Oct–Apr	bisexual (?)
Cunoniaceae			
Schizomeria ovata	3	Sept–Nov	bisexual
Ebenaceae			
Diospyros australis	1,2,3	Oct–Dec	dioecious
Diospyros pentamera	2,3	Dec–Feb	dioecious
Elaeocarpaceae			
Elaeocarpus obovatus	1,2,3	Sept–Dec	bisexual, occas. polygamous
Elaeocarpus reticulatus	1,2	Oct–Jan	bisexual, occas. polygamous
Euphorbiaceae			
Alchornea ilicifolia	3	Nov	usually dioecious
Baloghia inophylla	2,3	Dec	monoecious (?)
Claoxylon australe	2,3	Oct–Nov	normally dioecious
Croton insularis	3	Sept–Jan	monoecious

Continued

Table A1.4. (*Continued*)

Family/Species	site	flowering	breeding system
Croton verreauxii	3	Nov–Jan	monoecious?
Homalanthus populifolius	1,2,3	Sept–Dec	monoecious
Mallotus philippensis	3		normally dioecious
Eupomatiaceae			
Eupomatia laurina	1,2,3	Nov–Feb	bisexual
Fabaceae			
Pararchidendron pruinosum	1,2,3	Oct–Jan	bisexual
Flacourtiaceae/Salicaceae			
Scolopia braunii	1,2,3	Sept–Nov	bisexual
Lamiaceae/Verbenaceae			
Clerodendrum tomentosum	2,3	Oct–Jan	bisexual
Gmelina leichardtii	2	Nov–Jan	bisexual
Lauraceae			
Beilschmiedia obtusifolia	2	Oct–Nov	bisexual
Cryptocarya microneura	3	Sept–Nov	bisexual
Cryptocarya rigida	3	June–Oct	bisexual
Endiandra discolor	1,2	Oct–Dec	bisexual
Endiandra sieberi	1,3	Jun–Oct	bisexual
Litsea reticulata	1,2,3	May–July	dioecious
Neolitsea australiensis	1,2	Mar–Sept	dioecious
Malvaceae			
Hibiscus heterophyllus	3	Oct–Feb	bisexual
Malvaceae/Sterculiaceae			
Brachychiton acerifolius	3	Oct–Jan	monoecious
Commersonia fraseri	2,3	Sep–Mar	bisexual
Meliaceae			
Dysoxylum fraserianum	1,2,3	Apr–Jun	bisexual, or unisexual
Melia azedarach var. *australasica*	1	Oct	bisexual, or unisexual
Synoum glandulosum	1,2,3	Feb–Aug	bisexual
Monimiaceae			
Wilkiea huegeliana	1,2,3	Sept–Feb	dioecious, unisexual
Moraceae			
Ficus fraseri	1,2,3	Nov	dioecious, monoecious
Ficus henneana	1,2,3		dioecious, monoecious
Ficus obliqua	2,3	Mar	dioecious, monoecious
Ficus rubiginosus	1,2,3		dioecious, monoecious
Ficus watkinsiana	1,2		dioecious, monoecious

Continued

Table A1.4. (*Continued*)

Family/Species	site	flowering	breeding system
Myrtaceae			
Backhousia sciadophora	3	June–July	bisexual
Gossia bidwilli	3	Oct–Dec	bisexual
Lophostemon confertus	3	Oct–Dec	bisexual
Rhodamnia rubescens	3	Aug–Oct	bisexual
Rhodomyrtus psidioides	1,2,3	Nov–Dec	bisexual
Syzygium oleosum	1,2,3	Nov–Apr	bisexual
Syzygium smithii	1,2,3	Nov–Mar	bisexual
Tristaniopsis laurina	3	Dec–Feb	bisexual
Myrsinaceae/Primulaceae			
Myrsine variabilis	1,2,3	July–Aug, Dec	bisexual (?)
Nyctaginaceae			
Pisonia umbellifera	1	Sept–Jan	bisexual, or unisexual
Oleaceae			
Notelaea longifolia	1,2,3	Apr–Oct	usually bisexual
Notelaea venosa	1	Oct–Dec	bisexual
Olea paniculata	1	Oct–Dec	usually bisexual
Phyllanthaceae/Euphorbiaceae			
Bredelia exaltata	2,3	Dec	monoecious
Breynia oblongifolia	1,2	Mar–Apr	monoecious
Cleistanthus cunninghamii	3	Aug–Sept	monoecious
Glochidion ferdinandi	1,2,3	Sept–Dec	monoecious
Pittosporaceae			
Pittosporum multiflorum	2,3	'spring-summer'	bisexual
Pittosporum revolutum	1,3	'spring'	bisexual
Pittosporum undulatum	1,3	Aug–Oct	bisexual
Proteaceae			
Banksia integrifolia	1,2,3	Jan–June	usually bisexual (?)
Stenocarpus salignus	3	Oct–Jan	usually bisexual (?)
Putranjivaceae/Euphorbiaceae			
Drypetes deplanchei	2,3	Nov–Dec	dioecious
Rhamnaceae			
Alphitonia excelsa	1,2,3	Nov–Apr	bisexual
Rubiaceae			
Cyclophyllum coprosmoides	1,2,3	Jan–Mar	bisexual (?)
Psydrax odorata	1,3	Nov–Dec	bisexual (?)
Ixora beckleri	3	Nov–Jan	bisexual
Psychotria loniceroides	1,2,3	Dec–Mar	bisexual (?)

Continued

Table A1.4. *(Continued)*

Family/Species	site	flowering	breeding system
Rutaceae			
Acronychia imperforata	1,2,3	Feb–Apr	bisexual
Acronychia oblongifolia	2,3	Jan–Mar, May	bisexual
Acronychia wilcoxiana	2,3	Jan–May	bisexual
Geijera salicifolia	3	Sept–Oct	bisexual
Melicope micrococca	3	Nov–Mar	bisexual, or functionally unisexual
Sarcomelicope simplicifolia	2,3	Mar–Sept	functionally unisexual
Sapindaceae			
Alectryon coriaceous	1,2,3	Dec–Feb	monoecious
Arytera divaricata	1,2,3	Nov–Apr	monoecious
Cupaniopsis anacardioides	1,2,3	Apr–July	dioecious, or monoecious
Elattostachys nervosa	2,3	Sept–Nov	monoecious
Guoia semiglauca	1,2,3	Sept–Jan	monoecious
Jagera pseudorhus	3	Mar–May	normally dioecious
Rhysotoechia bifoliolata	3	Aug–Sept	usually monoecious
Sapotaceae			
Pouteria australis	1,2,3	Oct–Jan	bisexual (?)
Scrophulariaceae/Myoporaceae			
Myoporum acuminatum	3	Oct–Feb	bisexual (?)
Solanaceae			
Duboisia myoporoides	2,3	Feb–Nov	bisexual
Solanum stelligerum	3	'spring–summer'	bisexual (?)
Symplocaceae			
Symplocus thwaitesii	3	Sept–Dec, May–June	bisexual, occas. unisexual
Urticaceae			
Dendrocnide photinophylla	2,3	Nov–Mar, June	dioecious
Zingiberaceae			
Alpinia caerulea	1	'spring'	bisexual

Of the mass-flowering littoral rainforest tree and shrub species studied in the Manning region of northern New South Wales, Adam and Williams (2001) cited a high incidence of dioecy; species so characterised being obligate out-crossers. However, there were significant differences in the incidence of dioecy between sites. Of the ~50 species of native trees, shrubs and large herbs recorded in a littoral rainforest remnant at Camden Haven (G. Williams 2000), to the immediate north of the Manning Valley, ~11 were dioecious, or partly dioecious. Such high levels of dioecy assume greater significance with respect to the management and conservation of individual stands, and the preservation of species, especially where these exist at low population levels, or where individuals within populations are hyper-dispersed within stands; such that pollen flow between male and female plants may be uncertain, or not be achieved at all.

Appendix 2. Large insects and their place in the scheme of things

POLLEN LOADS CARRIED BY LARGE INSECTS IN AUSTRALIAN RAINFORESTS

Most insects that visit Australian rainforest trees are less than 6 mm in length. They can transport pollen and act as pollinators of a great number of Australian tropical, subtropical and temperate zone rainforest plants. But small-sized insects are not the only size class of invertebrates found on the flowers of rainforest plants.

Although large ecologically specialised bees are pollinators of some Australian rainforest plants, the capacity of ecologically unspecialised larger insects (arbitrarily characterised as species >9 mm in body length) to transfer pollen in Australian rainforest communities has received little attention. The point at which insects can be considered large is subjective, but beyond 9 mm individual insects become conspicuous to the casual human observer.

Large insects constitute only a small proportion (generally much less than 10%) of the number of individuals and species of anthophilous insects that visit Australian rainforest trees (Basset and Kitching 1991; G. Williams 1995; Williams and Adam 1998). However, their potential for relatively long-distance inter-plant movements may have important consequences for the reproductive success of disjunct or isolated plants. This may be particularly important for plants possessing small entomophilous flowers – such plants are not regularly visited by birds (which might act as long-distance pollen vectors) and rarely, if ever, by bats.

Historically, bees and hover-flies have been considered more efficient and specialised collectors of pollen and, consequently, pollinators of flowering plants, than other insects. This is because of the perceived nature of their foraging behaviour and obligate adaptation to flowers as a source of food. The foraging behaviour of eusocial apid bees is well documented but the pollination contributions of other bee families are varied and the many 'cousin' Hymenoptera of large aculeate wasps (Vespoidea and Spheciformes) that visit numerous mass-flowering rainforest trees (Williams and Adam 1995) largely remain outside the guild of animals popularly thought of as pollinators.

Although several beetle families possess dense or long vestiture, beetles (Coleoptera) have been characterised as smooth-bodied and so are unlikely to be effective pollinators. The alleged smooth body surface of many Coleoptera was assumed to make them unlikely pollinators; the absence of coarse surface sculpture and profuse hair would not permit the ready adherence of pollen grains. Also, their clumsy movements were thought to restrict them to open, bowl-shaped blossoms which offered little in the way of nectar or pollen resources. This assumption failed to examine the ability of fine surface sculpturing and micro-hairs to snare pollen, and the collective pollen and nectar rewards that the inflorescences of many mass-flowering rainforest trees can offer.

Increased surface vestiture, pollen-collecting hairs (palynophilic setae) and minute sculpturing (punctation and foveae) of the face, forebody and undersurface that retain pollen, and possibly pits and grooves on the head and elongated mouthparts (as in members of the family Buprestidae) are considered derived (apomorphic) evolutionary characters in flower-visiting beetles. Such pollen-retaining structures in pollen-feeding beetles may have evolved as food-collecting surfaces. As an unintended evolutionary outcome (on the part of the beetle) they also permit transport of pollen between plants. Thus beetles and other insects that possess these structures have the potential to act as pollinators. Individual grains of pollen can be found almost anywhere on the body of a flower-visiting insect, but most pollen lodged on large insects is located as aggregations or discrete masses. This may be because large beetles and many wasps, and potentially other insect groups, alight on a flowering tree then clamber among the available flowers, their bodies moving heavily over open anthers and haphazardly grading pollen into clusters. This premise requires testing.

Although most large anthophilous rainforest insects undertake frequent inter-plant movements and relatively long-distance inter-plant flights between widely spaced trees, several uncommon or episodic insect taxa (seasonal absences interspaced by occurrence of massed populations, as in some Cantharidae), particularly in the Coleoptera (Tenebrionidae-Alleculinae, Cantharidae and

Oedemeridae), undertake less frequent inter-plant movements. The species could still be important carriers of pollen between flowers in individual inflorescences, but these more or less sedentary insects reduce the pollen available for carriage by insects that undertake more frequent and relatively longer inter-plant movements.

Few, if any, of the large insects attracted to the flowers of rainforest plants can be considered specialist foragers in the sense of participating in a one-to-one relationship, or even being restricted to specific plant families. Rather, the limiting factors are whether the flower's architecture permits feeding and whether the flower offers an appropriate food reward (or other attribute, such as a refuge site). Thus Lepidoptera are not expected at flowers that do not manufacture and exude nectar and most large beetles normally do not feed from deep gullet and tube-shaped blossoms, although several cetoniine beetles use brute force to reach any nectaries, no matter how deeply hidden. Nevertheless, large insects may exhibit temporal fidelity or constancy to a particular species in flower but, unlike bees and some behaviourally advanced large wasps, very few large insects undertake cleaning behaviour (which would reduce their pollen load). If a plant continues to provide pollen and or nectar sufficient to meet the requirements of foragers, there is no incentive (other than satiation) to move on. Unless diurnal and nocturnal behavioural cycles require flight from a particular flowering tree to a site of temporary daily refuge, many large insects will remain on the flowering tree, even if hidden among the inflorescence mass.

EXAMPLES OF LARGE POLLEN-CARRYING INSECT TAXA

Coleoptera-Buprestidae

Normally brightly coloured or of brilliant metallic hue, the buprestid fauna (commonly called jewel beetles) includes several genera and many species restricted to Australian rainforests. Some, as adults, are exclusively associated with foliage (phyllophagous) or are anthophilous (flower associates, feeding on pollen or, more usually, nectar) (Bellamy *et al.* 2013; G. Williams 2020). The principal larger-sized anthophilous buprestid genera encountered in rainforest include *Castiarina*, *Metaxymorpha*, *Calodema* (Plates 100, 101, 102) and sometimes *Stigmodera* and *Temognatha*. Many species of *Castiarina* are known only from the flowers of rainforest canopy trees but anthophilous rainforest Buprestidae exhibit no obligate association with single plants or even families of plants, although they favour many of the canopy-flowering Myrtaceae (e.g. Volkovitsh *et al.* 2003). The restriction is largely to nectar-producing flowers exhibiting the general entomophilous syndrome of open flower structure. Anthophilous rainforest buprestids, large and small, will gather to these and remain constantly attracted to them until floral rewards are depleted or another flowering tree or species takes their attention. The combination of fidelity to a temporally discrete resource and promiscuity in relation to whatever appropriate flowering plants are available, means that pollen types found on large buprestids may be mixed or of a singular nature.

Although vestiture in several non-Australian buprestid genera is long and dense (e.g. southern African *Acmaeodera* and *Julodis*) most Buprestidae possess short setae. Placement of pollen in anthophilous Australian rainforest Buprestidae is principally ventral and on upper front (anterodorsal) areas, with major sites of pollen deposition being the frontal cavity of the head, areas of the thorax (thoracic hypomera, prosternum and metasternum), antennal cavities and base of the legs. Surface punctation and vestiture on the head, upper surface of the thorax and undersurfaces generally permits the retention and transport of potentially large pollen loads. Vestiture on the undersurface of the abdomen (the abdominal ventrites) and sutures between individual abdominal segments often carry significant loads. However, vestiture is vulnerable to wear and abrasion, which may reduce pollen-carrying capacity.

Coleoptera-Cerambycidae

Members of this beetle family are commonly called 'longicorns' and, while many are found only on fallen timber and leaves, they represent a taxonomically and ecologically diverse group that includes several obligate flower-frequenting species (Ślipiński and Escalona 2013, 2016). Large Cerambycidae are infrequent or spatially and temporally localised flower visitors in rainforest; distinctive large and resplendent species include *Tropocalymma dimidiata* and *Distichocera superba* (Plates 103, 104). Several species are thought restricted to rainforests, though several (e.g. *Aridaeus thoracicus* and *Tragocerus spencei*) are also found on the flowers of numerous sclerophyllous Myrtaceae. As with the Buprestidae, principal placement of pollen in Cerambycidae is on the ventral and anterodorsal regions of the body, with areas of the head, antennal and leg cavities, and the lower thorax surface (mesosternum) being major deposition sites. In Cerambycidae, however, the prosternum (the central area of the undersurface of the thorax) is generally reduced to a smooth convex structure without a prominent frontal lobe or vestiture. This reduces its capacity to function as a pollen deposition site.

Coleoptera–Scarabaeidae–Cetoniinae

The subfamily Cetoniinae, of the family known as scarab beetles, are often called rose-chafers; owing to their association with various flowering plants (not solely roses) in Europe. Adult Cetoniinae constitute a flower-adapted group within the otherwise predominantly phyllophagous Scarabaeidae. They are unique among all beetles in that their wing covers (modified and hardened fore-wings called elytra, sing. elytron) are held down against the abdomen in flight (rather than being raised) with the flight wings articulating out at the sides. This is achieved by a curvature on the margin of each elytron that permits the wing to swing outwards and allows rapid flight. The consequent ability to move over relatively long inter-plant distances within plant metapopulations is unique in the world of insect pollinators.

The Australian temperate rainforest cetoniine fauna is relatively poor, but upon the entomophilous flowers of subtropical rainforest trees are found *Bisallardiana gymnopleura, Chondropyga dorsalis, Eupoecila australasiae, Glycyphana brunnipes* (Plate 105), *Neorrhina octopunctatum, Neorrhina punctatum* and *Storeyus fasciculatus*. These are ecological generalists, for they are also just as typical of the large anthophilous beetle fauna found in adjacent eucalypt-dominated sclerophyllous forests. In the more northern subtropical rainforests and the rainforests of tropical Queensland many additional genera occur (e.g. *Lenosoma, Lomaptera* and *Dilochrosis*), including species wholly restricted to flowers of the rainforest environment. However, rainforest-associated cetoniines forage at a variety of flowers, for example Hasenpusch and Moeseneder (2010) record *Storeyus fasciculatus* at flowering *Barkyla* species (Caesalpiniaceae), *Backhousia myrtifolia* and *Syzygium floribundum* (both Myrtaceae), and *Lenosoma fulgens* at the flowers of *Melicope micrococca* (Rutaceae) and *Barklya syringifolia*; both in southern Queensland.

Although Cetoniinae generally lack the prominent body punctation of Buprestidae several species possess long, dense setal clusters and surface vestiture which readily snare pollen grains. Some structures reflect specialised modifications that increase the beetle's ability to carry pollen, for example, the hooked setae noted on the lateroventral surface of *Bisallardiana gymnopleura* (Plate 113) (Williams and Adam 1998); a feature of Cetoniinae previously only recorded for the European cetoniine *Cetonia aurata* (Englund 1993). Although modified plumose hairs are known from other scarab-related taxa (e.g. from the scarabaeoid genera *Amphicoma* and *Lichnanthe* [Glaphyridae]) (Crowson 1981), and may occur in overlooked related species, the presence of specialised pollen-collecting hairs on *Bisallardiana gymnopleura* (and *Cetonia aurata*) represents a faculty otherwise described only from bees, hover-flies and more recently bee-flies (Bombyliidae) (Neff *et al.* 2003).

The head, and lateral and ventral surfaces of the thorax of Cetoniinae carry heavier pollen loads than other parts of the body. Unlike Buprestidae, in the Cetoniinae the upper forehead area (frons) of the head capsule does not generally possess a pronounced frontal or vertical cavity or depression. Consequently, pollen clusters are not normally associated with this part of the body. The clypeus (roughly analogous to an upper lip), however, frequently possesses large and readily discernible pollen loads. The importance of the clypeus as a site of pollen deposition probably reflects the pronounced development of this structure in the family Scarabaeidae more generally, although some members of the scarab subfamilies Dynastinae and Rutelinae possess a reduced clypeus. However, in Australia these are usually not anthophilous groups.

Coleoptera potentially carry large pollen loads, and feeding behaviour in anthophilous Buprestidae, Cerambycidae and Cetoniinae is similar. Species characteristically feed on nectar by inserting the head and mouthparts into the perianth or hypanthium. Short-distance inter-flower movements are undertaken without the aid of flight by clambering across and through paniculate and corymbose inflorescences. Long-distance inter-plant flights are undertaken either without preceding visual warning, as in Cetoniinae, or as a response to disturbance (e.g. in the buprestid genus *Castiarina*). The similarities in sites of pollen deposition on these taxonomically unrelated families is not unexpected, given their generally similar foraging behaviour.

Hymenoptera: wasps (Ichneumonoidea, Spheciformes and Vespoidea)

Among the larger wasps occurring on the flowers of subtropical rainforest trees are species of the families

Crabronidae, Thynnidae, Scoliidae, Sphecidae, Pompilidae and Vespidae (aculeate wasps), and Ichneumonidae. Of these, Crabronidae, Thynnidae, Vespidae, and occasionally Sphecidae (*Sceliphron*, *Sphex*), are the most abundant larger-sized wasps frequently carrying significant pollen loads, yet as a hymenopteran guild only a small proportion (e.g. <20% of individuals) may carry large pollen loads. This is lower than the proportion of Coleoptera individuals from New South Wales subtropical rainforest that have been recorded carrying abundant quantities of pollen and is probably due, apart from any influence of actual time spent foraging (and therefore time available to accrue pollen loads), to the cleaning behaviour periodically undertaken by many hymenopteran species.

Sites of pollen deposition on large wasps appear to be more varied than on Coleoptera. The head, basal antennomeres and dorsal surface of the mesosoma (the 'thorax', for lack of a more lengthy and accurate description) are commonly major pollen deposition sites. The legs, particularly the femora and tibiae, carry pollen loads to an extent not generally paralleled in the Coleoptera. Species of *Anthobosca* (Thynnidae) possess specialised 'pollen combs' on the forelegs (Plates 89, 112) – functionally analogous structures are found in Apiformes and floricolous Scarabaeidae, which can carry the largest proportion of the pollen load. In *Dimorphothynnus dimidiatus* (Thynnidae) (Plate 90) pollen is associated with mesepisternal grooves located on the sides of the thorax, while in *Sphex fumipennis*, *Sphex ephippium* (Sphecidae) and species of *Bembicinus* (Crabronidae), *Radumeris* (Scoliidae), *Rhagigaster* (Plate 91) and *Zeleboria* (Thynnidae) pollen can be found deposited at the base of the forewings. In *Echthromorpha intricatoria* (Ichneumonidae) pollen may adhere to sites adjacent to the mouth (the prementum and malar region of the head), and the forelegs.

Robust, setose Hymenoptera-Scoliidae and some Vespidae-Eumeninae move across and between inflorescences in a similar manner to that of large Coleoptera. This clambering method of inter-flower movement maximises physical contact between the flower and insect's body. There is potential for heavy pollen loads, deposited widely upon the insect.

Sphecidae and Crabronidae undertake frequent interplant and inter-flower movements and remain on individual inflorescences or flowers for short periods (usually <10 s). For example, various species of *Bembix* generally make foraging bouts of less than three seconds at individual flowers. Sphecidae and large Crabronidae frequently feed with their body relatively raised above most of the flower structure, reducing contact with stamens. Although closely related to bees, sphecid and crabronid behaviour differs from that of bees, which frequently undertake foraging bouts involving prolonged and repeated contact with stamens. In Australian subtropical rainforest, Sphecidae and Crabronidae can be frequently found on *Alphitonia excelsa*.

Diptera

Although small-sized flies (<6 mm) are a very abundant component of the insect fauna visiting rainforest flowers, very few large species are observed or collected. The pool of larger-sized Diptera in temperate, subtropical and tropical rainforests, from which pollinators can be drawn, appears to be small both in terms of species and individuals (though tabanids may be important in the pollination of Tasmanian species of *Eucryphia*). However, entomophilous rainforest trees flowering on rainforest margins (Plate 121) are able to recruit large anthophilous flies, such as Bombyliidae, from adjoining vegetation types. Large-sized members of the Gondwanan fly family Pelechorhynchidae likewise may move from the blossoms of montane sclerophyllous plants to flowering angiosperms in adjacent cool temperate rainforests. Syrphidae are generally considered to be variously specialised oligolectic flower visitors. However, few ecologically specialised Syrphidae are recorded from tree blossoms although several generalist semi-adapted genera occur in rainforest (e.g. *Psilota*, *Dideopsis*, *Eristalinus* and *Mesembrius*).

The genus *Trichophthalma* (Nemestrinidae) visits various mass-flowering tree species in rainforest (e.g. *Cuttsia viburnea* [Escalloniaceae/Rousseaceae] and *Syzygium floribundum* [Myrtaceae]) and individuals, although relatively uncommon, can carry significant amounts of pollen (>300 grains) with minimal contamination (0–<5%) by foreign pollen, indicating that members exhibit a level of foraging constancy (fidelity) to a species when in flower. Similarly, species of *Paramenia* (Calliphoridae) and *Scaptia* (Tabanidae) have been found to carry heavy pollen loads (>200 grains, >90% home pollen). In large flies the anteroventral surface generally, and the head, femora and tibiae, are the major pollen deposition sites. In species of *Trichophthalma* the ventral surface of the abdomen and lateral areas of the thorax carries pollen masses. In general, however, the dense pilose surface of large Diptera precludes pollen accretion on the body surface.

Table A2.1. Examples of pollen loads from large insects that undertake frequent inter-canopy movements in Australian subtropical rainforest (after Williams and Adam 1998).

Insect taxa	Plant species	Pollen load (%)	Home pollen (%)
COLEOPTERA			
Buprestidae			
Castiarina acuminata	*Alphitonia excelsa*	6–54(1)	98–100
C. acuminata	*Cuttsia viburnea*	A	<30
C. acuminata	*Guioa semiglauca*	M–A	>95
C. delta	*Tristaniopsis laurina*	A	>99
C. insignis	*Syzygium floribundum*	A	100
C. insignis	*Syzygium smithii*	A	100
C. insignis	*Tristaniopsis laurina*	H	>99
C. neglecta	*Alphitonia excelsa*	M	100
C. producta	*Guioa semiglauca*	A	>70
C. producta	*Syzygium floribundum*	A	100
C. producta	*Syzygium smithii*	A	100
C. producta	*Tristaniopsis laurina*	M–A	<70–?100
C. pulchripes	*Cuttsia viburnea*	A	>95
C. ?sexcavata	*Syzygium smithii* var. *minor*	A	>95
C. ?vicina	*Syzygium floribundum*	A	>99
Curis aurifera	*Syzygium floribundum*	A	>70
C. splendens	*Syzygium smithii* var. *minor*	M	<50
C. splendens	*Syzygium floribundum*	A	>99
Metaxymorpha grayi	*Tristaniopsis laurina*	A	>99
Torresita cuprifera	*Syzygium floribundum*	M	>95
Cerambycidae			
Aridaeus thoracicus	*Tristaniopsis laurina*	H	>95
A. thoracicus	*Syzygium floribundum*	A	>99
Distichocera superba	*Syzygium floribundum*	A	>99
D. superba	*Tristaniopsis laurina*	H	100
Hesthesis sp.	*Tristaniopsis laurina*	M	<30
Tragoceras spencei	*Syzygium floribundum*	A	>90
Tropocalymma dimidiata	*Syzygium smithii*	H	>99
Scarabaeidae			
Bisallardiana gymnopleura	*Syzygium floribundum*	M	100
Chondropyga dorsalis	*Euroschinus falcatus*	18(1)	94
C. dorsalis	*Syzygium floribundum*	A	>99
Eupoecila australasiae	*Alphitonia excelsa*	M	100
E. australasiae	*Syzygium floribundum*	A	>99
Glycyphana brunnipes	*Alphitonia excelsa*	M	<70
G. brunnipes	*Euroschinua falcatus*	21(7)	33

Continued

Table A2.1. (*Continued*)

Insect taxa	Plant species	Pollen load (%)	Home pollen (%)
G. brunnipes	*Tristaniopsis laurina*	M	>99
Neorrhina punctatum	*Alphitonia excelsa*	H	>99
N. punctatum	*Syzygium floribundum*	A	>99
N. punctatum	*Tristaniopsis laurina*	A	>95
DIPTERA			
Calliphoridae			
Paramenia sp.	*Alphitonia excelsa*	H	>90
Nemestrinidae			
Trichophthalma nigrovittata	*Cuttsia viburnea*	A	100
T. ?punctata	*Syzygium floribundum*	H	>95
Syrphidae			
Dideopsis sp.	*Alphitonia excelsa*	3	100
Tabanidae			
Cydistomyia nigripicta	*Tristaniopsis laurina*	9(1)	89
Scaptia auriflua	*Syzygium floribundum*	M	>95
S. ?quadrimacula	*Cuttsia viburnea*	H	>95
HYMENOPTERA			
Crabronidae			
Bembicinus sp.	*Alphitonia excelsa*	3(1)	66
Bembix promontorii	*Alphitonia excelsa*	4	100
Bembix sp.	*Alphitonia excelsa*	8(@4)	@50
Bembix sp.	*Syzygium smithii*	19(8)	58
Tachysphex sp.	*Alphitonia excelsa*	96(0)	100
Ichneumonidae			
Echthromorpha intricatoria	*Alphitonia excelsa*	H	100
Pompilidae			
Chryptocheilus sp.	*Tristaniopsis laurina*	A	100
Scoliidae			
Radumeris ?tasmaniensis	*Alphitonia excelsa*	83(3)	96
R. ?tasmaniensis	*Diospyros australis*	A	<1
Pseudotrielis zonata	*Tristaniopsis laurina*	A	>99
Scolia verticollis	*Alphitonia excelsa*	H	>95
S. ?verticollis	*Tristaniopsis laurina*	M	>95
Sphecidae			
Sceliphron laetum	*Tristaniopsis laurina*	M	>90
Sphex ephippium	*Alphitonia excelsa*	7(4)	43
S. fumipennis	*Alphitonia excelsa*	48(3)	94

Continued

Table A2.1. *(Continued)*

Insect taxa	Plant species	Pollen load (%)	Home pollen (%)
Thynnidae			
Anthobosca ?signata	*Alphitonia excelsa*	M	>95
A. ?signata	*Alphitonia excelsa*	A	100
A. ?signata	*Tristaniopsis laurina*	M	>95
Diamma bicolor	*Alphitonia excelsa*	73(7)	>90
Dimorphothynnus dimidiatus	*Syzygium smithii*	83(@50)-H	<40–>95
D. dimidiatus	*Guioa semiglauca*	M	100
Hemithynnus apterus	*Syzygium smithii*	22(?)	>50
H. rufiventris	*Tristaniopsis laurina*	H	>99
Zaspilothynnus sp.	*Alphitonia excelsa*	38(@9)	@76
Vespidae			
Abispa splendida	*Tristaniopsis laurina*	A	>99
Bidentodynerus bicolor	*Tristaniopsis laurina*	H	80
?Epiodynerus sp.	*Tristaniopsis laurina*	M	<60
Polistes humilis	*Alectryon coriaceus*	8(3)	63
P. tepidus	*Alphitonia excelsa*	M	100
Pseudabispa confusa	*Tristaniopsis laurina*	A	>99

Approximate pollen grain counts in three abundance classes: M = moderate, >100 grains; H = heavy, >200 grains; A = abundant, >300 grains. Total counts given for <100 grains; number of foreign grains given in brackets; >99 = probably all home pollen with several unassignable grains; @ = pollen count indeterminate due to grain deformation of similarity of type; sp. = species; home pollen = pollen from plant which insect was visiting; ? = species uncertain.

SUMMARY

Although large insects comprise only a small proportion of the insect visitors to rainforest flowers, they are capable of transporting quantities of pollen potentially beneficial to the reproductive ecology of rainforest plants. Large insects collected from subtropical rainforest Myrtaceae generally carry heavy pollen loads which consist principally of home pollen (Table A2.1). The examination of pollen loads on insects indicates temporal foraging constancy by individual taxa, which would minimise pollen loss to non-conspecific flowering plants. Sites of pollen deposition on the body surface of individual insect taxa are diverse and pollen loads are variable, influenced by time spent foraging, foraging behaviour, body morphology and insect size, as well as quantity of pollen presented by individual flowers and plants. For example, the relatively small amounts of pollen produced individually by the flowers of *Alphitonia excelsa* and the method of pollen presentation, which requires depression of the stamens and enveloping petals to release pollen masses, may restrict the pollen loads on large insects regardless of the time they spend foraging and the number of flowers they visit. In general, Coleoptera seem to carry large amounts of pollen, contrary to the suggestion that pollen does not adhere to their smooth bodies. It has been suggested that large Coleoptera are pollinators of trees and shrubs in several Australian open forest and woodland communities. This function can be applied equally to rainforest communities.

Based upon the development of body vestiture, two broad groupings of taxonomically unrelated large anthophilous insects on mass-flowering rainforest plants can be recognised:

1 species with sparse or poorly developed vestiture (e.g. Buprestidae, Cerambycidae and Vespidae) which generally spend individually long periods foraging on inflorescences and make inter-flower movements upon individual trees by clambering from blossom to blossom. In so doing, large quantities of pollen may adhere to the integument;

2 species with dense, often long, setae (e.g. Crabronidae and some Diptera) that are unlikely to carry heavy pollen loads directly on the integument. Rather,

individual pollen grains are ensnared by setae or small pollen clusters may aggregate in spatially discrete regions. This group contains species characteristically considered specialised in their foraging behaviour but these generally transport small pollen loads, potentially reducing their contribution to reproductive success in mass-flowering rainforest plants.

The flowers of most Australian rainforest trees, especially those found in subtropical and temperate latitudes, conform to the entomophilous floral syndrome. Birds and bats are important inter-plant and long-distance pollinators in many ecosystems but are relatively rare visitors to plants with entomophilous flowers. Consequently, larger-sized insects that tend to undertake flights to widely separated flowering plants (even though their movements may be short compared with those of many birds and bats) may be important in the reproductive ecology of fragmented rainforest plant populations, because isolated plants can suffer greatly reduced levels of pollen receipt.

The daily movements by small insects (<6 mm), particularly those of small Coleoptera and brachyceran Diptera, in rainforests are predominantly within the crowns of individual trees. The majority of inter-plant movements are to near neighbours, and thus are likely to promote autogamous and geitonogamous pollination in self-compatible species. In contrast, the frequent daily movements of large insects between trees are likely to be important for pollen transfer in self-incompatible and dioecious plant species and to promote out-crossing, rather than geitonogamous pollination. Even if inter-plant movements occur less often, they can result in successful pollination of highly dispersed plants.

Appendix 3. Generalised pollen groups based on exine sculpture

Family	Species	1	2	3	4
Acanthaceae	*Pseuderanthemum variabile*	H			
Alstroemeriaceae/Luzuriagaceae	*Drymophila moorei*	H			
Amarylliidaceae/Liliaceae	*Crinum pedunculatum*	H			
Anacardiaceae	*Euroschinus falcatus*	S			
Apocynaceae	*Parsonsia straminea*		V		
Araceae	*Alocasia brisbanensis*				H
	Gymnostachys anceps	H			
Araliaceae	*Polyscias sambucifolia*		S/T		
Arecaceae	*Archontophoenix cunninghamiana*	T			
	Linospadix monostachyos		S/T		
Asparagaceae/Laxmanniaceae/Dracaenaceae	*Cordyline stricta*	S			
Asphodelaceae/Liliaceae	*Dianella* sp.	H			
Asteraceae	*Senecio minimus*				H
Bignoniaceae	*Pandorea pandorana*		V		
Boraginaceae	*Ehretia acuminata*		T		
Cannabaceae/Ulmaceae	*Aphananthe philippinensis*	T			
	Trema tomentosa	S			
Celastraceae	*Elaeodendron australe*	T			
	Denhamia celastroides	T			
Colchiaceae/Liliaceae	*Tripladenia multiflora*		H		
Commelinaceae	*Aneilema acuminatum*		H		
	Commelina cyanea	H			
	Pollia crispata		H		
Convolvulaceae	*Convolvulus erubescens*		H		
	Dichondra repens		H		
	Stephania japonica				V
Cunoniaceae	*Aphanopetalum resinosum*		V		
	Ackama paniculosa	T			
	Callicoma serratifolia	S/T			
	Schizomeria ovata		T		
Cyperaceae	*Cyperus* sp.		H		
Dilleniaceae	*Hibbertia scandens*		V		
Ebenaceae	*Diospyros australis*	T			
Elaeocarpaceae	*Elaeocarpus obovatus*	T			
	Elaeocarpus reticulatus	S/T			

Continued

Family	Species	1	2	3	4
Ericaceae/Epacridaceae	*Trochocarpa laurina*		S/T		
Escalloniaceae/Rousseaceae	*Abrophyllum ornans*	T			
	Anopterus macleayanus		S/T		
	Cuttsia viburnea	T			
Euphorbiaceae	*Aleurites moluccana*				T
	Croton insularis			T	
	Croton verreauxii		S/T		
	Mallotus philippensis	T			
Eupomatiaceae	*Eupomatia laurina*		S/T		
Fabaceae	*Austrosteenisia blackii*	V			
	Castanospermum australe		T		
	Desmodium acanthocladum	S			
	Glycine sp.		V		
Flacourtiaceae/Saliciceae	*Scolopia braunii*		T		
Flagellariaceae	*Flagellaria indica*	V			
Geraniaceae	*Geranium ?homeanum*				H
Lamiaceae	*Prunella vulgaris*		H		
Lamiaceae/Verbenaceae	*Clerodendrum tomentosum*		S/T		
Lauraceae	*Cryptocarya glaucescens*	T			
	Cryptocarya ?laevigata			S/T	
	Cryptocarya meissneriana		T		
	Cryptocarya rigida		T		
	Endiandra discolor			T	
	Endiandra muelleri			T	
Lobeliaceae	*Lobelia trigonocaulis*	H			
	Pratia ?purpurascens	H			
Malvaceae	*Hibiscus heterophyllus*				S/T
	Hibiscus splendens				S/T
Malvaceae/Sterculiaceae	*Brachychiton acerifolius*		T		
	Brachychiton discolor			T	
	Commersonia fraseri		S/T		
Meliaceae	*Dysoxylum fraserianum*	T			
	Dysoxylum rufum	T			
	Synoum glandulosum	T			
Monimiaceae/Atherospermataceae	*Daphnandra micrantha*	T			
	Wilkiea huegeliana	S/T			

Continued

Family	Species	1	2	3	4
Moraceae	*Ficus coronata*	T			
	Ficus rubiginosus	T			
	Maclura cochinchinensis	V			
	Trophis scandens	V			
	Streblus brunonianus	T			
Myrsinaceae/Primulaceae	*Aegiceras corniculatum*	T			
	Myrsine howittiana	T			
Myrtaceae	*Archirhodomyrtus beckleri*		S/T		
	Backhousia sciadophora	T			
	Rhodamnia rubescens	T			
	Rhodomyrtus psidioides		T		
	Syncarpia glomulifera	T			
	Syzygium smithii	T			
	Syzygium floribundum	T			
	Syzygium oleosum	T			
	Tristaniopsis laurina	T			
Oleaceae	*Jasminium volubile*				V
	Notelaea longifolia			S/T	
Orchidaceae	*Microtis* sp.			H	
	Pterostylis sp.				H
Philesiaceae	*Geitonoplesium cymosum*	V			
Phyllanthaceae/Euphorbiaceae	*Breynia oblongifolia*			S	
Pittosporaceae	*Hymenosporum flavum*		S/T		
	Pittosporum revolutum	S			
	Pittosporum undulatum	T			
Podocarpaceae	*Podocarpus elatus*			T	
Proteaceae	*Banksia integrifolia*	T			
	Macadamia integrifolia	T			
Putranjivaceae/Euphorbiaceae	*Drypetes deplanchei*		T		
Ranunculaceae	*Ranunculus plebeius*		H		
Rhamnaceae	*Alphitonia excelsa*	T			
Ripogonaceae	*Ripogonum album*		V		
Rosaceae	*Rubus hillii*		V		
	Rubus rosifolius		V		
Rubiaceae	*Psydrax ?odorata*			S/T	
	Morinda jasminoides		V		
	Psychotria ?daphnoides		S		

Continued

Family	Species	1	2	3	4
Rutaceae	*Acradenia euodiiformis*	T			
	Acronychia imperforata		T		
	Geijera salicifolia		T		
Sapindaceae	*Alectryon coriaceus*	S/T			
	Alectryon subcinereus		T		
	Alectryon subdentatus			T	
	Cupaniopsis anacardioides	T			
	Guioa semiglauca		T		
	Harpullia pendula		T		
	Mischocarpus pyriformis	T			
Sapotaceae	*Pouteria australis*	T			
Smilacaceae	*Smilax australis*		V		
	Smilax glyciphylla			V	
Solanaceae	*Duboisia myoporoides*	S/T			
	Solanum aviculare	S			
	Solanum stelligerum	S			
Violaceae	*Viola* sp.			H	
Vitaceae	*Cayratia clematidea*		V		
	Cissus antarctica		V		
	Cissus hypoglauca			V	
Winteraceae	*Tasmannia insipida*				S/T
Zingiberaceae	*Alpinia caerulea*		S		
	Alpinia sp.	S			

Sculpture types after Faegri *et al.* (1992). 1 = sculpture indistinct or smooth; 2 = sculpture weakly scabrate-verrucate, striate; 3 = sculpture strongly scabrate-verrucate, striate; 4 = sculpture echinate, baculate or clavate. Habit: H = herb, S = shrub, S/*t* = shrub/small tree, T - tree, V = vine.
Source: Williams and Adam (1999).

Appendix 4. Captions to photographs

Photographs by Geoff Williams unless otherwise credited; additional photographs by Hugh Nicholson and Jack Hasenpusch.

1. Emergent *Araucaria cunninghamii* (Araucariaceae), Richmond Range, northern New South Wales. This is a large to very large wind-pollinated conifer in subtropical, warm temperate, dry and littoral rainforests north from the Macleay River and extending to North Queensland. The genus *Araucaria* dates to the Jurassic. Araucariaceae are an ancient group, now largely confined to the Southern Hemisphere. Fossil Araucarians are recorded from the Upper Triassic but are apparently best represented and most widely distributed during the Jurassic (e.g. *Agathoxylon, Brachyphyllum, Agathis mirabilis*). However, fossil evidence suggests that they have declined since the Cretaceous. *Araucarioides* is a foliage type described from the Eocene of Tasmania. Foliage remains attributed to *Agathis* have also been described from the Eocene of Australia (Taylor *et al.* 2009). All extant Araucariaceae are trees. Three genera survive; *Araucaria, Agathis* and *Wollemia*.

2. *Macrozamia* sp. (Zamiaceae) in dry rainforest on the mid north coast, New South Wales. *Macrozamia* is more commonly considered a genus associated with open eucalypt forests and woodlands, but several species occur in rainforest and wet sclerophyll forest. Cycads are usually dioecious (have separate male and female cone-bearing plants), so pollen has to be transferred between sexes for pollination to occur. However, there are instances of sex change, 'sequential hermaphroditism', in several species and asexual reproduction by the production of bulbils arising as off-shoots of the stem, has also been recorded. Cycads were first thought to be wind-pollinated but more recent studies, while not wholly displacing wind as a possible agent of pollen transport, have demonstrated that insects (beetles and occasionally thrips) are primary pollinators, and that wind-borne cycad pollen has short range carriage such that the cones of widely spaced plants are unlikely to be reached by pollen grains mobilised in the air column.

3. Female dioecious *Lepidozamia peroffskyana* (Zamiaceae) in wet sclerophyll forest, northern New South Wales. This is a palm-like shrub in habit, producing a trunk >5 m in height (though most plants are shorter), and is distributed from northern New South Wales to near Nambour in south-eastern Queensland. Plants are pollinated by long-snouted weevils of the genus *Tranes* (see Plate 4). *Lepidozamia peroffskyana* occurs in subtropical rainforest and wet sclerophyll forest, but is more often found growing prolifically in wet sclerophyll forest-dry sclerophyll forest ecotones. This zone can be subject to frequent fires to which cycads respond post-fire with vigorous new frond growth. Cone production throughout cycad populations may be episodic following fire. *Lepidozamia* is endemic to eastern Australia but has a highly disjunct distribution, with a second extant species, *Lepidozamia hopei*, restricted to north-eastern Queensland, this growing from near sea level (Cape Tribulation) to ~700–1000 m in elevation. *Lepidozamia hopei* is the tallest of Australian cycads (and possibly the tallest in the world), reaching potentially to a height of ~20 m, but unlike *Lepidozamia peroffskyana L. hopei* is principally an inhabitant of shaded rainforest understoreys, and rainforest margins. Fossil species assigned to *Lepidozamia* are also known (Hill 1980), with the current distribution of living *Lepidozamia* suggesting that the two extant species are relics of a once more extensive taxon. *Lepidozamia* was at one time included in *Macrozamia* but the leaflets of *Lepidozamia* are inserted differently on the rhachis, lack the coloured callus of the leaf fronds of *Macrozamia* and have different sporophyll characteristics (Bogler and Francisco-Ortega 2004).

4. Pollinating *Tranes* weevils (Curculionidae) massing on a male cone of the cycad *Lepidozamia peroffskyana*. Although the weevils consume pollen, grains gather about the body surface. This is available for pollination as the weevils have no difficulty entering the spaces formed when female *Lepidozamia* cones are in the receptive stage. The weevils lay their eggs in the mature, spirally splitting, male cones, and although a few larvae develop in the small amount of decaying cone residue sometimes left on the apex of the trunk, or in the central cone core itself, the greater pulpy mass of the cone falls away to the ground carrying many hundreds, if not thousands, of *Tranes* larvae with it. Most larvae enter the soil profile, there to mature and later emerge (upwards of five weeks later; G. Williams pers. obs.). Some adults also enter the microsporophylls but larvae are generally not found in these structures. Ornduff (1989) cites *Tranes lyterioides* as the species associated with the cones of *Lepidozamia peroffskyana* in the Dorrigo–Coffs Harbour region, however, the species associated with *Lepidozamia peroffskyana* is different (Rolf Oberprieler cited in G. Williams 2020).

5. *Drymophila moorei* (Alstroemeriaceae/Luzuriagaceae) is a perennial herb found growing in rainforests at higher elevations (e.g. warm temperate rainforest, Tapin National Park, north from the Manning River, northern New South Wales). The genus is endemic to Australia. The species occurs on shaded forest margins as well as in the dim understorey below closed canopies, flowers mainly during spring. The flowers are solitary or borne in few-flowered cymes. Anthers dehisce by longitudinal slits. Pollinators include halictid bees and syrphid flies; in similar fashion to the pollinator cohort found visiting the mesic forest understorey herb *Pollia crispata* (Commelinaceae). Chrysomelid beetles and gryllid crickets are recorded feeding on flowers (Conran 1988).

6. *Crinum pedunculatum* (Amaryllidaceae/Liliaceae) is a large summer-flowering herbaceous shrub of littoral rainforest margins and sunlit understoreys, damp conditions on the margins of lowland subtropical rainforest, but also inhabits swamps and stream banks. Flowers are bisexual, with the anthers atop long thin filaments suggesting adaptation to morphologically specialised pollinators. Though not discounting birds, these appear to be large butterflies (e.g. *Papilio aegeus*), but principally hawkmoths such as *Theretra nessus* – the only insects that can contact both anthers and stigmas. Smaller visiting insects are resource thieves, their size and foraging behaviour precluding them from the role of pollinators. *Crinum* has a mainly tropical-subtropical distribution, with *Crinum pedunculatum* occurring in eastern mainland Australia, Northern Territory, New Caledonia, Fiji and Timor (Harden 1993). *Crinum pedunculatum* has been variably placed in the Amaryllidaceae, a family with a poor fossil record (Friis *et al.* 2011).

7. *Euroschinus falcatus* (Anacardiaceae) (plate shows female flowers) occurs in subtropical, dry and littoral rainforest formations, and also in wet sclerophyll forest, north from southern coastal New South Wales. It is a small to medium-sized dioecious tree possessing separate male and female plants; pollen must be carried between trees of the two sexes for pollination and reproduction to occur. To enhance pollen transfer, the flowers of male trees open before females so that pollen has been mobilised before female anthesis; anthesis in male flowers occurring over longer periods than that of females. This minimises waste of female flowers and maximises reproductive outcomes. Such early opening may also 'train' pollinators, before availability of synchronously opening female flowers. Flower visitors recorded from subtropical littoral rainforests are predominantly small Diptera, Coleoptera and Hymenoptera, the majority being ecological generalists. Larger-sized Hymenoptera include the crabronid genera *Acanthostethus, Bembicinus, Sphodrotes and Tachysphex*. Bees are dominated by Colletidae (e.g. *Leiopructus irroratus, Amphylaeus nubilosellus, Hylaeus microphenax, H. perhumilis, H. primulipictus*) and Halictidae (*Homalictus* spp., *Lasioglossum* spp.). However, the suite of flower visitors, and pollinators, can differ between individual stands (G. Williams 1995).

8. Floret cluster of the bisexual woody climber *Melodinus australis* (Apocynaceae), an endemic species distributed from southern New South Wales to Queensland. Flowers summer to autumn, and is common on lowland subtropical rainforest margins and also the margins of wet sclerophyll forest. Vines may straggle high into the canopy. Individual flowers are coloured orange-yellow to white. Florets are ~5 mm in length, and comprised of five fused sepals. Anthers are located at the base of the floret corolla, denying access to all but long-tongued visitors; though thrips potentially are also able to gain entry. Worldwide, members of the Apocynaceae are mainly distributed in the tropics but include numerous species occurring in Australian subtropical rainforests (e.g. *Alyxia buxifolia, Cynanchum elegans, Gymnanthera oblonga, Heterostemma acuminatum, Hoya australis, Marsdenia flavescens, Parsonsia straminea, Secamone elliptica, Tylophora barbata*).

9. *Alocasia brisbanensis* (Araceae) in wet sclerophyll forest. Plants grow in sunlit conditions, frequently in damp locations on forest margins and below canopy gaps. As the canopy is restored individual *Alocasia* plants are displaced, along with their attendant insects, both insect and plant populations facing potential extinction within individual rainforest stands. This is an unlikely scenario in large areas of rainforest where disturbance results in numerous opportunities for colonisation, but population extinction is a prospect where rainforest stands are small and isolated from other rainforest; as with many small lowland rainforest remnants.

10. *Alocasia brisbanensis* (Araceae). The 'flower' of the monocot *Alocasia brisbanensis* consists of an elongate spadix, this comprising discrete masses of male and female flowers, and an enveloping collar called a spathe. Male flowers are located nearer the apex of the spadix and females are restricted to the base of the column. Flowers of *Alocasia brisbanensis* attract native *Tetragonula* bees, ants and flies (e.g. Neurochaetidae) but the true pollinators are mainly small beetles – generally only these can gain access to the secluded female flowers. The spadices of numerous species of Araceae are known to produce heat. At the top of the spadix is a large sterile appendage, which becomes hot when the flower is emitting odours that are attractive to its beetle pollinators. Araceae have a fossil history that dates, based on fossil flowers, back to the Early Cretaceous (Friis *et al.* 2011).

11. *Archontophoenix cunninghamiana* (Arecaceae) palms establishing in the subcanopy of subtropical rainforest, Lansdowne Escarpment, northern New South Wales. Plants are vulnerable to extreme drought events. *Archontophoenix cunninghamiana* is a monoecious species, the flowers unisexual. Pollinators include numerous small nematoceran and brachyceran flies, small beetles, wasps, and solitary and social nesting bees. The species occurs in subtropical, and sometimes littoral, rainforest, north from southern New South Wales. The related *Archontophoenix alexandrae* occurs north from Eurimbula, Queensland. However, it has been extensively planted in subtropical regions as an ornamental garden tree. In such situations it can be invasive in adjacent lowland subtropical rainforest and may re-direct pollination services from co-occurring/co-flowering *Archontophoenix cunninghamiana* populations. Its seeds are dispersed by native Wompoo fruit-dove and Topknot pigeons. Arecaceae have an extensive fossil record through the Cenozoic, however, it is difficult to assign relationships of fossil pollen to extant genera (Friis *et al.* 2011).

12. *Calamus caryotoides* (Arecaceae) is an endemic slender climbing palm occurring in the rainforests of lowland and highland north-eastern Queensland. Worldwide, there are more than 370 species in the genus, with eight recorded from Australia (Harden 1993). Species can occur as dense vine tangles armed with whip-like barbed flagella. However, only *Calamus muelleri* and *C. australis* occur at subtropical latitudes. Of these *Calamus muelleri* is the sole representative found in New South Wales; being restricted in distribution from the Macksville area, north to near Maryborough in southern Queensland. Insects are associated with the flowers of *Calamus radicalis*, but in general the pollination ecology of Australian species is poorly studied. However, studies undertaken of species to the north of Australia may offer guidance as to the nature of pollination strategies more broadly found within the genus. Species of *Calamus* in Malaysia and southern Thailand are pollinated by small, nocturnally active noctuid and pyralid moths, and day-active Hymenoptera as well as other insect visitors (Fah 1995; Lee *et al.* 1995). *Calamus manan* from Sabah is known to flower at night and this may serve to filter those flower visitors (i.e. moths) which are most likely to provide effective pollination services (Alloysius 1999). Floral mimicry is also recorded in southern Thailand *Calamus castaneus* (Kidyoo and Mckey 2012). This dioecious species produces early flowering male flowers that offer nectar and pollen rewards. These flower longer than females and are more abundant, and mobilise pollen availability before female anthesis. Female plants each have a pistillate flower accompanied by a sterile male flower that appears to contribute to insect attraction. Female flowers are quite similar in morphology and function to fertile male flowers, they produce similar visual and olfactory cues, but produce no pollen (only nectar); pollination success in *Calamus castaneus* being ensured by the combination of intersexual phenological differences and floral mimicry.

13. The small solitary palm *Linospadix monostachyos* (Arecaceae) in the understorey of a mixed subtropical-warm temperate rainforest, Coorabakh National Park, northern New South Wales. Small weevils (possibly *Amorphoidea linospadicis*) and thrips are likely pollinators. *Linospadix monostachyos* inhabits shaded understorey environments where it may be the only palm found in warm temperate rainforest. It is distributed from Bulahdelah, New South Wales, north to Cooloola and the vicinity of Gympie in south-eastern Queensland. Related species (e.g. *Linospadix minor, L. microcarya*) are endemic to north-eastern Queensland. The genus is restricted to eastern Australia, but mainly in the tropical north-east, and New Guinea.

14. *Normanbya normanbyi* (Arecaceae) is an endemic feather-leaved palm restricted to north-eastern Queensland. *Normanbya normanbyi* is monoecious but has flowers that are temporarily dioecious, and some androdioecious plants (with males and hermaphrodites) possess hermaphrodites with non-functional pollen (Kolesik *et al.* 2006).

15. *Pararistolochia praevenosa* (Aristolochiaceae) is a large woody climber growing in subtropical rainforest in north-east New South Wales and south-eastern Queensland. It is a larval foodplant of the endangered Richmond birdwing butterfly *Ornithoptera richmondia* (Papilionidae). The flowers are bisexual and clustered in axils on leaves. Flowers are pollinated by species of *Forcipomyia,* small midges of the nematoceran fly family Ceratopogonidae. (Photo: Hugh Nicholson)

16. *Cordyline stricta* (Asparagaceae; but previously in Laxmanniaceae, Dracaenaceae, Asteliaceae and Agavaceae) is a small to medium-sized shrub associated with numerous rainforest types, as well as wet sclerophyll forest (Chase *et al.* 2009). Plants usually grow on sunlit margins and in sunny forest openings. The species occurs from the Sydney region, north to south-eastern Queensland. Flowers are actinomorphic and bisexual, readily accessible and primarily visited and pollinated by small bees.

17. *Neoastelia spectabilis* (Asteliaceae) is a mass-flowering rhizomatous perennial herb occurring in New England National Park, northern New South Wales. It is the only member of the genus. Flowers are actinomorphic and bisexual. Flowers November–December and occurs along stream lines and near waterfalls in cool temperate rainforest above 900 m. The species is endemic and is listed as vulnerable under the Federal *Environmental Protection and Biodiversity Conservation Act, 1999.*

18. *Balanophora fungosa* (Balanophoraceae) is a low-growing herbaceous root parasite of Queensland rainforest trees. Potential pollinators include bush rats (*Rattus* spp.), Musky rat kangaroo (*Hypsiprymnodon moschatus*), Giant White-tailed rat (*Uromys caudimaculatus*), Long-nosed bandicoot (*Parameles nasuta*) and four honeyeater species (i.e. *Meliphaga notata, M. gracilis, Xanthotis macleayana, Myzomela obscura*) (Irvine and Armstrong 1990; Pierce and Ogle 2017). Elsewhere ants have been considered possible pollinators of the related Japanese species *Balanophora kuroiwa* (Kawakita and Kato 2002), and the related New Zealand *Dactylanthus taylori* is pollinated by the Lesser short-tailed bat *Mystacina tuberculata* (Czenze and Thurley 2018). (Photo: Hugh Nicholson)

19. *Pandorea pandorana* (Bignoniaceae), a woody climber found on the margins of many rainforest types, including vine thickets, and wet sclerophyll forest. Shade limits occupation of sites. The species ranges from southern Victoria, where it can be conspicuous in isolated coastal stands, to Central Queensland. The flowers are bisexual and deeply tubular. Flower structure and colour suggest adaptation to ecologically specialised pollinators such as meliphagid honeyeaters and long-tongued bees.

20. *Capparis arborea* (Capparaceae), a tree particularly of dry rainforests and some mangrove margins, north from the Hunter River, New South Wales. The conspicuous white flowers are visited by rosechafer beetles, bees and butterflies. However, the principal pollinators of plants growing adjacent estuary zones appear to be long-billed Brown honeyeaters (*Lichmera indistincta*). Capparaceae are mainly tropical and subtropical in distribution. Fossil Capparaceae are known from the Late Cretaceous (*Dressiantha bicarpellata*) (Friis *et al.* 2011). *Dressiantha bicarpellata* flowers are bisexual and slightly zygomorphic, are small in size, but show no obvious adaptation to pollination by birds.

21. *Pollia crispata* (Commelinaceae), a low straggling herbaceous plant that colonises mesic forests, often found densely growing along forest margins, seepage zones and sunlit gaps below open canopies. Plants do not persist in dense shade. Flowers are bisexual. Pollinators include halictid bees and syrphid flies. The genus *Pollia* is known from Miocene leaf and fruit fossils collected in Kenya. Extant Commelinaceae mostly have a tropical and subtropical distribution.

22. *Pollia crispata* occupying a sunlit gap in subtropical rainforest. Colonising growth can be considerably denser, with plant height exceeding 50 cm. Such population densities initially restrict the recruitment of other rainforest species.

23. *Diplocyclos palmatus* (Cucurbitaceae) is a monoecious climbing herb distributed north from northern New South Wales to central Queensland. Male and female flowers occur in mixed clusters. Pollen is slightly sticky. The fleshy anthers appear to exude a clear fluid. The greenish-white corollas are densely hairy, and although such structures in plants have been interpreted as an adaptation to deter ants, these readily enter deep within the corolla of this species. Though found in a variety of rainforest formations colonisation and persistence at sites by *Diplocyclos palmatus* is associated with disturbance events such as canopy openings caused by tree fall, or fire. Such opportunistic recruitment strategies seemingly would preclude the establishment of specialised pollinator relationships. Cucurbitaceae are monoecious or dioecious. In addition to *Diplocyclos palmatus* the family includes several other trailing herbs and climbers that colonise ephemeral sunlight gaps below the rainforest canopy, or rainforest margins following fire (*e.g. Neoachmandra cunninghamii, Neoalsomitra capricornia, Sicyos australis, Trichosanthes subvelotina*); invasions of the latter can be profuse. Fossil evidence for Cucurbitaceae is sparse, however, pollen is recorded from the Oligocene of Africa (Friis *et al.* 2011).

24. *Eucryphia jinksii* (Cunoniaceae/Eucryphiaceae), a rare species with a highly restricted range in south-eastern Queensland (Springbrook Plateau) and north-eastern New South Wales (Numinbah Nature Reserve) (Floyd 2007). Individuals produce coppice growth, can grow as multi-stemmed shrub-like plants, but can also achieve heights of ~25 m (Harden *et al.* 2006). The species was discovered in 1993 by botanist David Jinks and is now listed as *endangered* on the Queensland *Nature Conservation Act, 1992. Eucryphia* has an apparent Gondwanan, or 'Southern', distribution occurring in Chile, Tasmania, New South Wales, south-eastern Queensland, and with an additional shrub-like species (*E. wilkiei*) occurring in the cloud forests on mountain tops of the Wet Tropics (Forster and Hyland 1997). (Photo: Hugh Nicholson)

25. *Schizomeria ovata* (Cunoniaceae), a small- to large-sized tree in subtropical, warm temperate and littoral rainforest, and occasionally in wet sclerophyll forest, from southern New South Wales to North Queensland. The species can be common in wet sclerophyll–dry sclerophyll forest ecotones of northern New South Wales. Flowers in spring. Recorded flower visitors from plants growing in wet sclerophyll forest are predominantly numerous species of small beetles (e.g. Chrysomelidae, Elateridae, Lycidae), wasps and ants. Cunoniaceae include evergreen trees and shrubs from the Southern Hemisphere. Flowers are generally bisexual. Fossil leaves, flowers and fruit are reported from the Cretaceous and Cenozoic. Fossil discoveries indicate that several modern Cunoniaceae lineages, now found mostly in the Southern Hemisphere, were more widely distributed early in their fossil history (Taylor *et al.* 2009).

26. *Vesselowskya ?rubifolia* (Cunoniaceae). The genus *Vesselowskya* consists of two endemic small tree species (*V. rubifolia, V. venusta*) that occupy a similar range (Barrington Tops-vic. Ebor, New South Wales), and are found mainly associated with *Nothofagus moorei*-dominated cool temperate rainforest, growing often along drainage lines. Flowers mainly in spring. Plants flowering in sunlit openings and forest margins are visited by bees and flies.

27. The giant rosette plant *Doryanthes palmeri* (Doryanthaceae) has a restricted range in north-east New South Wales (e.g. Mt Warning) and south-eastern Queensland. Grows on exposed rock outcrops, and flowers late winter–spring, the tall conspicuous inflorescences attracting meliphagid honeyeaters. Flowers are actinomorphic or slightly zygomorphic and bisexual. Each rosette flowers only once in its lifetime. The family Doryanthaceae is endemic, with the genus otherwise only represented by *D. excelsa*, this growing in dry sclerophyll forest and open woodland, from Karauh south to Mt Keira, and with several outlying populations further north near Glenreagh and Corindi, northern New South Wales (Harden 1993). (Photo: Hugh Nicholson)

28. Minute thrips species on a flower of *Diospyros australis* (Ebenaceae). Thrips (Thysanoptera) are implicated as pollinators of the earliest of flowering plants, and are pollinators of many rainforest angiosperms. They are also pollen feeders, thus a proportion of the pollen crop produced by flowers to which they are attracted is lost to them. *Diospyros* is a large genus, worldwide numbering some 700 species, various recorded pollinators including birds, butterflies bees and beetles (Rose *et al.* 2018). The pollinators of *Diospyros australis* also include several bees (e.g. *Hemirhiza melliceps, Palaeorhiza* sp. – Colletidae-Hylaeinae), which appear to be specialists in the rainforest understorey. Ebenaceae are dioecious (flowers unisexual) and in the subtropical rainforest of eastern Australia include several widespread tree species (*Diospyros australis, D. fasciculosa, D. pentamera*) as well as species with restricted distributions (*D. mabacca, D. ellipticifolia*). Male flowers of *Diospyros australis* are occasionally clustered in leaf axils and begin flowering sequentially several weeks before the synchronous flowering of females. Female flowers are either solitary or clustered. Both male and female flowers are generally obscured by foliage.

29. *Elaeocarpus holopetalus* (Elaeocarpaceae), a small tree in elevated cool temperate and warm temperate rainforest from East Gippsland, Victoria, to northern New South Wales. Australia has the greatest diversity of living Elaeocarpaceae (e.g. *Elaeocarpus, Sloanea*) with most species being found in moist environments. The Elaeocarpaceae are one of the families that exemplify the post-Cretaceous diversification of the Australian rainforest flora. The family appears to have arisen in the

Cretaceous, being widespread and diverse in the early Tertiary. Fossil leaves and pollen of Elaeocarpaceae are known in Australia from the Eocene and onwards (Friis *et al.* 2011).

30. *Abrophyllum ornans* (Escalloniaceae/Rousseaceae) is a large shrub-small tree in subtropical and warm temperate rainforest. Plants are usually found in association with gullies and stream banks. *Abrophyllum ornans* occurs from the south coast of New South Wales, north to Queensland. Plants flower October–December. Flowers are bisexual, but unlike the related *Cuttsia viburnea*, which is largely self-incompatible, those of *Abrophyllum ornans* are highly self-compatible (see Table 4.1). Potential pollinators include *Paralastor* sp. (Vespidae), *Trigonula carbonaria* (Apidae), *Megachile pictiventris* (Megachilidae), and *Callomelitta ?fulvicornis* (Colletidae-Colletinae). These undertake inter-plant flights. Small Coleoptera also frequent flowers (e.g. *Meriphus* sp. [Curculionidae], *?Hoshihananomia* [Mordellidae]) (G. Williams 1995). Although small beetles appear to forage only within the confines of single plants or inflorescences, nevertheless, given that *Abrophyllum ornans* is highly self-compatible, their contribution to pollination cannot be dismissed.

31. *Cuttsia viburnea* (Escalloniaceae/Rousseaceae) is an endemic mass-flowering small tree of sunlit subtropical and warm temperate rainforest margins and streamlines. The species is distributed from the Manning Valley, northern New South Wales, to about Rockhampton. Flowers are bisexual, slightly protogynous but self-incompatible. The stigma is 3-lobed and fleshy, and appears receptive at anthesis.The open, readily accessible, flowers attract a diversity of beetles, flies, thrips, bees and wasps. Although most of these are <5 mm in size there are several large beetles that undertake relatively long distance inter-plant flights (e.g. *Calodema regale* [Buprestidae], *Bisallardiana gymnopleura, Chondropyga dorsalis, Eupoecila australasiae* [Scarabaeidae-Cetoniinae]) that contribute to out-crossing. The three Cetoniinae species are habitat generalists, occurring in open forest and woodland, in addition to lowland subtropical rainforest.

32. *Aleurites moluccana* (Euphorbiaceae), is a self-compatible tree native to North Queensland, and also recorded from New Zealand, New Caledonia, Polynesia, India and China (Harden 1990). Occasionally planted as an ornamental in New South Wales. Male and female flowers are mixed in a terminal panicle. The female flowers are larger and placed at the base of the inflorescence. The species is potentially invasive in lowland subtropical rainforest owing to high levels of viable seed produced by single isolated trees (G. Williams pers. obs.). Fossil Euphorbiaceae are present in the Early Cenozoic of Australia.

33. *Baloghia inophylla* (Euphorbiaceae). This is a medium-sized tree occurring in subtropical, dry and littoral rainforest north of Mt Dromedary, southern New South Wales. It usually occupies the subcanopy stratum, and regenerates from basal coppice following low-medium fire impacts. Plants expel their mature seeds ballistically. *Baloghia* is a genus of monoecious small to medium-sized trees. The related *Baloghia marmorata* is a very rare shrub or small tree known only from a few sites north from the Lismore district (northern New South Wales) to Mt Pinbarren, south-east of Gympie (Queensland) (Harden *et al.* 2006).

34. *Croton insularis* (Euphorbiaceae) is a deciduous or partly deciduous small tree (to ~15 m high) occurring in subtropical rainforest, dry rainforest, vine thickets, and more rarely littoral rainforest (Black Head, Redhead south-east of Taree), north from the Sydney region to Carnarvon National Park (central Queensland). Flowers lack petals. Inflorescences are raceme-like. Flowers September–January. Mass-flowering events attract a diversity of insects, particularly bees and beetles.

35. The evolutionarily conservative small tree *Eupomatia laurina* (Eupomatiaceae) is one of the few known instances of close plant–pollinator relationships in Australian rainforests. The genus *Eupomatia* is endemic to Australia and is pollinated solely by small weevils in the genus *Elleschodes* (here emerging from flower [the synandrium]). There are three species in the genus *Eupomatia*; *E. laurina* (East Gippsland to Eurimbula north of Bundaberg, central coastal Queensland, New Guinea), *E. bennettii* (northern New South Wales to south-eastern Queensland) and *E. barbata* (Wet Tropics, Cooktown to Ingham Queensland) (Cooper and Cooper 2004; Jessup 2002). *Eupomatia* possess actinomorphic and bisexual flowers but these are functionally protogynous in which the carpels mature before the anthers. Fossil evidence suggests that the Eupomatiaceae may have been more widely distributed (Friis *et al.* 2011).

36. *Eupomatia laurina* showing the excision of the fused operculum at anthesis. After fertilisation the fused floral body (synandrium) falls away carrying the eggs of pollinating *Elleschodes* weevils with it to the ground. Weevil larvae develop within the soil profile, later emerging.

37. *Cynometra lenticellata* (Fabaceae) illustrating conspicuous recruitment contrast between flowers and foliage. *Cynometra lenticellata* occurs in North Queensland and New Guinea. Trees generally flower in spring and may grow to more than 20 m in height; though most are much smaller (10–12 m). (Photo: Jack Hasenpusch, Australian Insect Farm)

38. *Cynometra lenticellata* (Fabaceae), showing flower detail and visiting vespid wasp. Flowers are bisexual and are probably pollinated by marsupials and bats, with insects as secondary pollinators. (Photo: Jack Hasenpusch, Australian Insect Farm)

39. Unlike many mass-flowering rainforest plants which recruit numerous small-sized and ecologically unspecialised pollinators, the shrub or small tree *Clerodendrum tomentosum* (Lamiaceae/Verbenaceae) is principally pollinated by Lepidoptera. Its long tubular flowers obstruct entry, and only the long proboscises of some moths and butterflies are able to reach the nectar within. Small insects do not make contact with stigmas and anthers – only Lepidoptera can reliably do this – and so cannot perform pollen transfer. *Clerodendrum tomentosum* ranges north from southern New South Wales and occurs in subtropical and littoral rainforest, and mesic forest regrowth. Following fire plants are able to regenerate from basal trunk coppice growth.

40. *Endiandra muelleri* is a member of the basal Lauraceae, a family that originated in the Early Cretaceous and was widely established by the Mid Cretaceous. The flowers of Lauraceae are typically small, bisexual and actinomorphic. In Australia, similarly basal families include the Annonaceae, Austrobaileyaceae, Eupomatiaceae, Monimiaceae, Trimeniaceae and Winteraceae. The diminutive flowers of *Endiandra muelleri* allow entry only to small insects such as thrips and ants. The species occurs in subtropical, warm temperate and littoral rainforest, north from the Hunter Valley (New South Wales) to Kroombit Tops, near Gladstone (Queensland).

41. *Argyrodendron actinophyllum* (Malvaceae/Sterculiaceae) has earlier been known under the name *Heritera actiniphylla*. It is a large to very large buttressed tree occurring in warm temperate rainforest and subtropical rainforest, distributed north from Gloucester to near Gympie. The species can be characteristic of complex subtropical rainforest growing at intermediate elevations. Populations can synchronously mass flower, their white-flowered tall crowns conspicuous in rainforest mosaics within eucalypt-dominated forest matrices. Flowers are unisexual, the mass-flowering episodes facilitating the optimal availability of pollen. Flowering events can occur in autumn when pollinators are less abundant and active, suggesting the possibility of wind-pollination, in combination with biotic agents, as a pollen dispersal strategy.

42. *Brachychiton acerifolius* (Malvaceae/Sterculiaceae) is a medium- to large-sized partly deciduous tree growing in subtropical, dry and warm temperate rainforest north from southern coastal New South Wales. The species occurs in tropical and subtropical complex rainforest, as well as semi-deciduous vine forests. However, trees may not be deciduous every year, and are often widely spaced within stands. Flowers are unisexual, but there is no difference in the size of male and female flowers. The corolla is absent and the coloured perianth is usually treated as a calyx. Nectaries are absent (Guymer 1988). Flowers are pollinated mainly by bees. Populations often mass-flower in spring–summer, their distinctive red crowns conspicuous within an otherwise green landscape.

43. *Brachychiton discolor* (Malvaceae/Sterculiaceae) is a medium- to large-sized partly deciduous tree growing in dry and subtropical rainforest north from the lower north coast of New South Wales. Flowers in summer. The species is known to hybridise with *Brachychiton acerifolius* and *B. bidwilli*, the latter a deciduous shrub on the margins of dry rainforests and vine thickets north from Mt Tamborine (south-eastern Queensland) to central Queensland (Harden *et al.* 2006).

44. *Commersonia fraseri* (Malvaceae/Sterculiaceae) is a large colonising shrub or small tree occurring on the margins of subtropical and littoral rainforest, and also wet sclerophyll forest at low elevations. Distribution ranges from Victoria, north to south-eastern Queensland. Flowers are bisexual. The species usually flowers reliably each year but few insects are seen in association. Plants coppice vigorously following fire, forming dense low mats in the initial stages of vegetative regeneration. However, trees are soft-wooded and relatively short-lived, and when entangled with rampant *Cissus* vines collapsing dead trunks and vines may smother and deform any native plants attempting to regenerate underneath.

45. *Hibiscus splendens* (Malvaceae), a shrub to ~6 m high, but generally much shorter. The species is found on the margins of subtropical and dry rainforest regrowth, and lowland wet sclerophyll forest, north from the central and lower north coast of New South Wales. However, populations exhibit often highly disjunctive distributions and are replaced by *H. heterophyllus* in adjacent forests; though usually not co-occurring in the same stand. Flowers open for a single day, and are visited and pollinated by small beetles of the family Nitidulidae and flies of the family Drosophilidae. Nitidulids and drosophilids form part of the small insect assemblage of pollinators and visitors recorded from many rainforest flowers.

46. In subtropical rainforests of New South Wales, most angiosperms flower between spring and late summer. However, several trees (e.g. *Synoum glandulosum* [Meliaceae]) have flowering patterns that correspond with the cooler months. During such periods the diversity and abundance of potential pollinators is greatly reduced, and plants that depend on biotic vectors for pollination must rely on suites of generally ecologically unspecialised small insects, since (at least in higher latitudes) larger insects are no longer available.

47. *Daphnandra micrantha* (Monimiaceae/Atherospermataceae), a small protogynous understorey tree restricted in range to the subtropical rainforests of the Manning and Hastings valleys, northern New South Wales. It is possibly pollinated by minute gnats in the family Cecidymiidae. Flowering opening is generally nocturnal and sequential. Flowers lack obvious fragrance and are without nectar guides. There is no obvious nectar. The gynoecium bears approximately nine 'fleshy' stigmatic papillae, the apical half of each papilla tufted, and producing a clear gelatinous exudate; whose purpose remains enigmatic. Related species are found in the subtropical and warm temperate rainforests of the Tweed Caldera region of far northern New South Wales and south-eastern Queensland (*Daphnandra tenuipes*) and undescribed species elsewhere from southern New South Wales to southern Queensland (Harden *et al.* 2006). Flowers are small and coloured cream-white, except for *Daphnandra tenuipes* in which the flowers are reddish.

48. *Wilkiea huegeliana* (Monimiaceae) is a dioecious (sometimes monoecious) shrub of rainforest understoreys, and occasionally also found in wet sclerophyll forest. The species normally flowers annually. Male flowers are usually in ramiflorous clusters, female flowers often borne on terminal or apical branch sections. The flowers are pollinated solely by thrips (Thysanoptera). These tiny insects gain access to each urn-shaped flower via a small hole (ostiole) at the apex. *Wilkiea huegeliana* is a larval food plant of the Regent skipper butterfly *Euschemon rafflesia* (Nymphalidae). Monimiaceae are sometimes broken up into several smaller families, such as Atherospermataceae and Siparunaceae (Renner 1998). Flowers of Monimiaceae are usually actinomorphic, rarely bisexual (*Hortonia*) but usually unisexual. Flowers are generally inconspicuous but vary in form. Three species of *Wilkiea* occur in subtropical south-eastern Australia: *Wilkiea huegeliana* in subtropical, warm temperate, dry and littoral rainforest formations north from Mt Dromedary to central Queensland and an outlier in North Queensland; *W. macrophylla* in subtropical, dry and littoral rainforests north from Alstonville; and *W. austroqueenslandica* in subtropical rainforest in north-eastern New South Wales to south-eastern Queensland. The fossil record, and that of Siparunaceae, dates to the Late Cretaceous (Friis *et al.* 2011).

49. Cauliflorous presentation of *Ficus coronata* (Moraceae) synconia. *Ficus coronata* (Moraceae) is a small understorey tree of numerous rainforest types (especially at lower elevations), but also found along stream margins, in numerous rainforest types north of East Gippsland, Victoria. The fig fruit is a highly specialised structure that encloses small male and female flowers. Figs are pollinated by small wasps of the subfamily Agaoninae (family Agaonidae), the relationship normally involving a single fig wasp as the pollinator of a single fig species. Such one-to-one plant–pollinator mutualisms are rare; even among the figs there are instances of breakdown of reliance upon a single wasp species as the pollinator.

50. *Ficus henneana* (Moraceae), a medium-sized to large buttressed strangling fig (sometimes deciduous) occurring in lowland rainforest, north from southern New South Wales. The genus *Ficus* contains ~900 species and is generally dependent on wasps of the family Agaonidae, subfamily Agaoninae ('fig wasps'), for pollination. *Ficus* species are dioecious (or functionally dioecious) or monoecious. Responses to crown and trunk damage by fire includes the production of basal trunk shoots. Large *Ficus* trees act as focal points for the attraction of fruit-eating birds, these facilitating the recruitment (via the seed they void) of additional plants to stands.

51. *Streblus brunonianus* (Moraceae) is a dioecious tall shrub to medium-sized tree in dry rainforest and subtropical rainforest, north from southern coastal New South Wales to Queensland. Flowers expel pollen ballistically. The inflexed stamens are held in tension before the opening of flower buds, following which the anther-carrying filaments fling out, releasing small clouds of pollen into the air column. Pollen release appears to be cued by direct sunlight and attendant temperature increase. However, pollen release is not especially synchronised, rather appearing to be randomly expelled from individual flowers. The genus *Streblus* occurs in Madagascar, South-east Asia to Australia.

52. *Trophis scandens* (Moraceae), previously *Malaisia scandens*, is a woody vine, or sometimes a scandent shrub, in numerous rainforest types, including vine thickets. Occurs from southern New South Wales to central Queensland, being usually encountered in forest gaps or along margins. Plants are dioecious, the dust-like dry pollen grains expelled ballistically into the surrounding air column, and then dispersed by wind – thus no biotic agent is employed (unlike *Ficus*), though many grains are wasted as movement by wind to receptive stigmas is uncertain. The Moraceae number around 1000 species of trees, shrubs and vines. The family worldwide is mainly distributed in tropical and subtropical regions, however, there is considerable variation in flower structure and modes of pollination. The family has an extensive Cenozoic fossil record (Friis *et al.* 2011).

53. *Myrsine variabilis* (Myrsinaceae/Primulaceae) is a shrub-small tree associated with numerous rainforest types, and is distributed from southern New South Wales to Carnarvon National Park in central Queensland. The species also grows in wet sclerophyll forest. The small massed pale flowers of *Myrsine variabilis* (here), and those of the related *M. howittiana*, develop on lateral branches (ramiflorous) and are pollinated by thrips (several discernable in this photo) (e.g. *Thrips setipennis*). Flower opening is synchronous. Like the enclosed urn-shaped flowers of *Wilkiea* entry by thrips can only be achieved through a small opening at the apex. Fossils of *Myrsine* are reported from the later Cenozoic of Europe suggesting that the obligate plant-pollination relationship with Thysanoptera is an ancient one.

54. *Archirhodomyrtus beckleri* (Myrtaceae), a small tree found in warm temperate and subtropical rainforest, rarely in wet sclerophyll forest, north from Barrington Tops to south-east Queensland. Flowers in spring, the numerous anthers surround the stigma which is expanded to form a broad surface, the floral structure suggesting an adaptation to insect and wind pollination. Note presence of immature orthopteran. Orthoptera are often consumers of floral parts. Myrtaceae have an extensive fossil record and are particularly well represented in the Cenozoic floras of the Southern Hemisphere.

55. *Backhousia myrtifolia* (Myrtaceae), a small, densely flowering, dry fruited tree occurring from southern New South Wales to central coastal Queensland. It is widespread in subtropical, warm temperate and dry rainforest, as well as floristically depauperate mesic gullies within dry sclerophyll forest, and along stream margins, where it can be dominant. In southern New South Wales *Backhousia myrtifolia*, in combination with *Syzygium smithii* (Myrtaceae), can form fringing *gallery rainforest* stands along the banks of streams. The species occurs in the understorey of well developed rainforest on the sandmass of Fraser Island, southern Queensland. Flowers mainly December–January. Here the flowers are not fully open.

56. *Metrosideros* sp. (Myrtaceae). *Metrosideros* is an example of extinction in the Tertiary of a group of plants on the Australia mainland; the genus now being generally restricted to landmasses of the Pacific such as Lord Howe Island, New Zealand and Hawaii. However, fossils assigned to *Metrosideros* are known from the Eocene-Oligocene of Tasmania, providing evidence for a possible Australian origin of the genus (Tarran *et al.* 2016). Throughout its extant natural range individual species are variously pollinated by birds, insects and lizards, but various species and cultivars are widely planted in gardens of the Australian eastern seaboard, the showy bright flowers commonly attracting a variety of native birds (e.g. Rainbow lorikeet *Trichoglossus haematodus,* Little wattlebird *Anthochaera chrysoptera*).

57. *Rhodamnia rubescens* (Myrtaceae) is a shrub or small tree found growing in a variety of rainforests types north from southern New South Wales to central coastal Queensland. The species is sometimes encountered in eucalypt-dominated wet sclerophyll forest. *Rhodamnia rubescens* is threatened by the introduced fungal pathogen Myrtle rust (*Austropuccinia psidii*). Flower visitors include native bees (Halictidae, Colletidae).

58. *Rhodomyrtus psidioides* is a threatened species of Myrtaceae, occurring in lowland subtropical and littoral rainforests, and ranging from Gosford (New South Wales) to near Maryborough, Queensland. The plant faces extinction in the wild owing to high vulnerability to the introduced fungal pathogen Myrtle rust *Austropuccinia psidii.* Produces numerous root suckers which are prone to Myrtle rust re-infection (see Plate 149). Although pollinated by insects, flowers of *Rhodomyrtus psidioides* possess a relatively large and broad stigma that suggests partial adaptation to being pollinated by wind. Populations flower synchronously but flowers on individual trees open sequentially. Flowers are highly perfumed, in cymes or raceme-like inflorescences. Pollen is slightly sticky but is readily expelled from dehiscent anthers (G. Williams 1995).

59. *Syzygium cormiflorum* (Myrtaceae) is a mass-flowering tree of north Queensland tropical rainforests. It has a flowering period that can last from July to mid-November, and flowering events are distinguished by the conspicuous cauliflorous and ramiflorous presentation of flowers. Ramiflorous and cauliflorous positioning of flowers and associated large, fleshy, animal-dispersed (zoochorous) fruit reflects selection influences from both pollinators and dispersers. It exhibits varying degrees of self-compatibility, the degree of self-compatibility conferring resilience in the absence of major pollinators. Birds, hawkmoths, and particularly blossom bats (which include the genera *Macroglossus* and *Syconycteris*), are effective pollinators (Crome and Irvine 1986). *Szyzygium cormiflorum* produces large quantities of weak nectar (this conforming to the bat-pollinated floral syndrome). (Photo: Hugh Nicholson)

60. *Syzygium floribundum* (Myrtaceae), previously *Waterhousea floribunda,* is a moderate-sized tree that grows along the stream lines of subtropical rainforest at low elevation. There it is often the dominant canopy-forming tree. Occurs north from the Hunter Valley. Flowers are bisexual but are largely self-incompatible. Floral structure and presentation is similar to that of *Eucalyptus* species. Like other members of the genus its flowers have numerous, conspicuous long stamens. The stigma appears receptive at flower opening, but freshly opened flowers do not possess nectar. Flowers are strongly and sweetly scented, and attract large numbers of insects, the beetle fauna being especially numerous and diverse. The beetles are species of *Phyllotocus* (Scarabaeidae-Melolonthinae), a genus often prolific on mass-flowering non-rainforest Myrtaceae, but also encountered on rainforest Myrtaceae (e.g *Rhodomyrtus psidioides, Tristaniopsis laurina*) as well as other families in rainforest (e.g.*Cuttsia viburnea* – Escalloniaceae/Rousseaceae, *Euroschinus falcatus* – Anacardiaceae, *Guioa semiglauca* – Sapindaceae. *Scolopia braunii* – Flacourtiaceae/Salicaceae).

61. *Xanthostemon chrysanthus* (Myrtaceae), a tree distributed in north-eastern Queensland from Cardwell to Cape York Peninsula. Occurs in open forest and rainforest, often along the banks of streams and rivers. Flowers are bisexual with five petals attached to the rim of the calyx. Visitors include birds, bees, wasps and ants.

62. *Olea paniculata* (Oleaceae). This is a medium-sized tree occurring in dry, littoral and subtropical rainforest. It occurs north from the Hunter River, New South Wales, north to Queensland and also Lord Howe Island and New Caledonia. Often a co-dominant in the canopy of dry rainforest. Flowers bisexual, and flowers October–December. Oleaceae include several alien species that invade rainforest; *Olea europaea*, a native of the Mediterranean area, *Ligustrum lucidum, L. sinense*, natives to China, and *L. vulgare*, a native of southern Europe and North Africa (Harden 1992).

63. *Pseudovanilla foliata* (Orchidaceae) is a climbing saprophytic herb with stems up to 15 m long, found in subtropical rainforest north from the mid north coast of New South Wales to southern Queensland. Flowers October–January in the understorey, but also in sunlit openings. Grows on tree trunks, logs and rotting stumps. Flowers are bisexual.

64. *Rhinerrhiza divitiflora* (Orchidaceae) is an epiphytic, sometimes lithophytic, orchid endemic to northern New South Wales and Queensland. Grows in rainforest at lower altitudes, usually on tree trunks, less commonly on rocks. Flowers August–November. Individual flowers stay open for ~1–2 days. As with most Australian rainforest orchids little is known of its pollination ecology.

65. *Breynia oblongifolia* (Phyllanthaceae/Euphorbiaceae) is a small unisexual shrub commonly found growing on the margins of rainforest, and in wet sclerophyll forest. It occurs from southern New South Wales to central Queensland. Species of the moth genus *Epicephala*, as well as thrips, are considered to be pollinators.

66. The inconspicuous tiny flowers of *Glochidion ferdinandi* (Phyllanthaceae/Euphorbiaceae) are pollinated by small moths in the genus *Epicephala* (Gracillariidae). However, larval *Epicephala* are seed predators. *Glochidion ferdinandi* readily reproduces from root coppice. It tends to flower each year, but alternates with heavy and light crops (Floyd 1990). During protracted drought events trees suffer high levels of leaf drop, but tree and sapling mortality is low, with intense leaf flushes being initiated in response to renewed rainfall.

67. *Banksia integrifolia* (Proteaceae). Leaf forms vary from entire to serrate. The genus *Banksia* characterises several non-rainforest vegetation types, however, *Banksia integrifolia* occurs commonly in littoral rainforest and as an associate of high-elevation cool temperate rainforest (*Banksia integrifolia* var. *monticola*) in New South Wales. The flowers of *Banksia integrifolia* are principally adapted to pollination by vertebrates (e.g. specialised nectar- and pollen-feeding Queensland blossom bat *Syconycteris australis*). Proteaceae are widely distributed in tropical and subtropical regions of the Southern Hemisphere. The Proteaceae include several genera (e.g. *Grevillea, Telopea*) that in Australia are well known from open woodland and dry forest, and several genera (e.g. *Hicksbeachia, Macadamia, Stenocarpus*) which occur in rainforest. The family has a diverse and extensive fossil record with the earliest records being from the Late Cretaceous of Australia (reviewed in Friis *et al.* 2011), with fossil *Banksia archaecarpa* described on the basis of infructescences from the Early Eocene of north-western Australia. Fossil Proteaceae-type pollen from the Late Cretaceous is particularly abundant in the Otway Basin of south-eastern Australia. Fossil evidence suggests that pollination by birds, and possibly by small mammals, was already developed by the Early Cenozoic.

68. *Eidothea hardeniana* (Proteaceae) is a medium-sized to large tree restricted to warm temperate rainforest in the Nightcap Range, north-eastern New South Wales. The flowers are actinomorphic, but pollen is the only floral resource offered to flower visitors. Flowers are likely insect-pollinated, however, species observed include many groups (e.g. melolonthine scarab and chrysomelid beetles, orthopterans) that are known to eat pollen and flower parts (including reproductive structures). The genus *Eidothea* is known only from northern New South Wales and north-eastern Queensland (*E. zoeexylocarya*) (Rossetto and Kooyman 2005). (Photo: Hugh Nicholson)

69. Flowers of *Stenocarpus sinuatus* (Proteaceae), a medium-sized tree occurring from northern New South Wales to southern and northern Queensland. Often planted as an ornamental. Flowers are bisexual, zygomorphic and sometimes ramiflorous. It is one of the few trees encountered in subtropical rainforest that diverges from an otherwise dominating theme of small, white-coloured actinomorphic flowers adapted to pollination by small insects.

70. *Ripogonum discolor* (Ripogonaceae) is a robust armed vine found growing in the understorey and sheltered margins of subtropical and warm temperate rainforest, north from the lower north coast of New South Wales. The many-flowered racemes are conspicuous against the green foliage. Flowers are bisexual and strongly fragrant, however, they produce no obvious nectar reward. Pollen is dry and white in colour. Thrips and *Apis mellifera* are associated with the flowers.

71. *Guioa semiglauca* (Sapindaceae) is a common small- to medium-sized tree occurring in subtropical and littoral rainforests, as well as wet sclerophyll forest. It can be common in regrowth forest. The species ranges from southern New South Wales to Kroombit Tops in Queensland. Flowers are small, actinomorphic and usually monoecious (sometimes dioecious [G. Williams 1995]). Flower presentation is often ramiflorous. *Guioa semiglauca* is characteristic of the 'general entomophily' floral syndrome, possessing open, shallow flowers readily accessible to small, ecologically unspecialised, insects. Populations may flower episodically, flowering events punctuated by several seasons. Male anthesis is initiated 1–3 weeks before female trees flowering. Flower visitors and pollinators are small insects, however, the visiting assemblage varies each year and between individual flowering populations. Assemblages are rich in small Coleoptera, Diptera and Hymenoptera, the majority of which are morphologically and ecologically unspecialised nectar feeders (G. Williams 2020). Depending on how the family is defined the Sapindaceae comprises ~1500 species, most of which are trees and shrubs. Extant Sapindaceae have mainly a tropical and subtropical distribution, and are characterised by actinomorphic and zygomorphic flowers. Fossil flowers are recorded from the Eocene of British Colombia (*Wehrwolfea striata*) and the Eocene of Wyoming (*Landeenia*) (in Taylor *et al.* 2009). Pollen assigned to the extant genera *Cupania* and *Serjania* is reported from the Eocene of Argentina, however, the earliest records (*Sapindoxylon, Sapindospermum*) date to the Late Cretaceous of the Indian Deccan and Central Europe respectively (in Friis *et al.* 2011).

72. *Solanum aviculare* (Solonaceae) is an erect, ephemeral shrub found growing on the margins of numerous rainforest types north from the Otway Ranges, south-west Victoria, to central Queensland. The species may also colonise gaps in disturbed rainforest. The conspicuous flowers are predominantly visited by bees (e.g. species of *Trichocolletes* – Colletidae).

73. *Trimenia moorei* (*Piptocalyx moorei*) (Trimeniaceae) is a slender woody vine in subtropical, warm temperate and cool temperate north from the Barrington Tops to the McPherson Range. Plants may form dense growth in disturbed, open, sunlit areas. The flowers of *Trimenia* possess floral characters that grade between bracts and perianth. Terminal flowers are bisexual, the lower ones are usually male (apical male and lower female shown here), a condition that may favour out-crossing and minimise loading of self-pollen upon apically placed hermaphroditic flowers.

74. *Elatostema ?reticulatum* (Urticaceae). Prostrate plants of the genus *Elatostema* are often found in moist shady locations adjacent rainforest streams. Although their pollination ecology remains unstudied flower size and structure suggest the pollinators are small flies or bees. Worldwide the Urticaceae comprise ~2600 species of herbs, shrubs, trees and hemi-epiphytes. Flowers are small, unisexual or rarely bisexual. Urticaceae are known from the Late Cretaceous of Europe.

75. The large syrphid fly *Dideopsis aegrota* (Syrphidae), feeding from the solitary open flower of *Cayratia clematidea* (Vitaceae). *Cayratia clematidea* is a weakly climbing vine found growing on the margins of rainforest (here in littoral rainforest). Flowers are bisexual and produce nectar in combination with dehiscent anthers. Although flowers are individually small and inconspicuously coloured, nevertheless they attract numerous insects. These include thynnid, gasteruptid, pompilid and vespid wasps, pierid butterflies, tachinid and large syrphid flies. However, the composition of the visiting anthophilous fauna can change dramatically from site to site.

76. *Cissus antarctica* (Vitaceae), a rampant native vine common to coastal and elevated rainforests, north from southern New South Wales to Queensland. Flowers in spring and summer. Flowers bisexual, floral rewards minute. Often forms dense persisting growth in disturbed rainforest and eucalypt-dominated wet sclerophyll forest. Although this can be smothering, at littoral rainforests such growth on forest margins buffers the impact of salt-laden winds

77. *Cissus sterculifolia* (Vitaceae). This is a woody vine commonly encountered on sunlit margins of littoral rainforest. Its range extends from southern New South Wales to central coastal Queensland. Flowers are bisexual and diminutive, without apparent visually conspicuous pollinator recruitment cues. Nevertheless, flowers are visited by a diverse assemblage of native bees and large flies (e.g. *Dideopsis aegrota, Mesembrius hilaris* – Syrphidae).

78. *Tasmannia insipida* (Winteraceae), a dioecious shrub in various rainforest types such as subtropical, warm temperate, cool temperate and littoral rainforest. Usually grows *within* rainforest rather than on exposed margins. Flowers late winter to late spring. Male flowers (shown here) open differentially, the dry pollen falling readily away from dehiscent anthers. Pollen sculpture suggests adaptation for carriage by insects, however, the dry nature of the pollen indicates potential for transport by wind. The calyptra (hood-like) structures of both male and female flowers produce droplets that may function as a reward to potential pollinators. Winteraceae occur in the South Pacific, including Australia and New Guinea. One monotypic genus (*Takhtajania*) is known from Madagascar and *Drimys* occurs in South America. Micro- and macrofossils of Winteraceae are known from Laurasia and Gondwana (Grímsson *et al.* 2018). Flowers are recorded from the Late Cretaceous and pollen (*Walkeripollis*) is reported from the Lower Cretaceous of Israel (in Friis *et al.* 2011).

79. *Alpinia caerulea* (Zingiberaceae) is a yellow-white flowered rhizomatous perennial herb that is widespread in coastal rainforest and wet sclerophyll forest, north from central coastal New South Wales. Plants flower in spring. A second species, pink-red flowered *Alpinia arundelliana*, grows in similar habitats north from the same region. The inflorescences of many angiosperms open sequentially, as in *Alpinia caerulea*. The zygomorphic flowers of *Alpinia caerulea* are pollinated by large long-tongued bees which can land upon the specially adapted lower 'platform'. Many such bees 'trap-line', that is, they repeatedly undertake distinct foraging patterns between series of specific flowering plants. Only bees large enough (such as *Amegilla* spp.) to trip the modified anther structure, and so receive a dusting of pollen, are potential pollinators. Smaller bees, such as *Tetragonula*, generally do not pollinate and are thus thieves of floral resources upon which true pollinators feed. A potential threat to *Alpinia* is posed by the highly invasive bright red-yellow flowered ginger *Hedychium gardnerianum* (Zingiberaceae), a native of India, but locally naturalised in coastal habitats north from Sydney. This large and persistent rhizomatous plant occupies the same coastal forest microhabitats as southern *Alpinia* populations, can aggressively occupy the forest understorey and may disrupt pollination networks.

80. The ubiquitous introduced honey bee *Apis mellifera* (Apidae) foraging on flowers of *Acradenia euodiiformis* (Rutaceae), a small tree restricted to rainforests of north-eastern New South Wales, and south-eastern Queensland. *Apis mellifera* is common on mass-flowering rainforest trees, often outnumbering native species on flowers. When it occurs in large numbers it often disturbs native pollinators. Although *Apis mellifera* competes with native pollinators for floral resources, its impact on rainforest plant pollination and seed-set is still poorly understood. *Acradenia euodiiformis* is a small tree associated with warm temperate and subtropical rainforests north of the Hunter Valley to the McPherson Range. Flowers October–December. Flowers are bisexual, slightly protogynous, are highly self-compatible (Table 4.1) and produce nectar before fully open. However, flowering events can be punctuated by several years. Flowers in lowland subtropical rainforest attract large numbers of eusocial *Tetragonula carbonaria* and *Apis mellifera*. The Rutaceae are a large family of aromatic trees and shrubs and has an almost cosmopolitan distribution. Fossil Rutaceae date from the Cretaceous. Subtropical rainforests include numerous Rutaceae genera (*Acronychia, Bosistoa, Bouchardatia, Citrus/Microcitrus, Flindersia, Geijera, Medicosma, Melicope, Micromelum, Pentaceras, Pitaviaster, Sarcomelicope* and *Zanthoxylum*); many species of which are also associates of tropical Australian rainforests.

81. Long-tongued bee *Amegilla bombiformis* (Apidae) sonicating – buzzing – a *Senna acclinis* (Fabaceae) flower. The anthers of *Senna acclinis* are poricidal and must be buzzed (sonicated) by bees to release the pollen within. In addition to plants with zygomorphic flowers, such as *Senna acclinis, Amegilla bombiformis* has been recorded collecting nectar from rainforest trees with open actinomorphic flowers; for example small tubular blossoms (e.g. *Diospyros australis*), as well as species with little depth effect in the corolla (e.g. *Tristaniopsis laurina).* Species of *Amegilla* are known to 'trap line', repeatedly moving along set foraging paths, visiting particular plants when in flower. *Senna acclinis* is an endangered small shrub occurring in subtropical, littoral and, dry rainforest, as well as vine thickets, and ranges from the Gloucester area of New South Wales to the Expedition Range (Queensland). It grows on rainforest margins and internal sunlit gaps formed by damage to the rainforest canopy. Such gap phase species occupy temporal and spatial habitat mosaics formed by forest disturbance, upon which they are partly dependent. Suitable habitat may need manipulation so as to sustain *Senna acclinis* populations in individual rainforest stands; as at Black Head on the mid north coast of New South Wales. However, rainforests are subject to invasion by alien *Senna* species, such as *S. pendula* and *S. septemtrionalis*. These are widespread invasive shrubs often found growing profusely in, and on the margins of, lowland subtropical rainforest. By diverting pollinators (bees able to sonicate flowers) flowering alien *Senna* plants potentially threaten reproductive success in co-occurring native *Senna* (e.g. *S. acclinis, S. odorata*) populations.

82. *Tetragonula carbonaria* (Apidae) foraging on *Alectryon subcinereus* in lowland wet sclerophyll forest, and illustrating pollen collected in the corbiculae. Pollen collected in the corbiculae of social Apidae is not available for pollination. *Tetragonula carbonaria* is a eusocial species, and can be found in large numbers foraging on a diversity of trees, shrubs and herbs flowering on sunny margins of rainforest and wet sclerophyll forest. However, individuals tend to prefer plants flowering at lower forest levels and rarely enter shaded zones. Though frequently found in association with the introduced honey bee *Apis mellifera, Tetragonula carbonaria* colonies usually commence foraging latter in the morning than those of *Apis*, thus a level of nectar/pollen depletion by alien honey bees has already occurred.

83. *Leioproctus flavomaculatus* (Colletidae-Colletinae) foraging on flowering *Acronychia imperforata* (Rutaceae) in a littoral rainforest remnant. The dense yellow setae on the scutellum and metanotum resemble the yellow markings of several co-occurring bees of the genus *Hylaeus* and related genera (Colletidae-Hylaeinae) (Houston 2018). *Leioproctus flavomaculatus* forages at numerous mass-flowering trees and shrubs in lowland subtropical rainforests. Additional food plants include *Alectryon coriaceus* and *Guioa semiglauca* (Sapindaceae), *Euroschinus falcatus* (Anacardiaceae), and *Alphitonia excelsa* (Rhamnaceae); all species with open shallow flowers. Colletidae are short-tongued bees divided into five subfamilies (Michener 2007), of which the Euryglossinae, Colletinae and Hylaeinae occur in Australia. Colletids are often 'pollen thieves', harvesting pollen for their own nutritional needs but without any pollination benefit, or only a small one, for the plant. They visit numerous flowering plants in subtropical rainforest and open forest communities, and forage in all forest strata, but show a level of temporal foraging constancy to individual plants when in flower. Flower types visited range from those with little depth effect, tubular flowers, and flowers requiring sonication, or 'buzz-pollination', to release pollen (as in *Senna acclinis* – Fabaceae/Caesalpinioideae).

84. *Amphylaeus nubilosellus* (Colletidae-Hylaeinae) foraging on *Cupaniopsis anacardioides* (Sapindaceae). Hylaeinae carry pollen internally in the crop. In so doing, they harvest quantities of pollen for their own nutritional benefit, but this is a loss to the plants, as only the lesser numbers of grains incidentally caught on their bodies are potentially available for pollination. Hylaeinae are a diverse group in Australian lowland subtropical rainforests, with *Hylaeus* being especially diverse. *Cupaniopsis anacardioides* characterises littoral rainforest growing on sands in south-eastern Australia. It is a pioneer in heavily disturbed sites, and is a major contributor to the expansion of littoral rainforest boundaries. The species is monoecious and partly self-compatible, and flowers during cooler months. Fire-impacted trees coppice profusely from the base of the trunk.

85. Roosting *Megachile pictiventris* (Megachilidae), a large solitary bee with distinctive orange-coloured scopal hairs between which large pollen loads can be deposited. Dense setae elsewhere on the body may carry extraneous pollen grains, their adherence incidental to foraging activities. Nectar and pollen sources in rainforest visited by *Megachile pictiventris* include *Arytera divaricata* (Sapindaceae) and *Abrophyllum ornans* (Escalloniaceae/Rousseaceae). But the diversity of recorded food plants for the genus is quite broad. *Megachile pictiventris* exhibits foraging fidelity to mass-flowering trees, returning repeatedly to sunlit inflorescences. Megachilids are can be episodic foragers, their presence in large numbers often coinciding with peak flowering events, but absent in some seasons. The family Megachilidae is a large one, occurring worldwide.

86. *Lasioglossum* sp. (possibly *L. carbonarium*) (Halictidae) on flowering *Parsonsia purpurascens* (Apocynaceae) in warm temperate rainforest (Tapin Tops National Park). Halictidae are short-tongued bees that forage predominantly from flowers with open, shallow, floral structures. *Lasioglossum* is a large genus of >250 Australian species, with many species of similar form which, as a consequence, are difficult to distinguish. The genus occurs on all continents except Antarctica and is diverse in the rainforests and wet sclerophyll forests of eastern Australia. Adults forage at all levels of the forest, but can be particularly diverse at plants flowering at or near ground level (e.g. *Dianella caerulea, Pollia crispata, Senna acclinis*). *Lasioglossum turgicollaris* and *L. polygoni* remove pollen from the endangered rainforest shrub *Senna acclinis* (Fabaceae–Caesalpinioidae) by vibrating ('buzzing') anthers with their flight muscles.

87. Male 'fig wasp' (Agaonidae – Agaoninae) in the synconium of *Ficus henneana* (see Plate 50). The *Ficus*–agaonid association is thought to date from some time in the Cretaceous, greater than 100 mya. Fig–fig wasp relationships are generally considered host-specific, although breakdown in pollinator specificity has been recorded for several species. Fossil Agaonidae are recorded in Miocene Dominican amber (Grimaldi and Engel 2005).

88. *Piyuma* sp. (Crabronidae). The Crabronidae are a polyphyletic group of aculeate wasps. Fossil crabronids are known from the mid-Cretaceous, and it is from within Crabronidae that bees arose. Although predatory, Crabronidae commonly feed on nectar. Species of *Acanthostehus, Bembecinus, Bembix, Cerceris, Ectemnius, Larra, Pison, Sericophorus, Sphodrotes, Tachysphex* and *Williamsita* are recorded visiting mass-flowering trees and shrubs in lowland subtropical rainforests. Many of these genera have a dense cover of setae on their bodies to which pollen grains are readily captured, or pollen grains aggregate in discrete parts of the body, and so individuals can function as pollinators.

89. *Anthoboscа signata*, a large thynnid wasp, on the flowers of the rainforest pioneer tree *Alphitonia excelsa* (Rhamnaceae). *Anthobosca signata* possesses specialised pollen combs on the forelegs and individuals undertake frequent long-distance movements between flowering rainforest trees. Flowers of *Alphitonia excelsa* produce a fetid odour which can be unpleasant. Its flowers are small and readily accessible to numerous insects, many of which are true pollinators. A number are relatively inefficient in their ability to transfer pollen, but their sheer numbers substitute for loss of 'quality' of service. Worldwide, flowers of Rhamnaceae are mostly insect-pollinated, the flowers are usually bisexual, or more rarely unisexual, with a distinct nectary disc between the androecium and gynoecium. The Cenozoic fossil record is extensive, however, this is based mainly on leaves and fruits (Friis *et al.* 2011)

90. *Dimorphothynnus dimidiatus* (Thynnidae) on flowers of *Elaeocarpus obovatus* (Elaeocarpaceae). Thynnids are often called 'flower wasps', and are first known from the Early Cretaceous. Females are flightless, are generally smaller than males; which seek out and carry the females in flight. Thynnidae are predatory wasps, however, many species also feed on nectar, and in visiting flowers potentially serve as pollinators. Most thynnids are ecological generalists and visit a diversity of mass-flowering trees, including those of rainforests. Thynnids were previously grouped as a subfamily of Tiphiidae, however, molecular studies have indicated new relationships among taxa historically placed within this family. Worldwide, Thynnidae now comprise the subfamilies Anthoboscinae, Diamminae, Myzininae, Methochinae and Thynninae. Tiphiidae, as now constituted, contain only the subfamilies Brachycistidinae and Tiphiinae. Brachycistidinae and Methochinae do not occur in Australia, and Tiphiinae are restricted to northern Australia. Thynninae and Diamminae are restricted to Australia and South America. Of the four subfamilies listed from Australia by Naumann (1991) only the Anthoboscinae (*Anthobosca*) and Thynninae (numerous genera) occur in the subtropical eastern coastal zone.

91. *Rhagigaster ?mutatus* (Thynnidae) and *Gasterupton ?pallidicus* (Gasteruptiidae) foraging on fructescent *Cayratia clematidea* (Vitaceae). These are members of a diverse aculeate wasp assemblage occurring on mass-flowering lowland subtropical rainforest trees, shrubs and vines.

92. *Deuterodiscoelius ephippium* (Vespidae) foraging on *Trochocarpa laurina* – Ericaceae/Epacridaceae. Vespidae, and other large-sized wasps, are frequent visitors to flowering rainforest trees and shrubs with easily accessible 'open' flower structures. Many large wasps are important agents of pollen transfer between co-flowering plants. Additional Vespidae associated with mass-flowering subtropical rainforest trees and shrubs include *Polistes humilis, P. tepidus,Ropalidia plebiana, Bidentodynerus bicolor, Abispa splendida* and *Rhynchium mirabile* (G. Williams 1995). All undertake frequent inter-plant flights.

93. *Myrmecia nigrocincta* (Formicidae) foraging on flowering *Alectryon subcinereus* (Sapindaceae); the latter a widespread small tree encountered in various rainforest formations and occasionally wet sclerophyll forest, north from East Gippsland (Victoria) to central coastal Queensland. Flowers are monoecious and attract a suite of ecologically unspecialised small bees and flies. These include short-tongued halictid bees and long-tongued bombyliid flies. *Myrmecia nigrocincta* is a commonly encountered, and aggressive, inhabitant of lowland subtropical rainforests. The species forages by day on the flowers and leaves of numerous low-growing rainforest trees and shrubs, and though a general scavenger and predator of small soft-bodied invertebrates, it is often observed imbibing nectar.

94. *Polyrhachis ammon* (Formicidae). Various species of *Polyrhachis* ants are found foraging on the floral resources of *Alectryon coriaceous* (Sapindaceae); shown here. Ants are often considered thieves and robbers of nectar, or are dismissed as pollinators; owing to the antibiotic nature of metapleural glandular secretions that may interfere with pollen germination or stigma function. However, the metapleural glands of *Polyrhachis* appear to be non-functional, thus removing a potential obstacle to pollen transfer and subsequent pollination. *Alectryon coriaceus* is a bushy shrub or small tree often encountered in littoral rainforests, north from the mid north coast of New South Wales to central coastal Queensland. Ants are potential pollinators; the dense overlapping foliage and branches, which often touch the ground surface, facilitating their foraging movements to and from their nests. However, ants are participants within a larger cohort of flower-visiting insects to *Alectryon coriaceus*, these including euryglossine and halictid bees, lycaenid butterflies and small Diptera.

95. A large acrocerid fly *Mesophysa flavipes* on male flowers of the dioecious rainforest tree *Rhodosphaera rhodanthema* (Anacardiaceae). *Mesophysa flavipes* is a relatively rare, fast-flying species that undertakes rapid inter-plant flights. Pollen grains readily adhere to the dense body setae. Although many insect pollinators can occur in large numbers on individual mass-flowering rainforest trees, numerous other pollinators (especially larger-sized species) are often uncommon. *Rhodosphaera rhodanthema* is a small to medium-sized tree that grows in dry rainforest north from the Macleay River in northern New South Wales, north to the Bunya Mountains and Maryborough in southern Queensland. Worldwide the Anacardiaceae comprise more than 70 genera and ~1000 species of trees and shrubs occurring both in tropical and temperate regions. Fossil fruit are extensively represented in the Early Eocene of southern England and North America, however, Anacardiaceae leaf and pollen records from the Cretaceous are of uncertain assignation (in Friis *et al.* 2011).

96. Female *Bibio imitator* (Bibionidae) on the flowering mangrove *Avicennia marina* (Acanthaceae/Avicenniaceae). Pollen grains adhering to head and thorax. The integument is densely clothed with short, erect setae allowing large numbers of pollen grains to accumulate during foraging. Individual *Bibio* flies will spend extended periods (>30 s) feeding on nectar in individual flowers, their movements usually being to flowers of the same cluster. Mangrove communities can be considered as unique tidally influenced rainforest, their floristic diversity and structural complexity being best developed in tropical northern Australia.

97. *Geron* sp. (Bombyliidae) on *Alectryon subcinereus* (Sapindaceae). This is a genus of 'long-tongued' nectar feeders that are able to feed from small flowers of varying corolla depth and structure (e.g. *Alectryon* as here, *Lobelia, Pratia, Trochocarpa*). Though not abundant, *Geron* forms part of the diverse assemblage of small Diptera that may be encountered on mass-flowering shrubs, trees and vines, as well as herbaceous species of rainforest margins and below canopy gaps. Bombyliidae are members of the Lower Brachycera and based on fossil records were quite abundant in the Cenozoic (extending from ~70 mya), especially the Oligocene and Eocene.

98. *Stomorhina* sp. (Calliphoridae). *Stomorhina* is a genus of nectar-feeding Brachyceran flies found on numerous rainforest (e.g. *Alectryon coriaceus, Alphitonia excelsa, Argyrodendron actinophyllum, Avicennia marina, Clematis clematidea, Drypetes deplanchei*) and non-rainforest (e.g. *Baeckea* spp., *Leptospermum* spp.) flowering trees, shrubs and vines. Like other Calliphoridae the proboscis is not prolonged, for example as in Lower Brachyceran Bombyliidae, Nemestrinidae and Tabanidae, but nevertheless provides an efficient morphologically adaptive means of imbibing nectar from flowers with open, readily accessible flowers. Diptera are among the most ancient pollinators of angiosperms, and played an important role in the origin of co-evolutionary relationships between flowering plants and insects. They are a dominant group found on mass-flowering rainforest plants with small actinomorphic shallow-structured flowers.

99. *Trichophthalma costalis* (Nemestrinidae). Long-tongued Nemestrinidae are members of the fly fauna found occasionally on the flowers of mass-flowering rainforest trees and shrubs, but are more typical of open forests and woodlands (here on *Leptospermum* – Myrtaceae). They undertake relatively frequent interplant movements that consequently contribute to out-crossing. The Nemestrinidae are a rather primitive family of Brachyceran flies, were abundant during the Mesozoic, and likely one of the first groups to feed on the nectar of Early Cretaceous angiosperms. Nemestrinids are grouped with Tabanidae and Bombyliidae in the 'Lower Brachycera'. Modern flower-visiting Lower Brachyceran flies usually have long proboscises, though nearly all fossil forms (e.g. those from Mid-Cretaceous Burmese amber) have relatively short proboscises and have a labellum consisting of a broad, fleshy expansion probably used to feed on nectars (Zhang and Wang 2017).

100. *Calodema regale* (Buprestidae). This is a large species distributed from Queensland to northern coastal New South Wales, and is the mostly southerly member of the genus; one with a restricted Australasian distribution (G. Williams 2020). Individuals are restricted to mesic forests (though will also frequent 'hill-top' flowering Myrtaceae on rock outcrops), and undertake relatively long distance flights between flowering trees, these movements being punctuated by extended periods of feeding on nectar of individual flower clusters (here *Tristaniopsis laurina* – Myrtaceae). Many buprestid species are known only from the flowers of rainforest canopy trees but anthophilous, rainforest-associated, Buprestidae exhibit no obligate association with single plants or even families of plants. The flowers of *Tristaniopsis laurina* possess groups of stamens that are united at the base and fall off as discrete units when the flowers are senescent. The flowers are present in small cymes, massed but partially obscured by foliage. The pollen is slightly sticky. Freshly opened flowers may attract a diverse insect pollinator fauna, in some seasons being especially rich in bees of the family Megachilidae. The theme of yellow flowers is exhibited by several other trees and shrubs in subtropical rainforests, for example the emergent tree *Archidendropsis thozetian* (Fabaceae), the small subcanopy trees *Alangium villosum* (Cornaceae/Alangiaceae) and *Guilfoylia monostylis* (Surianaceae), and shrubs *Bedfordia arborescens* (Asteraceae), *Hymenosporum flavum, Pittosporum revolutum* (Pittosporaceae), *Senna acclinis* and *Cassia marksiana* (Fabaceae).

101. *Castiarina rollei* (Buprestidae). Many species of *Castiarina* are known only from the flowers of rainforest canopy trees, however, the genus is widely distributed throughout Australia, including Tasmania. *Castiarina* species are morphologically adapted to feed on nectar, in particular the prolongation of the mouthparts, and are a distinctive element of the anthophilous beetle fauna associated with eucalypt-dominated forests and woodlands. Nevertheless, many species frequent flowering tropical and subtropical rainforest trees, including species of Myrtaceae, Rhamnaceae, Rutaceae and Sapindaceae. (Photo: Jack Hasenpusch, Australian Insect Farm)

102. *Metaxymorpha gloriosa* and *M. hauseri* (Buprestidae) on flowering *Buckinghamia celsissima* (Proteaceae). These, and other large beetles, are important agents of out-crossing in tropical and subtropical rainforest communities. The genus *Metaxymorpha* is restricted in distribution to Australia, New Guinea, and some near-associated islands of Indonesia. *Metaxymorpha* is a member of the Gondwanan tribe Stigmoderini, this including the genera *Calodema, Calotemognatha, Castiarina, Stigmodera, Temognatha* and (not exhaustively) the Neotropical genera *Agrilozodes, Conognatha, Dactylozodes* and *Hiperantha* (Bellamy 2003). (Photo: Jack Hasenpusch, Australian Insect Farm)

103. Male *Distichocera superba* (Cerambycidae) on *Syzygium floribundum* (Myrtaceae). This species is generally uncommon, however, individuals often undertake relatively long distance flights between co-flowering plants. Members of the speciose beetle family Cerambycidae are frequently referred to as 'longicorns'. Although many species are only associated with fallen timber and leaves, the family is a taxonomically and ecologically diverse one that includes several obligate flower-frequenting species (e.g. *Anastetha raripila, Hesthesis* spp., *Stenoderus* spp., *Tragocerus* spp., *Tropocalymma dimidiata*). Fossil Cerambycidae (*Cerambycomima*) date to the Late Jurassic, with examples recorded from Miocene Dominican amber, and Eocene-Oligocene shales from Colorado (Grimaldi and Engel 2005).

104. *Tropocalymma dimidiata* (Cerambycidae). Large Cerambycidae are infrequent or spatially and temporally localised flower visitors in rainforest. However, *Tropocalymma dimidiata* (here associated with *Acronychia imperforata* – Rutaceae) commonly occurs episodically in some seasons on a variety of mass-flowering trees or shrubs; including the mangrove *Avicennia marina* (G. Williams unpubl. records). These events are often punctuated by seasons of apparent absence. Individuals will spend long periods feeding on nectar from single flower clusters, before flying to adjacent flowering congeners.

105. *Glycyphana brunnipes* (Scarabaeidae), a member of the subfamily Cetoniinae, feeding on the nectar of the Grey mangrove *Avicennia marina* (Acanthaceae/Avicenniaceae) a common tree in south-eastern Australian mangrove rainforest formations. *Glycyphana brunnipes* undertakes long-distance inter-plant flights. Such species are useful in transferring pollen to co-flowering plants over relatively large distances – cetoniines commonly fly many metres before landing. Flight orientation is well developed in many anthophilous Coleoptera so that transfer of pollen is facilitated. It is not uncommon to observe cetoniines, and other large beetles, hovering above a chosen canopy inflorescence before deciding to land. Though adapted to feed on nectar *Glycyphana brunnipes* is a habitat generalist, occurring in heathlands, woodland and various forest types wherever suitable floral resources occur. It can be common on the flowers of dry-fruited Myrtaceae (e.g. *Leptospermum, Angophora, Eucalyptus*) and in rainforest is recorded from *Alphitonia excelsa* (Rhamnaceae) and *Euroschinus falcatus* (Anacardiaceae). Individuals undertake prolonged feeding bouts at individual flower clusters, punctuated by relatively long distance flights to other co-flowering plants.

106. *Amata* sp. (Erebidae–Arctiinae) feeding on the nectar of *Acronychia oblongifolia* (Rutaceae). The erebid subfamily Arctiinae includes several genera (*Amata, Asura, Eressa*) of day-active moths commonly associated with flowering rainforest trees and shrubs. Adults typically spend several minutes feeding from single flowers or individual inflorescence masses, only undertaking infrequent interplant movements. Additional records of *Amata* feeding on nectar from flowering subtropical rainforest trees include *Acronychia imperforata, Alectryon coriaceus, Alphitonia excelsa, Diospyros australis, Elaeocarpus reticulatus, Euroschinus falcatus, Guioa semiglauca, Syzygium floribundum, S. smithii* and *Tristaniopsis laurina* (G. Williams 2020).

107. *Oxyodes tricolor* (Noctuidae) larvae feeding on flower buds of *Arytera divaricata* (Sapindaceae) in littoral rainforest. About 50% of insect species are fully or partially herbivorous, their feeding on flower parts and foliage impacting adversely on plant reproduction. About 99% of Thysanoptera, 98% of Lepidoptera, 90% of Orthoptera, 47% of

Coleoptera and 25% of Diptera are phytophagous (Hendrix 1990). By their combined actions assemblages of insect herbivores can drive rainforest plant diversity and composition (Bagchi *et al.* 2014).

108. *Hypocysta metirius* (Nymphalidae) is a small butterfly commonly encountered in a range of forest and woodland communities along the east coast, its range extending from northern Queensland to southern New South Wales. Although larval food plants are grasses and *Gahnia* (Cyperaceae), adults feed on the nectar of several small flowering plants on the margins of rainforest and wet sclerophyll forest. These include shrubs (e.g. *Trochocarpa laurina* – Ericaceae/Epacridaceae as here), herbs (*Lobelia trigonocaulis* – Lobeliaceae) and epiphytes (*Dendrobium gracilicaule* – Orchidaceae). The Nymphalidae are poorly recorded from flowering subtropical rainforests plants, this mainly the preserve of species of Hesperiidae, Papilionidae, Pieridae and Lycaenidae, however, *Danaus affinis* is often associated with flowering *Avicennia marina* (Acanthaceae/Avicenniaceae) at higher latitudes.

109. *Appias paulina* (Pieridae) feeding on the nectar of the native vine *Cayratia clematidea* (Vitaceae). *Appias paulina* is a wide ranging butterfly distributed from India, through to eastern Australia and the south-west Pacific. In Australia its range reflects that of its preferred larval food plant *Drypetes deplanchei* (Putranjivaceae/Euphorbiace), but also *Capparis canescens* (Capparaceae); the availability of *Appias paulina* as a pollinator being dependent upon the rainforest and monsoon thickets in which the food plant grows. Adults exhibit foraging fidelity, tending to feed on particular flowering plants to the exclusion of adjacent co-flowering species. They undertake short feeding bouts at individual blossom clusters before moving rapidly on to adjacent clusters. Clusters are frequently revisited as the butterflies search for newly opened florets. In North Queensland adults are on the wing year round (Braby 2000) but in subtropical regions their occurrence is often episodic, being much reduced in number or absent in many years.

110. *Oncopeltus sordidus* (Lygaeidae) on *Avicennia marina* (Acanthaceae/Avicenniaceae). Movements are mainly to flowers in the same cluster or adjacent clusters on the same tree. The integument is clothed with long setae on the ventral surface, and stout setae on the legs. Most adults carry significant pollen loads (>100 grains), these scattered on the ventral surface, pronotum and forelegs. Though adults undertake prolonged feeding bouts at individual flowers and flower clusters their movements are usually to adjacent clusters rather than adjacent or distant trees.

111. *Pristhesancus plagipennis* (Reduviidae) is an ambush hunter commonly encountered on flowers and foliage of subtropical rainforest trees (including flowering mangrove species). Many predators, such as assassin bugs, wasps, spiders and insectivorous birds, also frequent flowers and so pollinator populations may suffer losses through predation during foraging movements. This may then result in the loss of offspring (e.g. the young of solitary bees).

112. Pollen grains on a section of the specialised comb structure on the foreleg of *Anthobosca signata* (Thynnidae). Bees and numerous large wasps possess structures for preening their bodies or collection of pollen grains. These are later ingested, or are relocated to other areas of their bodies to allow carriage to nest sites.

113. Detail of thoracic setae, *Bisallardiana gymnopleura* (Scarabaeidae-Cetoniinae). Several Cetoniinae, commonly called rosechafers, possess long, dense hair-like setae and surface vestiture which snare pollen grains. Some structures reflect specialised modifications that increase the beetle's ability to carry pollen, for example, the hooked setae shown here on the lateroventral surface of *Bisallardiana gymnopleura* (Williams and Adam 1998). Although modified plumose hairs are known from other scarab-related taxa (e.g. from the scarabaeoid genera *Amphicoma* and *Lichnanthe* [Glaphyridae]) (Crowson 1981), and may occur in overlooked related species, the presence of specialised pollen-collecting hairs on *Bisallardiana gymnopleura* represents a faculty otherwise described only from bees, hover-flies and more recently bee-flies (Bombyliidae) (Neff *et al.* 2003).

114. *Belenois java* (Pieridae) feeding on the nectar of lantana, *Lantana camara* (Verbenaceae). Lantana is a native of tropical South America, now extensively naturalised in eastern Australia. It is an invasive shrub commonly encountered as dense infestations that have colonised disturbed rainforest and wet sclerophyll forest. In such situations infestations can persist for decades allowing minimal recruitment of native plants. Lantana can also invade relatively undisturbed mesic forest stands, adding high fuel loads. This can pose a significant fire threat during periods of prolonged drought. However, lantana flowers offer a nectar source to several rainforest-associated butterflies; such as the birdwing species *Ornithoptera priamus* and *O. richmondia*, species of *Graphium* (Papilionidae), and varius Hesperiidae and Pieridae. The genus *Belenois* has a predominatly Afrotropical distribution, with only *B. java* occurring in Australia, where it has an extensive distribution, but with few records from Tasmania (Braby 2000). Larval food plants are members of the family Capparaceae. *Belenois java* is a migratory species, but populations can vary dramatically from season to season, with individual populations often concentrated about densely flowering nectar sources. Populations can exhibit foraging constantly, not visiting different flowering species even when these are adjacent.

115. Emergent *Araucaria* conifers in low vine thicket, central Queensland. Conifers are pollinated by wind and thus subject to the uncertain mobilisation of pollen in the air column. The occurrence and direction of seasonal winds are not always predictable, so individual plants, especially distant outliers, may be subject to inadequate levels of pollen transfer. The central-southern Queensland region, what is commonly referred to as the 'brigalow belt', has been subject to heavy clearance for agriculture (Young and McDonald 1987), and this continues. In addition to the more complex subtropical evergreen forests (of high floristic and structural diversity) of the extreme south-east/McPherson Range area there are inland araucarian vine forests and low semi-evergreen vine thickets extending south from Rockhampton to the New South Wales border; though vine thickets (e;g. Planchonella Nature Reserve) extend beyond this (Curran *et al.* 2008; McDonald 2010). These inland rainforests have disjunctive distributions. Within araucarian vine forest there is latitudinal sifting of plant families south to north with some families disappearing altogether from rainforest stands or becoming poorly represented (e.g. Meliaceae, Monimiaceae, Proteaceae) (Young and McDonald 1987).

116. Boab tree, *Adansonia gregorii* (Malvaceae/Bombacaceae), west of Timber Creek, Northern Territory. This is a conspicuous deciduous tree of savannah woodlands, sometimes occurring as small closed-canopy copses, and along stream lines. However, relatively few Australian rainforest-inhabiting angiosperms are truly deciduous. Adam (1992) notes *Adansonia* is elsewhere restricted to sub-Saharan Africa (*A. digitata;* widespread there) and with seven species in Madagascar. *Adansonia gregorii* is more closely related to *A. digitata* than to the Madagascan species, and in the absence of fossil records from Australia its occurrence here may be an instance of recent long-distance dispersal (Adam 1992).

117. Mosaic-like distribution of subtropical and warm temperate rainforest (illustrated by the dark green patches of vegetation) within a matrix of eucalypt-dominated forest types, northern New South Wales. The rainforests of Australia's subtropical south-east exist as disjunct mesic-adapted vegetation pockets within a landscape otherwise comprised of 'wet' and 'dry' sclerophyll forest. The so-called 'wet sclerophyll forests' can occur as narrow transitional zones between rainforest and more open, xeric-adapted, plant communities such as dry sclerophyll forest and woodland.

118. Dry rainforest dominated by *Backhousia sciadophora* (Myrtaceae), near Taree, New South Wales. In floristically simple rainforests there is greater potential for wind to achieve efficient transfer of pollen between co-flowering plants of the same species, simply because there are fewer other species to act as physical obstacles to the movement of pollen. *Backhousia sciadophora* is a small to medium-sized tree mainly occurring, often as a co-dominant in the canopy, in dry rainforest north from the Hunter Valley to southern Queensland. An exceptional occurrence is that of populations in headland littoral rainforest at Black Head and Redhead in northern New South Wales.

119. Rainforest understorey, Werrikimbe National Park, New South Wales. Understorey strata of rainforests normally offer limited floral resource opportunities for pollinators, with foraging in subcanopy strata mainly occurring in gaps below disturbed and open canopies. Relatively few pollinators frequent such dimly lit environments, the flowering plants found there often relying on specialist pollinators. Werrikimbe National Park is a focal point of the Gondwana Rainforests/Central Eastern Rainforest Reserves of Australia (CERRA) World Heritage Area (Adam 1987). It is one of the major regions of rainforest in New South Wales. The vegetation is extremely diverse, comprised of subtropical, warm temperate, cool temperate (with *Nothofagus moorei*) and dry rainforest types, these within a eucalypt-dominated matrix.

120. Subtropical rainforest–wet sclerophyll forest margin, northern New South Wales. In the mesic hinterland forests of northern New South Wales and southern Queensland rainforest margins are commonly ill-defined floristic and structural intergradations of species and canopy formation with that of eucalypt-dominated sclerophyll forest. Here, a well developed rainforest is extending into eucalypt-dominated wet sclerophyll forest, this with no evidence of eucalypt recruitment on the forest floor. The large eucalypts to the right foreground are *Eucalyptus microcorys*, the grey-trunked species discernable in the background are *Eucalyptus grandis*. Here, rainforest trees include, *Ackama paniculosa, Alectryon subcinereus, Archontophoenix cunninghamiana, Cryptocarya microneura, Daphnandra micrantha, Diploglottis australis, Eupomatia laurina, Ficus coronata, Gmelina leichhardtii, Karrabina benthamiana, Litsea reticualta, Polyscias murrayi, Schizomeria ovata, Sloanea woollsii, Syzygium smithii,* in addition to the cycad *Lepidozamia peroffskyana* (centre) (G. Williams 1993). Wet sclerophyll forests can carry high fuel loads, the accumulating fallen leaves, branches and bark representing an increasing fire risk to adjacent rainforest during extreme drought conditions. Exfoliated bark often builds up as a 'skirt' around the base of *Eucalyptus grandis*, a species often dominant in mesic eucalypt forests. Owing to the smooth-barked trunk fire does not normally carry up into the tree crown, however, long strips of loose bark are retained on the trunk for some time, thus adding to flammability. In addition, eucalypt trees annually drop quantities of small twigs and branches which collectively add to availability of fuel on the forest floor.

121. Forest margin formed by dissecting trail, League Scrub, northern New South Wales. Few plants flower in the dim rainforest understorey, however, many species (especially vines) are able to take advantage of sunlit openings and margins, and within such environments may flower profusely at lower levels.

122. Understorey, *Nothofagus moorei*-dominated cool temperate rainforest, Barrington Tops, New South Wales. The spatial and structural complexities of floristically diverse rainforest, as well as climatic conditions, may hinder the evolution and deployment of wind pollination (anemophily) strategies. But it may be difficult to determine whether this is because of physical barriers to dispersal imposed by dense vegetation or because air humidity decreases the dispersal of pollen.

123. *Nothofagus moorei* (Nothofagaceae)-dominated canopy, cool temperate rainforest. *Nothofagus moorei* often occurs in almost pure stands, and is an example of species gregariousness. Rainforests with lower diversity of woody taxa, such as *Nothofagus*-dominated rainforest, are likely to favour wind pollination, or mixed wind–biotic agent pollination strategies. Thus a greater prevalence of anemophily in Australian rainforests may be anticipated in floristically and structurally simple formations. There may also be a heightened incidence of wind pollination in gap phase rainforests because the initial open structure favours less impeded pollen transfer to conspecific plants.

124. Warm temperate rainforest, Boorganna Nature Reserve, northern New South Wales. Here occurs the Australian endemic *Akania bidwillii* (= *A. lucens*) (Akaniaceae), a small pinnate-leaved tree in subtropical and warm temperate rainforest north from the Comboyne Plateau to the Blackall Range in south-eastern Queensland (Harden *et al.* 2006). Boorganna Nature Reserve was proclaimed early in the 20th century to protect one of the remaining stands of warm temperate and subtropical rainforest on the Comboyne Plateau. This was once an extensive area of rainforest, but by the 1920s much had been cleared for agriculture (Chisholm 1925). Isolated pockets now remain.

125. Cool temperate rainforest understorey, Melba Gully State Park, Otway Ranges, south-west Victoria. Cool temperate rainforests of Victoria extend from East Gippsland to the Otway Ranges south-west of Melbourne. The Otways represent the westernmost occurrence of cool temperate rainforest in Australia. *Nothofagus cunninghamii* (Nothofagaceae), usually with *Acacia melanoxylon* (Fabaceae/Mimosaceae) as a co-dominant, forms a closed forest community in wet montane gullies and moist protected slopes. Understorey associates include *Hedycarya angustifolia* (Monimiaceae), *Tasmannia lanceolata* (Winteraceae) and the tree fern *Dicksonia antarctica* (Dicksoniaceae). However, cool temperate rainforests of the Otways have been subject to clearing for agriculture and forestry plantations.

126. Cool temperate rainforest canopy, Otway Ranges, south-west Victoria. Here, cool temperate rainforest canopies are dominated by *Nothofagus cunninghamii*, a large tree occurring extensively in Tasmania, but on mainland Australia it is restricted to the Otway Ranges, Wilsons Promontory and the Central Highlands and South Gippsland (Harden *et al.* 2006). However, *Nothofagus cunninghamii* is vulnerable to attack from the fungal pathogen *Chalara australis*, or 'Myrtle wilt' disease, the pathogen attacking trees with damaged limbs or trunks.

127. The 'Balfour Track' showing *Nothofagus cunninghamii* (Nothofagaceae) and *Dicksonia antarctica* (Dicksoniaceae) tree ferns beside Stephens Rivulet, a tributary of the Arthur River in north-west Tasmania (H. Nicholson pers. comm.). (Photo: Hugh Nicholson)

128. Victoria Pass near Queenstown showing the canopy of mixed *Atherosperma moschatum* (Monimiaceae/Atherospemataceae), *Nothofagus cunninghamii* (Nothofagaceae) and *Phyllocladus aspleniifolius* (Podocarpaceae/Phyllocladaceae). Cool temperate rainforest in Tasmania is characterised by the dominance of the angiosperm genera *Atherosperma, Eucryphia, Nothofagus,* and the gymnosperms *Athrotaxis, Lagarostrobos* and *Phyllocladus* or *Diselma*. The most extensive areas of rainforest in Tasmania are found in the north-west. Elsewhere it is fragmented to various degrees. The Tasmanian endemic pine *Phyllocladus aspleniifolius* is mainly restricted to western and south-western Tasmania and is a common component of cool temperate rainforest, though also occurring frequently in wet sclerophyll forests. The widespread *Nothofagus cunninghamii* is a large spreading tree occurring in sheltered gullies, and has a limited, disjunct distribution in Victoria. The range of *Atherosperma moschatum* extends to Victoria, and the Snowy Mountains, Barrington Tops, and Tia Gorge near Armidale in northern New South Wales (Harden *et al.* 2006). (Photo: Hugh Nicholson)

129. The forest canopy of Tahune Forest Reserve on the Huon River in south-eastern Tasmania showing *Atherosperma moschatum, Dicksonia antarctica* and *Eucryphia lucida*. (Photo: Hugh Nicholson)

130. Wet sclerophyll forest, northern New South Wales. *Livistona australis* palm in a forest stand dominated by 'Flooded gum' *Eucalyptus grandis*; demonstrating the heaving fuel load of fallen bark, leaves and branches that can build up in the understorey of such forests. Also known as 'tall open forest' such eucalypt-dominated mesic forest formations can form transitional zones between rainforest and more open forest communities. However, fire is a strong limiting determinant of wet sclerophyll forest – rainforest boundaries and a barrier against the ability of rainforest to expand.

131. Gallery rainforest dominated by *Tristaniopsis laurina* (Myrtaceae), northern New South Wales. At sites experiencing lower rainfall gallery rainforest typically occurs as narrow stream-associated communities within a eucalypt-dominated forest matrix. These are prone to frequent fires which act to halt the advance of the rainforest boundary. At moderate elevations (>150 m) *Backhousia myrtifolia* and *Lophostemon confertus* (Myrtaceae) are often associates of *Tristaniopsis laurina*. In lowland stands (<70 m) *Syzygium floribundum* (Myrtace) is often dominant, generally replacing *Tristaniopsis laurina*. However, where stands experience higher rainfall gallery forest may grade into adjacent, and more expansive, subtropical and warm temperate rainforest types. At lower altitudes such rainforest stands are typically restricted to stream banks within otherwise eucalypt-dominated sclerophyll forest. These are subject to frequent fires such that rainforest is unable to expand away from the stream zone.

132. *Avicennia marina* (Acanthaceae/Avicenniaceae)-dominated mangrove forest, northern New South Wales. Traditionally, mangroves have been considered a distinct vegetation type, however, on structural grounds (that of a 'closed canopy' in more developed stands), their placement within the broader conceptual notion of 'rainforest' may be appropriate (Adam 1992; G. Williams 2020). Mangrove communities vary structurally from that of a closed canopy forest to open woodland, being floristically diverse at lower latitudes and most extensively developed in protected estuary zones. Mangroves are usually restricted to the inter-tidal zone, and in Australia they are distributed along the eastern and northern coasts; but are essentially tropical and subtropical in distribution (Adam 1992; Duke 2006). At higher latitudes insects are the primary pollinators of mangrove flowers, but in tropical regions birds may be more important in the role of pollinator.

133. Littoral rainforest, Redhead, northern New South Wales, illustrating the wind shearing of the frontal canopy, and resultant uniform canopy, caused by frequent winds. The extent of littoral rainforest in Australia has been significantly reduced since European occupation, and remnants are now recognised as *critically endangered* ecological communities. Surviving stands extend from tropical Australia to eastern Victoria. However, most are small in size and are subject to ongoing edge effects and invasion by alien plants. Many are subject to repeated fires which restrain regeneration and expansion of remnants.

134. Littoral rainforest understorey, Seal Rocks, New South Wales. The seaward margins of littoral rainforest on sandy soil are low in response to wind impacts, however, plants protected by beach dunes and frontal vegetation are able to grow to greater heights (e.g. as at Fraser Island in south-east Queensland, Iluka northern New South Wales). Structural characters though, show considerable uniformity with usually only 1–2 strata comprising canopy and sub-canopy layers. The understorey is usually open, with little diversity of shrubs or herbs. However, several tree species characterise subtropical littoral rainforest on sand; *Cupaniopsis anacaradioides, Arytera divaricata, Alectryon coriaceus* (Sapindaceae), *Alphitonia excelsa* (Rhamnaceae), *Acronychia imperforata* (Rutaceae), *Syzygium hemilamprum, S. oleosum, S. smithii* (Myrtaceae), *Elaeodendron australe* (Celastraceae), *Ficus fraseri* (Moraceae).

135. Canopy of complex tropical rainforest, Wooroonooran National Park, North Queensland. (Photo: Hugh Nicholson)

136. Understorey, complex tropical rainforest, Tully Gorge, North Queensland. Species include *Pullea stutzei* (Cunoniaceae) with red leaf growth, endemic to the Wet Tropics; *Musa banksii* (Musaceae), a species restricted to Queensland and New Guinea; and over-branching *Homalanthus novoguineensis* (Euphorbiaceae), occurring widely in northern Australia (including vine thickets), and also nearby islands. The most extensive continuous areas of rainforest in Australia are the evergreen closed forests extending from Ingham to Cooktown in north-eastern Queensland; the so named 'Wet Tropics' rainforests, and World Heritage Area. In size this area of rainforest covers nearly 8000 km^2, with an additional ~2500 km^2 in the Cape York Peninsula region to the north. Mean annual rainfall can exceed 2500 mm (Daintree-Cape Tribulation, Tully-Babinda areas). In altitude rainforest extends from sea level (Cape Tribulation), to hinterland ranges of 900–1200 m, with Mt Bartle Frere, the highest peak, being 1622 m. (Photo: Hugh Nicholson)

137. Understorey, lowland rainforest, Mission Beach, North Queensland. Species here include: *Epipremnum pinnatum* (Araceae), climbing trunk, a species with a distribution including Australia, Malaysia, Indochina, China, Japan, Melanesia; *Davidsonia pruriens* (Davidsoniaceae/Cunoniaceae) with new bright leaf colour, endemic to North Queensland; and the vigorous climbing endemic palm *Calamus moti* (Arecaceae) at centre foreground (H. Nicholson pers. comm.). *Calamus moti* forms a prominent feature of lowland and highland rainforest in north-east Queensland. The species occurs at heights above 1000 m on the Atherton Tableland. (Photo: Hugh Nicholson)

138. Canopy of upland tropical rainforest on Bellenden Ker, Wet Tropics World Heritage Area, North Queensland. At 1593 m Mt Bellenden Ker is the second highest peak in Queensland, the highest being the adjacent Mt Bartle Frere (1622 m). Mean annual rainfall on Mt Bellenden Ker can be in excess of 9000 mm. It is named after the English botanist John Bellenden Ker Gawler (1764–1842), is located south of Cairns near Babinda, and forms part of the Wooroonooran Range (syn. Bellenden Ker Range). (Photo: Hugh Nicholson)

139. Small pockets of vine thicket and tree clusters, Katherine River, Northern Territory. Rainforest stands in the Northern Territory and Western Australia are relatively biotically and structurally simple, are highly fragmented, and are mostly small in size, frequently less than several hectares (Adam 1992; Russell-Smith and Dunlop 1987). Seemingly minute stands (e.g. <0.01 ha), though, may have as many as 40 monsoon vine forest species present, though these generally are represented by individually small populations. However, in western Arnhem Land single stands can be extensive (~100 ha). Soil type (or rock features), moisture regimes, presence of streams and fire are factors in the extension and survival of individual stands, with sites on rock outcrops being particularly vulnerable to fire (Russell-Smith and Dunlop 1987). In the Kimberley region of Western Australia vine thickets are found on stabilised sand dunes, and in the Northern Territory a high proportion of the obligate monsoon vine forest flora occurs on recent landforms; this highlighting the dispersal capacity of the flora (Russell-Smith and Dunlop 1987). Canopy height in northern Australian evergreen stands can be >20 m, however, in seasonally dry lowland habitats rainforest may have a significant component of deciduous plants (Adam 1992; Kenneally and Beard 1987). Russell-Smith and Dunlop (1987) considered that the flora comprised two floristic elements; 1. a large number of dispersible species with a capacity to exploit new ecological opportunities, and 2. species that are confined to precarious refugia, such as spring habitats with wide disjunctions.

140. Isolated 'bottle tree' *Brachychiton ruprestris* (Malvaceae/Sterculiaceae) tree in remnant semi-evergreen vine thicket, Goodnight Scrub, central Queensland. Semi-evergreen vine thickets are typified by emergent *Brachychiton* species, with associated species that include *Alectryon connatus, Backhousia angustifolia, Bridelia leichhardtii, Ehretia membranifolia, Flindersia collina* and *F. australis* (Young and McDonald 1987). More widely, however, many rainforest trees occur as sparsely distributed individuals within rainforest stands. This situation is common in tiny remnant rainforest patches where populations are often represented by single plants. In such circumstances, if species are obligate outcrossers (and where agamospermy or vegetative growth offers no alternative opportunity for reproduction), then for all practical purposes they may be functionally extinct – their relatively long life spans masking their inability to reproduce.

141. Canopy gap following collapse of a large fig tree in Victoria Park Nature Reserve, New South Wales. Such ephemeral *gap phase* environments not only allow colonisation by plants that will re-establish a closed rainforest canopy, but permit expression of pollination strategies, such as wind-pollination, that are otherwise impeded in the understorey of floristically complex rainforests. Victoria Park Nature Reserve is one of several small rainforest remnants (now isolated within an otherwise agricultural and residential landscape) surviving from the once-extensive 'Big Scrub' rainforest in north-eastern New South Wales (Parkes *et al.* 2012).

142. Destruction of the canopy caused by flying-foxes, at a small lowland subtropical rainforest remnant, Maclean, northern New South Wales, 1998. Large populations progressively destroyed foliage and branches, opening the understorey to high light levels and subsequent invasion by smothering vines, these acting as a barrier to further tree recruitment and loss of much of the native plants that otherwise would have been present at lower strata. The clearing of most of the lowland subtropical rainforest since European occupation in 1788 has reduced roost sites available for flying-foxes (predominantly *Pteropus poliocephalus, P. scapulatus,* and increasingly southward-extending *P. alecto* [Pteropodidae]). These form large concentrated camps of many thousands, which remaining lowland subtropical stands are unable to sustain. Lowland rainforest remnants are characterised by small size (typically <4–8 ha) and minimal 'core' area, within which a significant number of tree species survive at very low population levels (commonly just 1or 2 mature individuals); such that persistent or frequent seasonal fly-fox residency leads step-wise to destruction of inflorescences, plant reproductive failure, and ultimately extinction of plant populations. A classic example of alien plant invasion and recurring canopy damage by flying-foxes in lowland rainforest remnants is provided by Wingham Brush Nature Reserve, also northern New South Wales. Prior to the commencement of a forest regeneration project in 1980 (Harden *et al.* 2004), this small lowland rainforest remnant was severely infested with aggressive exotic vines and understorey herbs (e.g.*Anredera cordifolia* [Basellaceae], *Cardiospermum grandiflorum* [Sapindaceae], *Dolichandra unguis-cati* (Bignoniaceae], *Tradescantia fluminensis* [Commelinaaceae]). Much of the canopy had been damaged, with many limbless 'vine towers' resulting, and a significant portion of the forest floor was smothered by *Tradescantia fluminensis,* such that the reproduction of native plants, and the long-term survival of the stand as a functioning rainforest ecosystem, was threatened. Despite a seemingly successful restoration program the site regularly hosts a large flying-fox population (comprising 3 species), their residency resulting in ongoing canopy damage.

143. Rampant invasive growth of the native vine *Cissus antarctica* in disturbed mixed subtropical rainforest-wet sclerophyll forest gully, northern New South Wales. Such infestations can persist for decades smothering native seedlings and saplings.

144. 'Mast' flowering *Argyrodendron actinophyllum* (Malvaceae/ Sterculiaceae) in a burnt mixed rainforest-wet sclerophyll forest gully, northern New South Wales, 2020. The dead vine tangles and open foreground indicate the extent of invasive native *Cissus antarctica* (Vitaceae) before the fire event. *Argyrodendron actinophyllum* is a large to very large palmate-leaved tree often common in montane warm temperate-subtropical rainforest (northern New South Wales to south-eastern Queensland). Regional populations may flower synchronously.

145. Weed invasion in the understorey of a dry rainforest stand (Kiwarrak State Forest) following fire (2019), New South Wales; illustrating 4 month post-fire response to increased light penetration by invasive alien plants following damage to the forest canopy. In addition, adjacent roadside margins were colonised by numerous minor alien plants (e.g. *Bidens pilosa, Tagetes minuta* [Asteraceae], as well as *Lantana camara*, this a pre-existing invasive weed regenerating from fire damage to its woody stems. Two native vines, *Cayratia clematidea* (Vitaceae) and *Sicyos australis* (Curcubitaceae), had also vigorously colonised margins, but had not penetrated into the sunlit understorey.

146. Although the photograph is taken from a different direction to that of Plate 145 this illustrates the nature of the dry rainforest understorey before the impact of fire and subsequent weed invasion.

147. Dense *Asparagus aethiopicus* (Asparagaceae) infestation in littoral rainforest, Old Bar, northern New South Wales (1996). Such carpeting forest floor growth by invasive alien plants suppresses germination of native seeds, smothers herbaceous species restricted to the forest floor, and disrupts pollination networks. *Asparagus aethiopicus* is pollinated by the invasive honey bee *Apis mellifera*, the liaison between the two alien species facilitating further invasion. This leads progressively to recruitment failure and extinction of native plants, and overall loss of ecosystem integrity. Various Asparagaceae seem able to invade coastal rainforest with relative intact canopies, suggesting they have a degree of inherent or developed shade tolerance (see discussion in Godoy *et al.* 2011 of plant adaptive characters that may facilitate plant invasiveness in shaded conditions).

148. Fire damage to rainforest, littoral rainforest, northern New South Wales (2019). The photo partly demonstrates the initial (4 months post-fire) regeneration response by plants. Surviving canopy plants on the margin are mainly *Acronychia imperforata* (Rutaceae), *Cupaniopsis anacardioides* and *Arytera divaricata* (Sapindaceae). Margins have been aggressively colonised by suckering native *Cissus antarctica* vines and *Breynia oblongifolia* shrubs in concert with minor alien weeds. Further in, native trees, shrubs, herbs and vines are resprouting from the base of trunks or root stock (see Table 9.1). Numerous *Alphitonia excelsa* seedlings have germinated from soil seed stock. However, propagules of alien Asparagaceae, *Lantana camara* and *Chrysanthemoides monilifera* have germinated. These are persistent invasive species that can smother understorey strata. Fires, and associated extreme drought conditions, threaten plant–pollinator interactions by removing floral resources and habitat. Regeneration of the plant community and the preservation of pollinator assemblages in such circumstances cannot be predicted.

149. Regeneration of *Rhodomyrtus psidioides* (Myrtaceae) from root stock in a post-fire littoral rainforest remnant (see Plate 148). New leaf growth is infected with the fungal pathogen Myrtle rust, *Austropuccinia psidii.* The impact of post-fire infection, and repeated later pathogen attack, has critical implications for the conservation of threatened Myrtaceae and the ecological integrity of rainforest stands.

150. Coppicing *Mischocarpus pyriformis* (Sapindaceae), Harrington, northern New South Wales, following an extreme fire event and drought in 2019–2020. Most medium to large rainforest trees that were impacted by the fire produced basal coppice (see Table 9.1). Although rainforest structure is likely to re-establish (in the absence of future fires) a critical issue is the length of time that will ensue before trees are able to bud and set flowers; and thus sexually reproduce. *Mischocarpus pyriformis* is monoecious, flowers are unisexual, and occurs in lowland rainforest from northern New South Wales to Queensland.

Appendix 5. Divisions of geologic time

(adapted from W. Hansen 1991; Barlow 1981)

Era	Period	Epoch	elapsed time my*
Cenozoic	Quaternary	Holocene	11 400 yrs
		Pleistocene	1.8
	Tertiary	Pliocene	5.3
		Miocene	23.3
		Oligocene	33.9
		Eocene	55.8
		Palaeocene	65.5
Mesozoic	Cretaceous	Late/Upper	99.6
		Early/Lower	145.5
	Jurassic	Late/Upper	161.2
		Middle	175.6
		Early/Lower	199.6
	Triassic	Late/Upper	228
		Middle	245
		Early/Lower	251
Palaeozoic	Permian		299
	Carboniferous	Upper	318
		Lower	359.2
	Devonian		416
	Silurian		443.7
	Ordovician		488
	Cambrian		542

* elapsed time in millions of years since beginning of the Period or Epoch

Glossary

abiotic pollination: pollination without the aid of biological agents, thus by wind, gravity or water.

abscission/absciss: to separate or shed an organ from a plant or parent part when aged or mature (e.g. petal, fused corolla, fruit).

abundance: the relative frequency of a species within a community.

actinomorphic: radially symmetrical; a flower that can be bisected in a number of planes, with sepals and petals of the same or largely similar shape. cf. **zygomorphic**.

Aculeata: clade of Hymenoptera containing stinging wasps, ants and bees.

aculeate wasps: species of vespoid and spheciform wasps. See Naumann (1991).

adaptation: a modification in condition or behaviour, usually in relation to environment or ecological influences, from which a survival benefit is obtained.

adaxial: adjacent to or turned towards the axis, as the upper surface of a leaf in relation to the stem.

adventitious: arising from a source incidental or abnormal to the regular source; roots arising from shoots.

agamospermy, agamospermous: production of an embryo without fertilisation (meiosis does not occur), thus all progeny possess an identical genetic constitution.

alkaloids: group of nitrogen-containing, basic organic compounds; possibly end products of nitrogen metabolism. Important because of their poison and medicinal properties.

allogamy, allogamous: reproducing by cross-fertilisation. cf. **autogamy**.

allopatric: originating from or occupying different areas.

allotropic: poorly adapted for utilisation of flowers; food forms part of a mixed diet.

amber: produced by plants; in the restricted sense of Langenheim (2003), a lipid-soluble mixture of terpenoid or phenolic compounds, distinguishing it from gums, waxes, plant mucilage, oils and latex.

ambophily, ambophilous: facultatively wind-pollinated.

androecium: the collective male part of the flower (i.e. stamens, anthers, filaments).

androdioecy, androdioecious: where normally hermaphroditic populations also possess some functionally male members.

andromonoecy, andromonoecious: the condition where inflorescences include bisexual flowers and some that are functionally male.

anemophily, anemophilous: pollination, dispersal of pollen, by wind.

angiosperm: a flowering, seed-bearing, vascular plant characterised by ovules enclosed in ovaries; distinct from gymnosperms. See Bateman *et al.* (2006).

angiospermy: in which the ovules of flowering, fruit-bearing, plants develop in a closed ovary.

Antarctic element: part of the present-day flora postulated as originating in the Southern Hemisphere.

antepetalous: in which the stamens are equal to, and aligned with, the petals.

anther: the pollen producing part of the stamen.

anthesis: the timing and process of flowering; also the timing of the shedding of pollen from the anthers.

anthocyanin: a water-soluble glycoside occurring as a pigment in flowering plants; may appear red, purple or blue according to ph.

anthophyte: a clade of plants comprising Gnetophyta (Gnetales), angiosperms, Bennettitales and Pentoxylales, but also the Caytoniales; see the 'pseudanthial' and 'anthophyte' hypotheses (Doyle and Donoghue 1992).

anthophile, anthophilous: 'flower loving', used in reference to animals (including insects) that visit and usually feed on flowers; but may not be true pollinators.

anthoxanthins: water soluble floral pigments (flavenoids); usually clear, white to yellow. Counterpart to **anthocyanins**.

apetalous: without a corolla.

Apiformes: bees.

apocarpous: a gynoecium which consists of two or more carpels, these free and distinct from one another. cf. **syncarpous**.

apogamy, apogamous: in general asexually reproductive.

apomixis, apomicts: reproduction usually by vegetative means, or by agamospermy through seeds, but in which meiosis does not occur. Thus offspring are genetically identical to the parent plant.

arboreal: living in trees.

aril: a fleshy structure fully or partly covering the seed coat (e.g. as in Sapindaceae).

asexual reproduction: reproduction without the formation of gametes.

assemblage: in pollination ecology, the various taxa (not necessarily related) associated with a flowering plant. These may include species highly adapted for flower visitation and a suite of ecological-generalist or opportunistic species exhibiting no particular morphological or behavioural adaptation to visiting flowers. Participating taxa within the assemblage may not exhibit spatial or temporal constancy to the flowering resource. See also **guild**.

asterids: a large clade of flowering plants. See Angiosperm Phylogeny Group III (2009), Angiosperm Phylogeny Group IV (2016).

autochthonous: being indigenous to or originating from a particular region.

autogamy: process of self-fertilisation.

axis: the central stem of a plant or inflorescence; the elongated part of the flower receptacle.

behavioural plasticity: in this sense feeding from many plant species.

biogeography: the study of the distribution of organisms, their habits and historical and biological factors that produced them.

biota: the flora or fauna of a particular area or biogeographic region.
biotic: relating to the living process.
biotic pollination: where pollination is achieved through the actions of animal vectors such as birds, bats, non-flying mammals, and insects and other invertebrates.
bipinnate: twice-divided pinnate leaves.
bisexual: flower, reproductive structure with both functional male and female parts.
boreal forest: biome characterised by coniferous forests; also **boreal ecosystem**, an ecosystem with subarctic climate in the Northern Hemisphere; **boreal climate**, characterised by long winters and short, cool to mild summers.
bract: in angiosperms, a leaf at the base of the peduncle; often modified in shape and form, sometimes brightly coloured so as to superficially appear flower-like.
bracteole: an accessory bract, such as a pedicel.
bryophytes: photosynthetic, non-vascular or partially vascular plants; e.g. liverworts, hornworts, mosses. Reproduce sexually by fusion of male and female gametes produced in multcellular sex organs.
bucca, buccal: the mouth in insects, of the mouth.
bulbils: a small bulb-like structure arising (by asexual production) from a stem, leaf or bract, and so allowing the plant to reproduce vegetatively.
calyptra: hood or lid-like structure of certain buds and fruits.
calyx: the sepals of a flower; sterile appendage of a floral receptacle composed of sepals. cf. **corolla**.
calyx tube: a tube formed by the bases of sepals.
cambium: the growing sheath-like region (meristem) of woody stems and fruits that forms the woody tissue.
canopy: the upper foliage and structural layer of a plant or plant community.
cantharophily, cantharophilous: of a flower adapted to, and pollinated by, beetles (Coleoptera).
carbohydrate: a group of chemically related compounds (such as sugars) basically comprised of carbon, hydrogen and oxygen.
carnivorous: of a plant, when utilising animal tissue for nutrients, by the action of enzymes that are able to reduce proteins to amino acids. Usually more strictly considered **insectivorous**.
carotenoid: a water-insoluble pigment mainly as a carotene, responsible for orange, red and yellow colours (in leaves, some flowers and fruits).
carpel: the ovule-bearing part of the female part of the flower (**gynoecium**).
cauliflory, cauliflorous: the development of flowers directly from the trunk; see also **ramiflory** (production of flowers from branches).
Cenozoic: geological Era; including Tertiary and Quaternary periods (also Caenozoic, Cainozoic); 65.5 mya to present.
chasmogamy, chasmogamous: flowers that are pollinated when open. cf. **cleistogamous**.
chiropterophily, chirophilous: pollination by blossom-bats, fruit-bats and flying foxes.
cilia: fine hairs.
circadian rhythmicity: a biological rhythm having a periodicity of 1 day (24 hours); exhibiting particular periods of activity with respect to daily periods.
clade, clades: branch/branches of a devised phylogenetic 'tree', their hypothetical membership formed by sequential loss or acquisition of particular morphological characters or genetic attributes.
classification: in systematic botany, and zoology, the arrangement of taxa in such a way as to express their relationships, structure and evolutionary origin.
cleistogamy, cleistogamous: of flowers when they are fully or partly closed so as to suppress cross-pollination. cf. **chasmogamous**.
climax community: usually refers to a plant community in equilibrium with its environment, but in which the presence of individual species may vary such that there are episodes of localised extinction and recruitment, and fluctuating abundance (as in complex rainforests). Thus in tropical and floristically complex subtropical rainforest equilibrium may be unattainable at the species level.
closed community/forest: forming a continuous and unbroken cover, as in rainforest, mangroves and some other vegetation types such as dense heaths and coastal shrubland.
commensalism: in which different species share the same environment but in which there is no obvious nutritional dependence or antagonism.
community: interactive group or system of plants and animals, these considered in relation to each other and the resources of their environment.
cone: the reproductive structure of cycads and most conifers (**gymnosperms**), and several other non-flowering plants (e.g. lycopods); also the woody fruit of flowering (**angiosperm**) Casuarinaceae.
conifer: seed-bearing tree, sometimes a shrub (i.e. Araucariaceae, Cupressaceae, Pinaceae, Podocarpaceae, Taxodiaceae).
constancy: although often used as a synonym of **fidelity**, it is more correctly applied to visitation to a resource *in time* (i.e. of limited temporal duration), rather than having an obligate association with a particular plant species, resource-related group of plants or forest stratum.
coppicing: producing shoots from dormant buds of a trunk, particularly when reduced to a stump.
corbicula, corbiculae: 'pollen baskets', concave, smooth space, edged by a fringe of hairs arising from the hind tibia of some bees. Coxal corbiculae are possessed by some agaonid fig-wasps.
corolla: the second whorl of sterile appendages of a complete flower; composed of petals, these free at their margins (**polypetalous**, **choripetalous**) or united (**sympetalous**) to form a partial **tube**. cf. **calyx**.
cosexual species: those with bisexual or monoecious breeding systems.
cotyledon: first leaf or leaves of a mature embryo; at germination may become photosynthetic.
crepuscular: active during the twilight hours of dawn and/or dusk.

Cretaceous: Period of geologic time; c. 145.5–65.5 mya.
cryptogams: algae, bacteria, fungi, pteridophytes, bryophytes.
cycad: gymnospermous cone-bearing seed plant usually considered fully dioecious.
cyme: an inflorescence in which the main axis ends in a flower. Further growth is by additional branches which also end in a flower.
deceitful attraction: attracting pollinators without offering a reward, by mimicking rewarding species.
deciduous: shedding leaves at the end of the growing season, but here also applied to the loss of floral parts during the period of flower presentation or at end of receptivity.
decussate: arranged in pairs of opposite members, each pair turned at right-angles to the succeeding pair.
dehiscent, dehisce: of an anther, rupturing or opening to release pollen.
Devonian: Period of geologic time, c. 416–359 mya.
dichogamy, dichogamous: protandry (in which flower is first male), protogyny (in which flower is first functionally female); also 'sequential hermaphroditism'.
dichotomous: branching.
diel: daily; pertaining to a 24 hour cycle.
dimethyl sulphide: an organosulphur compound; a breakdown product of many biological sulphur compounds.
dioecy, dioecious: of a species separate male or female plants.
dipterocarp: member of the large tropical family Dipterocarpaceae.
distichous: arranged in two ranks and in the same vertical plane.
diurnal: active or flying during the hours of daylight.
double fertilisation: the process in which two sperm nuclei of a pollen grain unite with nuclei of the embryo sac (female gametophyte) of an angiosperm; unique to angiosperms.
dystropic: visitor showing no adaptation to the blossom; may be destructive yet able to undertake pollination.
echinate: surface with spine-like projections; of pollen.
ecological generalist: exhibiting no preference to a particular taxon, or food source.
ecological release: exhibiting flexibility (switching) away from native species when exotic plants are available.
elaiophores: small oil secreting glands.
emboli: gaseous bubbles in vascular plants that impact on water balance.
endemic: having a natural distribution confined to a particular area or region.
energetics: broadly, the caloric reward and energy cost of foraging.
entomophily, entomophilous: pollinated by insects.
entomophilous species: plant whose flowers are adapted to being pollinated by insects.
Eocene: Epoch of geologic time; c. 56–34 mya.
epicormic buds: buds formed on the old wood, or bole, of a tree; usually applied to shoots that arise from dormant buds after fire or general injury, especially in eucalypt trees.
epidermis: outermost tissue of primary growth; forms a continuous covering to flowers, fruit and seeds.
epiphyte: plant dependent on another for mechanical support but not usually nutritionally dependent.
estragole: methyl chavicol; a liquid colourless ether, its chemical structure consists of a benzene ring substituted with a methoxy group and a propenyl group.
eucalypt: common name for members of the genus *Eucalyptus* (Myrtaceae) and allied genera such as *Corymbia* more recently derived from it.
eudicot: taxa with tricolpate and derived pollen.
eugenol: colourless, aromatic compound, slightly water soluble liquid in certain plant oils.
eumagnoliids: an informal cluster of plant orders that comprise about 1/3 of all angiosperm species.
eusocial: organisms (e.g. bees) whose members are fully integrated and cooperate in the tending of young.
eutropic: fully adapted, obligate blossom visitors.
exine: generally the outermost structural layer of the pollen grain, 2nd layer if pollenkitt considered as a layer.
extrafloral nectaries: nectar-producing glands outside of the flower; not within the flower; the specialised nectar-excreting glands of some plant phyllodes, leaves and stems.
facultative inbreeders: self-compatible and apomictic plants.
fecundity: potential reproductive capacity of an organism (or population).
femur, femora: 'thigh', usually the stoutest segment of the insect leg, bearing the tibia.
fertilisation: union of male and female gamete to form a zygote (a diploid cell formed by the fusion of two haploid gametes).
fidelity: of a pollinator or flower visitor to a particular plant species, group of floristically- or resource-related group of plants, or to a particular forest stratum/synusia. Also **constancy**.
filaments: the bodies that carry the pollen-bearing anthers.
flavinoids: group of yellow or orange pigments usually found in association with anthocyanins.
flavones: colourless, crystalline compounds that are part of white and yellow plant pigments.
floral reward: nectar, pollen, or other substances such as oils and resins, offered by a plant to a pollinator.
floral tube: a cup or tube formed by the fusion of the basal parts of sepals, petals and stamens.
florets: small massed flowers, particularly ones of composite inflorescences, as in Asteraceae/Compositae.
floricolous: adapted to feed from floral resources or strongly associated with flowers in some way.
frontoclypeus: frons and clypeus (loosely the 'foreface'), frequently delineated by a suture.
fructescent: maturing state of fruit following pollination and fertilisation of the flower.
gametophyte: haploid phase of a plant which exhibits an alternation of generations from which the gametes are produced (usually by miotic division). **Pollen grains** microgametophytes, **ovules** are macrogametophytes.
gender dimorphism: differential expression of structure, or some other trait, by each sex.
generalised entomophily: pollinated by ecologically unspecialised insects.

geoflory, geoflorous: flowers presented at or near ground level.
glycoside: a colourless crystalline derivative of sugar, which is water soluble and dissolved in cytoplasm or sap of certain plants, which yields sugar and non-sugar components on hydrolysis with acid or an enzyme.
Gondwana, Gondwanan: historically incorrectly called Gondwanaland, the landmass once comprising Africa, Madagascar, India, Australia, Antarctica, South America, New Zealand, New Caledonia, and other associated islands and island arcs. Formed from the break-up of Pangaea.
guild: a group of species having similar ecological resource requirements and foraging strategies, and therefore undertaking similar roles within a community.
gynodioecy, gynodioecious: a species having both bisexual (hermaphrodite) and female flowers on different plants. See **gynomonoecious**.
gynoecium: the ovule-bearing structure of the flower, made up of one or more carpels.
gynomonoecious: a plant bearing inflorescences of mixed bisexual and female flowers.
gymnosperm: seed-bearing vascular plant characterised by the non-enclosure of its ovule by a structure resembling the ovary formed by carpels of an angiosperm, but exposed except for one or two integuments on the surface of the sporophylls. Lacks true flowers. Condition of **gymnospermy**.
haploid: having a single set of unpaired chromosomes in each nucleus.
hermaphrodite, hermaphroditism: bisexual: comprising both male and female reproductive organs; these may be separated in time (**dichogamous**), or individually non-function, and thus obligate out-crossers. cf. **unisexual**.
heterozygosity, heterozygous: having two different allelomorphs in two corresponding loci of a pair of chromosomes. cf. **homozygous**.
Holocene: synoym **Recent**; Epoch of geologic time comprising approximately the last 12 000 years.
homologous: a trait or condition possessed by two or more taxa, especially when suggesting a common ancestry.
homozygosity, homozygosity: having identical genes (not having different allelomorphs) in the two corresponding loci of a pair of chromosomes. cf. **heterozygous**.
horsetails: Sphenopsida, *Equisetum*.
hypanthium, hypanthia: a floral cup or tube formed above the base or top of the ovary, with the stamens and perianth inserted on the rim.
ichnospecies (also ichnotaxon): species name assigned to trace fossils (e.g. nest burrows); taxon based on the fossilised work of an organism.
inflorescence: general term for the flower-bearing system of the plant, portions with the system separated from one another by vegetative sections; the flower contained thereon usually small (**florets**).
insectivore: fully or principally adapted to feed on insects, but also may forage on other invertebrates, nectar and fruit.
intine: 2nd structural layer of a pollen grain.
Jurassic: Period of geologic time; c. 200–145.5 mya.
kairomones: fragrance stimuli emitted by flowers.
keystone species: any species (pollinator, seed disperser, predator etc.), either plant or animal, that functions for the environmental balance of a habitat, area or community.
labellum: the distinctive, usually enlarged, median petal of orchids.
lamellate: composed of thin plates or scales (lamellae).
leaf flush: synchronous presentation, growth, of leaves owing to physiological or environmental influences.
lignotubers: woody swelling, partly or wholly underground; particularly of eucalypts.
ligula: of insects (Hymenoptera); also referred to as the glossa, or 'tongue', more accurately the central sclerite of the labium.
lipids: 'fats'; molecular group comprising fatty acids, sterols, oils, waxes, terpenes.
lithophyte (epilithic): a plant adapted to the substrate of a rock.
littoral: on or growing near the sea.
longevity, of flowers: the period between opening and senescence, though in a more restricted sense, to the end of receptivity.
malacophily: pollination by snails and slugs.
Malesia: the biogeographic region encompassing Peninsula Malaysia, Indonesia, Philippines and Papua New Guinea); overlaps boundaries of the Indomalayan and Australasian ecozones. Has a shared flora mostly from Asia but with numerous elements of the Antarctic flora.
manose sugar: a hexose sugar, or fermentable monosaccharide.
mass-flowering: simultaneous flowering episodes of individual species, usually restricted to flowering events by canopy - dominant trees.
maxillary lobes: also the galea; structures attached to the paired maxillae that form a major part of the 'tongue' or proboscis in nectar-feeding insects.
megasporophyll/s: female structure, cone units, cones scales that bear the ovules. When male a **microsporophyll.** Collectively called **sporophylls**.
megathermal vegetation/rainforest: tropical rainforests.
meiosis: the process of a diploid cell duplicating its chromosomes, sequential splitting into four haploid cells, each then becoming a gamete.
mesic-adapted: relating to conditions of moderate moisture or water supply; of a plant (**mesophyte**) thriving under intermediate environmental conditions of moderate moisture and temperature.
Mesozoic: geologic Era; including Triassic, Jurassic and Cretaceous periods; c. 251–65.5 mya.
metapleural glands: functional(or non-functional), paired, glands possessed by most ants, opening through a small orifice of the mesosoma. Otherwise unknown among the Hymenoptera.
microhabitat: a small specialised habitat; a **microenvironment**.
micropyle: the small pore in an ovule, allowing entry of the pollen tube for penetration of the nucleus.
micropyle drop: an exudate thought to function in the reproduction of cycads, by gamete induction or as a nutritive inducement to pollinators.

microspore: pollen; the spore of heterosporous (producing different spores – the micro- and macro-spores of seed plants) plants which develops into the male gametophyte.
Miocene: geologic Era of time; c. 23–5.3 mya.
monoecy, monoecious: having separate female and male flowers on the same plant.
monophilic: pollinated by one single or closely related species only.
monophyletic: of a taxon; having a shared ancestor; the possession of shared, derived, characteristics.
monotropic, monolectic: visiting single or closely related plant species.
morphology: the form and structure of an organism, with particular emphasis on external structure.
morphometry: the measurement of external form and structure; the characterisation of form for quantitative analysis.
motile: capable of movement.
mutualism: a shared, usually interdependent, relationship; as of a particular pollinator to a single or restricted suite of flowering plants.
myophily, myophilous: pollination by flies, adapted to being pollinated by flies.
myrmechory, myrmechorous: seed dispersal by ants.
necrophagy: attraction to the products of animal decay.
nectary, nectaries: a nectar-producing gland or group of cells.
nectar guides: marks, or structures, on flowers that indicate the position of the nectary.
nectariferous: producing nectar.
nectarivores: animals that are obligate feeders upon nectar, or principally feed on nectar but that may include other items (e.g. insects, fruit) in their diet.
Neogene: a unit of geologic time, collectively both the Pliocene and Miocene Epochs; c. 23–5 mya.
Neotropics: South America, but also loosely including Mesoamerica.
niche: a trophic or ecological division within any available habitat.
offspring vigour: the general metabolic activity and intensity of growth of an individual organism.
Oligocene: geologic Epoch; c. 34–23 mya.
oligophilic: pollinated by some related visiting taxa only.
oligotropic, oligolectic: visiting some related plant taxa only.
operculum: cap-like covering.
order: a taxonomic rank usually consisting of a number of families (e.g. Coleoptera, Hymenoptera).
ordination: methods of numerical analysis used for summarising similarities between communities or species.
ornithophily, ornithophilous: pollinated by birds.
orthognathous mouthparts: of insects, often Coleoptera, in which the mouthparts are positioned perpendicular to the head.
osmophores: specialised floral scent glands, particularly in Orchidaceae.
osmosis: diffusion of a solute or water through a membrane that is selectively permeable to it, from a region of low concentration of solute to one of higher concentration.
out-crossing, out-breeding: interbreeding via pollen transfer to individuals of the same species that are not genetically closely related.
ovary: female organ; of a flower the lower part of the carpel; comprised of one or more ovaries.
oviposition: egg-laying, as in flowers.
ovule: the component/structure of a seed-bearing plant in which the reproductive cycle of the female is carried out.
ovule parasites: organisms (including developing larvae) that feed upon ovules; may be an inherent component of the pollination cycle of some plant–pollinator relationships, as in fig wasps (Agaoninae).
Palaeozoic, Paleozoic: geologic Era; including Cambrian, Ordivician, Silurian, Devonian, Carboniferous and Permian periods; c. 542–251 mya.
Palaeocene, Paleocene: Epoch of geologic time; c. 65.5–56 mya.
Palaeogene, Paleogene: a unit of geologic time, collectively Oligocene, Eocene, Palaeocene Epochs; c. 65.5–23 mya.
panicle: a compound inflorescence with a main axis and lateral branches which are further branched, and in which each axis ends in a flower or flower bud.
pantropical: distributed, occurring, throughout regions of the Tropics.
paracarpels (pistillodes): sterile vestigial pistil remaining in a staminate flower; floral appendages interpreted as non-functional pistil-like structures.
paraphyletic: a group of taxa derived from a single ancestral taxon, but one which does not contain all the descendents of the most recent common ancestor.
parasite: an animal or plant that feeds on another organism.
perianth: the outer structures of a flower (sepals, petals etc.), but does not include the reproductive structures.
Permian: Period of geologic time; c. 299–251 mya.
petal: if present, a leaf-like component of the inner whorl of the corolla, located inside the sepal whorl but outside of the gynoecium and androecium.
petaloid: resembling a petal, petal-like.
phalaenophily, phalaenophilous: pollination by moths.
phenology: the periodic timing of life cycle events, such as flowering episodes; related to seasonal and/or annual cycles.
phenological event: leaf fall, leaf flush, flower budding/opening; usually highly synchronous or within a discrete seasonal/climatic cycle.
phloem: vascular plant tissue that conducts synthesised foods such as sugars, proteins, through the plant. Characterised by presence of sieve tubes.
photoperiod: the light phase of a 24 hour light/dark cycle.
photosynthesis: the biochemical process that uses radiant energy from sunlight to synthesise carbohydrates/sugars from carbon dioxide and water in the presence of chlorophyll.
phyllophagous (phyllophagy): organisms, particularly insects, adapted to feed on leaves, as in larvae of Lepidoptera.
phyllotaxy, phyllotaxis: the arrangement of floral parts (and leaves) on their axes in whorls or spiral formation.

phylogeny: the history of the genealogical development of a group.
phytochemicals: the chemotaxonomy of plants; hormones, secondary defence compounds etc.
phytochemical stimuli: plant chemical that elicits responses, normally recruitment/attraction, by pollinators to flower resources.
phytochemical syndrome: suites of plant chemicals that define relationships or traits.
phytophagous (phytophagy): adapted to eating plant tissue; feeding upon plants.
pioneer species: plant species, and other organisms, well-adapted to colonise vacant habitats, disturbed areas following logging etc.
pistillate: the female flower, or female phase, of a bisexual plant.
pistillodes: see paracarpels.
plasticity: see behavioural plasticity.
Pleistocene: Epoch of geologic time; c. 1.8 mya–11 800 years ago.
Pliocene: Epoch of geologic time; 5.3–1.8 mya.
pollenkitt: an oily, sometimes pigmented, coating of pollen grains, but not always present; sometimes considered the outermost layer of the pollen grain.
pollen tube: a projection, developed from the intine of a pollen grain as a response to its reception by a stigma or nucleus, through which passes the grain's content to form the male gametophyte.
pollination: of angiosperms, the transfer by biotic or abiotic means of pollen to the stigma.
pollination limitation: pollination below an optimum achievable level; due to inadequate pollen receipt or some other barrier.
pollination syndrome: the grouping of flowers (may be artificial and not necessarily taxonomically related) that have evolved or exhibit shared traits that attract visitation by a particular pollinator or suite of pollinators.
pollinium, pollinia: complex bound pollen grain assemblages such as polyads, found in orchids and asclepiads.
pollinivorous: pollen-feeding.
polygamous: having bisexual and unisexual flowers on the same plant.
Polyneoptera: such as: stoneflies, webspinners, zorapterans, walking-sticks, crickets, grasshoppers, cockroaches, mantises and termites.
polyphenols: class of organic chemicals characterised by the presence of large multiples of of phenol structural units.
polytropic, polylectic: visiting many different plant taxa.
proboscis: any extended mouth structure, but normally is used to describe the extensile mouth of Diptera, the elongated beak of Hemiptera, the 'tongue' of Lepidoptera and the mouth of long-tongued bees.
prognathous mouthparts: of insects, often Coleoptera, in which the head is horizontal with the jaws directed downward.
progymnosperm: ancestral fossil plant types from which modern gymnosperms are thought to have been derived. See Taylor *et al.* (2009).
propagule: structures that are able to give rise to a new plant; seed, or a vegetative structure that if detached from the parent body is able to reproduce as a new plant.
promiscuous: adapted to or capable of being pollinated by many kinds of animals.
protandry, protandrous: a bisexual flower that sheds pollen (male phase) prior to the stigma becoming receptive (female phase).
protogyny, protogynous: a bisexual flower in which the stigma is receptive, and loses receptivity, prior to pollen being shed.
pseudocopulation: copulation-like behaviour, normally by male wasps, responding to sexually deceitful plant mimicry.
psychophily, psychophilous: pollination by butterflies.
Quaternary: a Period of geological time, collectively comprising the Pleistocene and the Holocene Epochs; c. 1.8 mya to the present.
ramiflory, ramiflorous: formation of flowers and fruit on lateral woody branches.
Recent: see **Holocene**.
receptacle: the part of the flower on which the organs are situated.
refugium, refugia: an area that has escaped major climate changes typical of a larger region and acts as a refuge for species once more widely distributed; an isolated habitat that retains environmental conditions previously widespread.
rhizome: a main shoot that is more or less horizontal, subterranean or prostrate and from which adventitious shoots and buds develop.
rosids: a large clade of flowering plants. See Angiosperm Phylogeny Group III (2009), Angiosperm Phylogeny Group IV (2016).
sapromyophily, sapromyophilous: pollination by flies that are attracted to carrion, by mimicry.
saprophyte, saprophytic: deriving nourishment from dead and decaying organic matter.
saurochory, sauchorous: dispersal of seeds by reptiles.
saurophily, saurophilous: pollination by reptiles.
sclerophyll, sclerophyllous: a plant with hard/stiff leaves.
scopa, scopae: a brush or series of hairs, usually on the ventral surface of the abdomen, also in Hymenoptera the thick hair covering the hind tibia.
seed ferns: includes rhacopterids, Pteridospermophyta, Glossopteridales, Caytoniales, Corystospermales and others. See Taylor *et al.* (2009).
self-compatible: flowers that are able to self-pollinate and self-fertilise. cf. **self-incompatible**.
self-incompatible: self-sterile, unable to self-fertilise.
sensilla: of insects, a sense organ.
sepal: if present, individual leaf-like structures, collectively forming the outer whorl of the corolla.
sessile: closely and permanently seated or attached, incapable of movement; without a stalk.
seta, setae: of insects, hairs (macrotrichia).

sexual dimorphism: marked, phenotypic, differences (usually structural) between males and females of the same species.
Silurian: Period of geologic time; c. 448–416 mya.
sonication (buzz pollination): vibration by an insect of the anthers to allow release of pollen.
spadix, spadices: the elongate, thicked inflorescence of Araceae, bearing male and female flowers.
spathe: a large bract that encloses the spadix.
spathulate: spoon-shaped.
speciation: the formation of new species.
species gregariousness: tending to aggregate into groups or clusters.
speciose: consisting of many species; species-rich, diverse.
spermatazoids: sperm cells capable of movement.
Spheciformes: sphecoid wasps (Ampulicidae, Sphecidae, Crabronidae).
sphecophily, sphecophilous: pollinated by wasps.
sphingophily, sphingophilous: pollinated by hawkmoths.
sporangium, sporangia: a hollow unicellular or multicellular structure in which spores are produced.
sporophylls: modified leaf or leaf-like organ that bears sporangia; applied to stamens and carpels of angiosperms, and to the fertile fronds of ferns and some other groups. see **megasporophyll/microsporophyll**.
sporopollenin: a tough substance from which the pollen exine, or outer structural wall of the pollen grain, is composed.
stamens: that part of the flower that produces the pollen; usually composed of an anther and a connective filament; collectively, stamens comprise the **androecium**.
staminate: relating to a flower having stamens but no functional carpel.
staminode, staminodia: a sterile stamen, often enlarged, but also reduced or modified.
starch: complex insoluble carbohydrate; the principal substance in which plants store their food.
sternum: of an insect, the entire ventral surface of any segment; in general usage, the underside of the thorax, between the coxal cavities. See Naumann (1991).
sterols: an unsaturated alcohol of complex ring structure, universally distributed in plants.
stigma: the region of the carpel which serves as the receptive surface for pollen grains, and on which they germinate.
stigma receptivity: the time/period at which the stigma is reproductive receptive (of pollen).
stochastic: simplistically 'chance' event; the property of probability so that a given input may produce a range of possible outcomes due to chance alone.
stomata: structures in the leaf surface allowing gaseous exchange during respiration and photosynthesis, and the transpiration of water.
stratum, strata: distinct vertical layers of vegetation in rainforest, expressed in relation to structure and often with regard to life form and floristic composition.
strobilus, strobili: a reproductive structure that consists of modified sporophylls (leaves) or ovule-bearing scales grouped terminally on a stem; cone.
style: the elongate sterile column of the carpel, through which the tube of the germinating pollen grain, carrying the sperm, moves from stigma to the ovary.
sucker: a vegetative shoot formed from an underground source (e.g. root).
sugar: one of the carbohydrates; glucose, fructose, sucrose.
sulcate (as in pollen): of a surface when grooved or furrowed.
Sundaland: a biogeographic region essentially consisting of the Malay Peninsula, Borneo, Java, Sumatra and adjacent islands (the Sunda Shelf), the eastern boundary delineated by Wallace's Line.
synandrium: the fused flower structure of Eupomatiaceae that is shed as a single unit following fertilisation. See Armstrong and Irvine (1990).
syncarpous: of a multicarpellate gynoecium when the carpel walls are fused to form a single ovary.
synconium, synconia: a 'fig'; the fruit formed in figs (*Ficus*) by the invagination of the floral axis such that the flowers (which are minute) and the fruits are inside the swollen stem of the inflorescence.
syndrome: floral suites of convergent floral characters that are adapted to specific classes of pollinators.
synusia, synusiae: stratified layers/levels within forest community; can possess broadly defined characteristics of floral morphology, colour and scent cues, structure, floral rewards and breeding systems.
taxon, taxa: 'taxon' being a plant or animal of a certain taxonomic rank, e.g. a species, genus, family.
tepal: the free segment of the perianth that is not differentiated into a calyx or a corolla.
Tertiary: Period of geologic time; c. 65.5–1.8 mya.
testa: the seed coat.
tetrachromatic: able to perceive multiple (4) colours; visual colour discrimination that is based on four differently tuned types of photoreceptors.
tetrad: a group of four pollen grains formed by meiotic divisions of a spore mother cell.
thermogenesis: the endogenous production of heat.
thripophily, thripophilous: pollinated by thrips.
tibia, tibiae: of insects; the fourth division of the leg, articulated, distal and attached to the femur, and bearing the tarsal segments (tarsi).
trap-lining: movements by a pollinator between flowers (the resource being 'trapped') of the same species, which may be widely spaced. The fidelity of the pollinator/visitor is not necessarily fixed to a particular flowering taxon, but may differ with respect to temporal availability and spatial distribution of available flowering plants.
Triassic: Period of geologic time; c. 251–200 mya.
trichromatic: able to perceive multiple (3) colours; visual colour discrimination that is based on three differently tuned types of photoreceptors.
tricolpate: having 3 colpi; these being elongate grooves or furrows (as in Eudicots).
trimerous: as in flowers; a flower whorl in which the number of parts (sepals, petals, stamens) is divisible by 3.

triungulin: early mobile larval stage of certain meloid, ripiphorid and rhipicerid beetles.

ultraviolet, UV: light wavelengths between 10 and 400 nm, beyond the human visual spectrum.

unisexual: of a flower that is either male or female. cf. **hermaphrodite**, **bisexual**.

vanillin: an organic compound; its functional groups include aldehyde, ether and phenol.

vascular plants: the more highly evolved plants above mosses and liverworts; pertains to any plant with conducting tissue, e.g. xylem, phloem, vascular cambium.

vector: usually an animal, but more broadly physical influences such as wind and water, that facilitate pollen (or seed) transfer.

vegetative: non-sexual (asexual) form of reproduction/ growth without the use of seeds or spores.

volatile chemicals: chemicals that will readily disperse into vapour under normal environmental conditions.

volatilisation: dispersal of chemicals into the air column.

wet forest: a collective term to include rainforest and wet sclerophyll forest; the latter comprised of a canopy generally dominated by or completely composed by eucalypts, but frequently with a well developed understorey of typically rainforest plant species.

wet sclerophyll forest: synonym **tall open forest**; tall forest in which the canopy layer is dominated by eucalypts (usually exclusively, but may also include *Lophostemon* and *Syncarpia*) and in which there is a distinctive understorey of typically rainforest-type species; this rainforest understorey is absent in dry sclerophyll forest.

whorl: the ring of floral parts (also bracts) borne on the same level of a stem or axis.

xylem: the water and mineral conducting tissue of vascular plants.

zoochory, zoochorous: dispersed by animals, usually relates to seeds.

zoophily: pollination by animals.

zygomorphic (bilaterally symmetrical): pertaining to a flower in which the petals and sepals differ in their shape, size and/ or number such the flower can only be bisected in a single plane. cf. **actinomorphic**.

Bibliography

Abe H, Ueno S, Takahashi T, Tsumura Y, Hasegawa M (2013) Resilient plant–bird interactions in a volcanic island ecosystem: pollination of Japanese Camellia mediated by the Japanese White-eye. *PLoS One* **8**, e62696 doi:10.1371/journal.pone.0062696.

Abrahamczyk S, Kluge J, Gareca Y, Reichle S, Kessler M (2011) The influences of climatic seasonality on the diversity of different tropical pollinator groups. *PLoS One* doi:10.1371/journal.pone.0027115.

Abrahamczyk S, Lozada-Gobilard S, Ackermann M, Fischer E, Krieger V, Redling A, Weigend M (2017) A question of data quality – testing pollination syndromes in Balsaminaceae. *PLoS One* **12**, e0186125 doi:10.1371/journal.pone.0186125.

Ackerman JD (2000) Abiotic pollen and pollination: ecological, functional, and evolutionary perspectives. *Plant Systematics and Evolution* **222**, 167–185. doi:10.1007/BF00984101

Acosta MC, Mathiasen P, Premoli AC (2012) Predominant regeneration strategy results in species specific genetic patterns in sympatric *Nothofagus s.s* congenors (Nothofagaceae). *Australian Journal of Botany* **60**, 319–327. doi:10.1071/BT11277

Adam P (1987) *New South Wales Rainforests: The Nomination for the World Heritage List.* National Parks and Wildlife Service, NSW.

Adam P (1992) *Australian Rainforests.* Oxford University Press, Oxford.

Adam P, Williams G (2001) Dioecy, self-compatibility and vegetative reproduction in Australian subtropical rainforest trees and shrubs. *Cunninghamia* **7**, 89–100.

Adams PB, Bartareau T, Walker KL (1992) Pollination of Australian orchids by *Trigona* (Tetragona) Jurine bees (Hymenoptera: Apidae). *Australian Entomological Magazine* **19**, 97–101.

Aguilar R, Martén-Rodríguez S, Avila-Sakar G, Ashworth L, Lopezaraiza-Mikel M, Quesada M (2015) A global review of pollination syndromes: a response to Ollerton *et al.* 2015. *Journal of Pollination Ecology* **17**, 126–128. doi:10.26786/1920-7603 (2015)21

Ahmed S, Compton SG, Butlin RK, Gilmarten PM (2009) Wind-borne insects mediate directional pollen transfer between desert fig trees 160 kilometres apart. *Proceedings of the National Academy of Sciences of the United States of America* **106**, 20342–20347. doi:10.1073/pnas.0902213106

Ainsworth C (2000) Boys and girls come out to play: the molecular biology of dioecious plants. *Annals of Botany* **86**, 211–221. doi:10.1006/anbo.2000.1201

Aira MJ, Jato V, Iglesias I (1998) *Alnus* and *Betula* pollen content in the atmosphere of Santiagao de Compostela, North-Western Spain. *Aerobiologia* **14**, 135–140. doi:10.1007/BF02694197

Aizen MA, Ezcurra C (1998) High incidence of plant-animal mutualisms in the woody flora of the temperate forest of southern South America: a biogeographical origin and present ecological significance. *Ecología Austral* **8**, 217–236.

Aizen MA, Ashworth L, Galetto L (2002) Reproductive success in fragmented habitats: do compatibility systems and pollination specialization matter? *Journal of Vegetation Science* **13**, 885–892. doi:10.1111/j.1654-1103.2002.tb02118.x

Albuquerque AAE, de Lima HA, Gonçalves-Esteves V, Benevides CR, Rodarte AT (2013) *Myrsine parvifolia* (Primulaceae) in sandy coastal plains marginal to Atlantic rainforest: a case study of pollination by wind or by both wind and insects? *Brazilian Journal of Botany* **36**, 65–73. doi:10.1007/s40415-013-0007-4

Alloysius D (1999) Reproductive biology of *Calamus manan* Miquel. Msc. thesis, Faculty of Forestry, Universiti Putra Malaysia.

Allsopp PG (1992) Volatile compounds as attractants for *Campsomeris tasmaniensis* (Saussure) (Hymenoptera: Scoliidae). *Australian Entomological Magazine* **19**, 107–110.

Alonso-Rodríguez A, Finegan B, Fiedler K (2017) Neotropical moth assemblages degrade due to oil palm expansion. *Biodiversity and Conservation* **26**, 2295–2326. doi:10.1007/s10531-017-1357-1

Altenburger R, Matile P (1990) Further observations on rhythmic emission of fragrance in flowers. *Planta* **180**, 194–197. doi:10.1007/BF00193995

Altshuler DL (2003) Flower colour, humming-bird pollination, and habitat irradiance in four neotropical forests. *Biotropica* **35**, 344–355. doi:10.1111/j.1744-7429.2003.tb00588.x

Aluri JSR, Reddi CS, Das KR (1998) Temporal dioecism and pollination by wasps and bees in *Allophyllus serratus* (Sapindaceae). *Plant Species Biology* **13**, 1–5. doi:10.1111/j.1442-1984.1998.tb00242.x

Anderson RS (1995) An evolutionary perspective on diversity in Curculionoidea. *Memoirs of the Entomological Society of Washington* **14**, 103–114.

Anderson AB, Overal WL, Henderson A (1988) Pollination ecology of a forest-dominant palm (*Orbignya phalerata* Mart.) in northern Brazil. *Biotropica* **20**, 192–205. doi:10.2307/2388234

Angiosperm Phylogeny Group III (2009) An update of the Angiosperm Phylogeny Group classification for the orders and families of flowering plants: APG III. *Botanical Journal of the Linnean Society* **161**, 105–121. doi:10.1111/j.1095-8339.2009.00996.x

Angiosperm Phylogeny Group IV (2016) An update of the Angiosperm Phylogeny Group classification for the orders and families of flowering plants: APG IV. *Botanical Journal of the Linnean Society* **181**, 1–20. doi:10.1111/boj.12385

Antoń S, Komoń-Janczara E, Denisow B (2017) Floral nectary, nectar production dynamics and chemical composition in five nocturnal *Oenothera* species (Onagraceae) in relation to flower visitors. *Planta* **246**, 1051–1067. doi:10.1007/s00425- 017-2748-y

Antonelli A, Sanmartín I (2011) Mass extinction, gradual cooling, or rapid radiation? Reconstructing the spatiotemporal evolution of the ancient angiosperm genus *Hedyosmum* (Chloanthaceae) using empirical and simulated approaches. *Systematic Biology* **60**, 596–615. doi:10.1093/sysbio/syr062

Antropov AV (2016) On the taxonomic state of a digger wasp of the tribe Crabronini (Hymenoptera, Crabronidae, Crabroninae) from the Dominican amber. *Paleontological Journal* **50**, 612–615. doi:10.1134/S0031030116060022

Appanah S (1982) Pollination of androdioecious *Xerospermum intermedium* Radlk. (Sapindaceae) in a rainforest. *Biological Journal of the Linnean Society* **18**, 11–34. doi:10.1111/ j.1095-8312.1982.tb02031.x

Appanah S (1990) Plant–pollinator interactions in Malaysian rainforests. In *Reproductive Ecology of Tropical Forest Plants.* Vol. 7. (Eds KS Bawa and M Hadley) pp. 85–101. Unesco. Parthenon, Carnforth.

Appanah S, Chan HT (1981) Thrips: the pollinators of some dipterocarps. *The Malaysian Forester* **44**, 234–252.

Apple JL, Faener DH (2001) Ant visitation of extrafloral nectaries of *Passiflora*: The effects of nectary attributes and ant behavior on patterns in facultative ant–plant mutualisms. *Oecologia* **127**, 409–416. doi:10.1007/s004420000605

Archibold OW (1995) *Ecology of World Vegetation.* Chapman and Hall, London.

Armbruster WS, Howard JJ, Clausen TP, Debevec EM, Loquvam JC, Matsuki M, Cerndolo B, Andel F (1997) Do biochemical exaptations link evolution of plant defense and pollination success? Historical hypotheses and experimental tests with *Delechampia* vines. *American Naturalist* **149**, 461–484. doi:10.1086/286000

Armstrong JA (1979) Biotic pollination mechanisms in the Australian flora: a review. *New Zealand Journal of Botany* **17**, 467–508. doi:10.1080/0028825X.1979.10432565

Armstrong JE, Drummond BA, III (1986) Floral biology of *Myristica fragrans* (Myristicaceae), the nutmeg of commerce. *Biotropica* **18**, 32–38. doi:10.2307/2388359

Armstrong JE, Irvine AK (1989) Floral biology of *Myristica insipida* (Myristicaceae), a distinctive beetle pollination syndrome. *American Journal of Botany* **76**, 86–94. doi:10.1002/j.1537-2197.1989.tb11288.x

Armstrong JE, Irvine AK (1990) Functions of staminodia in the beetle-pollinated flowers of *Eupomatia laurina. Biotropica* **22**, 429–431. doi:10.2307/2388563

Aronne G, Giovanetti M, Guarracino MR, Micco V (2012) Foraging rules of flower selection applied by colonies of *Apis mellifera*: ranking and associations of floral resources. *Functional Ecology* **26**, 1186–1196. doi:10.1111/j.1365-2435.2012.02017.x

Arroyo-Correa B, Burkle LA, Emer C (2019) Alien plants and flower visitors disrupt the seasonal dynamics of mutualistic networks. *Austral Ecology* **108**, 1475–1486.

Arroyo MTK, Uslar P (1993) Breeding systems in a temperate mediterranean-type climate montane sclerophyllous forest in central Chile. *Botanical Journal of the Linnean Society* **111**, 83–102. doi:10.1006/bojl.1993.1008

Ashman T-L (2009) Sniffing out patterns of sexual dimorphism in floral scent. *Functional Ecology* **23**, 852–862. doi:10.1111/j. 1365-2435.2009.01590.x

Ashton DH, Frankenberg J (1976) Ecological studies of *Acmena smithii* (Poir.) Merrill and Perry with special reference to Wilson's Promontory. *Australian Journal of Botany* **24**, 453–487. doi:10.1071/BT9760453

Ashton PS (1969) Speciation among tropical forest trees: some deductions in the light of recent evidence. *Biological Journal of the Linnean Society* **1**, 155–196. doi:10.1111/j.1095-8312.1969.tb01818.x

Ashton PS, Givnish TJ, Appanah S (1988) Staggered flowering in the Dipterocarpaceae: new insights into floral induction and the evolution of mast fruiting in the aseasonal tropics. *American Naturalist* **132**, 44–66. doi:10.1086/284837

Atwell BJ, Kriedemann PE, Turnbull CGN (1999) *Plants in Action. Adaptation in Nature. Performance in Cultivation.* Macmillan Education Australia, Melbourne.

Augspurger CK (1980) Mass-flowering of a tropical shrub (*Hybanthus prunifolius*): influence on pollinator attraction and movement. *Evolution* **34**, 475–488.

Baba Y (2013) Evolution, systematics and taxonomy of *Elaeocarpus* (Elaeocarpaceae) in Australia. PhD thesis. James Cook University, Cairns.

Bach CS (2002) Phenological patterns in monsoon rainforests in the Northern Territory, Australia. *Austral Ecology* **27**, 477–489. doi:10.1046/j.1442-9993.2002.01209.x

Bagchi R, Gallery RE, Gripenberg S, Gurr SJ, Narayan L, Addis CE, Freckleton RP, Lewis OT (2014) Pathogens and insect herbivores drive rainforest plant diversity and composition. *Nature* **506**, 85–88. doi:10.1038/nature12911

Baker HG (1961) The adaptation of flowering plants to nocturnal and crepuscular pollinators. *The Quarterly Review of Biology* **36**, 64–73. doi:10.1086/403276

Baker HG (1975) Sugar concentrations in nectars from hummingbird flowers. *Biotropica* **7**, 37–41. doi:10.2307/2989798

Baker HG (1986) Yuccas and yucca moths: a historical commentary. *Annals of the Missouri Botanical Garden* **73**, 556–564. doi:10.2307/2399193

Baker HG, Baker I (1990) The predictive value of nectar chemistry to the recognition of pollinator types. *Israel Journal of Botany* **39**, 157–166.

Bale CL, Williams JB (1994) Lost and found: *Nothofagus moorei* at Comboyne. *Cunninghamia* **3**, 529–533.

Balvanera P, Quijás S, Pérez-Jiménez A (2011) Distribution patterns of tropical dry forest along a mesoscale water availability gradient. *Biotropica* **43**, 414–422. doi:10.1111/j. 1744-7429.2010.00712.x

Banza P, Macgregor CJ, Belo ADF, Fox R, Pocock MJO, Evans DM (2019) Wildfire alters the structure and season dynamics of nocturnal pollen-transport networks. *Functional Ecology* **33**, 1882–1892. doi:10.1111/1365-2435.13388

Bänziger H (1996) The mesmerising wart: the pollination strategy of epiphytic lady slipper orchid *Paphiopedilum villosum* (Lindl.) Stein (Orchidaceae). *Botanical Journal of the Linnean Society* **121**, 59–90.

Bänziger H (2002) Smart alecks and dumb flies: natural pollination of some wild lady slipper orchids (*Paphiopedilum* spp., Orchidaceae). In *Proceedings of the 16th World Orchid*

Conference 1999. (Eds J Clark, WM Elliott, G Tingley and J Biro) pp. 165–169. Vancouver Orchid Society, Vancouver.

Bänziger H, Disney HL (2006) Scuttle flies (Diptera: Phoridae) imprisoned by *Aristolochia baenzigeri* (Aristolochiaceae) in Thailand. *Mitteilungen der Schweizerische Entomologische Gesellschaft* **79**, 29–61.

Bänziger H, Hansen B (2000) A new taxonomic revision of a deceptive flower, *Rhizanthes* Dumortier (Rafflesiaceae). *Natural History Bulletin of the Siam Society* **48**, 117–143.

Bänziger H, Pape T (2004) Flowers, faeces and cadavers: natural feeding and laying habits of flesh flies in Thailand (Diptera *Sarcophaga* spp.). *Journal of Natural History* **38**, 1677–1694. doi:10.1080/0022293031000156303

Barfod AS, Hagen M, Borchsenius F (2011) Twenty-five years of progress in understanding pollination mechanisms in palms (Arecaceae). *Annals of Botany* **108**, 1503–1516. doi:10.1093/aob/mcr192

Barker PCJ (1995) *Phyllocladus aspleniifolius*: phenology, germination, and seedling survival. *New Zealand Journal of Botany* **33**, 325–337. doi:10.1080/0028825X.1995.10412960

Barlow BA (1981) The Australian flora: its origin and evolution. In *Flora of Australia.* Vol. 1, Introduction. Australian Government Publishing Service, Canberra.

Barnes RW, Hill RS (1999) *Ceratopetalum* fruits from Australian cainozoic sediments and their significance for petal evolution in the genus. *Australian Systematic Botany* **12**, 635–645. doi:10.1071/SB98014

Barrett SCH (2008) Major evolutionary transitions in flowering plant reproduction: an overview. *International Journal of Plant Sciences* **169**, 1–5. doi:10.1086/522511

Basilio AM, Medan D, Torretta JP, Bartoloni NJ (2006) A year-long plant–pollinator network. *Austral Ecology* **31**, 975–983. doi:10.1111/j.1442-9993.2006.01666.x

Basinger JF, Christophel Basinger JF, Greenwood DR, Wilson PG, Christophel DC (2007) Fossil flowers and fruits of capsular Myrtaceae from the Eocene of South Australia. *Canadian Journal of Botany* **85**, 204–215. doi:10.1139/B07-001

Baskorowati L, Moncur MW, Cunningham SA, Kanowski PJ (2010) Reproductive biology of *Melaleuca alternifolia* (Myrtaceae) 2. Incompatibility and pollen transfer in relation to the breeding system. *Australian Journal of Botany* **58**, 384–391. doi:10.1071/BT10036

Basset Y, Kitching RL (1991) Species number, species abundance and body length of arboreal arthropods associated with an Australian rainforest tree. *Ecological Entomology* **16**, 391–402. doi:10.1111/j.1365-2311.1991.tb00232.x

Basukriadi A, Sjamsuridzal W, Putra BB (2010) Molecular identification and diversity of yeasts associated with *Apis cerana* foraging on flowers of *Jatropha integerrima. Microbiology Indonesia* **4**, 44–48. doi:10.5454/mi.4.1.9

Bateman AJ (1951) The taxonomic discrimination of bees. *Heredity* **5**, 271–278. doi:10.1038/hdy.1951.24

Bateman RM, Hilton J, Rudall PJ (2006) Morphological and molecular phylogenetic context of the angiosperms: contrasting the 'top-down' and 'bottom-up' approaches used to infer the likely characteristics of the first flowers. *Journal of Experimental Botany* **57**, 3471–3503. doi:10.1093/jxb/erl128

Batley M, Hogendoorn K (2009) Diversity and conservation status of native Australian bees. *Apidologie* **40**, 347–354. doi:10.1051/apido/2009018

Baum DA (1995) The comparative pollination and floral biology of baobabs (*Adansonia* – Bombacaceae). *Annals of the Missouri Botanical Garden* **82**, 322–348. doi:10.2307/2399883

Bawa KS (1990) Plant–pollinator interactions in tropical rainforests. *Annual Review of Ecology and Systematics* **21**, 399–422. doi:10.1146/annurev.es.21.110190.002151

Bawa KS, Crisp JE (1980) Wind-pollination in the understorey of a rainforest in Costa Rica. *Journal of Ecology* **68**, 871–876. doi:10.2307/2259462

Bawa KS, Webb CL (1983) Floral variation and sexual differentiation in *Muntingia calabura* (Elaeocarpaceae), a species with hermaphrodite flowers. *Evolution* **37**, 1271–1282. doi:10.2307/2408847

Beard JS (1990) *Plant Life of Western Australia.* Kangaroo Press, Kenthurst.

Beardsell DV, Clements MA, Hutchinson JF, Williams EG (1986) Pollination of *Diuris maculata* R.Br. (Orchidaceae) by floral mimicry of native legumes *Daviesia* spp. and *Pultenaea scabra* R.Br. *Australian Journal of Botany* **34**, 165–174. doi:10.1071/BT9860165

Beardsell DV, O'Brien SP, Williams EG, Knox RB, Calder DM (1993) Reproductive ecology of Australian Myrtaceae. *Australian Journal of Botany* **41**, 511–526. doi:10.1071/BT9930511

Beattie AJ (1982) Ants and gene dispersal in flowering plants. In *Pollination and Evolution.* (Eds JA Armstrong, JM Powell and AJ Richards) pp. 1–8. Royal Botanic Gardens, Sydney.

Bell D, Lin Q, Gerelle WK, Joya S, Chang Y, Taylor ZN, Rothfels CJ, Larsson A, *et al.* (2020) Organellonic data sets confirm a cryptic consensus on (unrooted) land-plant relationships and provide new insight into bryophyte molecular evolution. *American Journal of Botany* **107**, 91–115. doi:10.1002/ajb2.1397

Bellamy CL (2003) An Illustrated Summary of the Higher Classification of the Superfamily Buprestoidea (Coleoptera). *Folia Heyrovskyana* **10**(Supplementum), 1–197.

Bellamy CL, Williams GA, Hasenpusch J, Sundholm A (2013) A summary of the published data on host plants and morphology of immature stages of Australian jewel beetles (Coleoptera: Buprestidae) with additional new records. *Insecta Mundi* **0293**, 1–172.

Bello MA, Bruneau A, Forest A, Hawkins JA (2009) Elusive relationships within order Fabales: phylogenetic analysis using *matK* and *rbcL* sequence data. *Systematic Botany* **34**, 102–114. doi:10.1600/036364409787602348

Bennett KF, Ellison AM (2009) Nectar, not colour, may lure insects to their death. *Biological Letters* **5**, 469–472. doi:10.1098/rsbl.2009.0161

Benson J (1991) The effect of 200 years of European Settlement on the vegetation and flora of New South Wales. *Cunninghamia* **2**, 343–370.

Benson J, Allen C (2007) Vegetation associated with *Wollemia nobilis* (Araucariaceae). *Cunninghamia* **10**, 255–262.

Bernhardt P (2000) Convergent evolution and adaptive radiation of beetle-pollinated angiosperms. *Plant Systematics and Evolution* **222**, 293–320. doi:10.1007/BF00984108

Bernhardt P, Burns-Balogh P (1986) Floral mimesis in *Thelymitra nuda* (Orchidaceae). *Plant Systematics and Evolution* **151**, 187–202. doi:10.1007/BF02430274

Bernhardt P, Thien LB (1987) Self-isolation and insect pollination in the primitive angiosperms: new evaluations of older hypotheses. *Plant Systematics and Evolution* **156**, 159–176. doi:10.1007/BF00936071

Bernhardt P, Sage T, Weston P, Azuma H, Lam M, Thien L, Bruhl J (2003) The pollination of *Trimenia moorei* (Trimeniaceae): floral volatiles, insect/wind pollen vectors and stigmatic self-incompatibility in a basal angiosperm. *Annals of Botany* **92**, 445–458. doi:10.1093/aob/mcg157

Berry PE, Calvo RN (1989) Wind pollination, self-incompatibility, and altitudinal shifts in pollination systems in the high Andean genus (*Espeletia*) (Asteraceae). *American Journal of Botany* **76**, 1602–1614. doi:10.1002/j.1537-2197.1989.tb15145.x

Berry EJ, Gorchov DL (2004) Reproductive biology of the dioecious understorey palm *Chamaedora radicalis* in a Mexican cloud forest: pollination vector, flowering phenology and female fecundity. *Journal of Tropical Ecology* **20**, 369–376. doi:10.1017/S0266467404001397

Bestmann HJ, Winkler L, von Helversun O (1997) Headspace analysis of volatile flower scent constituents of bat-pollinated plants. *Phytochemistry* **46**, 1169–1172. doi:10.1016/S0031-9422 (97)80004-0

Bezemer N, Krauss SL, Phillips RD, Roberts DG, Hopper SD (2016) Paternity analysis reveals wide pollen dispersal and high multiple paternity in a small isolation population of bird-pollinated *Eucalyptus caesia* (Myrtaceae). *Heredity* **117**, 460–471. doi:10.1038/hdy.2016.61

Bhatnagar S (1986) On insect adaptations for pollination in some asclepiads of Central India. In *Pollination Biology – An Analysis*. (Ed. RP Kapil) pp. 37–57. Inter-India Publications, New Delhi, India.

Biffin E, Lucas EJ, Craven LA, da Costa IR, Harrington MG, Crisp MD (2010) Evolution of exceptional species richness among lineages of fleshy-fruited Myrtaceae. *Annals of Botany* **106**, 79–93. doi:10.1093/aob/mcq088

Biffin E, Brobribb TJ, Hill RS, Thomas P, Lowe AJ (2011) Leaf evolution of Southern Hemisphere conifers tracks the angiosperm ecological radiation. *Proceedings. Biological Sciences* **279**, 1727.

Bischoff M, Lord JM, Robertson AW, Dyer AG (2013) Hymenoptera pollinators as agents of selection on flower colour in the New Zealand mountains: salient chromatic signals enhance flower discrimination. *New Zealand Journal of Botany* **51**, 181–193. doi:10.1080/0028825X.2013.806933

Blüthgen N, Reifenrath K (2003) Extrafloral nectaries in an Australian rainforest: structure and distribution. *Australian Journal of Botany* **51**, 515–527. doi:10.1071/BT02108

Boberg E, Ågren J (2009) Despite their apparent integration, spur length but not perianth size affects reproductive success in the moth-pollinated orchid *Plantanthera bifolia*. *Functional Ecology* **23**, 1022–1028. doi:10.1111/j.1365-2435.2009.01595.x

Bogarín D, Fernández M, Borkent A, Heemskerk A, Pupilin F, Ramírez S, Smets E, Gravendeel B (2018) Pollination of *Trichosalpinx* (Orchidaceae: Pseudothallidinae) by biting midges (Diptera: Ceratopogonidae). *Botanical Journal of the Linnean Society* **186**, 510–543. doi:10.1093/botlinnean/box087

Bogler DJ, Francisco-Ortega J (2004) Molecular systematic studies in cycads: evidence from *trnL* Intron and ITS2 rDNA sequences. *Botanical Review* **70**, 260–273. doi:10.1663/0006-8101(2004)070[0260:MSSICE]2.0.CO;2

Boland DJ, Brooker MIH, Chippendale GM, Hall N, Hyland BPM, Johnston RD, Kleinig DA, Turner JD (1984) *Forest Trees in Australia*. Nelson/CSIRO Publishing, Melbourne.

Bond WJ (1994) Do mutualisms matter? Assessing the impact of pollinator and disperser disruption on plant extinction. *Philosophical Transactions of the Royal Society of London. Series B, Biological Sciences* **344**, 83–90. doi:10.1098/rstb.1994.0055

Bond WJ, Midgley JJ (2012) Fire and angiosperm revolutions. *International Journal of Plant Sciences* **173**, 569–583. doi:10.1086/665819

Bond WJ, Scott AC (2010) Fire and the spread of flowering plants in the Cretaceous. *New Phytologist* **188**, 1137–1150. doi:10.1111/j.1469-8137.2010.03418.x

Borchert R (1983) Phenology and control of flowering in tropical trees. *Biotropica* **15**, 81–89. doi:10.2307/2387949

Borges RM, Somanathan H, Kelber A (2016) Patterns and processes in nocturnal and crepuscular pollination services. *The Quarterly Review of Biology* **91**, 389–418. doi:10.1086/689481

Boucek Z (1988) *Australian Chalcidoidea (Hymenoptera)*. CAB International, Wallingford.

Boulter SL, Kitching RL, Howlett BG, Goodall K (2005) Any which way will do: the pollination biology of a northern Australian rainforest canopy tree (*Syzygium sayeri*; Myrtaceae). *Botanical Journal of the Linnean Society* **149**, 69–84. doi:10.1111/j.1095-8339.2005.00430.x

Boulter SL, Kitching RL, Gross CL, Goodall K, Howlett BG (2008) Floral morphology, phenology and pollination in the Wet Tropics. In *Living in a Dynamic Tropical Forest Landscape*. (Eds NE Stork and SM Turton) pp. 224–239. Blackwell Publishing, Malden, US.

Boulter SL, McDonald WJF, Kitching RL, Zaluki JM, Jessup LW (2011) Plant reproduction, phenology and floral resources of an Australian subtropical rainforest. *Memoirs of the Queensland Museum – Nature* **55**, 463–479.

Bower C, Towle B, Bickel D (2015) Reproductive success and pollination of the threatened Tuncurry Midge orchid (*Genoplesium littorale*) by chloropid flies. *Telopea* **18**, 43–55. doi:10.7751/telopea8127

Bowman DMJS (1991) Recovery of some Australian monsoon forest tree species following fire. *Proceedings of the Royal Society of Queensland* **101**, 21–25.

Bowman DMJS, Harris S (1995) Conifers of Australia's dry forests and open woodlands. In *Ecology of Southern Conifers.* (Eds NJ Enright and RS Hill) pp. 252–270. Melbourne University Press, Carlton.

Braby MF (2000) *The Butterflies of Australia: Their Identification, Biology and Distribution.* CSIRO Publishing, Melbourne.

Braby MF (2019) Are insects and other invertebrates in decline in Australia? *Austral Entomology* **58**, 471–477. doi:10.1111/aen.12411

Bradshaw CJA (2012) Little left to lose: deforestation and forest degradation in Australia since European colonization. *Journal of Plant Ecology* **5**, 109–120. doi:10.1093/jpe/rtr038

Branstetter MG, Danforth BN, Pitts JP, Brant C, Faircloth BC, Ward PS, Buffington ML, Gates MW, Kula RR, Brady SG (2017) Phylogenomic insights into the evolution of stinging wasps and the origins of ants and bees. *Current Biology* **27**, 1019–1025. doi:10.1016/j.cub.2017.03.027

Braun M, Gottsberger G (2012) Reproduction of beetle-pollinated *Anaxagorea dolichocarpa* (Annonaceae) is resilient to habitat disturbance in rainforest fragments. *Nordic Journal of Botany* **30**, 453–460. doi:10.1111/j.1756-1051.2012.01256.x

Braun M, Dötterl S, Gottsberger G (2011) Absence of pollinators and apomictic fruit production in an Atlantic rainforest population of *Cymbopetalum brasiliense* (Annonaceae). *Plant Systematics and Evolution* **296**, 265–273. doi:10.1007/s00606-011-0493-4

Breeden S, Breeden K (1970) *Tropical Queensland.* Collins, Sydney.

Bremer K, Friis E, Bremer B (2004) Molecular phylogenetic dating of asterid flowering plants shows Early Cretaceous diversification. *Systematic Botany* **53**, 496–505.

Briggs BG, Johnson LAS (1979) Evolution in the Myrtaceae: evidence from inflorescence structure. *Proceedings of the Linnean Society of New South Wales* **102**, 157–256.

Briggs D, Walters SM (1984) *Plant Variation and Evolution.* Cambridge University Press, Cambridge.

Briscoe AD, Chittka L (2001) The evolution of colour vision in insects. *Annual Review of Entomology* **46**, 471–510. doi:10.1146/annurev.ento.46.1.471

Brockington SF, Walker RH, Glover BJ, Soltis PS, Soltis DE (2011) Complex pigment evolution in the Caryophyllales. *New Phytologist* **190**, 854–864. doi:10.1111/j.1469-8137.2011.03687.x

Brodmann J, Twele R, Francke W, Holzer G, Zhang Q-H, Ayasse M (2008) Orchids mimic green-leaf volatiles to attract prey-hunting wasps for pollination. *Current Biology* **18**, 740–744. doi:10.1016/j.cub.2008.04.040

Brodribb TJ, Feild TS, Sack L (2010) Viewing leaf structure and evolution from a hydraulic perspective. *Functional Plant Biology* **37**, 488–498. doi:10.1071/FP10010

Bronstein JL (1989) A mutualism at the edge of its range. *Experientia* **45**, 622–637. doi:10.1007/BF01975679

Bronstein JL (1992) Seed predators as mutualists: ecology and evolution of the fig-pollinator interaction. In *Insect-Plant Interactions.* (Ed. EA Bernays) pp. 1–47. CRC Press, Boca Raton, USA.

Brooks JG (1948) Some North Queensland Coleoptera and their food plants. *North Queensland Naturalist* **15**, 26–29.

Brooks JG (1965) North Queensland Coleoptera – their food and host plants. Part III. *North. Queensland Naturalist* **32**, 29–30.

Brown ED, Hopkins MJG (1996) How New Guinea rainforest flower resources vary in time and space: implications for nectarivorous birds. *Austral Ecology* **21**, 363–378. doi:10.1111/j.1442-9993.1996.tb00623.x

Brown J, York A, Christie F, McCarthy M (2017) Effects of fire on pollinators and pollination. *Journal of Applied Ecology* **54**, 313–322. doi:10.1111/1365-2664.12670

Brown KS, Hutchings RW (1997) Disturbance, fragmentation, and the dynamics of diversity in Amazonian forest butterflies. In *Tropical Forest Remnants.* (Eds WF Laurance and RO Bierregaard) pp. 91–110. University of Chicago Press, Chicago.

Brown R (1814) General remarks, geographical and systematical, on the botany of Terra Australis. In *A Voyage to Terra Australis.* (MA Flinders) Appendix III, pp. 65–66. William Bulmer, St. James's.

Buchmann SL (1983) Buzz pollination in angiosperms. In *Handbook of Experimental Pollination Biology.* (Eds CE Jones and JR Little) pp. 73–113. Van Nostrand Reinhold, New York.

Bullock SH (1994) Wind pollination of neotropical dioecious trees. *Biotropica* **26**, 172–179. doi:10.2307/2388806

Bullock SH, Bawa KS (1981) Sexual dimorphism and the annual flowering pattern in *Jacaratia dolichaula* (D. Smith) Woodson (Caricaceae) in a Costa Rican rain forest. *Ecology* **62**, 1494–1504. doi:10.2307/1941506

Bullock SH, Martinez del Rio C, Ayala R (1989) Bee visitation rates to trees of *Prockia crusis* differing in flower number. *Oecologia (Berl.)* **78**, 389–393. doi:10.1007/BF00379114

Burgett DM, Sukumalanand P (2000) Flight activity of *Xylocopa (Nyctomelitta) tranquebarica*: a night flying carpenter bee (Hymenoptera: Apidae). *Journal of Apicultural Research* **39**, 75–83. doi:10.1080/00218839.2000.11101024

Burleigh JG, Mathews S (2004) Phylogenetic signal in nucleotide data from seed plants: implications for resolving the seed plant tree of life. *American Journal of Botany* **91**, 1599–1613. doi:10.3732/ajb.91.10.1599

Burnham DA (2007) Archaeopteryx – a re-evaluation suggesting an arboreal habitat and an intermediate stage in *trees down* origin of flight. *Neues Jahrbuch für Geologie und Palaontologie. Abhandlungen* **245**, 33–44. doi:10.1127/0077-7749/ 2007/0245-0033

Burr B, Barthlott W (1993) Untersuchungen zur Ultraviolettreflexion von Angiospermen blüten. II Magnoliidae, Ranunculidae, Hameneliddidae, Caryophyllidae, Rosidae. *Tropische und Subtropische Pflanzenwelt* **87**, 1–193.

Butler J, Barrett PM, Penn MG, Kenrick P (2010) Testing coevolutionary hypotheses over geological timescales: interactions between Cretaceous dinosaurs and plants. *Biological Journal of the Linnean Society* **100**, 1–15. doi:10.1111/j.1095-8312.2010.01401.x

Byrne M, Steane DA, Joseph L, Yeates DK, Jordan GJ, Crayn D, Aplin K, *et al.* (2011) Decline of a biome: evolutionary contraction, fragmentation, extinction and invasion of the Australian mesic zone biota. *Journal of Biogeography* **38**, 1635–1656. doi:10.1111/j.1365-2699.2011.02535.x

Byron N, Shepherd G (1998) Indonesia and the 1997–98 El Niño: fire problems and long term solutions. *Natural Resource Perspectives* No. 28. Overseas Development Institute, London.

Cai C, Escalona HE, Li L, Yin Z, Huang D, Engel MS (2018) Beetle pollination of cycads in the Mesozoic. *Current Biology* **28**, 2806–2812. doi:10.1016/j.cub.2018.06.036

Cameron DG (1992) A portrait of Victorian rainforests: distribution, diversity and definition. In *Victoria's Rainforests: Perspectives on Definition, Classification and Management.* (Eds P Gell and D Mercer) pp. 13–96. Monash University, Melbourne.

Campbell ML, Clarke PJ (2006) Response of montane wet sclerophyll forest understorey species to fire: evidence from high and low intensity fires. *Proceedings of the Linnean Society of New South Wales* **127**, 63–73.

Canela MB, Sazima M (2003) Florivory by the crab *Armases angustipes* (Grapsidae) influences hummingbird visits to *Aechmea pectinata* (Bromeliaceae). *Biotropica* **35**, 289–294. doi:10.1111/j.1744-7429.2003.tb00287.x

Canny MJ (1981) A universe comes into being when a space is severed: some properties of boundaries in open systems. *Proceedings of the Ecological Society of Australia* **11**, 1–11.

Cardale JC (1993) Hymenoptera: Apoidea. In *Zoological Catalogue of Australia.* Vol. 10. (Eds WWK Houston and GV Maynard) pp. 1–406. AGPS, Canberra.

Carothers JH (1982) Effects of trophic morphology and behaviour on foraging rates of three Hawaiian honeycreepers. *Oecologia* **55**, 157–159. doi:10.1007/BF00384481

Carpenter FL (1976) Plant–pollinator interactions in Hawaii: pollination energetics of *Metrosideros collina* (Myrtaceae). *Ecology* **57**, 1125–1144. doi:10.2307/1935040

Carpenter RJ, Macphail MK, Jordan GJ, Hill RS (2015) Fossil evidence for open Proteaceae-dominated heathlands and fire in the Late Cretaceous of Australia. *American Journal of Botany* **102**, 2092–2107. doi:10.3732/ajb.1500343

Cause H, Foote D, Kremen C (2012) Invasive species management restores a plant–pollinator mutualism in Hawaii. *Journal of Applied Ecology* **50**, 147–155.

Chadwick CE (1993) The roles of *Tranes lyteroides* and *T. sparsus* Boh (Col., Curculionidae) in the pollination of *Macrozamia communis* (Zamiaceae). In *Proceedings of CYCAD 90, the Second International Conference of Cycad Biology.* (Eds DW Stevenson and NJ Norstog) pp. 77–88. Palm and Cycad Societies of Australia Ltd, Brisbane.

Chairul, Yoneda T, Yamauchi J (2010) Phenological study on a tropical secondary forest in West Sumatra, Indonesia. *Tropics* **18**, 211–220. doi:10.3759/tropics.18.211

Chalcoff VR, Aizen MA, Ezcurra C (2012) Erosion of a pollination mutualism along an environmental gradient in a south Andean treelet *Embothrium coccineum* (Proteaceae). *Oikos* **121**, 471–480. doi:10.1111/j.1600-0706.2011.19663.x

Chan YM, Saw LG (2011) Notes on the pollination ecology of the palm genus *Johannesteijsmannia* (Arecaceae). *Journal of Pollination Ecology* **6**, 108–117. doi:10.26786/1920-7603(2011)19

Charlesworth B, Charlesworth D (1978) A model for the evolution of dioecy and gynodioecy. *American Naturalist* **112**, 975–997. doi:10.1086/283342

Chase MW, Reveal JL, Fay MF (2009) A subfamilial classification for the expanded asparagalean families Amaryllidae, Asparagaceae and Xanthorrhoeaceae. *Botanical Journal of the Linnean Society* **161**, 132–136. doi:10.1111/j.1095-8339.2009.00999.x

Chazdon RL (1988) Sunflecks and their importance to forest understory plants. *Advances in Ecological Research* **18**, 1–63. doi:10.1016/S0065-2504(08)60179-8

Chazot N, Wahlberg N, Freitas AVL, Mitter C, Labandeira N, Sohn JC, Sahoo RK, Seraphim N, de Jong R, Heikkilä M (2019) Priors and posteriors in Bayesian timing of divergence analyses: the age of butterflies revisited. *Systematic Biology* **68**, 797–813. doi:10.1093/sysbio/syz002

Cheke RA, Mann CF, Allen R (2001) *Sunbirds: A Guide to the Sunbirds, Flowerpeckers, Spiderhunters and Sugarbirds of the World.* Yale University Press, New Haven.

Chen C, Song Q, Proffit M, Bessière J-M, Li Z, Hossaert-McKay M (2009) Private channel: a single unusual compound assures specific pollinator attraction in *Ficus semicordata. Functional Ecology* **23**, 941–950. doi:10.1111/j.1365-2435.2009.01622.x

Chisholm EC (1925) The Comboyne Plateau. Its general confirmation and flora. *Proceedings of the Linnean Society of New South Wales* **50**, 284–298.

Chittka L, Waser NM (1997) Why red flowers are not invisible to bees. *Israel Journal of Plant Sciences* **45**, 169–183. doi:10.1080/07929978.1997.10676682

Chittka L, Dyer AG, Bock F, Dornhaus A (2003) Bees trade off foraging speed for accuracy. *Nature* **424**, 388. doi:10.1038/424388a

Chmel K, Riegert J, Paul L, Novotný V (2016) Vertical stratification of an avian community in New Guinea tropical rainforest. *Population Ecology* **58**, 535–547. doi:10.1007/s10144-016-0561-2

Choat B, Jansen S, Brodribb TJ, Chochard H, Delzon S, Bhaskar R, *et al.* (2012) Global convergence in the vulnerability of forests to drought. *Nature* **491**, 752–755. doi:10.1038/nature11688

Christenhusz MJM, Byrng JW (2016) The number of known plant species in the world and its annual increase. *Phytotaxa* **261**, 201–217. doi:10.11646/phytotaxa.261.3.1

Christophel DC, Greenwood DR (1987) A megafossil flora from the Eocene of Golden Grove, South Australia. *Transactions of the Royal Society of South Australia* **111**, 155–162.

Christophel DC, Lys SD (1986) Mummified leaves of two new species of Myrtaceae from the Eocene of Victoria, Australia. *Australian Journal of Botany* **34**, 649–662. doi:10.1071/BT9860649

Clarke D, Whitney H, Sutton G, Robert D (2013) Detection and learning of floral electric fields by bumblebees. *Science* **340**, 66–69. doi:10.1126/science.1230883

Clifford HT, Drake WE (1981) Pollination and dispersal in eastern Australian heathlands. In *Ecosystems of the World. Heathlands and Related Shrublands. Analytical Studies*. (Ed. RL Specht) pp. 39–49. Elsevier Scientific Publications, Amsterdam.

Coiro M, Doyle JA, Hilton J (2019) How deep is the conflict between molecular and fossil evidence on the age of angiosperms. *New Phytologist* **223**, 83–99. doi:10.1111/nph.15708

Collins BG, Rebelo T (1987) Pollination biology of the Proteaceae in Australia and Southern Africa. *Australian Journal of Ecology* **12**, 387–421. doi:10.1111/j.1442-9993.1987.tb00958.x

Collins BG, Newland C, Briffa P (1984) Nectar utilization and pollination by Australian honeyeaters and insects visiting *Calothamnus quadrifidus* (Myrtaceae). *Austral Ecology* **9**, 353–365. doi:10.1111/j.1442-9993.1984.tb01373.x

Common IFB (1990) *Moths of Australia*. Melbourne University Press, Melbourne.

Compton SG, Ball AD, Collinson ME, Hayes P, Rasnitsyn AP, Ross AJ (2010) Ancient fig wasps indicate at least 34 myr of stasis in their mutualism with fig trees. *Biological Letters* **6**, 6.

Conran JG (1988) Observations on the pollination ecology of *Drymophila moorei* Baker (Luzuriagaceae) in Southeast Queensland. *Victorian Naturalist* **105**, 43–47.

Conran JG, Bradbury JH (2007) *Aspidistras*, amphipods and Oz: niche opportunism between strangers in a strange land. *Plant Species Biology* **22**, 41–48. doi:10.1111/j.1442-1984.2007.00174.x

Cook LG, Crisp MD (2005) Not so ancient: the extent of crown group of *Nothofagus* represents a post-Gondwanan radiation. *Proceedings. Biological Sciences* **272**, 1580.

Cooley JR (1995) Floral heat rewards and direct benefits to insect pollinators. *Annals of the Entomological Society of America* **88**, 576–579. doi:10.1093/aesa/88.4.576

Cooper W, Cooper WT (2004) *Fruits of the Australian Tropical Rainforest*. Nokomis Editions Pty. Ltd., Melbourne.

Corlett RT (2001) Pollination in a degraded tropical landscape: a Hong Kong case study. *Journal of Tropical Ecology* **17**, 155–161. doi:10.1017/S0266467401001109

Corlett RT (2004) Flower visitors and pollination in the Oriental (Indomalayan) Region. *Biological Reviews of the Cambridge Philosophical Society* **79**, 497–532. doi:10.1017/S1464793103006341

Corner EJ (1952) *Wayside Trees of Malaya*. Vol. 1. Government Printing Office, Singapore.

Corner EJ (1954) The evolution of tropical forest. In *Evolution as a Process*. (Eds JS Huxley, AC Hardy and EB Ford) pp. 34–46. Allen and Unwin, London.

Coutinho GIAC, Valente VMM, Strozi RM, Meira A (2010) Ontogenetic, anatomical and histochemical study of the extrafloral nectaries of *Sapium biglandulosum* (Euphorbiaceae). *Australian Journal of Botany* **58**, 224–232. doi:10.1071/BT09200

Cox PA, Elmqvist T, Pierson ED, Rainey WE (1991) Flying foxes as strong interactors in South Pacific Island ecosystems: a conservation hypothesis. *Conservation Biology* **5**, 448–454. doi:10.1111/j.1523-1739.1991.tb00351.x

Crane PR (1989) Patterns of evolution and extinction in vascular plants. In *Evolution and the Fossil Record*. (Eds KC Allen and DEG Briggs) pp. 153–187. Smithsonian Institution Press, Washington.

Crayn DM, Rossetto M, Maynard DJ (2006) Molecular phylogeny and dating reveals an Oligo-Miocene radiation of dry-adapted shrubs (former Tremandraceae) from rainforest tree progenitors (Elaeocarpaceae) in Australia. *American Journal of Botany* **93**, 1328–1342. doi:10.3732/ajb.93.9.1328

Crepet WL, Niklas KJ (2009) Darwin's second "abominable mystery": why are there so many angiosperm species? *American Journal of Botany* **96**, 366–381. doi:10.3732/ajb.0800126

Crepet WL, Taylor DW (1985) The diversification of the Leguminoseae: first fossil evidence of Mimosoideae and Papilionoideae. *Science* **228**, 1087–1089. doi:10.1126/science.228.4703.1087

Crichton A, Francis N, Doherty S, Tuiwawa C, Hayes S, Stevens MI, Schwarz MP (2019) Low endemic bee diversity and very wide host range in lowland Fiji: support for the pollinator super-generalist hypothesis in island biogeography. *Pacific Conservation Biology* **25**, 135–142. doi:10.1071/PC18037

Crisp MD, Cook LG (2011) Cenozoic extinctions account for the low diversity of extant gymnosperms compared with angiosperms. *New Phytologist* **192**, 997–1009. doi:10.1111/j.1469-8137.2011.03862.x

Crisp MD, Cook LG (2013) How was the Australian flora assembled over the last 65 million years? A molecular phylogenetic perspective. *Annual Review of Ecology Evolution and Systematics* **44**, 303–324. doi:10.1146/annurev-ecolsys-110512-135910

Crome FHJ, Irvine AK (1986) 'Two bob each way': the pollination and breeding system of the Australian rainforest tree *Syzygium cormiflorum* (Myrtaceae). *Biotropica* **18**, 115–125. doi:10.2307/2388754

Cromie WJ (1999) Oldest known flowering plants identified by genes. *Harvard University Gazette*, 16 December 1999.

Cronk Q, Ojeda I (2008) Bird-pollinated flowers in an evolutionary and molecular context. *Journal of Experimental Botany* **59**, 715–727.

Cronquist A (1988) *The Evolution and Classification of Flowering Plants*. 2nd edn. New York Botanical Gardens, Bronx, NY.

Crous KY (2019) Plant responses to climate warming: physiological adjustments and implications for plant functioning in a future, warmer world. *American Journal of Botany* **106**, 1049–1051.

Crowson RA (1981) *Biology of the Coleoptera*. Academic Press, London.

Cullen PJ (1987) Regeneration patterns in populations of *Athrotaxis selaginoides* D. Don from Tasmania. *Journal of Biogeography* **14**, 39. doi:10.2307/2844785

Culley TM, Weller SG, Sakai AK (2002) The evolution of wind pollination in angiosperms. *Trends in Ecology & Evolution* **17**, 361–369. doi:10.1016/S0169-5347(02)02540-5

Cunningham SA, Schellhorn NA, Marcora A, Batley M (2013) Movement and phenology of bees in a subtropical Australian agricultural landscape. *Austral Ecology* **38**, 456–464.

Curran TJ, Clarke PJ, Bruhl JJ (2008) A broad typology of dry rainforests on the western slopes of New South Wales. *Cunninghamia* **10**, 381–405.

Czenze ZJ, Thurley T (2018) Weather and demographics affect *Dactylanthus* flower visitation by New Zealand lesser short-tailed bats. *New Zealand Journal of Ecology* **42**, 1–5.

Dafni A (1992) *Pollination Ecology: A Practical Approach*. Oxford University Press, Oxford.

Dafni A (1994) Note on side advertisement in flowers. *Functional Ecology* **8**, 136–138.

Dafni A, Calder DM (1987) Pollination by deceit and floral mimesis in *Thelymitra antennifera* (Orchidaceae). *Plant Systematics and Evolution* **158**, 11–22. doi:10.1007/BF00936140

Dafni A, Lehrer M, Kevan PG (1997) Spatial flower parameters and insect spatial vision. *Biological Reviews of the Cambridge Philosophical Society* **72**, 239–282. doi:10.1017/S0006323196005002

Dahlgren RMT (1980) A revised system of classification of the angiosperms. *Botanical Journal of the Linnean Society* **80**, 91–124. doi:10.1111/j.1095-8339.1980.tb01661.x

Dainese M, Riedinger V, Holzschuh A, Kleijn D, Scheper J, Steffan-Dewenter I (2017) Managing trap-nesting bees as crop pollinators: Spatiotemporal effects of floral resources and antagonists. *Journal of Applied Ecology* **55**, 195–204. doi:10.1111/1365-2664.12930

Danforth B, Poinar G (2011) Morphology, classification and the antiquity of *Melittosphex burmensis* (Apoidea: Melittosphecidae) and implications for early bee evolution. *Journal of Paleontology* **85**, 882–891. doi:10.1666/10-130.1

Danforth BN, Sipes S, Fang J, Brady SG (2006) The history of early bee diversification based on five genes plus morphology. *Proceedings of the National Academy of Sciences of the United States of America* **103**, 15118–15123. doi:10.1073/pnas.0604033103

Davies SJ (1999) A new myrmecophytic, thrips-pollinated species of *Macaranga* from the highlands of Sarawak. *Harvard Papers in Botany* **4**, 433–437.

Dayanandan S, Attygalla DNC, Abeygunasekera AWWL, Gunatilleke IAUN, Gunatill CVS (1990) Phenology and floral morphology in relation to pollination of some Sri Lankan dipterocarps. In *Reproductive Ecology of Tropical Forest Plants*. Vol. 7. (Eds KS Bawa and M Hadley) pp. 103–133. Unesco, Parthenon, Carnforth.

de Jager ML, Ellis AG (2012) Gender-specific pollinator preference for floral traits. *Functional Ecology* **26**, 1197–1204. doi:10.1111/j.1365-2435.2012.02028.x

De Luca PA, Vallejo-Marín M (2013) What's the 'buzz' about? The ecology and evolutionary significance of buzz-pollination. *Current Opinion in Plant Biology* **16**, 429–435. doi:10.1016/j.pbi.2013.05.002

de Souza CS, Aoki C, Ríbas A, Pott A, Sigrist MR (2016) Floral traits as indicators of pollination vs. theft. *Rodriguésia* **67**, 309–320. doi:10.1590/2175-7860201667203

de Vega C, Herrera CM (2013) Micro-organisms transported by ants induce changes in floral nectar composition of an ant-pollinated plant. *American Journal of Botany* **100**, 792–800. doi:10.3732/ajb.1200626

Debevec AH, Cardinal S, Danforth BN (2012) Identifying the sister group to the bees: a molecular phylogeny of Aculeata with an emphasis on the superfamily Apoidea. *Zoologica Scripta* **41**, 527–535. doi:10.1111/j.1463-6409.2012.00549.x

DEC (2003) Recovery plan for Davidson's Plum (*Davidsonia jerseyana*). Department of Environment and Conservation (NSW).

DEC (2004a) Recovery plan for *Randia moorei* (Spiny Gardenia). Department of Environment and Conservation (NSW).

DEC (2004b) Recovery plan for *Davidsonia johnsonii* (Smooth Davidsonia). Department of Environment and Conservation (NSW).

DEC (2004c) Recovery plan for *Endiandra floydii* (Crystal Creek Walnut). Department of Environment and Conservation (NSW).

DEC (2004d) Recovery plan for *Elaeocarpus williamsianus*. Department of Environment and Conservation (NSW).

DEC (2004e) Recovery plan for *Fontainea oraria*. Department of Environment and Conservation (NSW).

Del-Claro K, Marullo R, Mound LA (1997) A new Brazilian species of *Heterothrips* (Insecta: Thysanoptera) co-existing with ants in the flowers of *Peixotoa tomentosa* (Malpighiaceae). *Journal of Natural History* **31**, 1307–1312. doi:10.1080/00222939700770731

Del Tredici P (2007) The phenology of sexual reproduction in *Ginkgo biloba*: ecological and evolutionary implications. *Botanical Review* **73**, 267–278. doi:10.1663/0006-8101(2007)73[267:TPOSRI]2.0.CO;2

DellaSala DA (Ed.) (2011) *Temperate and Boreal Rainforests of the World: Ecology and Conservation*. Island Press, Washington.

Delmas CEL, Kooyman RM, Rossetto M (2020) Evolutionary constraints and adaptation shape the size and colour of rain forest fruits and flowers at continental scale. *Global Ecology and Biogeography* **29**, 830–841. doi:10.1111/geb.13065

Delph LF, Lively CM (1989) The evolution of floral colour change: pollinator attraction versus physiological constraints in *Fuchsia excorticata*. *Evolution* **43**, 1252–1262. doi:10.1111/j.1558-5646.1989.tb02572.x

Department of Environment and Energy (2019) *National Recovery Plan for the Littoral Rainforests and Coastal Vine Thickets of Eastern Australia Ecological Community*. Department of Environment and Energy, Canberra.

Dettmann ME (1990) The Antarctic Australian rift-valley – Late Cretaceous cradle of northeastern Australasian relics. *Review of Palaeobotany and Palynology* **65**, 131–144. doi:10.1016/0034-6667(90)90064-P

Dettmann ME, Clifford HT (2001) The fossil record of *Elaeocarpus* L. fruits. *Memoirs of the Queensland Museum* **46**, 461–497.

Dettmann ME, Jarzen DM (2000) Pollen of extant *Wollemia* (Wollemi Pine) and comparisons with pollen of other extant and fossil Araucariaceae. In *Pollen and Spores: Morphology and Biology*. (Eds MM Harley, CM Morton and S Blackmore) pp. 197–203. Royal Botanic Gardens, Kew.

Dettmann ME, Clifford HT, Peters M (2009) *Lovellea wintonensis* gen. et sp. nov. – Early Cretaceous (late Albian), anatomically preserved, angiosperm flowers and fruits from the Winton Formation, western Queensland. *Cretaceous Research* **30**, 339–355. doi:10.1016/j.cretres.2008.07.015

Devoto M, Montaldo NH, Medan D (2006) Long-probiscid fly pollination in 'ornithophilous' *Embothrium coccineum* (Proteaceae) along a rainfall gradient in Patagonia, Argentina. *Austral Ecology* **31**, 512–519. doi:10.1111/j.1442-9993.2006.01601.x

Devoto M, Medan D, Roig-Alsina A, Montaldo NH (2009) Patterns of species turnover in plant-pollinator communities along a precipitation gradient in Patagonia (Argentina). *Austral Ecology* **34**, 848–857. doi:10.1111/j.1442-9993.2009.01987.x

Devy MS, Davidar P (2003) Pollination systems of trees in Kakachi, a mid-elevation wet evergreen forest in Western Ghats, India. *American Journal of Botany* **90**, 650–657. doi:10.3732/ajb.90.4.650

Dick CW (2001) Genetic rescue of remnant tropical trees by an alien pollinator. *Proceedings. Biological Sciences* **268**, 2391–2396. doi:10.1098/rspb.2001.1781

Dirzo R, Young HS, Galetti M, Ceballos G, Isaac NJB, Collen B (2014) Defaunation in the Anthropocene. *Science* **345**, 401–406. doi:10.1126/science.1251817

Dixon DJ, Jackes BR, Bielig LM (2001) Figuring out the figs: the *Ficus rubiginosa* complex (Moraceae: Urostigma sect. Malvanthera). *Australian Systematic Botany* **14**, 133–154. doi:10.1071/SB99029

Dobson HEM (2006) Relationship between floral fragrance composition and type of pollinator. In *Biology of Floral Scent*. (Eds N Dudareva and E Pichersky) pp. 147–198. CRC Press, Boca Raton.

Dodd ME, Silverton J, Chase MW (1999) Phylogenetic analysis of trait evolution and species diversity variation among angiosperms. *Evolution* **53**, 732–744. doi:10.1111/j.1558-5646.1999.tb05367.x

Donaldson JS (1997) Is there a floral parasite mutualism in cycad pollination? The pollination biology of *Encephalartos villosus*. *American Journal of Botany* **84**, 1398–1406. doi:10.2307/2446138

Donoghue MJ, Doyle JA (2000) Seed plant phylogeny: demise of the anthophyte hypothesis? *Current Biology* **10**, R106–R109. doi:10.1016/S0960-9822(00)00304-3

Doust AN, Drinnan AN (2004) Floral development and molecular phylogeny support the generic status of *Tasmannia* (Winteraceae). *American Journal of Botany* **91**, 321–331. doi:10.3732/ajb.91.3.321

Doyle JA (2008) Integrating molecular phylogenetic and paleobotanical evidence on the origin of the flower. *International Journal of Plant Sciences* **169**, 816–843.

Doyle JA (2012) Molecular and fossil evidence on the origin of angiosperms. *Annual Review of Earth and Planetary Sciences* **40**, 301–326. doi:10.1146/annurev-earth-042711-105313

Doyle JA, Donoghue MJ (1992) Fossils and seed plant phylogeny reanalyzed. *Brittonia* **44**, 89–106. doi:10.2307/2806826

Drinnan AN, Chambers TC (1985) A reassessment of *Taeniopteris daintreei* from the Victorian Early Cretaceous: a member of the *Pentoxylales* and a significant Gondwanaland plant. *Australian Journal of Botany* **33**, 89–100. doi:10.1071/BT9850089

Duchen P, Renner SS (2010) The evolution of *Cayaponia* (Cucurbitaceae): repeated shifts from bat to bee pollination and long-distance dispersal to Africa 2–5 million years ago. *American Journal of Botany* **97**, 1129–1141. doi:10.3732/ajb.0900385

Dukas R, Dafni A (1990) Buzz-pollination in three nectariferous Boraginaceae and possible evolution of buzz-pollinated flowers. *Plant Systematics and Evolution* **169**, 65–68. doi:10.1007/BF00935985

Duke NC (2006) *Australia's Mangroves: The Authoritative Guide to Australia's Mangrove Plants*. University of Queensland and Norman C. Duke, Brisbane.

Duncan CJ, Worth JRP, Jordan GJ, Jones RC, Vaillancourt RE (2016) Genetic differentiation in spite of high gene flow in the dominant rainforest tree of southeastern Australia, *Nothofagus cunninghamii*. *Heredity* **116**, 99–106. doi:10.1038/hdy.2015.77

Duncan DH, Nicotra AB, Cunningham SA (2004) High self-pollen transfer and low fruit set in buzz-pollinated *Dianella revoluta* (Phormiaceae). *Australian Journal of Botany* **52**, 185–193. doi:10.1071/BT03139

Dutton EM, Fredrickson ME (2012) Why ant pollination is rare: new evidence and implications of the antibiotic hypothesis. *Arthropod-Plant Interactions* **6**, 561–569. doi:10.1007/s11829-012-9201-8

Dyer AG, Whitney HM, Arnold SEJ, Glover BJ, Chittka L (2006) Bees associate warmth with floral colour. *Nature* **442**, 525. doi:10.1038/442525a

Dyer AG, Boyd-Gerny S, McLoughlin S, Rosa MGP, Simonov V, Wong BBM (2012) Parallel evolution of angiosperm colour signals: common evolutionary pressures linked to hymenopteran vision. *Proceedings of the Royal Society B. Biological Sciences* **279**, 3606–3615. doi:10.1098/rspb.2012.0827

Eby P (1991) "Finger-winged night workers": managing forests to conserve the role of the Grey-headed Flying Foxes as pollinators and seed dispersers. In *Conservation of Australia's Forest Fauna*. (Ed. D Lunney) pp. 91–100. Royal Zoological Society of New South Wales, Mosman.

Eckhart VM (1999) Sexual dimorphism in flowers and inflorescences. In *Gender and Sexual Dimorphism in Flowering Plants*. (Eds MA Geber, TE Dawson and LF Delph) pp. 123–148. Springer, Berlin.

Edwards W, Gadek P, Weber E, Worboys S (2001) Idiosyncratic phenomenon of regeneration from cotyledons in the idiot fruit tree, *Idiospermum australiense*. *Austral Ecology* **26**, 254–258. doi:10.1046/j.1442-9993.2001.01110.x

Eliott FG, Shepherd M, Rossett M, Bundock P, Rice N, Henry RJ (2014) Contrasting breeding systems revealed in the rainforest genus *Davidsonia* (Cunoniaceae): can polyembryony turn tables on rarity? *Australian Journal of Botany* **62**, 451–464. doi:10.1071/BT14063

Elmqvist T, Cox PA, Rainey WE, Pierson ED (1992) Restricted pollination on Oceanic islands: pollination of *Ceiba pentandra* by flying foxes in Samoa. *Biotropica* **24**, 15–23. doi:10.2307/2388469

Eltz T, Brühl CA, Imiyahir Z, Linsenmair KE (2003) Nesting and nest trees of stingless bees (Apidae: Meliponini) in lowland dipterocarp forests of Sabah, Malaysia, with implications for forest management. *Forest Ecology and Management* **172**, 301–313. doi:10.1016/S0378-1127(01)00792-7

Endress PK (1994) *Diversity and Evolutionary Biology of Tropical Flowers.* Cambridge University Press, Cambridge.

Endress PK (2001) The flowers in extant basal angiosperms and inferences on ancestral flowers. *International Journal of Plant Sciences* **162**, 1111–1140. doi:10.1086/321919

Endress PK (2003) Early floral development and nature of the calyptra in Eupomatiaceae (Magnoliales). *International Journal of Plant Sciences* **164**, 489–503. doi:10.1086/375319

Endress PK (2010) The evolution of floral biology in basal angiosperms. *Philosophical Transactions of the Royal Society of London. Series B, Biological Sciences* **365**, 411–421. doi:10.1098/rstb.2009.0228

Endress PK, Doyle JA (2007) Floral phyllotaxis in basal angiosperms: development and evolution. *Current Opinion in Plant Biology* **10**, 52–57. doi:10.1016/j.pbi.2006.11.007

Engel MS (1999) A new xeromelissine bee in Tertiary amber of the Dominican Republic. *Entomologica Scandinavica* **30**, 453–458. doi:10.1163/187631200X00561

Engel MS (2001) A monograph of the Baltic amber bees and evolution of the Apoidea (Hymenoptera). *Bulletin of the American Museum of Natural History* **259**, 1–192.

Englund R (1993) Movement patterns of *Cetonia* beetles (Scarabaeidae) among flowering *Viburnum opulus* (Caprifoliaceae): option for long-distance pollen dispersal in a temperate shrub. *Oecologia* **94**, 295–302. doi:10.1007/BF00341330

Enright NJ, Hill RS, Veblen TT (1995) The southern conifers – an introduction. In *Ecology of Southern Conifers.* (Eds NJ Enright and RS Hill) pp. 1–9. Melbourne University Press, Carlton.

Eriksson D, Bremer B (1992) Pollination systems, dispersal modes, life forms, and the diversification rates in angiosperm families. *Evolution* **46**, 258–266. doi:10.2307/2409820

Espeland M, Breinholt J, Willmott KR, Warren AD, Vila R, Toussaint EFA, Mannsell SC, Aduse-Poku K, Talavera DJ, Eastwood R, Jarzyna MA, Guralnick R, Lohman DJ, Pierce NE, Kawahara AY (2018) A comprehensive and dated phylogenomic analysis of butterflies. *Current Biology* **28**, 770–778. doi:10.1016/j.cub.2018.01.061

Espolador Leitão CA, Meira RSA, Azevedo AA, de Araújo JM, Silva KLF, Collevatti RG (2005) Anatomy of the floral, bract, and foliar nectaries of *Triumfetta semitriloba* (Tiliaceae). *Canadian Journal of Botany* **83**, 279–286. doi:10.1139/b05-001

Ettershank G, Ettershank JA (1990) Insects associated with flowers of Tasmanian leatherwoods (*Eucryphia* spp.). *Proceedings of the Tasmanian National Rainforest Conservation Program.* Report No. 1, pp. 65–71.

Evans TJ (1993) Aspects of the Structure and Floristics of rainforest remnants occurring within the Lower Manning Valley. Vols. 1 and 2. Master of Letters thesis, The University of New England, Armidale.

Evenhuis NL (Ed.) (2007) *Barefoot on Lava: the Journals and Correspondence of Naturalist R.C.L. Perkins in Hawai'i, 1892–1901.* Bishop Museum Press, Honolulu.

Faegri K, van der Pijl L (1979) *The Principles of Pollination Ecology.* Pergamon Press, Oxford.

Faegri K, Iversen J, Kaland PM, Krzywinski K (1992) *Textbook of Pollen Analysis.* John Wiley and Sons, Chichester.

Fah LY (1995) Genetic and ecological studies relevant to the conservation and management of some Bornean *Calamus* species. PhD thesis, University of Aberdeen.

Fancy SG, Ralph CJ (1998) 'I'iwi (*Vestiaria coccinea*). In *The Birds of North America.* No. 327. (Eds A Poole and F Gill). The Birds of North America, Inc., Philadelphia. doi:10.2173/bna.327

Fedorov AA (1966) The structure of the tropical rainforest and speciation in the humid tropics. *Journal of Ecology* **54**, 1–11. doi:10.2307/2257656

Feil JP (1992) Reproductive ecology of dioecious *Siparuna* (Monimiacae) in Ecuador: a case of gall midge pollination. *Botanical Journal of the Linnean Society* **110**, 171–203. doi:10.1111/j.1095-8339.1992.tb00290.x

Feild TS, Franks PJ, Sage TL (2003) Ecophysical shade adaptation in the basal angiosperm, *Austrobaileya scandens* (Austrobaileyaceae). *International Journal of Plant Sciences* **164**, 313–324. doi:10.1086/346167

Feinsinger P, Swarm LB (1978) How common are ant-repellant nectars? *Biotropica* **10**, 238–239. doi:10.2307/2387911

Fensham RJ, Fairfax RJ, Cannell RJ (1994) The invasion of *Lantana camara* L. in Forty Mile Scrub National Park, north Queensland. *Australian Journal of Ecology* **19**, 297–305. doi:10.1111/j.1442-9993.1994.tb00493.x

Fensham RJ, Fairfax RJ, Butler DW, Bowman DJMS (2003) Effects of fire and drought in a tropical eucalypt savanna colonized by rain forests. *Journal of Biogeography* **30**, 1405–1414. doi:10.1046/j.1365-2699.2003.00934.x

Fenster CB, Armbruster WS, Wilson P, Dudash MR, Thomson JD (2004) Pollination syndromes and floral specialization. *Annual Review of Ecology Evolution and Systematics* **35**, 375–403. doi:10.1146/annurev.ecolsys.34.011802.132347

Fernandez Winzer L, Cuddy W, Pegg GS, Carnegie AJ, Mauea A, Leishman MR (2020) Plant architecture, growth and biomass allocation effects of the invasive pathogen myrtle rust (*Austropuccinia psidii*) on Australian Myrtaceae species after fire. *Austral Ecology* **45**, 177–186. doi:10.1111/aec.12845

Ferreira PA, Boscolo D, Carvalheiro LG, Beismeijer JC, Rocha PLB, Viana BF (2015) Responses of bees to habitat loss in fragmented landscapes of Brazilian Atlantic rainforest. *Landscape Ecology* **30**, 2067–2078. doi:10.1007/s10980-015-0231-3

Fiala B, Meyer U, Hashim R, Maschwitz U (2011) Pollination systems in pioneer trees of the genus *Macaranga* (Euphorbiaceae) in Malaysian rainforests. *Biological Journal of the Linnean Society* **103**, 935–953. doi:10.1111/j.1095-8312.2011.01680.x

Finch JTD, Power SA, Welbergen JA, Cook JM (2018) Two's company, three's a crowd: co-occurring pollinators and parasite species in *Breynia oblongifolia* (Phyllanthaceae). *BMC Evolutionary Biology* **18**, 193. doi:10.1186/s12862-018-1314-y

Flannery T (1995) *Mammals of the South-West Pacific and Moluccan Islands.* Reed Books, Chatswood.

Floyd AG (1990) *Australian Rainforests in New South Wales.* Vol. 1, pp. 1–136, Vol. 2, pp. 1–180. Surrey Beatty and Sons, Sydney.

Floyd AG (2007) *Rainforest Trees of Mainland South-eastern Australia.* Terania Rainforest Publishing, The Channon, NSW.

Ford HA, Paton DC, Forde N (1979) Birds as pollinators of Australian plants. *New Zealand Journal of Botany* **17**, 509–519. doi:10.1080/0028825X.1979.10432566

Fordyce IR, Eamus D, Duff GA (2000) Episodic seedling growth in *Allosyncarpia ternata*, a lignotuberous, monsoon rainforest tree in northern Australia. *Austral Ecology* **25**, 25–35. doi:10.1046/j.1442-9993.2000.01029.x

Forster PI (1989) Pollination of *Marsdenia fraseri* (Asclepiadaceae) by *Metriorrhynchus lateralis* (Coleoptera: Lycidae). *Coleopterists Bulletin* **43**, 311–312.

Forster PI (1992a) Pollination of *Hoya australis* (Asclepiadaceae) by *Ocybadistes walkeri sothis* (Lepidoptera: Hesperiidae). *Australian Entomological Magazine* **19**, 39–43.

Forster PI (1992b) Insects associated with the flowers of *Marsdenia cymulosa* Benth. (Asclepiadaceae) and their possible role in pollination. *Australian Entomological Magazine* **19**, 45–47.

Forster PI, Hyland BPM (1997) Two new species of *Eucryphia* Cav. (Cunoniaceae) from Queensland. *Austrobaileya* **4**, 589–596.

Forster PI, Machin PJ, Mound LA, Wilson GW (1994) Insects associated with reproductive structures of cycads in Queensland and north-east New South Wales, Australia. *Biotropica* **26**, 217–222. doi:10.2307/2388812

Fox BJ, Taylor JE, Fox MD, Williams C (1997) Vegetation changes across edges of rainforest remnants. *Biological Conservation* **82**, 1–13. doi:10.1016/S0006-3207(97)00011-6

Fragnière Y, Bétrisey S, Cardinaux L, Stoffel M, Kozlowski G (2015) Fighting their last stand? A global analysis of the distribution and conservation status of gymnosperms. *Journal of Biogeography* **42**, 809–820. doi:10.1111/jbi.12480

Frame D (2003) The pollen tube pathway in *Tasmannia insipida* (Winteraceae): homology of the male gametophyte condition tissue in Angiosperms. *Plant Biology* **5**, 290–296. doi:10.1055/s-2003-40793

Frankie GW, Baker HG, Opler PA (1974) Comparative phenological studies of trees in tropical wet and dry forests in lowlands of Costa Rica. *Journal of Ecology* **62**, 881–919. doi:10.2307/2258961

Frankie GW, Opler PA, Bawa KS (1976) Foraging behaviour of solitary bees: implications for outcrossing of a neotropical forest tree species. *Journal of Ecology* **64**, 1049–1057. doi:10.2307/2258824

Franklin DC, Noske RA (2000) Nectar sources used by birds in monsoonal north-western Australia: a regional survey. *Australian Journal of Botany* **48**, 461–474. doi:10.1071/BT98089

Friedman J, Barrett SCH (2008) A phylogenetic analysis of the evolution of wind pollination in angiosperms. *International Journal of Plant Sciences* **169**, 49–58. doi:10.1086/523365

Friis EM, Crane PR, Pedersen KR (2011) *Early Flowers and Angiosperm Evolution.* Cambridge University Press, Cambridge.

Frohlich MW, Chase MW (2007) After a dozen years of progress the origin of angiosperms is still a great mystery. *Nature* **450**, 1184–1189. doi:10.1038/nature06393

Fu Q, Diez JB, Pole M, Ávila MG, Liu Z-J, Chu H, Hou Y, Yin P, Zang G-Q, Du K, Wang X (2018) An unexpected noncarpellate epigynous flower from the Jurassic of China. *eLife* **7**, e38827. doi:10.7554/eLife.38827

Fukuhara T, Tokumaru S (2014) Inflorescence dimorphism, heterdichogamy and thrips pollination in *Platycarya strobilacea* (Juglandaceae). *Annals of Botany* **113**, 467–476. doi:10.1093/aob/mct278

Fulton M, Hodges SA (1999) Floral isolation between *Aquilegia formosa* and *Aquilegia pubescens. Proceedings. Biological Sciences* **266**, 2247–2252. doi:10.1098/rspb.1999.0915

Gandolfo MA, Nixon KC, Crepet WL (2004) Cretaceous flowers of Nymphaceae and implications for complex insect entrapment pollination mechanisms in early angiosperms. *Proceedings of the National Academy of Sciences of the United States of America* **101**, 8056–8060.

Gardère ML (2018) Snails in the flowers of *Campanula jacobaea* C. Sm. ex Webb (Campanulaceae) from the island of Sao Nicolau, Cabo Verde archipelago. *Adansonia* **40**, 183–186. doi:10.5252/adansonia2018v40a14

Gardner S, Sidisunthorn P, Anusarnsunthorn V (2000) *A Field Guide to the Forest Trees of Northern Thailand.* Kobfai Publishing Project, Bangkok.

Gauli A, Vaillancourt RE, Steane DA, Bailey TG, Potts BM (2013) Effect of forest fragmentation and altitude on the mating system of *Eucalyptus pauciflora* (Myrtaceae). *Australian Journal of Botany* **61**, 622–632. doi:10.1071/BT13259

Geerts S, Malherbe SD, Pauw A (2012) Reduced flower visitation by nectar-feeding birds in response to fire in Cape fynbos vegetation, South Africa. *Journal of Ornithology* **153**, 297–301. doi:10.1007/s10336-011-0743-9

Gentle CB, Duggin JA (1997) *Lantana camara* L. invasions in dry rainforest – open forest ecotones: The role of disturbances associated with fire and cattle grazing. *Australian Journal of Ecology* **22**, 298–306. doi:10.1111/j.1442-9993.1997.tb00675.x

Gervasi DDL, Schiestl FP (2017) Real-time divergent evolution in plants driven by pollinators. *Nature Communications* **8**, 14691. doi:10.1038/ncomms14691

Ghazoul J (1997) The pollination and breeding system of *Dipterocarpus obtusifolius* (Dipterocarpaceae) in dry deciduous forests of Thailand. *Journal of Natural History* **31**, 901–916. doi:10.1080/00222939700770441

Ghazoul J (2001) Can floral repellants pre-empt potential ant–plant conflicts? *Ecology Letters* **4**, 295–299. doi:10.1046/j.1461-0248.2001.00229.x

Ghazoul J, McLeish M (2001) Reproductive ecology of tropical forest trees in logged and fragmented habitats in Thailand and Costa Rica. *Plant Ecology* **153**, 335–345. doi:10.1023/A:1017542124699

Gibbs P (2014) Late-acting self-incompatibility – the pariah breeding system in flowering plants. *New Phytologist* **203**, 717–734. doi:10.1111/nph.12874

Gilbert LE (1972) Pollen feeding and reproductive biology of *Heliconius* butterflies. *Proceedings of the National Academy of Sciences of the United States of America* **69**, 1403–1407. doi:10.1073/pnas.69.6.1403

Gilmore M (1934) *Old Days, Old Ways*. Angus and Robertson, Sydney.

Gilpin A-M (2017) Investigating the combined role of the introduced honeybee (*Apis mellifera*) and magnet plants on the pollination of both native and exotic plants in Australian ecosystems. PhD, University of Wollongong.

Godínez-Álvarez H (2004) Pollination and seed dispersal by lizards: a review. *Revista Chilena de Historia Natural* **77**, 569–577. doi:10.4067/S0716-078X2004000300015

Godoy O, Saldaña A, Fuentos N, Valladeres F, Gianoli E (2011) Forests are not immune to plant invasions: phenotypic plasticity and local adaptation allow *Prunella vulgaris* to colonize a temperate evergreen forest. *Biological Invasions* **13**, 1615–1625. doi:10.1007/s10530-010-9919-0

Goldblatt P, Bernhardt P, Manning JC (1998) Pollination of petaloid geophytes by monkey beetles (Scarabaeidae: Rutelinae: Hopliini) in southern Africa. *Annals of the Missouri Botanical Garden* **85**, 215–230. doi:10.2307/2992006

Goldingay RL (2000) Small dasyurid marsupials: are they effective pollinators? *Australian Journal of Zoology* **48**, 597–606. doi:10.1071/ZO00015

Gómez GM (2000) Effectiveness of ants as pollinators of *Lobularia maritima*: effects on main sequential fitness components of the host plant. *Oecologia* **122**, 90–97.

Gómez JM, Zamora R (1992) Pollination by ants: consequences of the quantitative effects on a mutualistic system. *Oecologia* **91**, 410–418. doi:10.1007/BF00317631

Gonzalez F, Poncy O (1999) A new species of *Aristolochia* (Aristolochiaceae) from Thailand. *Brittonia* **51**, 452–456. doi:10.2307/2666529

González-Varo JF, Biesmeijer JC, Bommar R, Potts SG, Sweiger O, Smith HG, Steffan-Dewenter I, Szentgyörgyi H, Woyciea-chowski M, Vila M (2013) Combined effects of global change pressures on animal-mediated pollinators. *Trends in Ecology & Evolution* **28**, 524–530. doi:10.1016/j.tree.2013.05.008

Gorelick R, Olson K (2011) Is lack of cycad (Cycadales) diversity a result of a lack of polyploidy? *Botanical Journal of the Linnean Society* **165**, 156–167. doi:10.1111/j.1095-8339. 2010.01103.x

Goto R, Yamakoshi G, Matsuzawa T (2012) A novel brood-site pollination mutualism?: the root holoparasite *Thonningia sanguinea* (Balanophoraceae) and an inflorescence-feeding fly in tropical rainforests of West Africa. *Plant Species Biology* **27**, 164–169. doi:10.1111/j.1442-1984.2011.00338.x

Gottsberger G (1974) The structure and function of the primitive angiosperm flower: a discussion. *Acta Botanica Neerlandica* **23**, 461–471. doi:10.1111/j.1438-8677.1974.tb00962.x

Gottsberger G (1977) Some aspects of beetle pollination in the evolution of flowering plants. *Plant Systematics and Evolution* **1**(Supp.), 211–226.

Gottsberger G (1999) Pollination and evolution in neotropical Annonaceae. *Plant Species Biology* **14**, 143–152. doi:10.1046/j. 1442-1984.1999.00018.x

Gottsberger G (2015) Generalist and specialist pollination in basal angiosperms (ANITA grade, basal monocots, magnoliids, Chloranthaceae and Ceratophyllaceae): what we know now. *Plant Diversity and Evolution* **131**, 263–362. doi:10.1127/pde/2015/0131-0085

Gottsberger G, Silberbauer-Gottsberger I, Ehrendorfer F (1980) Reproductive biology in the primitive relic angiosperm *Drimys brasiliensis* (Winteraceae). *Plant Systematics and Evolution* **135**, 11–39.

Gould E (1978) Foraging behaviour of Malaysian nectar-feeding bats. *Biotropica* **10**, 184–193. doi:10.2307/2387904

Goulson D, Derwent LC (2004) Synergistic interactions between an exotic honeybee and an exotic weed: pollination of *Lantana camara* in Australia. *Weed Research* **44**, 195–202. doi:10.1111/j.1365-3180.2004.00391.x

Goulson D, Wright NP (1998) Flower constancy in the hoverflies *Episyrphus balteatus* (Degeer) and *Syrphus ribesii* (L.) (Syrphidae). *Behavioral Ecology* **9**, 213–219. doi:10.1093/ beheco/9.3.213

Gowda V, Zacharias M, Vishveshwara N, Borges RM (2001) Do 'pollinator flags' aid in floral visitation in *Mussaenda frondosa*? In *Tropical Ecosystems: Structure, Diversity and Human Welfare*. (Eds KN Ganeshaiah, RU Shaankar and KS Bawa) pp. 207–210. Oxford-IBH, New Delhi, India.

Grant EL, Wallace HM, Trueman SJ, Reddell PW, Ogbourne SM (2017) Floral and reproductive biology of the medicinally significant rainforest tree, *Fontainea picrosperma* (Euphorbiaceae). *Industrial Crops and Products* **108**, 416–422. doi:10.1016/j.indcrop.2017.07.013

Gratshev VG, Zherikhin VV (2003) The fossil record of weevils and related beetle families (Coleoptera, Curculionoidea). *Acta Zoologica Craeoviensia* **46** (Suppl. – Fossil Insects), 129–138.

Greenwood DR (1987) Early Tertiary Podocarpaceae: mega fossils from Eocene Anglesea locality, Victoria, Australia. *Australian Journal of Botany* **35**, 111–133. doi:10.1071/BT9870111

Grehan JR (1991) A panbiogeographic perspective for Pre-Cretaceous angiosperm-Lepidoptera coevolution. *Australian Systematic Botany* **4**, 91–110. doi:10.1071/SB9910091

Gribel R (1988) Visits of *Caluromys lanatus* (Didelphidae) to flowers of *Psuedobombax tomentosum* (Bombacaceae): a probable case of pollination by marsupials in Central Brazil. *Biotropica* **20**, 344–347. doi:10.2307/2388329

Griffin AR, Hingston AB, Ohmart CP (2009) Pollinators of *Eucalyptus regnans* (Myrtaceae), the world's tallest flower plant species. *Australian Journal of Botany* **57**, 18–25. doi:10.1071/BT08168

Grimaldi DA, Engel MS (2005) *Evolution of the Insects*. Cambridge University Press, New York.

Grimaldi DA, Peñalver E, Barrón E, Herbold HW, Engel MS (2019) Direct evidence for eudicot pollen-feeding in a Cretaceous stinging wasp (Angiospermae-Hymenoptera, Aculeata) preserved in Burmese amber. *Communications Biology* **2**, 408. doi:10.1038/s42003-019-0652-7

Grímsson F, Grimm GW, Pott AJ, Zetter R, Renner SS (2018) A Winteraceae pollen tetrad from the early Paleocene of

Western Greenland, and the fossil record of Winteraceae in Laurasia and Gondwana. *Journal of Biogeography* **45**, 567–581. doi:10.1111/jbi.13154

Grison-Pigé L, Bessière J-M, Hossaert-McKey MM (2002) Specific attraction of fig-pollinating wasps: role of volatile compounds released by tropical figs. *Journal of Chemical Ecology* **28**, 283–295. doi:10.1023/A:1017930023741

Groffen J, Rethus G, Pettigrew J (2016) Promiscuous pollination of Australia's baobab, the boabab, *Adansonia gregorii*. *Australian Journal of Botany* **64**, 678–686. doi:10.1071/BT16049

Gross GL (1993) The breeding system and pollinators of *Melastoma affine* (Melostomataceae): a pioneer shrub in tropical Australia. *Biotropica* **25**, 468–474. doi:10.2307/2388870

Gross CL (2005) A comparison of the sexual systems in the trees from the Australian tropics with other tropical biomes: more monoecy but why? *American Journal of Botany* **92**, 907–919. doi:10.3732/ajb.92.6.907

Gross CL, Whitehead JD, de Souza CS, Mackay D (2017) Unsuccessful introduced biocontrol agents can act as pollinators of invasive weeds: Bitou Bush (*Chrysanthemoides monilifera* ssp. *rotundata*) as an example. *Ecology and Evolution* **7**, 8643–8656. doi:10.1002/ece3.3441

Guymer GP (1988) A taxonomic revision of *Brachychiton* (Sterculiaceae). *Australian Systematic Botany* **1**, 199–232. doi:10.1071/SB9880199

Hackett DJ, Goldingay RL (2001) Pollination of *Banksia* spp. By non-flying mammals in north-eastern New South Wales. *Australian Journal of Botany* **49**, 637–644. doi:10.1071/BT00004

Hajibabaei M, Xia J, Drouin G (2006) Seed plant phylogeny: Gnetophytes are derived conifers and a sister group to Pinaceae. *Molecular Phylogenetics and Evolution* **40**, 208–217. doi:10.1016/j.ympev.2006.03.006

Hall JA, Walter GH (2011) Does pollen aerodynamics correlate with pollination vector? Pollen settling velocity as a test for wind versus insect pollination among cycads (Gymnospermae: Cycadaceae: Zamiaceae). *Biological Journal of the Linnean Society* **104**, 75–92. doi:10.1111/j.1095-8312.2011.01695.x

Hall JA, Walter GH (2018) Pollination of the Australian cycad *Cycas ophiolitica* (Cycadaceae): the limited role of wind pollination in a cycad with beetle pollinator mutualists, and its ecological significance. *Journal of Tropical Ecology* **34**, 121–134. doi:10.1017/S0266467418000111

Hall JA, Walter GH, Bergstrom DM, Machin P (2004) Pollination ecology of the Australian cycad *Lepidozamia peroffskyana* (Zamiaceae). *Australian Journal of Botany* **52**, 333–343. doi:10.1071/BT03159

Hall L, Richards G (2000) *Flying Foxes: Fruit and Blossom Bats of Australia*. UNSW Press, Sydney.

Hamilton AG (1897) On the fertilization of *Eupomatia laurina*. *Proceedings of the Linnean Society of New South Wales* **22**, 48–55. doi:10.5962/bhl.part.12702

Hamilton AG (1919) Pollination of some Australian plants. *The Australian Naturalist* **4**, 75–81.

Handel SN, Peakall R (1993) Thynnine wasps discriminate among heights when seeking mates: tests with a sexually deceptive orchid. *Oecologia* **95**, 241–245. doi:10.1007/BF00323496

Hansen DM, Müller CB (2009) Invasive ants disrupt gecko pollination and seed dispersal of the endangered plant *Roussea simplex* in Mauritius. *Biotropica* **41**, 202–208. doi:10.1111/j.1744-7429.2008.00473.x

Hansen DM, Beer K, Müller CB (2006) Mauritian coloured nectar no longer a mystery: a visual signal for lizard pollinators. *Biology Letters* **2**, 165–168. doi:10.1098/rsbl.2006.0458

Hansen WR (Ed.) (1991) *Suggestions to authors of the reports of the United States Geological Survey. 7th edn (STA7)*. U.S. Geological Survey, Reston, Va.

Hansman DJ (2001) Floral biology of dry rainforest in north Queensland and a comparison with adjacent savanna woodland. *Australian Journal of Botany* **49**, 137–153. doi:10.1071/BT00017

Hao S, Xue J (2013) Earliest record of megaphylls and leafy structure, and thei initial diversification. *Chinese Science Bulletin* **58**, 2784–2793. doi:10.1007/s11434-013-5799-x

Harden GJ (Ed.) (1990) *Flora of New South Wales*. Vol 1. UNSW Press, Sydney.

Harden GJ (Ed.) (1991) *Flora of New South Wales*. Vol 2. UNSW Press, Sydney.

Harden GJ (Ed.) (1992) *Flora of New South Wales*. Vol 3. UNSW Press, Sydney.

Harden GJ (Ed.) (1993) *Flora of New South Wales*. Vol 4. UNSW Press, Sydney.

Harden GJ, Fox MD, Fox BJ (2004) Monitoring and assessment of restoration of a rainforest remnant at Wingham Brush, NSW. *Austral Ecology* **29**, 489–507. doi:10.1111/j.1442-9993.2004.01421.x

Harden GJ, McDonald WJF, Williams J (2006) *Rainforest Trees and Shrubs*. Gwen Harden Publishing, Nambucca Heads.

Harden GJ, McDonald WJF, Williams J (2007) *Rainforest Climbing Plants: A Field Guide to Their Identification*. Gwen Harden Publishing, Nambucca Heads.

Harden GJ, Nicholson H, McDonald WJF, Nicholson N, Tame T, Williams J (2014) *Australian Rainforest Plants (Rockhampton to Victoria) – Interactive Identification Key*. Gwen Harden Publishing, Nambucca Heads.

Harder LD, Johnson SD (2008) Function and evolution of aggregated pollen in angiosperms. *International Journal of Plant Sciences* **169**, 59–78. doi:10.1086/523364

Harrington GN, Irvine AK, Crome FHJ, Moore LA (1997) Regeneration of large-seeded trees in Australian rainforest fragments: a study of higher order interactions. In *Tropical Forest Remnants: Ecology, Management, and Conservation of Fragmented Communities*. (Eds WF Laurance and RO Bierregaard) pp. 292–303. University of Chicago Press, Chicago.

Harrison RD (2001) Drought and the consequences of El Niño in Borneo: a case study of figs. *Population Ecology* **43**, 63–75. doi:10.1007/PL00012017

Harriss FCL, Beattie AJ (1991) Viability of pollen carried by *Apis mellifera* L., *Trigona carbonaria* Smith and *Vespula germanica* (F.) (Hymenoptera: Apidae Vespidae).

Journal of the Australian Entomological Society **30**, 45–47. doi:10.1111/j.1440-6055.1991.tb02189.x

Hart MB (Ed.) (1996) *Biotic Recovery from Mass Extinction Events*. Geological Society Special Publication No. 102, London.

Hasenpusch JW, Moeseneder CH (2010) *Storeyus*, a new genus and *Storeyus pseudodipterus*, a new species of Cetoniinae (Coleoptera: Scarabaeidae) from Australia, with a redescription of *Lenosoma* Kraatz. *Australian Entomologist* **36**, 221–234.

Haslett JR (1983) A graphic account of pollen digestion by adult hoverflies. *Physiological Entomology* **8**, 167–171. doi:10.1111/j.1365-3032.1983.tb00345.x

Haston E, Richardson JE, Stevens PF, Chase MW, Harris DJ (2009) The Linear Angiosperm Phylogeny Group (LAPG) III: a linear sequence of the families in APG III. *Botanical Journal of the Linnean Society* **161**, 128–131. doi:10.1111/j.1095-8339.2009.01000.x

Hawkeswood TJ (1980) Jewel beetles as pollinators of *Melaleuca pauperiflora* F. Muell between Eucla (W.A.) and Koonalda (S.A.). *Western Australian Naturalist (Perth)* **14**, 238–239.

Hawkeswood TJ (1981) Notes on the pollination of *Nuytsia floribunda* (Labill.) R.Br. (Loranthaceae) and some literature reviewed. *Western Australian Naturalist (Perth)* **15**, 17–21.

Hawkeswood TJ (1987) Notes on some Coleoptera from *Baeckea stenophylla* F. Muell. (Myrtaceae) in New South Wales, Australia. *Giornale Italiano di Entomologia* **3**, 285–290.

Hawkeswood TJ (1990) Insect pollination of *Bursaria spinosa* (Pittosporaceae) in the Armidale area. New South Wales, Australia. *Giornale Italiano di Entomologia* **5**, 67–87.

Hawkeswood TJ (1993) Observations on insect pollination of *Melaleuca lanceolata* Otto and *Verticordia picta* Endl. (Myrtaceae), *Leucopogon parviflorus* (Andr.) Lindl. (Epacridaceae) and *Eryngium pinnatifidum* Bunge (Apiaceae) in Western Australia. *Giornale Italiano di Entomologia* **6**, 239–242.

Hawkins CP, MacMahon JA (1989) Guilds: the multiple meanings of a concept. *Annual Review of Entomology* **34**, 423–451. doi:10.1146/annurev.en.34.010189.002231

Hawthorne WD, Hughes CE (2008) Optimising linear taxon sequences derived from phylogenetic trees: a reply to Haston *et al. Taxon* **57**, 1–7.

Heard TA (1993) Pollinator requirements and flowering patterns of *Macadamia integrifolia. Australian Journal of Botany* **41**, 491–497. doi:10.1071/BT9930491

Heideman PD (1989) Temporal and spatial variation in phenology of flowering and fruiting in a tropical rainforest. *Journal of Ecology* **77**, 1059–1079. doi:10.2307/2260823

Heiling AM, Cheng K, Herbertstein ME (2004) Exploitation of floral signals by crab spiders (*Thomisus spectabilis*, Thomisidae). *Behavioral Ecology* **15**, 321–326. doi:10.1093/beheco/arh012

Heinrich B (1975) Energetics of pollination. *Annual Review of Ecology and Systematics* **6**, 139–170. doi:10.1146/annurev.es.06.110175.001035

Heithaus ER (1979) Community structure of neotropical flower visiting bees and wasps: diversity and phenology. *Ecology* **60**, 190–202. doi:10.2307/1936480

Henderson E, Missen M, Zaluki J (2019) To pupate or not to pupate: a case study of an obligate pollination mutualism in *Glochidion ferdinandi* (Phyllanthaceae) and *Epicephala colymbetella* (Gracillariidae). *Australian Journal of Botany* **67**, 473–479. doi:10.1071/BT18142

Hendrix SD (1990) Herbivory and its impact on plant reproduction. In *Plant Reproductive Ecology: Patterns and Strategies*. (Eds JL Doust and JL Doust) pp. 246–263. Oxford University Press, New York.

Herendeen PS, Crane PR, Drinnan AN (1995) Fagaceous flowers, fruits, and cupules from the Campian (Late Cretaceous) of Central Georgia, USA. *International Journal of Plant Sciences* **156**, 93–116. doi:10.1086/297231

Herendeen PS, Friss EM, Pedersen KR, Crane PR (2017) Palaeobotanical redux: revisiting the age of the angiosperms. *Nature Plants* **3**, 17015. doi:10.1038/nplants.2017.15

Hermansen TD, Ayre DJ, Minchinton TE (2014a) Effects of stand size on pollination of temperate populations of the mangrove *Avicennia marina. Plant Ecology* **215**, 1153–1162. doi:10.1007/s11258-014-0374-3

Hermansen TD, Britton DR, Ayre DJ, Minchinton TE (2014b) Identifying the real pollinators? Exotic honeybees are the dominant flower visitors and only effective pollinators of *Avicennia marina* in Australian temperate mangroves. *Estuaries and Coasts* **37**, 621–635. doi:10.1007/s12237-013-9711-3

Herrera CM (1997) Thermal biology and foraging responses of insect pollinators to the forest floor irradiance mosaic. *Oikos* **78**, 601–611. doi:10.2307/3545623

Herrera CM, Medrano M, Rey PJ, Sánchez-Lafuente AM, Garcia MB, Guitián J, Manzaneda AJ (2002) Interaction of pollinators and herbivores on plant fitness suggests a pathway for correlated evolution of mutualism – and antagonism-related traits. *Proceedings of the National Academy of Sciences of the United States of America* **99**, 16823–16828. doi:10.1073/pnas.252362799

Heslewood MM, Lowe AJ, Crayn DM, Rossetto M (2014) Contrasting levels of connectivity and localised persistence characterise the latitudinal distribution of a wind-dispersed rainforest canopy tree. *Genetica* **142**, 251–264. doi:10.1007/s10709-014-9771-8

Hesse M (1984) Pollenkitt is lacking in Gnetatae: *Ephedra* and *Welwitschia*: further proof for its restriction to the angiosperms. *Plant Systematics and Evolution* **144**, 9–16. doi:10.1007/BF00990796

Hetherington-Rauth MC, Ramírez SR (2016) Evolution and diversity of floral scent chemistry in the euglossine bee-pollinated orchid genus *Gongora. Annals of Botany* **118**, 135–148. doi:10.1093/aob/mcw072

Hickey JE (1994) A floristic comparison of vascular species in Tasmanian old growth mixed forest with regeneration resulting from logging and wildfire. *Australian Journal of Botany* **42**, 383–404. doi:10.1071/BT9940383

Higginson AD, Gilbert FS (2004) Paying for nectar with wingbeats: a new model of honeybee foraging. *Proceedings. Biological Sciences* **271**, 2595–2603. doi:10.1098/rspb.2004.2866

Higham RK, McQuillan PB (2000) *Cyathodes divaricata* (Epacridaceae) – the first record of a bird-pollinated dioecious plant in the Australian flora. *Australian Journal of Botany* **48**, 93–99. doi:10.1071/BT98027

Hilbert DW, Graham A, Hopkins MS (2007) Glacial and interglacial refugia within a long-term rainforest refugium: The Wet Tropics Bioregion of NE Queensland, Australia. *Palaeogeography, Palaeoclimatology, Palaeoecology* **251**, 104–118. doi:10.1016/j.palaeo.2007.02.020

Hill CJ (1988) The temporal distribution of butterfly species richness in a subtropical region of Australia. *Proceedings of the Ecological Society of Australia* **15**, 115–126.

Hill KD, Stevenson DW, Osborne R (2004) The world list of Cycads. *The Botanical Review* **70**, 274–298.

Hill RS (1980) Three new Eocene cycads from Eastern Australia. *Australian Journal of Botany* **28**, 105–122. doi:10.1071/BT9800105

Hill RS (1989) Early Tertiary leaves of the Menispermaceae from Nerriga, New South Wales. *Alcheringa* **13**, 37–44. doi:10.1080/03115518908619124

Hill RS (1994) The history of selected Australian taxa. In *History of Australian Vegetation*. (Ed. RS Hill) pp. 390–419. Cambridge University Press, Cambridge.

Hill RS (1995) Conifer origin, evolution and diversification in the Southern Hemisphere. In *Ecology of Southern Conifers*. (Eds NJ Enright and RS Hill) pp. 10–29. Melbourne University Press, Carlton.

Hill RS (2004) Origins of the southeastern Australian vegetation. *Philosophical Transactions of the Royal Society of London. Series B, Biological Sciences* **359**, 1537–1549. doi:10.1098/rstb.2004.1526

Hill RS, Dettmann ME (1996) Origin and diversification of the genus *Nothofagus*. In *The Ecology and Biogeography of Nothofagus Forests*. (Eds TT Veblen, RS Hill and J Read) pp. 11–24. Yale University Press, New Haven.

Hill RS, Read J (1984) Post-fire regeneration of rainforest and mixed forest in western Tasmania. *Australian Journal of Botany* **32**, 481–493. doi:10.1071/BT9840481

Hill RS, Truswell EM, McLoughlin S, Dettmann ME (1999) The evolution of the Australian flora: fossil evidence. In *Flora of Australia*. 2nd edn. (Eds AE Orchard and HS Thomson) pp. 251–320. Australian Biological Resources Study, Canberra.

Hill RS, Jordan GJ, McPhail MK (2015) Why we should retain "*Nothofagus sensu lato*". *Australian Systematic Botany* **28**, 190–193. doi:10.1071/SB15026

Hilton J, Bateman RM (2006) Pteridosperms are the backbone of seed-plant phylogeny. *The Journal of the Torrey Botanical Society* **133**, 119–168. doi:10.3159/1095-5674(2006)133[119:PATBOS]2.0.CO;2

Hilu KW, Borsch T, Müller K, Soltis DE, Soltis PS, Savolainen V, Chase MW, Powell MP, Alice LA, Evans R, Sauquet H, Neinhuis C, Slotla TAB, Rohwer JG, Campbell CS, Chatrou LW (2003) Angiosperm phylogeny based on *matK* sequence information. *American Journal of Botany* **90**, 1758–1776. doi:10.3732/ajb.90.12.1758

Hingston AB, McQuillan PB (2000) Are pollination syndromes useful predictors of floral visitors in Tasmania? *Austral Ecology* **25**, 600–609. doi:10.1111/j.1442-9993.2000.tb00065.x

Hingston AB, Wotherspoon S (2017) Introduced social bees reduce nectar availability during breeding season of the swift parrot (*Lathamus discolor*). *Pacific Conservation Biology* **23**, 52–62. doi:10.1071/PC16025

Hingston AB, Potts BM, McQuillan PB (2004a) Pollination services provided by various size classes of flower visitors to *Eucalyptus globulus* ssp. *globulus* (Myrtaceae). *Australian Journal of Botany* **52**, 353–369. doi:10.1071/BT03002

Hingston AB, Potts BM, McQuillan PB (2004b) The swift parrot *Lathamus discolor* (Psittacidae), social bees (Apidae) and native insects as pollinators of *Eucalyptus globulus* ssp. *globulus* (Myrtaceae). *Australian Journal of Botany* **52**, 371–379. doi:10.1071/BT03018

Hirthe G, Porembski S (2003) Pollination of *Nymphaea lotus* (Nymphaecaceae) by rhinoceros beetles and bees in the northeastern Ivory Coast. *Plant Biology* **5**, 670–676. doi:10.1055/s-2003-44717

Hjelmroos M (1991) Evidence of long-distance transport of *Betula* pollen. *Grana* **30**, 215–228. doi:10.1080/ 00173139109427802

Hnatiuk R (Co-ord.) (1990) *Census of Australian Vascular Plants*. Australian Flora and Fauna Series No. 11, Australian Biological Resources Study. AGPS, Canberra.

Hobbs RJ, Yates CJ (2003) Impacts of ecosystem fragmentation on plant populations: generalising the idiosyncratic. *Australian Journal of Botany* **51**, 471–488. doi:10.1071/BT03037

Hocking B, Sharplin CD (1965) Flower basking by Arctic insects. *Nature* **206**, 215. doi:10.1038/206215b0

Hölldobler B, Wilson EO (1990) *The Ants*. Belknap Press, Cambridge.

Homer LE (2009) Population structure and distance of gene flow in *Avicennia marina* (Forsk.) Vierh. (Avicenniaceae) on a local/regional scale in the Northern Rivers of New South Wales, Australia. PhD thesis, Southern Cross University, Lismore, NSW.

Honnay O, Jacquemyn H (2007) Susceptibility of common and rare plant species to the genetic consequences of habitat fragmentation. *Conservation Biology* **21**, 823–831. doi: 10.1111/j.1523-1739.2006.00646.x

Hopkins MS (1990) Disturbance – the forest transformer. In *Australian Tropical Rainforests*. (Eds LJ Webb and J Kikkawa) pp. 40–52. CSIRO Publishing, Melbourne.

Hopkins MS, Graham AW (1987) Gregarious flowering in a lowland tropical rainforest: a possible response to disturbance by Cyclone Winifred. *Austral Ecology* **12**, 25–29. doi:10.1111/j.1442-9993.1987.tb00924.x

Hopkins MS, Graham AW (1989) Community phenological patterns of a lowland tropical rainforest in north-eastern Australia. *Australian Journal of Ecology* **14**, 399–413. doi:10.1111/j.1442-9993.1989.tb01450.x

Hopper SD (1980) Pollination of the rainforest tree *Syzygium tierneyanum* (Myrtaceae) at Kuranda, northern Queensland. *Australian Journal of Botany* **28**, 223–237. doi:10.1071/BT9800223

Horvitz CC (1991) Light environments, stage structure, and dispersal syndromes of Costa Rican Marantaceae. In *Ant-Plant Interactions*. (Eds CR Huxley and DF Cutler) pp. 463–485. Oxford University Press, Oxford.

Hotchkiss AT (1958) Pollen and pollination in the Eupomatiaceae. *Proceedings of the Linnean Society of New South Wales* **83**, 86–91.

House SM (1985) Relationships between breeding and spatial pattern in some dioecious tropical rainforest trees. PhD thesis. Australian National University, Canberra.

House SM (1989) Pollen movement to flowering canopies of pistillate individuals of three rainforest tree species in tropical Australia. *Australian Journal of Ecology* **14**, 77–94. doi:10.1111/j.1442-9993.1989.tb01010.x

House SM (1992) Population density and fruit set in three dioecious tree species in Australian tropical rainforest. *Journal of Ecology* **80**, 57–69. doi:10.2307/2261063

House SM (1993) Pollination success in a population of dioecious rain forest trees. *Oecologia* **96**, 555–561. doi:10.1007/BF00320513

Houston TF (1975) A revision of the Australian Hylaeine bees (Hymenoptera: Colletidae). I. Introductory material and the genera *Heterapoides* Sandhouse, *Gephyrohylaeus* Michener, *Hyleoides* Smith, *Pharohylaeus* Michener, *Hemirhiza* Michener, *Amphylaeus* Michener and *Meroglossa* Smith. *Australian Journal of Zoology Supplementary Series* **23**, 1–135.

Houston TF (1981) A revision of the Australian Hylaeine bees (Hymenoptera: Colletidae). II. *Australian Journal of Zoology Supplementary Series* **29**, 1–128. doi:10.1071/AJZS080

Houston TF (2018) *A Guide to Native Bees of Australia*. CSIRO Publishing, Melbourne.

Houston TF, Ladd PG (2002) Buzz pollination in the Epacridaceae. *Australian Journal of Botany* **50**, 83–91. doi:10.1071/BT01020

Houston TF, Lamont BB, Radford S, Errington SG (1993) Apparent mutualism between *Verticordia nitens* and *V. aurea* (Myrtaceae) and their oil-ingesting bee pollinators (Hymenoptera: Colletidae). *Australian Journal of Botany* **41**, 369–380. doi:10.1071/BT9930369

Howell DJ (1974) Bats and pollen: physiological aspects of the syndrome of chiropterophily. *Comparative Biochemistry and Physiology A* **48**, 263–276. doi:10.1016/0300-9629(74)90707-5

Hu S, Dilcher DL, Jarzen DM, Taylor DW (2008) Early steps of angiosperm coevolution. *Proceedings of the National Academy of Sciences of the United States of America* **105**, 240–245. doi:10.1073/pnas.0707989105

Huang SQ, Fenster CB (2007) Absence of long-tongued proboscid pollinators for long-corolla-tubed Himalayan *Pedicularis* species: implications for the evolution of corolla length. *International Journal of Plant Sciences* **168**, 325–331. doi:10.1086/510209

Hugel S, Micheneau C, Fournel J, Warren BH, Gauvín-Bialecki A, Pailler T, Chase MW, Strasberg D (2010) *Glomeremus* species from the Mascarene islands (Orthoptera, Gryllacrididae) with the description of the pollinator of an endemic orchid from the island of Réunion. *Zootaxa* **2545**, 58–68. doi:10.11646/zootaxa.2545.1.6

Hunter RJ (2003) *World Heritage and Associative Natural Values of the Central Eastern Rainforest Reserves of Australia*. New South Wales National Parks and Wildlife Service, Sydney.

Ickert-Bond SM, Renner SS (2016) The Gnetales: recent insights on their morphology, reproductive biology, chromosome numbers, biogeography, and divergence. *Journal of Systematics and Evolution* **54**, 1–16. doi:10.1111/jse.12190

Igic B, Lande R, Kohn JR (2008) Loss of self-compatibility and its evolutionary consequences. *International Journal of Plant Sciences* **169**, 93–104. doi:10.1086/523362

Indsto JO, Weston PH, Clements MA, Dyer AG, Batley M, Whelan RJ (2006) Pollination of *Diuris maculata* (Orchidaceae) by male *Trichocolletes venustus* bees. *Australian Journal of Botany* **54**, 669–679. doi:10.1071/BT05146

Inkrot D, Sattler D, Geyer C, Morawetz W (2007) Flowering and fruiting phenology of *Normanbya normanbyi* (W. Hill) L.H. Bailey (Arecaceae), a palm endemic to lowland tropical rainforest of north-eastern Australia. *Austral Ecology* **32**, 21–28. doi:10.1111/j.1442-9993.2007.01737.x

Inoue T, Kato M, Kakutani T, Suka T, Itino T (1990) Insect–flower relationship in the temperate deciduous forest of Kibune Kyoto: an overview of the flowering phenology and the seasonal pattern of insect visits. *Contributions from Biological Laboratory, Kyoto University* **27**, 377–463.

Inouye DW, Pyke GH (1988) Pollination biology in the Snowy Mountains of Australia: comparisons with montane Colorado, USA. *Australian Journal of Ecology* **13**, 191–205. doi:10.1111/j.1442-9993.1988.tb00968.x

Inouye D, Morales M, Dodge G (2002) Variation in timing and abundance of flowering by *Delphinium barbeyi* Huth (Ranunculaceae): the roles of snowpack, frost and La Niña, in the context of climate change. *Oecologia* **130**, 543–550. doi:10.1007/s00442-001-0835-y

Ippolito A, Armstrong JE (1993) Floral biology of *Hornstedtia scottiana* (Zingiberaceae) in a lowland rain forest of Australia. *Biotropica* **25**, 281–289. doi:10.2307/2388786

Irvine AK, Armstrong JE (1988) Beetle pollination in Australian tropical rainforests. *Proceedings of the Ecological Society of Australia* **15**, 107–113.

Irvine AK, Armstrong JE (1990) Beetle pollination in tropical forests of Australia. In *Reproductive Ecology of Tropical Forest Plants*. Vol. 7. (Eds KS Bawa and M Hadley) pp. 135–149. Unesco, Parthenon, Carnforth.

Ishida C, Kono M, Sakai S (2009) A new pollination system: brood-site pollination by flower bugs in *Macaranga* (Euphorbiaceae). *Annals of Botany* **103**, 39–44. doi:10.1093/aob/mcn212

Itino T, Kato M, Hotta M (1991) Pollination ecology of the two wild bananas, *Musa acuminata* ssp. *halabanensis* and *Musa salaccensis*: chiropterophily and ornithophily. *Biotropica* **23**, 151–158. doi:10.2307/2388300

Jackson SM (2001) Foraging behaviour and food supply of the mahogany glider *Petaurus gracilis* (Petauridae: Marsupalia). *Journal of Zoology* **253**, 1–13. doi:10.1017/S0952836901000012

Jacques FMB, Wang W, Ortiz RDC, Li H-L, Zhou Z-H, Chen Z-D (2011) Integrating fossils in a molecular-based phylogeny and testing them as calibration points for divergence time estimates in Menispermaceae. *Journal of Systematics and Evolution* **49**, 25–49. doi:10.1111/j.1759-6831.2010.00105.x

James EA, Knox RB (1993) Reproductive biology of the Australian species of the genus *Pandorea* (Bignoniaceae). *Australian Journal of Botany* **41**, 611–626. doi:10.1071/BT9930611

Janzen DH (1969) Seed-eaters versus seed size, number, toxicity and dispersal. *Evolution* **23**, 1–27.

Janzen DH (1971) Euglossine bees as long-distance pollinators of tropical plants. *Science* **171**, 203–205. doi:10.1126/science.171.3967.203

Janzen DH (1977) Why don't ants visit flowers? *Biotropica* **9**, 252. doi:10.2307/2388142

Janzen DH (1979) How many babies do figs pay for babies? *Biotropica* **11**, 48–50. doi:10.2307/2388172

Jarman SJ, Brown MJ, Kantvilas G (1987) The classification, distribution and conservation status of Tasmanian rainforests. In *The Rainforest Legacy. Australian National Rainforests Study*. Vol. 1. (Eds G Werren and AP Kershaw) pp. 9–22. Australian Government Publishing Service, Canberra.

Jessup LW (1988) Australian Annonaceae in an Asian-Pacific context. *Proceedings of the Ecological Society of Australia* **15**, 249–257.

Jessup LW (2002) A new species of *Eupomatia* R.Br. (Eupomatiaceae) from Queensland. *Austrobaileya* **6**, 333–335.

Jevanandam N, Goh AGR, Corlett RT (2013) Climate warming and the potential extinction of fig wasps, the obligate pollinators of figs. *Biology Letters* **9**, doi:10.1098/rsbl.2013.0041.

Johanson LG, Hoffmann AA, Walker KL, Nash MA (2019) Bees of the Victorian Alps: network structure and interactions with introduced species. *Austral Ecology* **44**, 245–254. doi:10.1111/aec.12669

Johnson KA (2013) Are there pollination syndromes in the Australian epacrids (Ericaceae: Styphelioides)? A novel statistical method to identify key floral traits per syndrome. *Annals of Botany* **112**, 141–149. doi:10.1093/aob/mct105

Johnson KA, McQuillan PB (2011) Comparative floral presentation and bee-pollination in two *Sprengelia* species (Ericaceae). *Cunninghamia* **12**, 45–51.

Johnson KA, McQuillan PB, Kirkpatrick JB (2011) Nocturnal mammals, diurnal lizards, and the pollination ecology of the cryptic flowering *Acrotriche serrulata* (Ericaceae). *International Journal of Plant Sciences* **172**, 173–182. doi:10.1086/657280

Johnson SD, Anderson B (2010) Coevolution between food-rewarding flowers and their pollinators. *Evolution (New York)* **3**, 32–39. doi:10.1007/s12052-009-0192-6

Johnson SD, Nicolson SW (2008) Evolutionary associations between nectar properties and specificity in bird pollination systems. *Biology Letters* **4**, 49–52. doi:10.1098/rsbl.2007.0496

Johnson SD, Hargreaves AL, Brown M (2006) Dark, bitter-tasting nectar functions as a filter of flower visitors in a bird-pollinated plant. *Ecology* **87**, 2709–2716. doi:10.1890/0012-9658(2006)87[2709:DBNFAA]2.0.CO;2

Johnson SD, Moré M, Amorin FW, Haber WA, Frankie GW, Stanley DA, Cocucci AA, Raguso RA (2016) The long and the short of it: A global analysis of hawkmoth pollination niches and interaction networks. *Functional Ecology* **31**, 101–115. doi:10.1111/1365.2435.12753

Jolly WM, Cochrane MA, Freeborn PH, Holden ZA, Brown TJ, Williamson GJ, Bowman DM (2015) Climate-induced variations in global wildfire danger from 1979 to 2013. *Nature Communications* **6**, 7537. doi:10.1038/ncomms8537

Jones WG, Hill KD, Allen JM (1995) *Wollemia nobilis*, a new living Australian genus and species in the Araucariaceae. *Telopea* **6**, 173–176. doi:10.7751/telopea19953014

Jörgensen A, Rydin C (2015) Reproductive morphology in the *Gnetum cuspidatum* group (Gnetales) and its implications for pollination biology in the Gnetales. *Plant Ecology and Evolution* **148**, 387–396. doi:10.5091/plecevo.2015.1142

Joseph L, Toon A, Nyári AS, Longmore NW, Rowe KMC, Haryoko T, Trueman J, Gardner JL (2014) A new synthesis of the molecular systematics and biogeography of honeyeaters (Passeriformes: Meliphagidae) highlights biogeographical and ecological complexity of a spectacular avian radiation. *Zoologica Scripta* **43**, 235–248. doi:10.1111/zsc.12049

Joshi AR, Smith JLD, Cuthbert FJ (1995) Influence of food distribution and predation pressure spacing behavior in palm civets. *Journal of Mammalogy* **76**, 1205–1212. doi:10.2307/1382613

Judd WS, Campbell CS, Kellogg EA, Stevens PF, Donoghue MJ (2007) *Plant Systematics: A Phylogenetic Approach*. 3rd edn. Sinauer, Sunderland, MA.

Jürgens A (2009) The hidden language of flowering plants: floral odours as a key for understanding angiosperm evolution? *New Phytologist* **183**, 240–243.

Justiniano MJ, Fredericksen TS (2000) Phenology of tree species in Bolivian dry forests. *Biotropica* **32**, 276–281.

Kaczorowski RL, Seliger AR, Gaskett AC, Wigsten SK, Raguso RA (2012) Corolla shape vs. size in flower choice by a nocturnal hawkmoth pollinator. *Functional Ecology* **26**, 577–587. doi:10.1111/j.1365-2435.2012.01982.x

Kaiser CN, Hansen DM, Müller CB (2008) Habitat structure affects reproductive success of the rare endemic tree *Syzygium mamillatum* (Myrtaceae) in restored and unrestored sites in Mauritius. *Biotropica* **40**, 86–94.

Kalinganire A, Harwood CE, Slee MU, Simons AJ (2001) Pollination and fruit-set of *Grevillea robusta* in western Kenya. *Austral Ecology* **26**, 637–648. doi:10.1046/j.1442-9993. 2001.01139.x

Kanowski J, Kooyman RM, Catterall CP (2009) Dynamics and restoration of Australian tropical and subtropical rainforests. In *New Models of Ecosystem Dynamics and Restoration*. (Eds RJ Hobbs and KN Sading) pp. 206–220. Island Press, Washington.

Kato M (1996) Plant–pollinator interactions in the understorey of a lowland mixed forest in Sarawak. *American Journal of Botany* **83**, 732–743.

Kato M (2000) Anthophilous insect community and plant–pollinator interactions on Amami Islands in the Ryukyu Archipelago, Japan. *Contributions from Biological Laboratory, Kyoto University* **29**, 157–252.

Kato M, Miura R (1996) Flowering phenology and anthophilous insect community at a threatened natural lowland marsh at Nakaikemi in Tsuruga, Japan. *Contributions from Biological Laboratory, Kyoto University* **29**, 1–48.

Kato M, Kakutani T, Inoue T, Itino T (1990) Insect–flower relationship in the primary beech forest of Ashu Kyoto: an overview of the flowering phenology and seasonal pattern of insect visits. *Contributions from Biological Laboratory, Kyoto University* **27**, 309–375.

Kato M, Itino T, Hotta M, Inoue T (1991) Pollination of four Sumatran *Impatiens* species by hawkmoths and bees. *Tropics* **1**, 59–73. doi:10.3759/tropics.1.59

Kato M, Itino T, Nagamitsu T (1993) Melittophily and ornithophily of long-tubed flowers in Zingiberaceae and Gesneriaceae in West Sumatra. *Tropics* **2**, 129–142. doi:10.3759/tropics.2.129

Kato M, Takimura A, Kawakita A (2003) An obligate pollination mutualism and reciprocal diversification in the tree genus *Glochidion* (Euphorbiaceae). *Proceedings of the National Academy of Sciences of the United States of America* **100**, 5264–5267. doi:10.1073/pnas.0837153100

Kato M, Kosaka Y, Kawakita A, Okuyama Y, Kobayashi C, Phimminith T, Thongphan D (2008) Plant–pollinator interactions in tropical monsoon forests of Southeast Asia. *American Journal of Botany* **95**, 1375–1394. doi:10.3732/ajb.0800114

Kauffman JB (1991) Survival by sprouting following fire in tropical forests of the eastern Amazon. *Biotropica* **23**, 219–224. doi:10.2307/2388198

Kaul RB, Abbe EC, Abbe LB (1986) Reproductive phenology of the oak family (Fagaceae) in the lowland rainforests of Borneo. *Biotropica* **18**, 51–55. doi:10.2307/2388362

Kawakita A, Kato M (2002) Floral biology and unique pollination system of root holoparasites *Balanophora kuroiwai* and *B. tobiracola* (Balanophoraceae). *American Journal of Botany* **89**, 1164–1170. doi:10.3732/ajb.89.7.1164

Kawakita A, Takimura A, Terachi T, Sota T, Kato M (2004) Cospeciation analysis of an obligate pollination mutualism: have *Glochidion* trees (Euphorbiaceae) and pollinating *Epicephala* moths (Gracillariidae) diversified in parallel? *Evolution* **58**, 2201–2214.

Kay KM, Schemske DW (2003) Pollinator assemblages and visitation rates for 11 species of Neotropical *Costus* (Costaceae). *Biotropica* **35**, 198–207.

Kay KM, Reeves PA, Olmstead RG, Schemske DW (2005) Rapid speciation and the evolution of hummingbird pollination in Neotropical *Costus* subgenus *Costus* (Costaceae): evidence from nrDNA ITS and ET sequences. *American Journal of Botany* **92**, 1899–1910. doi:10.3732/ajb.92.11.1899

Kearns CA, Inouye DW (1993) *Techniques for Pollination Biologists*. University of Colorado Press, Colorado.

Keighery GJ (1975) Parallel evolution of floral structures in *Darwinia* (Myrtaceae) and *Pimelea* (Thymeleaceae). *Western Australian Naturalist (Perth)* **13**, 17–19.

Keith DA (Ed.) (2017) *Australian Vegetation*. Cambridge University Press, Cambridge.

Keller G, Mateo P, Punekar J, Khozyem H, Gertsch B, Spangenberg J, Bitchong A, Adatte T (2018) Environmental changes during the Cretaceous-Palaeogene mass extinction and Palaeocene-Eocene Thermal Maximum: Implications for the Anthropocene. *Gondwana Research* **56**, 69–89. doi:10.1016/j.gr.2017.12.002

Kellogg CA, Griffin DW (2006) Aerobiology and the global transport of desert dust. *Trends in Ecology & Evolution* **21**, 638–644. doi:10.1016/j.tree.2006.07.004

Kelly LT, Bennett AF, Clarke MF, McCarthy MA (2015) Optimal fire histories for biodiversity conservation. *Conservation Biology* **29**, 473–481. doi:10.1111/cobi.12384

Kenneally KF, Beard JS (1987) Rainforests of Western Australia. In *The Rainforest Legacy*. Vol. 1. (Eds GL Werren and AP Kershaw) pp. 289–304. Australian Government Publishing Service, Canberra.

Keppel G (2001) Notes on the natural history of *Cycas seemannii* (Cycadaceae). *South Pacific Journal of Natural Science* **19**, 35–41. doi:10.1071/SP01007

Kershaw AP, McGlone MS (1995) The Quaternary History of the southern conifers. In *Ecology of Southern Conifers*. (Eds NJ Enright and RS Hill) pp. 30–63. Melbourne University Press, Carlton.

Kessler A, Baldwin IT (2001) Defensive function of herbivore-induced plant volatile emissions in nature. *Science* **291**, 2141–2144. doi:10.1126/science.291.5511.2141

Kessler A, Halitschke R (2009) Testing the potential for conflicting selection on floral chemical traits by pollinators and herbivores: predictions and case study. *Functional Ecology* **23**, 901–912. doi:10.1111/j.1365-2435.2009.01639.x

Kevan PG (1978) Anthophilous springtails (Collembola) from the Alaskan north slope, from Signy Island, Antarctica, and from near Ottawa, Ontario. *Reviews in Ecology and Biological Science* **15**, 373–378.

Kevan PG, Backhaus WGK (1998) Color vision: ecology and evolution in making the best of the photopic environment. In *Color Vision: Perspectives from Different Disciplines*. (Ed. WGK Backhaus) pp. 163–183. Walter de Gruyter, Berlin.

Kevan PG, Lack AJ (1985) Pollination in a cryptically dioecious plant, *Decaspermum parviflorum* (Lam.) A.J.Scott (Myrtaceae), by pollen-collecting bees in Sulawesi, Indonesia. *Biological Journal of the Linnean Society* **25**, 319–330. doi:10.1111/j.1095-8312.1985.tb00398.x

Kevan PG, Chittka L, Dyer AG (2001) Limits to the salience of ultraviolet: lessons from colour vision in bees and birds. *The Journal of Experimental Biology* **204**, 2571–2580.

Kidyoo AM, Mckey D (2012) Flower phenology and mimicry of the rattan *Calamus castaneus* (Arecaceae) in southern Thailand. *Botany* **90**, 856–865. doi:10.1139/b2012-058

Kingston AB, McQuillan PB (1998) Does the recently introduced bumblebee *Bombus terrestris* (Apidae) threaten Australian ecosystems? *Austral Ecology* **23**, 539–549. doi:10.1111/j.1442-9993.1998.tb00764.x

Kinnaird MF (1992) Phenology of flowering and fruiting of an East African riverine forest system. *Biotropica* **24**, 187–194. doi:10.2307/2388672

Kirk WDJ (1985a) Effect of some floral scents on host finding by thrips (Insecta: Thysanoptera). *Journal of Chemical Ecology* **11**, 35–43. doi:10.1007/BF00987602

Kirk WDJ (1985b) Pollen-feeding and host specificity and fecundity of flower thrips (Thysanoptera). *Ecological Entomology* **10**, 281–289. doi:10.1111/j.1365-2311.1985.tb00725.x

Kirk WDJ (1988a) Assessing the effects of flower thrips. *Acta Phytopathologica et Entomologica Hungarica* **23**, 295–300.

Kirk WDJ (1988b) Thrips and pollination biology. In *Dynamics of Insect–Plant Interaction*. (Eds TN Ananthakrishnan and A Raman) pp. 129–135. Oxford and IBH, New Delhi.

Kitching RL, Bickel DJ, Boulter S (2005) Guild analyses of Diptera assemblages: a rationale and investigation of seasonality and stratification in selected rainforest faunas. In *The Evolutionary Biology of Flies*. (Eds DK Yeates and BM Wiegmann) pp. 388–416. Columbia University Press, New York.

Kitching RL, Boulter SL, Howlett BG, Goodall K (2007) Visitor assemblages at flowers in a tropical rainforest canopy. *Austral Ecology* **32**, 29–42. doi:10.1111/j.1442-9993.2007.01733.x

Kite GC, Hetterschield WLA (1997) Inflorescence odours of *Amorphophallus* and *Pseudodracontium* (Araceae). *Phytochemistry* **46**, 71–75. doi:10.1016/S0031-9422(97)00221-5

Knudsen JT, Andersson S, Bergman P (1999) Floral scent attraction in *Geonoma macrostachys*, an understorey palm of the Amazonian rain forest. *Oikos* **85**, 409–418. doi:10.2307/3546690

Knudsen J T, Erikkson R, Gershenzon J, Stähl B (2006) Diversity and distribution of floral scent. *The Botanical Review* **72**, 1. doi:10.1663/0006-8101

Kodela PG (2006) Pollen morphology of some rainforest taxa occurring in the Illawarra region of New South Wales, Australia. *Telopea* **11**, 346–389. doi:10.7751/telopea20065734

Koetz AH (2013) Ecology, behaviour and control of *Apis cerana* with a focus on relevance to the Australian incursion. *Insects* **4**, 558–592. doi:10.3390/insects4040558

Kolesik P, Geyer C, Morawetz W (2006) First known gall midge (Diptera: Cecidomyiidae) from Arecaceae *Normanbyomyia fructivora* gen. & sp. n. damaging fruit of black palm, *Normanbya normanbyi*, in tropical Australia. *Australian Journal of Entomology* **45**, 38–43. doi:10.1111/j.1440-6055. 2006.00497.x

Kondo T, Nishimura S, Naito Y, Tsumura Y, Okuda T, Ng KKS, Lee SL, Muhammad N, Nakagoshi N, Isagi Y (2011) Can tiny thrips provide sufficient pollination service during a general flowering period in tropical rainforest? In *Single-Pollen Genotyping*. (Eds Y Isagi and Y Suyama) pp. 66–81. Ecological Research Monographs. Springer, Tokyo.

Kono M, Tobe H (2007) Is *Cycas revoluta* (Cycadaceae) wind-or insect-pollinated? *American Journal of Botany* **94**, 847–855. doi:10.3732/ajb.94.5.847

Konrad W, Roth-Nebelsick A (2003) The dynamics of gas bubbles in conduits of vascular plants and implications for embolism repair. *Journal of Theoretical Biology* **224**, 43–61. doi:10.1016/S0022-5193(03)00138-3

Krahl AH, de Holanda ASS, Krahl DRP, Martucci MEP, Gobbo-Neto L, Webber AC (2019) Study of the reproductive biology of an Amazonian *Heterotaxis* (Orchidaceae) demonstrates the collection of resin-like material by stingless bees. *Plant Systematics and Evolution* **305**, 281–291. doi:10.1007/s00606-019-01571-9

Kramer AT, Ison JL, Ashley MV, Howe HF (2008) The paradox of forest fragmentation genetics. *Conservation Biology* **22**, 878–885. doi:10.1111/j.1523-1739.2008.00944.x

Kramer KU, Green PS, Götz E (Eds) (2013) *Pteridophytes and Gymnosperms*. Springer-Verlag, Berlin.

Krassilov VA, Rasnitsyn AP, Afonin SA (2007) Pollen eaters and pollen morphology: co-evolution through Permian and Mesozoic. *African Invertebrates* **48**, 3–11.

Krauss SL, Phillips RD, Karron JD, Johnson SD, Roberts DG, Hopper SD (2017) Novel consequences of bird pollination for plant mating. *Trends in Plant Science* **22**, 395–410. doi:10.1016/j.tplants.2017.03.005

Kress WJ (1985) Bat pollination of an Old World Heliconia. *Biotropica* **17**, 302–308. doi:10.2307/2388592

Kubitzki K (Ed.) (2013) *The Families and Genera of Vascular Plants*. Springer-Verlag, Berlin.

Kukalová-Peck J (1991) Fossil History and the Evolution of Hexapod Structures. In *The Insects of Australia*. Vol. I. (Ed. ID Naumann) pp. 141–179. CSIRO Publishing, Melbourne.

Kuriakose G, Sinu PA, Shivanna KR (2018a) Floral traits predict pollination syndromes in *Syzygium* species: a study on four endemic species of the Western Ghats, India. *Australian Journal of Botany* **66**, 575–582. doi:10.1071/BT18042

Kuriakose G, Sinu PA, Shivanna KR (2018b) Ant pollination of *Syzygium occidentale*, an endemic tree species of tropical rain forests of the Western Ghats, India. *Arthropod-Plant Interactions* **12**, 647–655. doi:10.1007/s11829-018-9613-1

Kuschel G, May BM (1990) Palophaginae, a new subfamily for leaf-beetles, feeding as adult and larva on Araucarian pollen in Australia (Coleoptera: Megalopodidae). *Invertebrate Taxonomy* **3**, 697–719. doi:10.1071/IT9890697

Labandeira C (2010) The pollination of Mid Mesozoic seed plants and the early history of long-proboscid insects. *Annals of the Missouri Botanical Garden* **97**, 469–513. doi:10.3417/2010037

Lacey CT, Jahnke R (1984) The occurrence and nature of lignotubers in *Notelaea longifolia* and *Elaeocarpus reticulatus*. *Australian Journal of Botany* **32**, 311–321. doi:10.1071/BT9840311

Lachance M-A, Starmer WT, Rosa CA, Bowles JM, Barker JSF, Janzen DH (2001) Biogeography of the yeasts of ephemeral flowers and their insects. *FEMS Yeast Research* **1**, 1–8. doi:10.1016/S1567-1356(00)00003-9

Lack AJ, Kevan PG (1984) On the reproductive biology of a canopy tree, *Syzygium syzygioides* (Myrtaceae), in a rainforest in Sulawesi, Indonesia. *Biotropica* **16**, 31–36. doi:10.2307/2387891

Lahondère C, Vinauger C, Okubo RP, Wolff GH, Chan JK, Akbari OS, Riffell JA (2020) The olfactory basis of orchid pollination by mosquitoes. *Proceedings of the National Academy of Sciences of the United States of America* **117**, 708–716. doi:10.1073/pnas.1910589117

Laidlaw MJ, McDonald WJF, Hunter RJ, Putland DA, Kitching RL (2011) The potential impacts of climate change on Australian subtropical rainforest. *Australian Journal of Botany* **59**, 440–449. doi:10.1071/BT10319

Lamont BB, He T (2012) Fire-adapted Gondwanan angiosperm floras evolved in the Cretaceous. *BMC Evolutionary Biology* **12**, 223. doi:10.1186/1471-2148-12-223

Langenheim JH (2003) *Plant Resins: Chemistry, Evolution, Ecology, Ethnobotany.* Timber Press, Portland, Oregon.

LaPolla JS, Dlussky GM, Perrichot V (2013) Ants and the fossil record. *Annual Review of Entomology* **58**, 609–630. doi:10.1146/annurev-ento-120710-100600

Laurance WF, Bierregaard RO (Eds) (1997) *Tropical Forest Remnants.* University of Chicago Press, Chicago.

Lautenbach S, Seppelt R, Liebscher J, Dormann CF (2012) Spatial and temporal trends in global pollination benefit. *PLoS One* **7**, e35954. doi:10.1371/journal.pone.0035954

Law BS (1992) Physiological factors affecting pollen use by Queensland blossom bats (*Syconycteris australis*). *Functional Ecology* **6**, 257–264. doi:10.2307/2389515

Law BS, Lean M (1999) Common blossom bats (*Syconycteris australis*) as pollinators in fragmented Australian tropical rainforest. *Biological Conservation* **91**, 201–212. doi:10.1016/S0006-3207(99)00078-6

Lawrence JF, Ślipiński A (2013) *Australian Beetles.* Vol. 1. CSIRO Publishing, Melbourne.

Lee YF, Jong K, Swaine MD, Chey VK, Chung AYC (1995) Pollination in the rattans *Calamus subcinereus* and *C. caesius* (Palmae: Calamoideae). *Sandakania* **6**, 15–39.

Leonardi S, Piovani P, Scalfi M, Piotti A, Giannini R, Menozzi P (2012) Effect of habitat fragmentation on the genetic diversity and structure of peripheral populations of Beech in Central Italy. *The Journal of Heredity* **103**, 408–417. doi:10.1093/jhered/ess004

Les DH (1993) Ceratophyllaceae. In *The Families and Genera of Vascular Plants.* Vol. 2. (Eds K Kubitzki, JG Rohwer and V Bittrich) pp. 246–250. Springer-Verlag, Berlin.

Les DH, Schneider EL, Padgett DJ, Soltis PS, Soltis DE, Zanis M (1999) Phylogeny, classification and floral evolution of water lilies (Nymphaeaceae; Nymphaeles): a synthesis of non-molecular, *rbcL, matK,* and 18S rDNA data. *Systematic Botany* **24**, 28–46. doi:10.2307/2419384

Li H-T, Yi T-S, Gao L-M, Ma P-F, Zhang T, Yang J-B, Gitzendanner MA, Fritsch PW, *et al.* (2019) Origin of angiosperms and the puzzle of the Jurassic gap. *Nature Plants* **5**, 461–470. doi:10.1038/s41477-019-0421-0

Lin H-Y, Tseng Y-H, Hsieh C-F, Hu J-M (2019) Geographical distribution of dioecy and its ecological correlates based on fine-scaled species distribution data from a subtropical island. *Ecological Research* **35**, 170–181. doi:10.1111/1440-1703.12068

Linder HP (1998) Morphology and the evolution of wind pollination. In *Reproductive Biology in Systematics, Conservation and Economic Botany.* (Eds SJ Owens and PJ Rudall) pp. 123–135. Royal Botanic Gardens, Kew.

Linhart YB, Mendenhall JA (1977) Pollen dispersal by hawkmoths in a *Lindenia rivalis* Benth. population in Belize. *Biotropica* **9**, 143.

Linskens HF (1996) No airborne pollen within tropical rain forests. *Proceedings of the Koninklijke Nederlandse Akademie van Wetenschappen* **99**, 175–180.

Linskens HF, Shrauwn J (1969) The release of free amino acids from germinating pollen. *Acta Botanica Neerlandica* **18**, 605–614. doi:10.1111/j.1438-8677.1969.tb00082.x

Listabarth C (1993) Insect-induced wind pollination of the palm *Chamaedorea pinnatifrons* and pollination of the related *Wendlandiella* sp. *Biodiversity and Conservation* **2**, 39–50. doi:10.1007/BF00055101

Little JK, Prior LD, Williamson GJ, Williams S, Bowman DMJS (2012) Fire weather risk differs across rain forest-savanna boundaries in the humid tropics of north-eastern Australia. *Austral Ecology* **37**, 915–925. doi:10.1111/j.1442-9993.2011.02350.x

Liu C-Q, Sun H (2019) Pollination in *Lilium sargentiae* (Liliaceae) and the first confirmation of long-tongued hawkmoths as a niche in Asia. *Journal of Systematics and Evolution* **57**, 81–88. doi:10.1111/jse.12419

Liu Z-J, Wang X (2017) *Yuhania*: a unique angiosperm from the Middle Jurassic of Inner Mongolia, China. *Historical Biology* **29**, 431–441. doi:10.1080/08912963.2016.1178740

Liu Z, Ślipiński A, Lawrence JF, Ren D, Pang H (2018) *Palaeoboganum* gen. nov. from the Middle Jurassic of China (Coleoptera: Cucujoidea: Boganiidae): the first cycad pollinators? *Journal of Systematic Palaeontology* **16**, 351–360. doi:10.1080/14772019.2017.1304459

Llandres AL, De Mas E, Rodríguez-Gironés A (2012) Response of pollinators to tradeoff between resource acquisition and predator avoidance. *Oikos* **121**, 687–696. doi:10.1111/j.1600-0706.2011.19910.x

Lloyd DG, Wells MS (1992) Stigmatic exudate rewards and self-incompatibility in a primitive angiosperm, *Pseudowintera colorata* (Winteraceae). *Newsletter of Australasian Pollination Ecologists Society* July.

Lopez-Vaamonde C, Dixon DJ, Cook JM, Rasplus J-Y (2002) Revision of the Australian species of *Pleistodontes* (Hymenoptera: Agaonidae) fig-pollinating wasps and their host-plant associations. *Zoological Journal of the Linnean Society* **136**, 637–683. doi:10.1046/j.1096-3642.2002.00040.x

Lorence DH (1985) A monograph of the Monimiaceae (Laurales) in the Malagasy region (southwest Indian ocean). *Annals of the Missouri Botanical Garden* **72**, 1–165. doi:10.2307/2399135

Lowe S, Browne M, Bondjelas S, De Poorter M (2000) *100 of the World's Worst Invasive Animal Species. A Selection from the Global Invasive Species Database.* Published by the Invasive Species Specialist Group (ISSG), a specialist group of the Species Survival Commission (SSC) of the World Conservation Union (IUCN), www.issg.org/booklet.pdf.

Lughadha E, Proença C (1996) A survey of the reproductive biology of the Myrtoideae (Myrtaceae). *Annals of the Missouri Botanical Garden* **83**, 480–503. doi:10.2307/2399990

Lumer C, Schoer RD (1986) Pollination of *Blakea austin-smithii* and *B. penduliflora* (Melastomataceae) by small rodents in Costa Rica. *Biotropica* **18**, 363–364. doi:10.2307/2388584

Lunau K, Ren Z-X, Fan X-Q, Trunscke J, Pyke GH, Wang H (2020) Nectar mimicry: a new phenomenon. *Scientific Reports* **10**, 7039. doi:10.1038/s41598-020-63997-3

Lynch AJJ, Barnes RW, Cambecèdes J, Vaillancourt RE (1998) Genetic evidence that *Lomatia tasmanica* (Proteaceae) is an ancient one. *Australian Journal of Botany* **46**, 25–33. doi:10.1071/BT96120

Ma S-B, Wang Y-H, Cui M-K (1997) A contribution to the reproductive biology of *Chloranthus holostegius* (Chloranthaceae) in Mile population. *Yunnan Zhi Wu Yan Jiu* **19**, 415–422.

MacGregor CJ, Pocock MJO, Fox R, Evans DM (2015) Pollination by nocturnal Lepidoptera, and effects of light pollution: a review. *Ecological Entomology* **40**, 187–198. doi:10.1111/een.12174

Machado IC, Lopez AV, Sazima M (2006) Plant sexual systems and a review of the breeding system studies in the Caatinga, a Brazilian tropical dry forest. *Annals of Botany* **97**, 277–287. doi:10.1093/aob/mcj029

Magallón S, Crane PR, Herendeen PS (1999) Phylogenetic pattern, diversity and diversification of eudicots. *Annals of the Missouri Botanical Garden* **86**, 297–372. doi:10.2307/2666180

Maia ACD, Gibernau M, Carvalho AT, Gonçalves EG, Schlindwein C (2013) The cowl does not make the monk: scarab beetle pollination of the Neotropical aroid *Taccarum ulei* (Araceae: Spathicarpeae). *Biological Journal of the Linnean Society* **108**, 22–34. doi:10.1111/j.1095-8312.2012.01985.x

Makinson RO (2018) Myrtle Rust in Australia – a draft Action Plan, presented at the Plant Biosecurity Cooperative Research Centre's National Science Exchange, Melbourne, 31 May 2018.

Mariani M, Connor SE, Theuerkauf M, Kuneš P, Fletcher M-S (2016) Testing quantitative pollen dispersal models in animal-pollinated vegetation mosaics: An example from Temperate Tasmania, Australia. *Quaternary Science Reviews* **154**, 214–225. doi:10.1016/j.quascirev.2016.10.020

Marloth R (1914) Note on the entomophilous nature of *Encephalartos*. *Transactions of the Royal Society of South Africa* **4**, 69–71. doi:10.1080/00359191409519515

Martén-Rodríguez S, Almarales-Castro A, Fenster CB (2009) Evaluation of pollination syndromes in Antillean Gesneriaceae: evidence for bat, hummingbird and generalized flowers. *Journal of Ecology* **97**, 348–359. doi:10.1111/j.1365-2745.2008.01465.x

Martin HA (1998) Tertiary climatic evolution and the development of aridity in Australia. *Proceedings of the Linnean Society of New South Wales* **119**, 115–136.

Massoni J, Couvreur TLP, Sauquet H (2015) Five major shifts of diversification through the long evolutionary history of Magnoliidae (angiosperms). *BMC Evolutionary Biology* **15**, 49. doi:10.1186/s12862-015-0320-6

Mast AR, Milton EF, Jones EH, Barker RM, Barker WR, Weston PH (2012) Time-calibrated phylogeny of the woody Australian genus *Hakea* (Proteaceae) supports multiple origins of insect-pollination among bird-pollinated ancestors. *American Journal of Botany* **99**, 472–487. doi:10.3732/ajb.1100420

Matallana G, Wendt T, Araujo DSD, Scarano FR (2005) High abundance of dioecious plants in a tropical coastal vegetation. *American Journal of Botany* **92**, 1513–1519. doi:10.3732/ajb.92.9.1513

Matheson A, Reid M (2011) *Practical Beekeeping in New Zealand*. Exisle Publications, Auckland.

McDonald WJF (2010) National recovery plan for the "semi-evergreen vine thickets of the Brigalow Belt (North and South) and Nandewar Bioregions' ecological community. Report for the Department of Environment, Water, Heritage and Arts, Canberra. Queensland Department of Environment and Resource Management, Brisbane.

McEvey SF, Barker JSF (2001) *Scaptodrosophila aclinata*: a new *Hibiscus* flower-breeding species related to *S. hibisci* (Diptera: Drosophilidae). *Records of the Australian Museum* **53**, 255–262. doi:10.3853/j.0067-1975.53.2001.1332

McKenzie GM (2002) The Late Quaternary vegetation history of south-central highlands of Victoria. II. Sites below 900 m. *Austral Ecology* **27**, 32–54. doi:10.1046/j.1442-9993.2002.01155.x

McLeish MJ, van Noort S (2012) Codivergence and multiple host species use by fig wasp populations of the *Ficus* pollination mutualism. *BMC Evolutionary Biology* **12**, 1. doi:10.1186/1471-2148-12-1

McLoughlin S (1992) Permian sphenophytes from the Collie and Perth Basins, Western Australia. *Review of Palaeobotany and Palynology* **75**, 153–182.

McLoughlin S (2001) The breakup history of Gondwana and its impact on pre-Cenozoic floristic provincialism. *Australian Journal of Botany* **49**, 271–300. doi:10.1071/BT00023

McMahon ARG (1987) *The Effects of the 1982–1983 Bushfires on Sites of Significance*. Environmental Studies Publication Series No. 411. Victorian Department of Conservation, Forests and Lands, Melbourne.

McQuillan PB, Hingston AB (1999) Displacement of Tasmanian native megachilid bees by the recently introduced bumblebee *Bombus terrestris* (Linnaeus, 1758) (Hymenoptera: Apidae). *Australian Journal of Zoology* **47**, 59–65. doi:10.1071/ZO98016

Melick DR (1990) Relative drought resistance of *Tristaniopsis laurina* and *Acmena smithii* from riparian warm temperate rainforest in Victoria. *Australian Journal of Botany* **38**, 361–370. doi:10.1071/BT9900361

Mellick R, Lowe A, Rossetto M (2011) Consequences of long- and short-term fragmentation on the genetic diversity and differentation of a late successional rainforest conifer. *Australian Journal of Botany* **59**, 351–362. doi:10.1071/BT10291

Melo GAR (1999) Phylogenetic relationships and classification of the major lineages of Apoidea (Hymenoptera), with emphasis on the crabronid wasps. *Scientific Papers Natural History Museum, University of Kansas* **14**, 1–55.

Melo GAR, Rosa BB (2018) New genus of apoid wasps (Hymenoptera, Apoidea) from the Cretaceous amber of Myanmar. *Revista Brasileira de Entomologia* **62**, 319–323. doi:10.1016/j.rbe.2018.09.004

Mendez M, Munzinger J (2010) *Planchonella*, first record of gynomonoecy for the family Sapotaceae. *Plant Systematics and Evolution* **287**, 65–73. doi:10.1007/s00606-010-0290-5

Menz MHM, Phillips RD, Winfree R, Kremen C, Aizen MA, Johnson SD, Dixon KW (2011) Reconnecting plants and pollinators: challenges in the restoration of pollination mutualisms. *Trends in Plant Science* **16**, 4–12. doi:10.1016/j.tplants.2010.09.006

Mesquita-Neto JN, Borges JPR, de Sá TFF, Oliveira Teixeira TP, Ferreira INM, Furtado MTR, Consolaro H, Franceschinelli EV (2018) Pollen flow and pollinator sharing among synchronopatric species of *Psychotria* (Rubiaceae). *Plant Systematics and Evolution* **304**, 943–953. doi:10.1007/s00606-018-1527-y

Miao Y, Liu Z-J, Wang M, Wang X (2017) Fossil and living cycads say "No more megasporophylls". *Journal of Morphology and Anatomy* **1**, 107.

Micheneau C, Fournel J, Warren BH, Hugel S, Gauvín-Bialecki A, Pailler T, Strasberg D, Chase MW (2010) Orthoptera, a new order of pollinator. *Annals of Botany* **105**, 355–364. doi:10.1093/aob/mcp299

Michener CD (2007) *The Bees of the World.* 2nd edn. Johns Hopkins University Press, Baltimore.

Michez D, Vanderplanck M, Engel MS (2012) Fossil bees and their plant associates. In *Evolution of Plant-Pollinator Relationships.* (Ed. S Patiny) pp. 103–164. Cambridge University Press, Cambridge.

Miles J, Kendall P (2006) *Endangered ecological communities survey and mapping, coastal vegetation communities – littoral rainforest and coastal saltmarsh.* New South Wales Department of Environment and Conservation, Sydney.

Miller BP, Murphy BP (2017) Fire and Australian Vegetation. In *Australian Vegetation.* (Ed. DA Keith) pp. 113–134. Cambridge University Press, Cambridge.

Milton K (1991) Leaf change and fruit production in six neotropical Moraceae species. *Journal of Ecology* **79**, 1–26. doi:10.2307/2260781

Minnaar C, Anderson B (2019) Using quantum dots as pollen labels to track the fates of individual pollen grains. *Methods in Ecology and Evolution* **10**, 604–614. doi:10.1111/2041-210X.13155

Mohr B, Friis EM (2000) Early angiosperms from the Aptian Crato Formation (Brazil): a preliminary report. *International Journal of Plant Sciences* **161**(Supplement), S155–S167. doi:10.1086/317580

Molbo D, Machado CA, Sevenster JG, Keller L, Herre EA (2003) Cryptic species of fig pollinating wasps: implications for the evolution of the fig-wasp mutualism, sex allocation, and precision of adaptation. *Proceedings of the National Academy of Sciences of the United States of America* **100**, 5867–5872. doi:10.1073/pnas.0930903100

Møller AP, Eriksson M (1995) Pollinator preference for symmetrical flowers and sexual selection in plants. *Oikos* **73**, 15–22. doi:10.2307/3545720

Momose K, Yumoto T, Nagamitsu T, Kato M, Nagamasu H, Sakai S, Harrison RD, Itioka T, Hamid AA, Inoue T (1998) Pollination biology in a lowland dipterocarp forest in Sarawak, Malaysia. I. Characteristics of the plant-pollinator community in a lowland dipterocarp forest. *American Journal of Botany* **85**, 1477–1501. doi:10.2307/2446404

Monteith GB (1986) Some curious insect–plant associations in Queensland. *Queensland Naturalist* **26**, 105–114.

Montero-Castaño A, Vilà M (2012) Impact of landscape alteration and invasions on pollinators: a meta analysis. *Journal of Ecology* **100**, 884–893. doi:10.1111/j.1365-2745.2012.01968.x

Moog U, Fiala B, Federle W, Maschwitz U (2002) Thrips pollination of the dioecious ant plant *Macaranga hullettii* (Euphorbiaceae) in Southeast Asia. *American Journal of Botany* **89**, 50–59. doi:10.3732/ajb.89.1.50

Moore C, Gross CL (2012) Great big hairy bees! Regulating the European bumblebee, 'Bombus Terrestris L.'. What does it say about the precautionary principle? *International Journal of Rural Law and Policy* **1**, 1–19. doi:10.5130/ijrlp.i1.2012.2627

Moragues E, Traveset A (2005) Effects of *Carpobrotus* spp. on the pollination success of native plant species of the Baleartic Islands. *Biological Conservation* **122**, 611–619. doi:10.1016/j.biocon.2004.09.015

Moreau CS, Bell CD, Vila R, Archibald SB, Pierce NE (2006) Phylogeny of ants: diversification in the age of angiosperms. *Science* **312**, 101–104. doi:10.1126/science.1124891

Morillo IMR, May FC, Fernández-Concha GC, Pat FM (2008) Reproductive biology of *Hechtia schottii*, a dioecious Bromeliaceae, in Mexico. *Revista de Biología Tropical* **56**, 279–289.

Morrant DS, Petit S, Schumann R (2010) Floral nectar sugar composition and flowering phenology of the food plants used by western pygmy possum, *Cercartetus concinnus*, at Innes National Park, South Australia. *Ecological Research* **25**, 579–589. doi:10.1007/s11284-009-0687-1

Morris RJ, Sinclair FH (2015) Food web structure changes with elevation but not rainforest stratum. *Ecography* **38**, 792–802. doi:10.1111/ecog.01078

Moulds MS, Tuttle JP, Lane DA (2020) *Hawkmoths of Australia: Identification, Biology and Distribution.* Monographs on Australian Lepidoptera. Volume 13. CSIRO Publishing, Melbourne.

Mound LA, Marullo R (1996) The thrips of Central and South America: an introduction. *Memoirs on Entomology, International* **6**, 1–488.

Mound LA, Marullo R (1998) Biology and identification of Aeolothripidae (Thysanoptera) in Australia. *Invertebrate Taxonomy* **12**, 929–950. doi:10.1071/IT97014

Mound LA, Palmer JM (1990) Two new Thripidae (Thysanoptera) from male flowers of *Araucaria* and *Casuarina* in Australia and Hawaii. *Entomologist's Monthly Magazine* **126**, 1–7.

Mound LA, Terry I (2001) Thrips pollination of the central Australian *Macrozamia macdonnellii* (Cycadales). *International Journal of Plant Sciences* **162**, 147–154. doi:10.1086/317899

Mound LA, den Hollander E, den Hollander L (1998) Do thrips help pollinate *Macrozamia* cycads? *Victorian Entomologist* **28**, 86–88.

Murawski DA, Hamrick JL, Hubbell SP, Foster RB (1990) Mating systems of two Bombacaceous trees of a neotropical moist forest. *Oecologia* **82**, 501–506. doi:10.1007/BF00319793

Murawski DA, Dayanandan B, Bawa KS (1994) Outcrossing rates of two endemic *Shorea* species from Sri Lankan tropical rain forests. *Biotropica* **26**, 23–29. doi:10.2307/2389107

Murphy HT, Metcalfe DJ, Bradford MG, Ford AF, Galway KE, Sydes TA, Westcott DJ (2008) Recruitment dynamics of invasive species in rainforest habitats following Cyclone Larry. *Austral Ecology* **33**, 495–502. doi:10.1111/j.1442-9993.2008.01904.x

Nagalingum NS, Marshall CR, Quental TB, Rai HS, Little DP, Mathews S (2011) Recent synchronous radiation of a living fossil. *Science* **334**, 796–799. doi:10.1126/science.1209926

Nagamitsu T, Inoue T (1997) Cockroach pollination and breeding system of *Uvaria elmeri* (Annonaceae) in a lowland mixed-dipterocarp forest in Sarawak. *American Journal of Botany* **84**, 208–213. doi:10.2307/2446082

Nakase Y, Kato M (2012) A nocturnal *Provespa* wasp species as the probable pollinator of the epiphytic orchid *Coelogyne fimbricata*. *Entomological Science* **15**, 253–256. doi:10.1111/j.1479-8298.2011.00504.x

Naumann ID (Ed.) (1991) *The Insects of Australia*. CSIRO Publishing, Melbourne.

Nayak KG, Davidar P (2010) Pollination and breeding systems of woody plant species in tropical evergreen forests, Southern India. *Flora – Morphology, Distribution, Functional Ecology of Plants* **205**, 745–753. doi:10.1016/j.flora.2009.12.041

Neal JM, Hardner CM, Gross CL (2010) Population demography and fecundity do not decline with habitat fragmentation in the rainforest tree *Macadamia integrifolia* (Proteaceae). *Biological Conservation* **143**, 2591–2600. doi:10.1016/j.biocon.2010.06.029

Neal PR, Dafni A, Giurfa M (1998) Floral symmetry and its role in plant-pollinator systems: terminology, distribution, and hypotheses. *Annual Review of Ecology and Systematics* **29**, 345–373. doi:10.1146/annurev.ecolsys.29.1.345

Neff JL, Simpson BB, Evenhuis NL, Dieringer G (2003) Character analysis of adaptations for tarsal pollen collection in the Bombyliidae (Insecta: Diptera): The benefits of putting your foot in your mouth. *Zootaxa* **157**, 1–14. doi:10.11646/zootaxa.157.1.1

Negrón-Juárez RI, Chambers JQ, Hurt GC, Annane B, Cocke S, Powell M, Stott M, Goosem S, Metcalfe DJ, Saatchi SS (2014) Remote sensing assessment of forest disturbance across complex mountainous terrain: the pattern and severity of impacts of Tropical Cyclone Yasi on Australian rainforests. *Remote Sensing of Environment* **6**, 5633–5649. doi:10.3390/rs6065633

Negrutiu I, Vyskot B, Barbacar N, Georgiev S, Moneger F (2001) Dioecious plants. A key to the early events of sex chrome evolution. *Plant Physiology* **127**, 1418–1424.

Neilan W, Catterall CP, Kanowski J, McKenna S (2006) Do frugivorous birds assist rainforest succession in weed-dominated oldfield regrowth of subtropical Australia. *Biological Conservation* **129**, 393–407. doi:10.1016/j.biocon.2005.11.007

NESP Earth Systems and Climate Change Hub (2019) *Climate change impacts on the Gondwana Rainforests of Australia (workshop report), Earth Systems and Climate Change Hub Report No. 8*. NESP Earth Systems and Climate Change Hub, Australia. Aspendale, Victoria.

Newstrom L, Robertson A (2005) Progress in understanding pollination systems in New Zealand. *New Zealand Journal of Botany* **43**, 1–59. doi:10.1080/0028825X.2005.9512943

Newstrom LE, Frankie GW, Baker HG (1994) A new classification for plant phenology based on flowering patterns in lowland tropical rain forest at La Selva, Costa Rica. *Biotropica* **26**, 141–159. doi:10.2307/2388804

Nicholson N, Nicholson H (1985–1994) *Australian Rainforest Plants*. Vols 1–6. Terania Rainforest Publishing, The Channon, NSW.

Nicolson SW (1994) Pollen feeding in the eucalypt nectar fly, *Drosophila flavohirta*. *Physiological Entomology* **19**, 58–60. doi:10.1111/j.1365-3032.1994.tb01074.x

Nicolson SW, Van Wyk B-E (1998) Nectar sugars in Proteaceae: patterns and processes. *Australian Journal of Botany* **46**, 489–504. doi:10.1071/BT97039

Niklas KJ (1985) The aerodynamics of wind pollination. *Botanical Review* **51**, 328–386. doi:10.1007/BF02861079

Niklas KJ, Crepet WL (2020) Morphological (and not anatomical or reproductive) features define early vascular plant phylogenetic relationships. *American Journal of Botany* **107**, 477–488. doi:10.1002/ajb2.1440

Niklas KJ, Norstog K (1984) Aerodynamics and pollen grain depositional patterns on cycad megastrobili: implications on the reproduction of three cycad genera (*Cycas, Dioon* and *Zamia*). *Botanical Gazette* **145**, 92–104. doi:10.1086/337431

Nishiyama T (2007) Evolutionary development biology of non-flowering land plants. *International Journal of Plant Sciences* **168**, 37–47. doi:10.1086/509609

Nixon KC, Crepet WL, Stevenson D, Friis EM (1994) A reevaluation of seed plant phylogeny. *Annals of the Missouri Botanical Garden* **81**, 484–533. doi:10.2307/2399901

Noble T, Bunbury N, Kaiser-Bunbury CN, Bell DJ (2011) Ecology and co-existence of two endemic day gecko (*Phelsuma*) species in Seychelles native palm forest. *Journal of Zoology* **283**, 73–80. doi:10.1111/j.1469-7998.2010.00751.x

Norgate M, Boyd-Gerny S, Simonov V, Rose RGP, Heard TA, Dyer AG (2010) Ambient temperature influences Australian native stingless bee (*Trigona carbonaria*) preference for warm nectar. *PLoS One* **5**, e12000. doi:10.1371/journal.pone.0012000

Norstog KJ, Fawcett KS (1989) Insect-cycad symbioisis and its relation to pollination of *Zamia furfuracea* by *Rhopalotria mollis*. *American Journal of Botany* **76**, 1380–1394. doi:10.1002/j.1537-2197.1989.tb15117.x

Norstog KJ, Nicholls TJ (1997) *The Biology of Cycads*. Cornell University Press, Ithaca, New York.

Nyhagen DF, Kragelund C, Olesen JM, Jones CG (2001) Insular interactions between lizards and flowers: flower visitation by an endemic Mauritian gecko. *Journal of Tropical Ecology* **17**, 755–761. doi:10.1017/S0266467401001560

O'Brien DM, Boggs CL, Fogel ML (2003) Pollen feeding in the butterfly *Heliconius charitonia*: isotopic evidence for essential amino acid transfer from pollen to eggs. *Proceedings. Biological Sciences* **270**, 2631–2636. doi:10.1098/rspb.2003.2552

Oberprieler RG (1995) The weevils (Coleoptera: Curculionoidea) associated with cycads 1. classification, relationships, and biology. In *Proceedings of the Third International Conference on Cycad Biology*. (Ed. P Voreter) pp. 295–334. Cycad Society of South Africa, Stellenbosch.

Ödeen A, Håstad O (2010) Pollinating birds differ in spectral sensitivity. *Journal of Comparative Physiology* **196**, 91–96. doi:10.1007/s00359-009-0474-z

Ohashi K, Makino TT, Arikawa K (2015) Floral colour change in the eyes of pollinators: testing possible contrasts and correlated evolution. *Functional Ecology* **29**, 1144–1155. doi:10.1111/1365-2435.12420

Okada T, Carson HL (1980) Drosophilidae associated with flowers in Papua New Guinea. II. *Alocasia* (Araceae). *Pacific Insects* **22**, 217–236.

Oldroyd BP, Lawler SH, Crozier RH (1994) Do feral honey bees (*Apis mellifera*) and regent parrots (*Polytelis anthopeplus*) compete for nest sites? *Australian Journal of Ecology* **19**, 444–450. doi:10.1111/j.1442-9993.1994.tb00510.x

Olesen JM, Valido A (2003) Lizards as pollinators and seed dispersers: an island phenomenon. *Trends in Ecology & Evolution* **18**, 177–181. doi:10.1016/S0169-5347(03)00004-1

Olesen JM, Eskildsen LI, Venkatasamy S (2002) Invasion of pollination networks on oceanic islands: importance of invader complexes and endemic super generalists. *Diversity & Distributions* **8**, 181–192. doi:10.1046/j.1472-4642.2002.00148.x

Ollerton J, Liede S (1997) Pollination systems in the Asclepiadaceae: a survey and preliminary analysis. *Biological Journal of the Linnean Society. Linnean Society of London* **62**, 593–610. doi:10.1111/j.1095-8312.1997.tb00324.x

Ollerton J, Alarcón R, Waser NM, Price MV, Watts S, Cranmer L, Hingston A, Peter CI, Rotenberger J (2009) A global test of the pollination syndrome hypothesis. *Annals of Botany* **103**, 1471–1480. doi:10.1093/aob/mcp031

Ollerton J, Winfree R, Tarrant S (2011) How many flowering plants are pollinated by animals? *Oikos* **120**, 321–326. doi:10.1111/j.1600-0706.2010.18644.x

Ollerton J, Rech AR, Waser NM, Price MV (2015) Using the literature to test pollination syndromes – some methodological cautions. *Journal of Pollination Ecology* **16**, 119–125. doi:10.26786/1920-7603(2015)17

Olsson M, Shine R, Ba'K-Olsson E (2000) Lizards as a plant's 'hired help': letting pollinators in and seeds out. *Biological Journal of the Linnean Society* **71**, 191–202.

Ondei S, Prior LD, Vigilante T, Bowman DMJS (2016) Post-fire resprouting strategies of rainforest and savanna saplings along the rainforest – savanna boundary in the Australian monsoon tropics. *Plant Ecology* **217**, 711–724. doi:10.1007/s11258-015-0531-3

Opedal ØH (2019) The evolvability of animal-pollinated flowers: towards predicting adaptation to novel pollinator communities. *New Phytologist* **221**, 1128–1135. doi:10.1111/nph.15403

Ornduff R (1989) Size distribution and coning behaviour of the Australian cycad *Lepidozamia peroffskyana*. *Australian Journal of Ecology* **14**, 241–245. doi:10.1111/j.1442-9993.1989.tb01432.x

Ornduff R (1990) Geographic variation in reproductive behaviour and size structure of the Australian cycad *Macrozamia communis* (Zamiaceae). *American Journal of Botany* **77**, 92–99. doi:10.1002/j.1537-2197.1990.tb13531.x

Osborne R, Gorelick R (2007) Sex change in cycads: cases, causes, and chemistry. *Memoirs of the New York Botanical Garden* **97**, 335–345.

Osunkoya O (1999) Population structure and breeding biology in relation to conservation in the dioecious *Gardenia actinocarpa* (Rubiaceae): a rare shrub of North Queensland rainforest. *Biological Conservation* **88**, 347–359. doi:10.1016/S0006-3207(98)00115-3

Packham JM, Elliott HJ, Bashford R (2008) Rate of spread of Myrtle wilt desease in undisturbed Tasmanian rainforests. *Australian Forestry* **71**, 64–69. doi:10.1080/00049158.2008.10676272

Parker KL (1905) *The Euahlayi Tribe: A Study of Aboriginal Life in Australia*. Archibold Constable, London.

Parkes T, Delaney M, Dunphy M, Woodford R, Bower H, Bower S, Bailey D, Joseph R, Nagle J, Roberts T, Lymburner S, McDonald T (2012) Big Scrub: a cleared landscape in transition back to forest? *Ecological Management & Restoration* **13**, 212–223. doi:10.1111/emr.12008

Parry-Jones KA, Augee ML (1991a) Food selection by grey-headed flying foxes (*Pteropus poliocephalus*) occupying a summer colony site near Gosford, New South Wales. *Wildlife Research* **18**, 111–124. doi:10.1071/WR9910111

Parry-Jones KA, Augee ML (1991b) The diet of flying-foxes in the Sydney and Gosford areas of New South Wales, based on sighting reports 1986–1990. *Australian Zoologist* **27**, 49–54. doi:10.7882/AZ.1991.006

Patiño S, Grace J, Bänzinger H (2000) Endothermy by flowers of *Rhizanthes lowii* (Rafflesiaceae). *Oecologia* **124**, 149–155. doi:10.1007/s004420050001

Patiny S (Ed.) (2012) *Evolution of Plant-Pollinator Relationships*. Cambridge University Press, Cambridge.

Paton DC (1981) The significance of pollen in the diet of the New Holland honeyeater *Phylidonyris novaehollandiae* (Aves: Meliphagidae). *Australian Journal of Zoology* **29**, 217–224. doi:10.1071/ZO9810217

Paton DC (1993) Honeybees in the Australian environment. *Bioscience* **43**, 95–103. doi:10.2307/1311970

Paton DC (2000) Disruption of bird-plant pollination systems in southern Australia. *Conservation Biology* **14**, 1232–1234.

Paton DC, Turner V (1985) Pollination of *Banksia ericifolia* Smith: Birds, mammals and insects as pollen vectors. *Australian Journal of Botany* **33**, 271–286. doi:10.1071/BT9850271

Paudel BR, Kessler A, Shresha M, Zhao JL, Li Q-J (2019) Geographic isolation, pollination syndromes, and pollinator generalisation in Himalayan *Roscoea* spp. (Zingiberaceae). *Ecosphere* **10**, e02943. doi:10.1002/ecs2.2943

Paulus HF (2019) Speciation, pattern recognition and the maximization of pollination: general questions and answers given by the reproductive biology of the orchid genus *Ophrys*. *Journal of Comparative Physiology. A, Neuroethology, Sensory, Neural, and Behavioral Physiology* **205**, 285–300. doi:10.1007/s00359-019-01350-4

Pauw A (1998) Pollen transfer on birds' tongues. *Nature* **394**, 731–732. doi:10.1038/29421

Pauw A (2006) Floral syndromes accurately predict pollination by a specialized oil-collecting bee (*Rediviva peringueyi*, Melittidae) in a guild of South African orchids (Coyciinae). *American Journal of Botany* **93**, 917–926. doi:10.3732/ajb.93.6.917

Pavlacky DC, Possingham HP, Goldizen AW (2015) Integrating life history traits and forest structure to evaluate the vulnerability of rainforest birds along gradients of deforestation and fragmentation in eastern Australia. *Biological Conservation* **188**, 89–99. doi:10.1016/j.biocon.2014.10.020

Peakall R (1990) Responses of male *Zaspilothynnus trilobatus* Turner wasps to females and the sexually deceptive orchid it pollinates. *Functional Ecology* **4**, 159–167. doi:10.2307/2389335

Peakall R, Whitehead MR (2014) Floral odour and reproductive isolation in sexually deceptive orchids. *Annals of Botany* **113**, 341–355. doi:10.1093/aob/mct199

Peakall R, Beattie AJ, James SH (1987) Pseudocopulation of an orchid by male ants: a test of two hypotheses accounting for the rarity of ant pollination. *Oecologia* **73**, 522–524. doi:10.1007/BF00379410

Peakall R, Handel SN, Beattie AJ (1991) The evidence for, and importance of, ant pollination. In *Ant–Plant Interactions*. (Eds CR Huxley and DF Cutler) pp. 421–429. Oxford University Press, Oxford.

Peakall R, Ebert D, Scott LJ, Meagher PF, Offord CA (2003) Comparative genetic study confirms exceptionally low genetic variation in the ancient and endangered relictual conifer, *Wollemia nobilis* (Araucariaceae). *Molecular Ecology* **12**, 2331–2343. doi:10.1046/j.1365-294X.2003.01926.x

Peakall R, Ebert D, Poldy J, Barrow RA, Francke W, Bower C, Schiestl P (2010) Pollinator specificity, floral odour chemistry and the phylogeny of Australian sexually deceptive *Chiloglottis* orchids: implications for pollinator-driven speciation. *New Phytologist* **188**, 437–450. doi:10.1111/j.1469-8137.2010.03308.x

Pederson HA (1995) Anthecological observations on *Dendrochilum longibracteatum* – a species pollinated by facultatively anthophilous insects. *Lindleyana* **10**, 19–28.

Peitsch D, Fietz A, Hertel H, de Souza J, Ventura DF, Menzel R (1992) The spectral input systems of hymenopteran insects and their receptor-based colour vision. *Journal of Comparative Physiology. A, Neuroethology, Sensory, Neural, and Behavioral Physiology* **170**, 23–40. doi:10.1007/BF00190398

Pellmyr O (1989) The cost of mutualism: interactions between *Trollius europaeus* and its pollinating parasite. *Oecologia* **78**, 53–59. doi:10.1007/BF00377197

Pellmyr O, Thompson JM (1992) Multiple occurrence of mutualism in the yucca moth lineage. *Proceedings of the National Academy of Sciences of the United States of America* **89**, 2927–2929. doi:10.1073/pnas.89.7.2927

Pellmyr O, Tang W, Groth I, Bergstrom G, Thien LBDA (1991) Cycad cone and angiosperm floral volatiles – inferences for the evolution of insect pollination. *Biochemical Systematics and Ecology* **19**, 623–627.

Pellmyr O, Kjellberg F, Herre EA, Kawakita A, Hembry DH, Holland JN, Terrazas T, Clement W, Segraves KA, Althoff DM (2020) Active pollination drives selection for reduced pollen-ovule ratios. *American Journal of Botany* **107**, 164–170. doi:10.1002/ajb2.1412

Pemberton RW (2013) Pollination of slipper orchids (Cypripedioideae): a review. *Lankesteriana* **13**, 65–73.

Peñalver E, Labandeira CC, Barrón E, Delclòs X, Nel P, Nel A, Tafforeau P, Soriano C (2012) Thrips pollination of Mesozoic gymnosperms. *Proceedings of the National Academy of Sciences of the United States of America* **109**, 8623–8628. doi:10.1073/pnas.1120499109

Penz CM, Krenn HW (2000) Behavioral adaptations to pollen-feeding in *Heliconius* butterflies (Nymphalidae, Heliconiinae): an experiment using lantana flowers. *Journal of Insect Behavior* **13**, 865–880. doi:10.1023/A:1007814618149

Perez-Mellado V, Ortega F, Martin-Garcia S, Perera A, Cortazar G (2000) Pollen load and transport by the insular lizard *Podarcis lilfordi* (Squamata, Lacertidae) in coastal inlets of Menorca (Balearic Islands, Spain). *Israel Journal of Zoology* **46**, 193–200. doi:10.1092/QMY9-PXWF-AG43-RP6F

Perret M, Chautems A, Spichger R, Peixoto M, Savolainen V (2001) Nectar sugar composition in relation to pollination syndromes in Sinningieae (Gesneriaceae). *Annals of Botany* **87**, 267–273. doi:10.1006/anbo.2000.1331

Perry DR, Starrett A (1980) The pollination ecology and blooming strategy of a neotropical emergent tree, *Dipteryx panamensis*. *Biotropica* **12**, 307–313. doi:10.2307/2387702

Petanidou T, Ellis WN (1993) Pollinating fauna of a phryganic ecosystem: composition and diversity. *Biodiversity Letters* **1**, 9–22. doi:10.2307/2999643

Peters RS, Krogmann L, Mayer C, Donath A, Gunkel S, Meusemann K, Kozlov A, Podsiadlowski L, Petersen M, *et al.* (2017) Evolutionary history of the Hymenoptera. *Current Biology* **27**, 1013–1018. doi:10.1016/j.cub.2017.01.027

Peterson JH, Roitberg BD (2006) Impact of resource levels on sex ratio and resource allocation in a solitary bee, *Megachile rotunda*. *Environmental Entomology* **35**, 1404–1410. doi:10.1093/ee/35.5.1404

Phillips RD, Hopper SD, Dixon KW (2009) Pollination ecology and the possible impacts of environmental change in the Southwest Australian Biodiversity Hotspot. *Philosophical Transactions of the Royal Society of London. Series B, Biological Sciences* **365**, 1539.

Pierce R, Ogle C (2017) Musky Rat Kangaroos and other vertebrates feeding from the flowers of the root parasite 'Balanophora fungosa'. *North Queensland Naturalist* **47**, 14–20.

Pimentel MRC, Antonini Y, Martins RP, Lachance M-A, Rosa CA (2005) *Candida riodocensis* and *Candida cellae* two new yeast species from the *Starmerella* clade associated with solitary bees in the Atlanta rain forest of Brazil. *FEMS Yeast Research* **5**, 875–879. doi:10.1016/j.femsyr.2005.03.006

Pires ND, Dolan L (2012) Morphological evolution in land plants: new designs with old genes. *Philosophical Transactions of the Royal Society of London. Series B, Biological Sciences* **367**, 508–518. doi:10.1098/rstb.2011.0252

Pisanu PC, Gross CL, Flood L (2009) Reproduction in wild populations of the threatened tree *Macadamia tetraphylla*: inter-population pollen enriches fecundity in a declining species. *Biotropica* **41**, 391–398. doi:10.1111/j.1744-7429.2008.00484.x

Poinar G (2009) *Melittosphex* (Hymenoptera: Melittosphecidae), a primitive bee and not a wasp. *Palaeontology* **52**, 483–484. doi:10.1111/j.1475-4983.2008.00840.x

Poinar G (2020) Disciscapidae fam. nov. (Hymenoptera: Apoidea), a new family of stem lineage bees with associated beetle triungulins in mid-Cretaceous Burmese amber. *Palaeodiversity* **13**, 1–19. doi:10.18476/pale.v13.a1

Ponomarenko AG (2004) Beetles (Insecta, Coleoptera) of the Late Permian and Early Triassic. *Palaeontology Journal* **38**, S185–S196.

Poska A, Pidek IA (2010) Pollen dispersal and deposition characteristics of *Abies alba, Fagus sylvatica* and *Pinus sylvestris*, Roztocze region (SE Poland). *Vegetation History and Archaeobotany* **19**, 91–101. doi:10.1007/s00334-009-0230-x

Powney GD, Carvel C, Edwards M, Morris RKA, Roy HE, Woodcock BA, Isaac NJB (2019) Widespread losses of pollinating insects in Britain. *Nature Communications* **10**, 1018. doi:10.1038/s41467-019-08974-9

Pratt Hd (2005) *The Hawaiain Honeyeaters*. Oxford University Press, New York.

Prendergast K (2018a) *Euryglossina* (*Euryglossina*) *perpusilla* (Hymenoptera: Colletidae: Euryglossinae) nesting in pre-formed cavities in *Banksia attenuata* (Proteaceae). *Journal of Melittology* **81**, 1–5. doi:10.17161/jom.v0i81.6936

Prendergast K (2018b) Nesting biology of 'Megachile ignita' Smith 1853 (Hymenoptera: Megachilidae) in artificial nesting blocks in urbanised southwestern Australia. *Australian Entomologist* **45**, 139–148.

Primack RB (1978) Variability in New Zealand montane and alpine pollinator assemblages. *New Zealand Journal of Ecology* **1**, 66–73.

Primack RB (1985) Longevity of individual flowers. *Annual Review of Ecology and Systematics* **16**, 15–37. doi:10.1146/annurev.es.16.110185.000311

Primack RB, Hall P (1992) Biodiversity and forest change in Malaysian Borneo. *Bioscience* **42**, 829–837. doi:10.2307/1312082

Primack RB, Duke NC, Tomlinson PB (1981) Floral morphology in relation to pollination ecology of five Queensland coastal plants. *Austrobaileya* **1**, 346–355.

Proctor M, Yeo P (1975) *The Pollination of Flowers*. Collins, London.

Proença CEB, Gibbs P (1994) Reproductive biology of eight sympatric Myrtaceae from Central Brazil. *New Phytologist* **126**, 343–354. doi:10.1111/j.1469-8137.1994.tb03954.x

Pulawski WJ (2018) *Catalog of Sphecidae* sensu lato. https://www.calacademy.org/scientists/projects/catalog.of.sphecidae (accessed 10.07.18).

Punekar J, Kellar G, Khozyem HM, Adatte T, Font E, Spangenberg J (2016) A multi-proxy approach to decode the end-Cretaceous mass extinction. *Palaeogeography, Palaeoclimatology, Palaeoecology* **441**, 116–136. doi:10.1016/j.palaeo.2015.08.025

Pyke GH, Kalman JRM, Bordin DM, Blanes L, Doble PA (2020) Patterns of floral nectar standing crops allow plants to manipulate their pollinators. *Scientific Reports* **10**, 1660. doi:10.1038/s41598-020-58102-7

Qiu Y-L, Dombrovska O, Lee J, Li L, Whitlock BA, Quadroni-Bernasconi F, Rest JS, Davis CC, Borsch T, Hilu KW, Renner SS, Soltis DE, Soltis PS, Zanis MJ, Cannone JJ, Gutell RR, Powell M, Savolainen V, Chatron LW, Chase MW (2005) Phylogenetic analyses of basal angiosperms based on nine plastid mitochondrial, and nuclear genes. *International Journal of Plant Sciences* **166**, 815–842. doi:10.1086/431800

Raguso RA (2004) Flowers as sensory billboards: progress towards an integrated understanding of floral advertisement. *Current Opinion in Plant Biology* **7**, 434–440. doi:10.1016/j.pbi.2004.05.010

Raguso RA (2020) Don't forget the flies: dipteran diversity and its consequences for floral ecology and evolution. *Applied Entomology and Zoology* **55**, 1–7. doi:10.1007/s13355-020-00668-9

Raju AJS, Jonathan KH (2010) Anemophily, accidental cantharophily, seed dispersal and seedling ecology of *Cycas sphaerica* Roxb. (Cycadaceae), a data-deficient red-listed species of northern Eastern Ghats. *Current Research* **99**, 1105–1111.

Raju AJS, Rao NG (2011) Taxonomic aspects and coning ecology of *Cycas circinalis* L. (Cycadales: Cycadaceae), a threatened species of India. *Journal of Threatened Taxa* **3**, 1425–1431. doi:10.11609/JoTT.o2372.1425-31

Raju AJS, Krishna JR, Chandra PH (2014) Reproductive ecology of *Syzygium alternifolium* (Myrtaceae), an endemic and endangered tropical tree species in the southern Eastern Ghats of India. *Journal of Threatened Taxa* **6**, 6153–6171. doi:10.11609/JoTT.o3768.6153-71

Ramírez SR, Gravendeel B, Singer RB, Marshall CR, Pierce NE (2007) Dating the origin of the Orchidaceae from a fossil orchid with its pollinator. *Nature* **448**, 1042–1045.

Ramírez VM, Ayala R, González HD (2016) Temporal variation in native bee diversity in the tropical sub-deciduous forest of the Yucatan Peninsula, Mexico. *Tropical Conservation Science* **9**, 718–734. doi:10.1177/194008291600900210

Ramírez W (1980) Evolution of the monoecious and dioecious habit in *Ficus* (Moraceae). *Brenesia* **18**, 207–216.

Ramírez-Barahona S, Sauquett H, Magallón S (2020) The delayed and geographically heterogeneous diversification of flowering plants. *Nature Ecology and Evolution* **4**, 1232–1238. doi:10.1038/s41559-020-1241-3

Ramsey M (1988) Differences in pollinator effectiveness of birds and insects visiting *Banksia menziesii* (Proteaceae). *Oecologia* **76**, 119–124. doi:10.1007/BF00379609

Ramsey M (1995) Ant pollination of the perenial herb *Blandfordia grandiflora* (Liliaceae). *Oikos* **74**, 265–272. doi:10.2307/3545656

Rathcke B, Lacey EP (1985) Phenological patterns of terrestrial plants. *Annual Review of Ecology and Systematics* **16**, 179–214. doi:10.1146/annurev.es.16.110185.001143

Raven PH, Axelrod DI (1974) Angiosperm biogeography and past continental movements. *Annals of the Missouri Botanical Garden* **61**, 539–637. doi:10.2307/2395021

Raven PH, Evert RF, Eichhorn S (1992) *Biology of Plants*. Worth, New York.

Read J (1989) Phenology and germination in some rainforest canopy species at Mt Field National Park, Tasmania. *Papers and Proceedings of the Royal Society of Tasmania* **123**, 211–221. doi:10.26749/rstpp.123.211

Real LA (1981) Nectar availability and bee foraging on *Ipomoea* (Convolulaceae). *Biotropica* **13**(Supp.), 64–69. doi:10.2307/2388072

Rech AR, Dalsgaard B, Sandel B, Sonne J, Svenning J-C, Holmes N, Ollerton J (2016) The macroecology of animal versus wind pollination: ecological factors are more important than historical climate stability. *Plant Ecology & Diversity* **9**, 253–262. doi:10.1080/17550874.2016.1207722

Regal PJ (1982) Pollination by wind and animals: ecology of geographical patterns. *Annual Review of Ecology and Systematics* **13**, 497–524. doi:10.1146/annurev.es.13.110182.002433

Reid JB, Hill RS (2005) *Vegetation of Tasmania*. Australian Biological Resources Study, Canberra.

Reiter N, Bohman B, Freestone M, Brown GR, Phillips RD (2019) Pollination by nectar-foraging thynnine wasps of the endangered *Caladenia arenaria* and *Caladenia concolor* (Orchidaceae). *Australian Journal of Botany* **67**, 490–500. doi:10.1071/BT19033

Remizowa MV, Sokoloff DD, Rudall PJ (2010) Evolutionary history of the monocot flower. *Annals of the Missouri Botanical Garden* **97**, 617–645. doi:10.3417/2009142

Renner SS (1998) Phylogenetic affinities of Monimiaceae based on cpDNA gene and spacer sequences. *Perspectives in Plant Ecology, Evolution and Systematics* **1**, 61–77.

Renner SS (1999) Circumscription and phylogeny of the Laurales: evidence from molecular and morphological data. *American Journal of Botany* **86**, 1301–1315. doi:10.2307/2656778

Renner SS, Ricklefs RE (1995) Dioecy and its correlates in the flowering plants. *American Journal of Botany* **82**, 596–606. doi:10.1002/j.1537-2197.1995.tb11504.x

Renner SS, Strijk JS, Strasberg D, Thébaud C (2010) Biogeography of the Monimiaceae (Laurales): a role for East Gondwana and long-distance dispersal, but not West Gondwana. *Journal of Biogeography* **37**, 1227–1238. doi:10.1111/j.1365-2699.2010.02319.x

Rensing SA (2018) Plant evolution: phylogenetic relationships between the earliest land plants. *Current Biology* **28**, R210–R213. doi:10.1016/j.cub.2018.01.034

Rentz D (2010) *A Guide to the Katydids of Australia*. CSIRO Publishing, Melbourne.

Rering CC, Beck JJ, Hall GW, McCartney MM, Vannette RL (2018) Nectar-inhibiting microorganisms influence nectar volatile composition and attractiveness to a generalist pollinator. *New Phytologist* **220**, 750–759. doi:10.1111/nph.14809

Reyes E (2016) Evolutionary history of floral key innovations in angiosperms. Doctoral thesis. Botanics. Université Paris-Saclay.

Ricklefs RE, Renner SS (1994) Species richness within families of flowering plants. *Evolution* **48**, 1619–1636. doi:10.1111/j.1558-5646.1994.tb02200.x

Rico-Gray V, Oliveira PS, Parra-Tabla V, Cuautle M, Díaz-Castelozo C (2004) Ant-plant interactions: their seasonal variation and effects on plant fitness. In *Coastal Dunes, Ecology and Conservation. Ecological Studies*. Vol. 171. (Eds MI Martínez and NP Psuty) pp. 219–239. Springer-Verlag, Berlin Heidelberg.

Riffell J, Lei H, Christensen T, Hildebrand JG (2009) Characterization and coding of behaviorally significant odor mixtures. *Current Biology* **19**, 335–340. doi:10.1016/j.cub.2009.01.041

Robertson AW, Ladley JJ, Kelly D (2005) Effectiveness of short-tongued bees as pollinators of apparently ornithophilous New Zealand mistletoes. *Austral Ecology* **30**, 298–309. doi:10.1111/j.1442-9993.2005.01474.x

Roemer R, Terry I, Chockley C, Jacobson J (2005) Experimental evaluation and thermo-physical analysis of thermogenesis in male and female cycad cones. *Oecologia* **144**, 88–97. doi:10.1007/s00442-005-0045-0

Roemer RB, Terry LI, Walter GH (2008) Unstable, self-limiting thermochemical temperature oscillations in *Macrozamia* cycads. *Plant, Cell & Environment* **31**, 769–782. doi:10.1111/j.1365-3040.2008.01795.x

Rohwer JG (2000) Towards a phylogenetic classification of the Lauraceae: evidence from *mat*K sequences. *Systematic Botany* **25**, 60–71. doi:10.2307/2666673

Ronse De Craene LP, Soltis PS, Soltis DE (2003) Evolution of floral structures in Basal Angiosperms. *International Journal of Plant Sciences* **164**, S329–S363. doi:10.1086/377063

Rønsted N, Weiblen GD, Clement WL, Zerega NJC, Savolainen V (2008) Reconstructing the phylogeny of figs (*Ficus*, Moraceae) to reveal the history of the fig pollination mutualism. *Symbiosis* **45**, 45–55.

Rosas-Guerrero V, Aguilar R, Martén-Rodriguez S, Ashworth L, Lopezaraiza-Mikel M, Bastida JM, Quesada M (2014) A quantitative review of pollination syndromes: do floral traits predict effective pollinators? *Ecology Letters* **17**, 388–400. doi:10.1111/ele.12224

Rose JP, Kleist TJ, Löfstrand SD, Drew BT, Schönenberger J, Sytsma KJ (2018) Phylogeny, historical biogeography, and diversification of angiosperm order Ericales suggest ancient Neotropical and East Asian connections. *Molecular Phylogenetics and Evolution* **122**, 59–79. doi:10.1016/j.ympev.2018.01.014

Rosenthal GA, Janzen DH (Eds) (1979) *Herbivores, Their Interaction with Secondary Plant Metabolites*. Academic Press, New York.

Rossetto M, Kooyman RM (2005) The tension between dispersal and persistence regulates the current distribution of rare palaeo-endemic rain forest flora: a case study. *Journal of Ecology* **93**, 906–917. doi:10.1111/j.1365-2745.2005.01046.x

Rossetto M, Gross C, Jones R, Hunter J (2003) Developing genetic-based management strategies for a rare clonal tree (*Elaeocarpus williamsianus*) in fragmented rainforest habitats. Unpublished report to NSW National Parks and Wildlife Service.

Rossetto M, Jones R, Hunter J (2004) Genetic effects of rainforest fragmentation in an early successional tree (*Elaeocarpus grandis*). *Heredity* **93**, 610–618. doi:10.1038/sj.hdy.6800585

Rossetto M, Kooyman R, Yap J-YS, Laffan SW (2015) From ratites to rats: the size of fleshy fruits shapes species distributions and continental rainforest assembly. *Proceedings. Biological Sciences* **282**, 20151998. doi:10.1098/rspb.2015.1998

Roth WE (1897) *Ethnological Studies among North-West-Central Queensland Aborigines.* Government Printer, Brisbane.

Roubik DW (1989) *Ecology and Natural History of Tropical Bees.* Cambridge University Press, New York.

Roulston TH, Cane JH (2000) Pollen nutritional content and digestibility for animals. *Plant Systematics and Evolution* **222**, 187–209. doi:10.1007/BF00984102

Rudall PJ, Sokoloff DD, Remizowa MV, Couran JG, Davis JI, Macfarlane TD, Stevenson DW (2007) Morphology of Hydatellaceae, an anomalous aquatic family recently recognized as an early-divergent angiosperm lineage. *American Journal of Botany* **94**, 1073–1092. doi:10.3732/ajb.94.7.1073

Rudder J (1977) *Introduction to Yolŋu Science.* Galiwinku Adult Education Centre, Galiwinku, NT.

Rugkhla A, McComb JA, Jones GK (1997) Intra- and Interspecific pollination of *Santalum specatum* and *S. album. Australian Journal of Botany* **45**, 1083–1095. doi:10.1071/BT96079

Rule S, Brook BW, Haberle SG, Turney CSM, Kershaw AP, Johnson CN (2012) The Aftermath of megafaunal extinction: ecosystem transformation in Pleistocene Australia. *Science* **335**, 1483–1486. doi:10.1126/science.1214261

Russell-Smith J, Dunlop C (1987) The status of monsoon vine forests in the Northern Territory: a perspectve. In *The Rainforest Legacy.* Vol. 1. (Eds GL Werren and AP Kershaw) pp. 227–288. Australian Government Publishing Service, Canberra.

Russell-Smith J, Lee AH (1992) Plant populations and monsoon rain forest in the Northern Territory, Australia. *Biotropica* **24**, 471–487. doi:10.2307/2389010

Rydin C, Bolinder K (2015) Moonlight pollination in the gymnosperm *Ephedra* (Gnetales). *The Royal Society. Biological Letters* **11**, 20140993. doi:10.1098/rsbl.2014.0993

Rydin C, Källersjö M, Friis EM (2002) Seed plant relationships and the systematic position of Gnetales based on nuclear and chloroplast DNA: conflicting data, rooting problems, and the monophyly of conifers. *International Journal of Plant Sciences* **163**, 197–214. doi:10.1086/338321

Rye BL (2009) A reduced circumscription of *Balaustion* and description of the new genus *Cheyniana* (Myrtaceae: Chamelaucieae). *Nuytsia* **19**, 129–148.

Rye BL (2013) A revision of the south-western Australian genus *Astartea* (Myrtaceae: Chamelaucieae). *Nuytsia* **23**, 189–269.

Saarela JM, Rai HS, Doyle JA, Endress PK, Mathews S, Marchant AD, Briggs BG, Graham SW (2007) Hydatellaceae identified as a new branch near the base of the angiosperm phylogenetic tree. *Nature* **446**, 312–315. doi:10.1038/nature05612

Saffer VM (2004) Are diel patterns of nectar production and anthesis associated with other floral traits in plants visited by potential bird and mammal pollinators? *Australian Journal of Botany* **52**, 87–92. doi:10.1071/BT02056

Sakai S (2000) Reproductive phenology of gingers in a lowland mixed dipterocarp forest in Borneo. *Journal of Tropical Ecology* **16**, 337–354. doi:10.1017/S0266467400001449

Sakai S (2001) Thrips pollination of androdioecious *Castilla elastica* (Moraceae) in a seasonal tropical forest. *American Journal of Botany* **88**, 1527–1534. doi:10.2307/3558396

Sakai S (2002a) A review of brood-site pollination mutualism: plants providing breeding sites for their pollinators. *Journal of Plant Research* **115**, 161–168. doi:10.1007/s102650200021

Sakai S (2002b) *Aristolochia* spp. (Aristolochiaceae) pollinated by flies breeding on decomposing flowers in Panama. *American Journal of Botany* **89**, 527–534. doi:10.3732/ajb.89.3.527

Sakai S, Inoue T (1999) A new pollination system: dung-beetle pollination discovered in *Orchidantha inouei* (Lowiaceae Zingiberales) in Sarawak, Malaysia. *American Journal of Botany* **86**, 56–61. doi:10.2307/2656954

Sakai S, Kato M, Inoue T (1999a) Three pollination guilds and variation in floral characteristics of Bornean gingers (Zingiberaceae and Costaceae). *American Journal of Botany* **86**, 646–658. doi:10.2307/2656573

Sakai S, Momose K, Yumoto T, Nagamitsu T, Nagamasu H, Hamid AA, Nakashizuka T (1999b) Plant reproductive phenology over four years, including an episode of general flowering, in a lowland dipterocarp forest, Sarawak, Malaysia. *American Journal of Botany* **86**, 1414–1436. doi:10.2307/2656924

Sakai S, Kato M, Nagamasu H (2000) *Artocarpus* (Moraceae)-gall midge pollination mutualism mediated by a male-flower parasitic fungus. *American Journal of Botany* **87**, 440–445. doi:10.2307/2656640

Saldaña-Vásquez RA, Ortega J, Guerrero JA, Aiza-Reynoso MI, McSwiney MC, Aguilar-Rodríguez PA, Ayala-Berdon J, Zamora-Gutierrez V (2020) Ambient temperature drives sex ratio and presence of pregnant females of *Anoura geoffroyi* (Phyllstomatidae) bats living in temperate forests. *Journal of Mammology* **101**, 234–240. doi:10.1093/jmammal/gyz186

Salomo K, Smith JF, Feild TS, Samain MS, Bond L, Davidson C, Zimmers J, Neinhuis C, Wanke S (2017) The emergence of earliest angiosperms may be earlier than fossil evidence indicates. *Systematic Botany* **42**, 607–619. doi:10.1600/036364417X696438

Sampson JF (1998) Multiple paternity in *Eucalyptus rameliana* (Myrtaceae). *Heredity* **81**, 349–355. doi:10.1046/j.1365-2540.1998.00404.x

Samuelson GA (1994) Pollen consumption and digestion by leaf beetles. In *Novel Aspects of the Biology of Chrysomelidae.* (Eds PH Jolivet, ML Cox and E Petitpierre) pp. 179–183. Kluwer Academic Publishers, Netherlands.

Samways MJ, Barton PS, Birkhofer K, Chichorro F, Deacon C, Fartmann T, Fukushima CS, Gaigher R, Habel JC, *et al.* (2020) Solutions for humanity on how to conserve insects. *Biological Conservation* **242**, 108427. doi:10.1016/j.biocon. 2020.108427

Sánchez-Bayo F, Wyckhuysb KAG (2019) Worldwide decline of the entomofauna: a review of its drivers. *Biological Conservation* **232**, 8–27. doi:10.1016/j.biocon.2019.01.020

Sanderson MJ, Doyle JA (2001) Sources of error and confidence intervals in estimating the age of angiosperms from *rbcL* and 18S *rDNA* data. *American Journal of Botany* **88**, 1499–1516. doi:10.2307/3558458

Sands DPA, House S (1990) Plant/insect interactions: food webs and breeding systems. In *Australian Tropical Rainforests.* (Eds LJ Webb and J Kikkawa) pp. 88–97. CSIRO Publishing, Melbourne.

Sands DPA, Scott SE, Moffatt R (1997) The threatened Richmond birdwing butterfly (*Ornithoptera richmondia* [Gray]): a community conservation project. *Memoirs of the Museum of Victoria* **56**, 449–453. doi:10.24199/j.mmv.1997.56.37

Sapir Y, Ghara M (2017) The (relative) importance of pollinator-mediated selection for evolution of flowers. *American Journal of Botany* **104**, 1787–1789. doi:10.3732/ajb.1700321

Sargent RD, Otto Sp (2004) A phylogenetic analysis of pollination mode and the evolution of dichogamy in angiosperms. *Evolutionary Ecology Research* **6**, 1183–1199.

Sarma K, Tandon R, Shivanna KR, Ram HYM (2007) Snail pollination in *Volvulopsis nummularium. Current Science* **93**, 826–831.

Sauquet H, le Thomas A (2003) Pollen diversity and evolution in Myristicaceae (Magnoliales). *International Journal of Plant Sciences* **164**, 613–628. doi:10.1086/375424

Sauquet H, Magallón S (2018) Key questions and challenges in angiosperm macroevolution. *New Phytologist* **219**, 1170–1187. doi:10.1111/nph.15104

Sauquet H, Doyle JA, Scharaschkin T, Borsch T, Hilu KW, Chatron LW, Thomas AL (2003) Phylogenetic analysis of Magnoliales and Myristicaceae based on multiple data sets: implications for character evolution. *Botanical Journal of the Linnean Society* **142**, 125–186. doi:10.1046/j.1095-8339. 2003.00171.x

Sauquet H, Cantrill DJ, Weston PH, Barker N, Mast A, Savolainen V (2007) Bringing together the living and the dead: integrating extant and fossil biodiversity in evolutionary studies of Proteaceae (Proteales). *Botany and Plant Biology 2007 Joint Congress, July 7–11,* Chicago, IL.

Sauquet H, Weston PH, Barker NP, Anderson CL, Cantrill DJ, Savolainen V (2009) Using fossils and molecular data to reveal the origins of the Cape proteas (subfamily Proteodeae). *Molecular Phylogenetics and Evolution* **51**, 31–43. doi:10.1016/j.ympev.2008.12.013

Sauquet H, von Balthazar M, Magallón S, Doyle JA, Endress PK, Bailes EJ, Barroso de Morais E, Bull-Hereñu K, Carrive L, Chartier M, Schönenberger J (2017) The ancestral flower of angiosperms and its early diversification. *Nature Communications* **8**, 16047. doi:10.1038/ncomms16047

Sauquet H, von Balthazar M, Doyle JA, Endress PK, Magallón S, Staedler Y, Schönenberger J (2018) Challenges and questions in reconstructing the ancestral flower of angiosperms: a reply to Sokoloff *et al. American Journal of Botany* **105**, 127–135. doi:10.1002/ajb2.1023

Sazima M, Sazima I (1978) Bat pollination of the passion flower *Passiflora mucronata* in south-eastern Brazil. *Biotropica* **10**, 100–109. doi:10.2307/2388012

Sazima I, Sazima C, Sazima M (2005) Little dragons prefer flowers to maidens: a lizard that laps nectar and pollinates trees. *Biota Neotropica* **5**, 185–192. doi:10.1590/S1676-06032005000100018

Schaefer HM, Rolshausen G (2005) Plants on red alert: do insects pay attention? *BioEssays* **27**, 1–7.

Schatz GE (1990) Some aspects of pollination biology in Central American forests. In *Reproductive Ecology of Tropical Forest Plants.* Vol. 7. (Eds KS Bawa and M Hadley) pp. 69–84. Unesco, Parthenon, Carnforth.

Schemske DW (1978) Evolution of reproductive characteristics in *Impatiens* (Balsaminaceae): the significance of cleistogamy and chasmogamy. *Ecology* **59**, 596–613. doi:10.2307/ 1936588

Schemske DW (1980) Floral ecology and hummingbird pollinations of *Combretum farinosum* in Costa Rica. *Biotropica* **12**, 169–181. doi:10.2307/2387968

Schimper AFW (1903) *Plant Geography upon a Physiological Basis.* English trans. Clarendon Press, Oxford.

Schlessman MA, Vary LB, Munzinger J, Lowry PP (2014) Incidence, correlates, and origins of dioecy in the island flora of New Caledonia. *International Journal of Plant Sciences* **175**, 271–286. doi:10.1086/674452

Schleuning M, Frund J, Schweiger O, Welk E, Albrecht J, Albrecht M, Beil M, Benadi G, *et al.* (2016) Ecological networks are more sensitive to plant than to animal extinction under climate change. *Nature Communications* **7**, 13965. doi:10.1038/ncomms13965

Schneemilch M, Williams C, Kokkinn M (2011) Floral visitation in the Australian native shrub genus *Acrotriche R. Br.* (Ericaceae): an abundance of ants (Formicidae). *Australian Journal of Entomology* **50**, 130–138. doi:10.1111/j.1440-6055. 2010.00805.x

Schneider EL, Williamson PS (1993) Nymphaeaceae. In *The Families and Genera of Vascular Plants.* Vol. 2. (Eds K Kubitzki, JG Rohwer and V Bittrich) pp. 486–493. Springer-Verlag, Berlin.

Schneider G, Moritz C (1999) Rainforest refugia and Australia's Wet Tropics. *Proceedings of the Royal Society B* **266**, 191–196. doi:10.1098/rspb.1999.0621

Schodde R, Mason IJ, Gill HB (1982) The avifauna of the Australian mangroves. A brief review of composition, structure and origin. In *Mangrove Ecosystems in Australia. Structure, Function and Management.* (Ed. BF Clough) pp. 141–150. AIMS/ANU Press, Canberra.

Schoene B, Eddy MP, Samperton KM, Keller B, Keller G, Adatte T, Khadri SFR (2019) U-Pb constraints on pulsed eruption of the Deccan Traps across the end-Cretaceous mass extinction. *Science* **363**, 862–866. doi:10.1126/science.aau2422

Schulte P, Alegret L, Arenillas I, Arz JA, Barton PJ, Brown PR, Bralower TJ, Christeson GL, Claeys P, Cockell CS, Collins GS, Deutsch A, Goldin TJ, Goto K, Grajales-Nishimura JM, *et al.* (2010) The Chicxulub asteroid impact and mass extinction at the Cretaceous-Paleogene Boundary. *Science* **327**, 1214–1218. doi:10.1126/science.1177265

Schulze CH, Linsenmair KE, Fiedler K (2001) Understorey versus canopy: patterns of vertical stratification and diversity among Lepidoptera in a Bornean rain forest. *Plant Ecology* **52**, 43–64.

Scriber JM, Larsen ML, Zalucki MP (2007) *Papilio aegeus* Donovan (Lepidoptera: Papilionidae) host plant range evaluated experimentally on ancient angiosperms. *Australian Journal of Entomology* **46**, 65–74. doi:10.1111/j.1440-6055. 2007.00580.x

Sedivy C, Müller A, Dorn S (2011) Closely related pollen generalist bees differ in their ability to develop on the same pollen diet: evidence for physiological adaptations to digest pollen. *Functional Ecology* **25**, 718–725. doi:10.1111/j.1365-2435.2010.01828.x

Senarath WTPSK (2008) Dioecy and monoecy in the flora of Sri Lanka and their evolutionary correlations to endemism, growth form, fruit type, seed number and flower size. *Bangladesh Journal of Plant Taxonomy* **15**, 13–19. doi:10.3329/bjpt.v15i1.907

Seymour RS, Matthews PGD (2006) The role of thermogenesis in the pollination biology of the Amazon waterlily *Victoria amazonica*. *Annals of Botany* **98**, 1129–1135. doi:10.1093/aob/mcl201

Seymour RS, Gibernau M, Ito K (2003) Thermogenesis and respiration of inflorescences of the dead horse arum *Hekicodiceros muscivorus*, a pseudothermoregulatory aroid associated with fly pollination. *Functional Ecology* **17**, 886–894. doi:10.1111/j.1365-2435.2003.00802.x

Seymour RS, Terry I, Roemer RB (2004) Respiration and thermogenesis by cones of the Australian cycad *Macrozamia machinii*. *Functional Ecology* **18**, 925–930. doi:10.1111/j.0269-8463.2004.00918.x

Shanahan M, Harrison RD, Yamuna R, Boen W, Thornton IWB (2001) Colonization of an island volcano, Long Island, Papua New Guinea, and an emergent island, Motmot, in its caldera lake. V. Colonization by figs (*Ficus* spp.), and their dispersers and pollinators. *Journal of Biogeography* **28**, 1365–1377. doi:10.1046/j.1365-2699.2001.00638.x

Shapcott A, Lamont R, O'Connor K, James H, Benwell A (2016) How is genetic variability in a threatened rainforest vine *Marsdenia longifolia* distributed at different geographical scales? *Botanical Journal of the Linnean Society* doi:10.1111/boj.12494.

Sharma MV, Armstrong JE (2013) Pollination of *Myristica* and other nutmegs in natural populations. *Tropical Conservation Science* **6**, 595–607. doi:10.1177/194008291300600502

Sharma N, Koul P, Koul AK (1993) Pollination biology of some species of genus *Plantago* L. *Botanical Journal of the Linnean Society* **111**, 129–138. doi:10.1111/j.1095-8339.1993.tb01895.x

Shaw DE, Cantrell BL (1982) A study of the pollination of *Alocasia macrorrhiza* (Araceae) in southeast Queensland. *Proceedings of the Linnean Society of New South Wales* **106**, 323–336.

Shoeva OY, Glagoleva AY, Khlestkina EK (2017) The factors affecting the evolution of the anthocyanin biosynthesis pathway genes in monocot and dicot plant species. *BMC Plant Biology* **17**(suppl. 2), 256. doi:10.1186/s12870-017-1190-4

Shrestha M, Dyer AG, Boyd-Gerny S, Wong BBM, Burd M (2013) Shades of red: bird-pollinated flowers target the specific colour discrimination abilities of avian vision. *New Phytologist* **198**, 301–310. doi:10.1111/nph.12135

Shrestha M, Garcia JE, Bukovac Z, Dorin A, Dyer AG (2018a) Pollination in a new climate: assessing the potential influence of flower temperature variation on insect pollinator behaviour. *PLoS One* **13**, e0200549. doi:10.1371/journal.pone.0200549

Shrestha M, Garcia JE, Bukovac Z, Dorin A, Dyer AG (2018b) Correction: Pollination in a new climate: assessing the potential influence of flower temperature variation on insect pollinator behaviour. *PLoS One* **13**, e0203158. doi:10.1371/journal.pone.0203153

Shuttleworth A, Johnson SD (2009) The importance of scent and nectar filters in a specialized wasp-pollination system. *Functional Ecology* **23**, 931–940. doi:10.1111/j.1365-2435.2009.01573.x

Silva RF, Tinoco NAB, Tsukui A, Koshnitzke C, Silva-Batista IC, Rezende CM, Bizzo HR (2019) Floral scent and nectar sugar composition of *Temnadenia odorifera* (Apocynoideae, Apocyneae). *Journal of the Brazilian Chemical Society* **30**, 2.

Simberloff DS (1978) Colonization of islands by insects: immigration, extinction and diversity. In *Diversity of Insect Faunas. Symposium of Royal Entomological Society, London.* No. 9. (Eds LA Mound and N Waloff) pp. 139–153. Blackwell Scientific, London.

Simon R, Holderied MW, Koch CU, von Helversen O (2011) Floral acoustics: conspicuous echoes of a dish-shaped leaf attract bat pollinators. *Science* **333**, 631–633. doi:10.1126/science.1204210

Simpson SR, Gross CL, Silberbauer LX (2005) Broom and honeybees in Australia: an alien liaison. *Plant Biology* **7**, 541–548. doi:10.1055/s-2005-865855

Singer RB (2002) The pollination mechanism in *Trigonidium obtusum* Lindl (Orchidaceae: Maxillariinae): sexual mimicry and trap-flowers. *Annals of Botany* **89**, 157–163. doi:10.1093/aob/mcf021

Sjöstrom A, Gross CL (2006) Life-history characters and phylogeny are correlated with extinction risk in Australian angiosperms. *Journal of Biogeography* **33**, 271–290. doi:10.1111/j.1365-2699.2005.01393.x

Skubatz H, Tang W, Meeuse BTD (1993) Oscillatory heat-production in the male cones of cycads. *Journal of Experimental Botany* **44**, 489–492. doi:10.1093/jxb/44.2.489

Slater AT (1991) Interaction of the stigma with the pollinium in *Dendrobium speciosum*. *Australian Journal of Botany* **39**, 273–282. doi:10.1071/BT9910273

Slater AT, Calder DM (1988) The pollination biology of *Dendrobium speciosum* Smith: a case of false advertising. *Australian Journal of Botany* **36**, 145–158. doi:10.1071/BT9880145

Slik JW (2004) El Niño droughts and their effects on tree species composition and diversity in tropical rain forest. *Oecologia* **141**, 114–120. doi:10.1007/s00442-004-1635-y

Ślipiński SA, Escalona HE (2013) *Australian Longhorn Beetles (Coleoptera: Cerambycidae). Volume 1. Subfamily Lamiinae.* CSIRO Publishing, Melbourne.

Ślipiński SA, Escalona HE (2016) *Australian Longhorn Beetles (Coleoptera: Cerambycidae). Volume 2. Subfamily Cerambycinae.* CSIRO Publishing, Melbourne.

Ślipiński SA, Lawrence JF (Eds) (2019) *Australian Beetles: Archostemata, Myxophaga, Adephaga, Polyphaga (part).* Vol. 2. CSIRO Publishing, Melbourne.

Smith JP, Heard TA, Beekman M, Gloag R (2017) Flight range of the Australian stingless bee *Tetragonula carbonaria* (Hymenoptera: Apidae). *Austral Ecology* **56**, 50–53.

Smith TJ, Mayfield MM (2015) Diptera species and functional diversity across tropical Australian countryside landscapes. *Biological Conservation* **191**, 436–443. doi:10.1016/j.biocon.2015.07.035

Smith TJ, Mayfield MM (2018) The effect of habitat fragmentation on the bee visitor assemblages of three Australian tropical rainforest tree species. *Ecology and Evolution* **8**, 8204–8216. doi:10.1002/ece3.4339

Smith-Ramírez C, Armesto JJ (2003) Foraging behaviour of bird pollinators on *Embothrium coccineum* (Proteaceae) trees in forest fragments and pastures in southern Chile. *Austral Ecology* **28**, 53–60. doi:10.1046/j.1442-9993.2003.01248.x

Smith-Ramírez C, Ramos-Jiliberto R, Valdovinos FS, Martínez P, Castillo JA, Armesto JJ (2014) Decadal trends in the pollinator assemblage of *Eucryphia cordifolia* in Chilean rainforests. *Oecologia* **176**, 157–169. doi:10.1007/s00442-014-3000-0

Smyth DR (2018) Evolution and genetic control of the floral ground plan. *New Phytologist* **220**, 70–86. doi:10.1111/nph.15282

Sobek S, Tscharntke T, Scherber C, Schiele S, Stephen-Dewenter I (2009) Canopy vs. understory: Does tree diversity affect bee and wasp communities and their natural enemies across forest strata? *Forest Ecology and Management* **258**, 609–615. doi:10.1016/j.foreco.2009.04.026

Sohn JC, Labandeira CC, Davis DR (2015) The fossil record and taphonomy of butterflies and moths (Insecta, Lepidoptera): implications for evolutionary diversity and divergence-time estimates. *BMC Evolutionary Biology* **15**, 12. doi:10.1186/s12862-015-0290-8

Sokoloff DD, Remizowa MV, Bateman PM, Rudall PJ (2018) Was the ancestral angiosperm whorled throughout? *American Journal of Botany* **105**, 5–15. doi:10.1002/ajb2.1003

Solé RV, Alenso D, Saldaña J (2004) Habitat fragmentation and biodiversity collapse in neutral communities. *Ecological Complexity* **1**, 65–75. doi:10.1016/j.ecocom.2003.12.003

Soltis DE, Smith SA, Cellinese N, Wurdack KJ, Tank DC, Brockington SF, Refulio-Rodriguez NF, *et al.* (2011) Angiosperm phylogeny: 17 genes, 640 taxa. *American Journal of Botany* **98**, 704–730. doi:10.3732/ajb.1000404

Somanathan H, Borges RM (2001) Nocturnal pollination by the carpenter bee *Xylocopa tenuiscapa* (Apidae) and the effect of floral display on fruit set of *Heterophragma quadriloculare* (Bignoniaceae) in India. *Biotropica* **33**, 78–89. doi:10.1111/j.1744-7429.2001.tb00159.x

Song B, Chen G, Stöcklin J, Peng D-L, Niu Y, Li Z-M, Sun H (2014) A new pollinating seed-consuming mutualism between *Rheum nobile* and a fly fungus gnat, *Bradysis* sp., by a specific flora compound. *New Phytologist* **203**, 1109–1118. doi:10.1111/nph.12856

Sonne J, Zanata TB, Martin González AM, Cumbicus Torres NL, Fjeldså J, Colwell RK, Tinoco BA, Rahbek C, Dalsgaard B (2019) The distributions of morphologically specialized hummingbirds coincide with floral trait matching across an Andean elevated gradient. *Biotropica* **51**, 205–218. doi:10.1111/btp.12637

Sousa VA, Hattemer HH (2003) Pollen dispersal and gene flow by pollen in *Araucaria angustifolia*. *Australian Journal of Botany* **51**, 309–317. doi:10.1071/BT02037

Spencer H, Menet A, Anderson B, de Ville d'Avray LT (2010) Assessment of coconut impacts on Wet Tropics littoral rainforests (coconuts as a "transformer" a species). Australian Tropical Research Foundation – Cape Tribulation Tropical Research Station, Cape Tribulation, Queensland.

Spigler RB, Kalisz S (2017) Persistent pollinators and the evolution of complete selfing. *American Journal of Botany* **104**, 1783–1786. doi:10.3732/ajb.1700332

Staedler YM, Weston PH, Endress PK (2007) Floral phyllotaxis and floral architecture in Calycanthaceae. *International Journal of Plant Sciences* **168**, 285–306. doi:10.1086/510417

Stangler ES, Hanson PE, Stefan-Dewenter I (2015) Interactive effects of habitat fragmentation and microclimate on trap-nesting Hymenoptera and their trophic interactions in small secondary rainforest remnants. *Biodiversity and Conservation* **24**, 563–577.

Stangler ES, Hanson PE, Stefan-Dewenter I (2016) Vertical diversity patterns and biotic interactions of trap-nesting bees along a fragmentation gradient of small secondary rainforest remnants. *Apidologie* **47**, 527–538. doi:10.1007/s13592-015-0397-3

Stephenson AG (1982) When does outcrossing occur in a mass-flowering plant? *Evolution* **36**, 762–767. doi:10.1111/j.1558-5646.1982.tb05442.x

Stilwell JD, Long JA (2011) *Frozen in Time: Prehistoric Life in Antarctica*. CSIRO Publishing, Melbourne.

Stilwell JD, Langendam A, Mays C, Sutherland LJM, Arillo A, Bickel DJ, De Silva WT, Pentland AH, Roghi G, Price GD, Cantrill DJ, Quinney A, Peñalver E (2020) Amber from the Triassic to Paleogene of Australia and New Zealand as exceptional preservation of poorly known terrestrial ecosystems. *Scientific Reports* **10**, 5703. doi:10.1038/s41598-020-62252-z

Stocker GC (1988) Tree species diversity in rainforests – establishment and maintenance. *Proceedings of the Ecological Society of Australia* **15**, 39–47.

Stork NE, Goosem S, Turton SM (2008) Australian rainforests in a global context. *Living in a Dynamic Tropical Landscape*. (Eds N Stork and SM Turton) pp. 4–20. Blackwell Publishing, Malden.

Stork NE, Stone M, Sam L (2016) Vertical stratification of beetles in tropical rainforests as sampled by light traps in North Queensland, Australia. *Austral Ecology* **41**, 168–178. doi:10.1111/aec.12286

Stpiczyńska M, Nepi M, Zych M (2012) Secretion and composition of nectar and the structure of perigonal nectaries in *Fritillaria meleagris* L. (Liliaceae). *Plant Systematics and Evolution* **298**, 997–1013. doi:10.1007/s00606-012-0609-5

Strahan R (Ed.) (1983) *The Australian Museum Complete Book of Australian Mammals*. Angus and Robertson, Sydney.

Stratton DA (1989) Longevity of individual flowers in a Costa Rican cloud forest: ecological correlates and phylogenetic constraints. *Biotropica* **21**, 308–318. doi:10.2307/2388281

Stride G, Thomas CD, Benedick S, Hodgson JA, Jelling A, Senior MJM, Hill JK (2018) Contrasting patterns of local richness of seedlings, saplings, and trees may have implications for regeneration in rainforest remnants. *Biotropica* **50**, 889–897. doi:10.1111/btp.12605

Stuessy TF (2004) A transitional-combinational theory for the origin of angiosperms. *Taxon* **53**, 3–16.

Sugden EA, Pyke GH (1991) Effects of honey bees on colonies of *Exoneura asimillima*, an Australian native bee. *Australian Journal of Ecology* **16**, 171–181. doi:10.1111/j.1442-9993.1991.tb01044.x

Suinyuy TN, Donaldson JS, Johnson SD (2010) Scent chemistry and patterns of thermogenesis in male and female cones of the African cycad *Encephalartos natalensis* (Zamiaceae). *South African Journal of Botany* **76**, 717–725. doi:10.1016/j.sajb.2010.08.014

Sullivan C, Xu X, O'Connor JK (2017) Complexities and novelties in the early evolution of avian flight, as seen in the Mesozoic Yanliao and Jehol Biotas of northeast China. *Palaeoworld* **26**, 212–229. doi:10.1016/j.palwor.2016.12.001

Summerbell G (1991) Regeneration of complex notophyll vine forest (humid subtropical rainforest) in eastern Australia: a review. *Cunninghamia* **2**, 391–410.

Sun M, Gross K, Schiestl FP (2014) Floral adaptation to local pollinator guilds in a terrestrial orchid. *Annals of Botany* **113**, 289–300. doi:10.1093/aob/mct219.

Suppiah R, Watterson IG, Macadam I, Collier MA, Bathols J (2010) Climate change projections for a tropical rainforest region of north Queensland. Final report on MTSRF activities, 2006–2010. Report for the Marine and Tropical Sciences Regional Faciluity (MTSRF). Research Report Series, Reef and Rainforest Centre Ltd., Cairns.

Susanti R, Sripanomyom S (2006) Role of calyx coloration in the attraction of pollinators in *Elaeocarpus* (Elaeocarpaceae). In *Proceedings of the Center for Tropical Forest Science – Arnold Arboretum, International Field Biology Course 2006*. (Eds MS Khoo, S Hong-Wa and RD Harrison) pp. 89–92. University of Peradenyia. Forest Department of Sri Lanka.

Sussman RW, Raven PH (1978) Pollination by lemurs and marsupials: an archaic co-evolutionary system. *Science* **200**, 731–736. doi:10.1126/science.200.4343.731

Svensson GP, Pellmyr O, Raguso RA (2011) Pollinator attraction to volatiles from virgin and pollinated host flowers in a yucca/moth obligate mutualism. *Oikos* **120**, 1577–1583. doi:10.1111/j.1600-0706.2011.19258.x

Takhtajan A (1997) *Diversity and Classification of Plants*. Columbia University Press, New York.

Tan MK, Artchawakom T, Wahib R, Lee C-Y, Belabut DM, Tan HTW (2017) Overlooked flower-visiting Orthoptera in Southeast Asia. *Journal of Orthoptera Research* **26**, 143–153. doi:10.3897/jor.26.15021

Tanaka H (2000) Size and dispersal of pollen grains in anemophilous angiosperms. *Shokubutsu Kenkyu Zasshi* **75**, 116–122.

Tang W (1987) Heat production in cycad cones. *Botanical Gazette* **148**, 165–174. doi:10.1086/337644

Tarran M, Wilson PG, Hill RS (2016) Oldest record of *Metrosideros* (Myrtaceae): fossil flowers, fruits, and leaves from Australia. *American Journal of Botany* **103**, 754–768. doi:10.3732/ajb.1500469

Taylor F, Hill RS (1996) A phylogenetic analysis of the Eucryphiaceae. *Australian Systematic Botany* **9**, 735–748. doi:10.1071/SB9960735

Taylor KJ, Lowe AJ, Hunter RJ, Ridgway T, Gresshoff PM, Rossetto M (2005) Genetic diversity and region identity in the Australian remnant *Nothofagus moorei*. *Australian Journal of Botany* **53**, 437–444. doi:10.1071/BT04215

Taylor TN, Taylor EL, Krings M (2009) *Paleobotany: The Biology and Evolution of Fossil Plants*. 2nd edn. Academic Press, Oxford.

Telford IRH (1993) Surianaceae. In *Flora of Australia*. Vol. 50. (Eds AS George, AE Orchards and HJ Hewson) pp. 187–188. Australian Government Publishing Service, Canberra.

Terborgh J (1990) Seed and fruit dispersal:– commentary. In *Reproductive Ecology of Tropical Forest Plants*. Vol. 7. (Eds KS Bawa and M Hadley) pp. 181–190. Unesco, Parthenon, Carnforth.

Terry I (2001) Thrips and weevils as dual specialist pollinators of the Australian cycad *Macrozamia communis* (Zamiaceae). *International Journal of Plant Sciences* **162**, 1293–1305. doi:10.1086/321929

Terry I, Walter GH, Donaldson JS, Snow E, Forster PI, Machin PJ (2005) Pollination of Australian *Macrozamia* cycads (Zamiaceae): effectiveness and behaviour of specialist vectors in a dependent mutualism. *American Journal of Botany* **92**, 931–940. doi:10.3732/ajb.92.6.931

Terry I, Walter GH, Moore C, Roemer R, Hull C (2007) Odor-mediated push–pull pollination in cycads. *Science* **318**, 70. doi:10.1126/science.1145147

Terry I, Forster PI, Moore CJ, Roemer RB, Machin PJ (2008) Demographics, pollination syndrome and conservation status of *Macrozamia platyrhachis* (Zamiaceae), a geographically restricted Queensland cycad. *Australian Journal of Botany* **56**, 321–322. doi:10.1071/BT06202

Teulon DAJ, Penman DR (1992) Colour preferences of New Zealand thrips (Terebrantia: Thysanoptera). *New Zealand Entomologist* **15**, 8–13. doi:10.1080/00779962.1992.9722621

Thalwitzer L, Kelly D, Smissen RD, Butler R, Suckling DM, El-Sayed A (2018) Species-specific male pollinators found for three native New Zealand greenhood orchids (*Pterostylis* spp.) suggest pollination by sexual deception. *Australian Journal of Botany* **66**, 243–254. doi:10.1071/BT17111

Thien LB, Azuma H, Kawano S (2000) New perspectives on the pollination biology of basal angiosperms. *International Journal of Plant Sciences* **161**, S225–S235. doi:10.1086/317575

Thomann M, Imbert E, Devaux C, Cheptou P-O (2013) Flowering plants under global pollinator decline. *Trends in Plant Science* **18**, 353–359. doi:10.1016/j.tplants.2013.04.002

Thomas CD (1994) Extinction, colonization, and metapopulations: environmental tracking of rare species. *Conservation Biology* **8**, 373–378. doi:10.1046/j.1523-1739.1994.08020373.x

Thompson JN (2014) *Interaction and Coevolution*. The University of Chicago Press, Chicago.

Thompson JN, Pellmyr O (1992) Mutualisms with pollinating seed parasites and co-pollinators: constraints on specialization. *Ecology* **73**, 1780–1791. doi:10.2307/1940029

Thurlby KAG, Wilson PG, Sherwin WB, Connelly C, Rossetto M (2012) Reproductive bet-hedging in a rare widespread tree, *Syzygium paniculatum* (Myrtaceae). *Austral Ecology* **37**, 936–944. doi:10.1111/j.1442-9993.2011.02353.x

Tng DY, Murphy BP, Weber E, Sanders G, Williamson GJ, Kemp J, Bowman DMJS (2012) Humid tropical rain forest has expanded into eucalypt forest and savanna over the last 50 years. *Ecology and Evolution* **2**, 34–45. doi:10.1002/ece3.70

Toledo-Hernández M, Wanger TC, Tscharntke T (2017) Neglected pollinators: can enhanced pollination services improve cocoa yields? A review. *Agriculture, Ecosystems & Environment* **247**, 137–148. doi:10.1016/j.agee.2017.05.021

Tosaki Y, Renner SS, Takahashi H (2001) Pollination of *Sarcandra glabra* (Chloranthaceae) in natural populations in Japan. *Journal of Plant Research* **114**, 423–427. doi:10.1007/PL00014007

Traveset A, Sáez E (1997) Pollination of *Euphorbia dendroides* by lizards and insects: spatio-temporal variation in patterns of flower visitation. *Oecologia* **111**, 241–248.

Trueman SJ, Pegg GS, King J (2007) Domestication for conservation of an endangered species: the case of the Wollemi Pine. *Tree and Forestry Science and Biotechnology* **1**, 1–10.

Truswell EM (1990) Australian rainforests: the 100 million year record. In *Australian Tropical Rainforests.* (Eds LJ Webb and J Kikkawa) pp. 7–22. CSIRO Publishing, Melbourne.

Tsugi K, Hasyim S, Nakamura HK (2004) Asian weaver ants, *Oecophylla smaragdina*, and their repelling of pollinators. *Biological Research* **19**, 669–673.

Turner V (1982) Marsupials as pollinators in Australia. In *Pollination and Evolution.* (Eds JA Armstrong, JM Powell and AJ Richards) pp. 55–66. Royal Botanic Gardens, Sydney.

Turner V (1984) *Banksia* pollen as a source of protein in the diet of two Australian marsupials, *Cercartetus nanus and Tarsipes rostratus. Oikos* **43**, 53–61. doi:10.2307/3544245

Turton SM (1988) Solar radiation regimes in a north Queensland rainforest. *Proceedings of the Ecological Society of Australia* **15**, 101–105.

Turton SM (2008) Landscape-scale impacts of Cyclone Larry on the forests of northeast Australia, including comparisons with cyclones impacting the region between 1858 to 2006. *Austral Ecology* **33**, 409–416. doi:10.1111/j.1442-9993. 2008.01896.x

Uhl NW, Moore HE (1977) Correlations in inflorescence, flower structure and floral anatomy with pollination in some palms. *Biotropica* **9**, 170–190. doi:10.2307/2387879

Ushio M, Osada Y, Kumagai T, Kume T, Pungga R, Nakashizuka T, Itioka T, Sakai S (2019) Dynamic and synergistic influences of air temperature and rainfall on general flowering in a Bornean lowland tropical rainforest. *Ecological Research* **35**, 17–29. doi:10.1111/1440-1703.12057

Vaknin Y, Gan-Mor S, Bechar A, Ronen B, Eisikowitch D (2000) The role of electrostatic forces in pollination. *Plant Systematics and Evolution* **222**, 133–142. doi:10.1007/BF00984099

Valenciá-Aguilar A, Cortés-Gómez AM, Ruíz-Aguledo C (2013) Ecosystem services provided by amphibians and reptiles in Neotropical ecosystems. *The International Journal of Biodiversity Science, Ecosystem Services & Management* **9**, 257–272. doi:10.1080/21513732.2013.821168

Vamosi JC, Otto SP (2002) When looks can kill: the evolution of sexually dimorphic floral display and the extinction of dioecious plants. *Proceedings. Biological Sciences* **269**, 1187–1194. doi:10.1098/rspb.2002.2004

Vanbergen A (2013) Threats to a ecosystem service: pressures on pollinators. *Frontiers in Ecology and the Environment* **11**, 251–259. doi:10.1890/120126

Vandelook F, Janssens SB, Gijbels P, Fischer E, Van den Ende W, Honnay O, Abrahamczyk S (2019) Nectar traits differ between pollination syndromes in Balsaminaceae. *Annals of Botany* **124**, 269–279. doi:10.1093/aob/mcz072

van der Kooi CJ, Wilts BD, Leertouwer HL, Staal M, Elzenga JTM, Stavenga DG (2014) Iridescent flowers? Contribution of surface structure to optical signaling. *New Phytologist* **203**, 667–673. doi:10.1111/nph.12808

Van der Niet T, Peakall R, Johnson SD (2014) Pollinator-driven ecological speciation in plants: new evidence and future perspectives. *Annals of Botany* **113**, 199–212. doi:10.1093/aob/mct290.

VanDerWal J, Shoo LP, Williams SE (2009) New approaches to understanding late Quaternary climate fluctuations and refugial dynamics in Australian wet tropical rain forests. *Journal of Biogeography* **36**, 291–301. doi:10.1111/j.1365-2699.2008.01993.x

Van Nieuwstadt MGL, Sheil D (2005) Drought, fire and tree survival in a Borneo rain forest, East Kalimantan, Indonesia. *Journal of Ecology* **93**, 191–201. doi:10.1111/j.1365-2745.2004.00954.x

Van Nuland ME, Haag EN, Bryant JAM, Read QD, Klein RN, Douglas MJ, Gorman CE, Greenwell TD, Busby MW, Collins J, LeRoy JT, Schuchmann G, Schweitzer JA, Bailey JK (2013) Fire promotes pollinator visitation. Implications for ameliorating declines in pollination services. *PLoS One* **8**, e79853. doi:10.1371/journal.pone.0079853

Van Schaik CPV (1986) Phenological changes in a Sumatran rain forest. *Journal of Tropical Ecology* **2**, 327–347. doi:10.1017/S0266467400000973

Vary LB, Gillen DL, Randrianjanahary M, Lowry PP, Sakai AK, Weller SG (2011) Dioecy, monoecy and their ecological correlates in the littoral forest of Madagascar. *Biotropica* **43**, 582–590. doi:10.1111/j.1744-7429.2010.00742.x

Vasconcelos TNC, Chartier M, Prenner G, Martins AC, Schönenberger J, Wingler A, Lucas E (2019) Floral uniformity through evolutionary time in a species-rich tree community. *New Phytologist* **221**, 1597–1608. doi:10.1111/nph.15453

Vincent BJ, Barretts S, Cochrone A, Plummer JA, Renton M (2015) Conservation biology of two endemic *Beyeria* (Euphorbiaceae) from southern Western Australia. *Australian Journal of Botany* **63**, 484–496. doi:10.1071/BT14310

Vislobokov NA (2017) Flowering biology of *Aspidistra* (Asparagaceae). *Plant Species Biology* **32**, 412–422. doi:10.1111/1442-1984.12175

Visscher PK, Seeley TD (1982) Foraging strategy of honeybee colonies in a temperate deciduous forest. *Ecology* **63**, 1790–1801. doi:10.2307/1940121

Vithanage V, Ironside DA (1986) The insect pollinators of *Macadamia* and their relative importance. *Journal of the Australian Institute of Agricultural Science* **52**, 155–160.

Vlasáková B (2015) Density dependence in flower visitation rates of cockroach-pollinated *Clusia blattophila* on the Nouragues inselberg, French Guina. *Journal of Tropical Ecology* **31**, 95–98. doi:10.1017/S0266467414000571

Vlasáková B, Kalinová B, Gustafsson MG, Teichert H (2008) Cockroaches as pollinators of *Clusia* aff. *sellowiana* (Clusiaceae) on Inselbergs in French Guina. *Annals of Botany* **102**, 295–304. doi:10.1093/aob/mcn092

Volkovitsh MG, Bílý S, Hasenpusch J (2003) Larvae of Australian Buprestidae (Coleoptera) Part 2. Genus *Metaxymorpha*. *Folia Heyrovskyana* **11**, 203–216.

von Helversen D, von Helversen O (1999) Acoustic guide in a bat-pollinated flower. *Nature* **398**, 759–760. doi:10.1038/19648

Vorster P (1995) Aspects of the reproduction of cycads. I. Pollinating mechanisms and the association of *Amophocerus* (Curculionidae) with *Encephalartos*. In *Proceedings of the Third International Conference on Cycad Biology*. (Ed. P Vorster) pp. 379–389. Cycad Society of South Africa, Stellenbosch.

Wagner RE, Mugnaini S, Sniezko R, Hardie D, Poulis B, Nepi M, Pacini E, von Aderkas P (2007) Proteomic evaluation of gymnosperm pollination drop proteins indicate highly conserved and complex biological functions. *Sexual Plant Reproduction* **20**, 181–189. doi:10.1007/s00497-007-0054-8

Wahlberg N, Wheat CW, Peña C (2013) Timing and patterns in the taxonomic diversification of Lepidoptera (butterflies and moths). *PLoS One* **8**, e80875. doi:10.1371/journal.pone.0080875

Walker KL (1986) Revision of the Australian species of the genus *Homalictus* Cockerell (Hymenoptera: Halictidae). *Memoirs of the Museum of Victoria* **47**, 105–200. doi:10.24199/j.mmv.1986.47.05

Walker KL (1995) Revision of the Australian native bee subgenus *Lasioglossum* (*Chilalictus*) (Hymenoptera: Halictidae). *Memoirs of the Museum of Victoria* **55**, 1–214. doi:10.24199/j.mmv.1995.55.01

Wan JSH, McDougall KL, Liew ECY (2019) The susceptibility of rare and threatened NSW species to the root-rot pathogen *Phytophthora cinnamoni* 1. Initial testing and the identification of key research questions. *Australian Journal of Botany* **67**, 510–516. doi:10.1071/BT19090

Wang H, Moore MJ, Soltis PS, Bell CD, Brockington SF, Alexandre R, Davis CC, Latvis M, Manchester SR, Soltis DE (2009) Rosid radiation and the rapid rise of angiosperm-dominated forests. *Proceedings of the National Academy of Sciences of the United States of America* **106**, 3853–3858. doi:10.1073/pnas.0813376106

Wang R-W, Sun B-F, Zheng Q (2010) Diffusive coevolution and mutualism maintenance mechanisms in a fig-fig wasp system. *Ecology* **91**, 1308–1316. doi:10.1890/09-1446.1

Wang X (2018) *The Dawn Angiosperms: Uncovering the Origin of Flowering Plants*. Springer-Verlag, Berlin.

Wappler T, Engel MS (2003) The middle Eocene bee faunas of Eckfeld and Messel, Germany (Hymenoptera: Apoidea). *Journal of Paleontology* **77**, 908–921. doi:10.1017/S0022336000044760

Wardhaugh CW (2015) How many species of arthropods visit flowers? *Arthropod-Plant Interactions* **9**, 547–565. doi:10.1007/s11829-015-9398-4

Ware AB, Compton SG (1992) Breakdown of pollinator specificity in an African fig tree. *Biotropica* **24**, 544–549. doi:10.2307/2389018

Watson S, Taylor R, Nimmo D, Kelly L, Clarke M, Bennett A (2012) The influence of unburnt patches and distance from refuges on post-fire bird communities. *Animal Conservation* **15**, 499–507. doi:10.1111/j.1469-1795.2012.00542.x

Webb CJ (1984) Flower and fruit movements in *Muntingia calabura*: a possible mechanism for avoidance of pollinator–disperser interference. *Biotropica* **16**, 37–42. doi:10.2307/2387892

Webb CJ, Kelly D (1993) The reproductive biology of New Zealand flora. *Trends in Ecology & Evolution* **8**, 442–447. doi:10.1016/0169-5347(93)90007-C

Webb CJ, Lloyd DG, Delph LF (1999) Gender dimorphism in indigenous New Zealand seed plants. *New Zealand Journal of Botany* **37**, 119–130. doi:10.1080/0028825X.1999.9512618

Webb GA (1984) Insect pollinators of *Hakea microcarpa* (Proteaceae) at Bombala, New South Wales. *Victorian Naturalist* **102**, 204.

Webb GA (1989) Insects as potential pollinators of *Micromyrtus ciliata* (Sm.) Druce, Myrtaceae. *Victorian Naturalist* **106**, 148–151.

Webb LJ, Kikkawa J (Eds) (1990) *Australian Tropical Rainforests*. CSIRO Publishing, Melbourne.

Webb LJ, Tracey JG, Williams WT (1984) A floristic framework of Australian rainforests. *Australian Journal of Ecology* **9**, 169–198. doi:10.1111/j.1442-9993.1984.tb01356.x

Webber BL, Curtis ASO, Cassis G, Woodrow IE (2008) Flowering morphology, phenology and flower visitors of the Australian rainforest tree *Ryparosa kurrangii* (Achariaceae). *Australian Entomologist* **35**, 7–17.

Weiblen CD (2002) How to be a fig wasp. *Annual Review of Entomology* **47**, 299–330.

Weller Sg, Sakai AK, Rankin AE, Golouka A, Kutcher B, Ashby KE (1998) Dioecy and the evolution of pollination systems in *Schiedea* and *Alsinidendron* (Caryophyllaceae: Alsinoideae) in the Hawaiian Islands. *American Journal of Botany* **85**, 1377–1388. doi:10.2307/2446396

Wetschnig W, Depisch B (1999) Pollination of *Welwitschia mirabilis* Hook. f. *(Welwitschiaceae, Gnetopsida)*. *Phyton* **39**, 167–183.

White DJ, Vesk PA (2019) Fire and legacy effects of logging on understorey assemblages in wet-sclerophyll forests. *Australian Journal of Botany* **67**, 341–357. doi:10.1071/BT18171

Whitehead DR (1969) Wind pollination in the angiosperms: evolutionary and environmental considerations. *Evolution* **23**, 28–35. doi:10.1111/j.1558-5646.1969.tb03490.x

Whitehead MR, Peakall R (2009) Integrating floral scent, pollination ecology and population genetics. *Functional Ecology* **23**, 863–874. doi:10.1111/j.1365-2435.2009.01620.x

Whitehead MR, Sherwin WB, Crayn D, Rossetto M (2019) A predominantly southern distribution conceals a northern reservoir of diversity in a wet sclerophyll tree. *Australian Journal of Botany* **67**, 455–462. doi:10.1071/BT19065

Wiebes JT (1979) Co-evolution of figs and their insect pollinators. *Annual Review of Ecology and Systematics* **10**, 1–12. doi:10.1146/annurev.es.10.110179.000245

Wiens D, Rourke JP (1978) Rodent pollination in southern African *Protea* spp. *Nature* **276**, 71–73. doi:10.1038/276071a0

Wilcock C, Neiland R (2002) Pollination failure in plants: why it happens and when it matters. *Trends in Plant Science* **7**, 270–277. doi:10.1016/S1360-1385(02)02258-6

Wilkins CF, Whitlock BA (2011) A revision of *Commersonia* including *Rulingia* (Malvaceae *s. l.* or Byttneriacaceae). *Australian Systematic Botany* **24**, 226–283. doi:10.1071/SB10031

Willemstein SC (1987) *An Evolutionary Basis for Pollination Ecology*. Vol. 11. Leiden Botanical Series. Brill, Leiden, The Netherlands.

Williams GA (1993) *Hidden Rainforests: Subtropical Rainforests and Their Invertebrate Biodiversity*. UNSW Press/Australian Museum, Sydney.

Williams GA (1995) Pollination ecology of lowland subtropical rainforests in New South Wales. PhD thesis. University of New South Wales, Sydney.

Williams GA (1998) Bee pollination in the threatened Australian shrub *Senna acclinis* (Caesalpinioidae). *Cunninghamia* **5**, 767–772.

Williams GA (2000) A survey of littoral rainforest, and a rainforest restoration plan for remnants on the western portion of Camden Head. Report for Laurieton Landcare, April 2000.

Williams GA (2018) Insects associated with flowering *Rhodomyrtus psidioides* (Myrtaceae): Is this a Myrtle Rust (*Austropuccinia psidii*) – induced plant–pollinator interaction extinction event? *Cunninghamia* **18**, 23–27.

Williams GA (2020) *The Invertebrate World of Australia's Subtropical Rainforests*. CSIRO Publishing, Melbourne.

Williams GA, Adam P (1991) Rainforest remnants on headlands in the Manning Valley: their composition and conservation significance. *Wetlands Australia* **11**, 21–30. doi:10.31646/wa.211

Williams GA, Adam P (1993) Ballistic pollen release in Australian members of the Moraceae. *Biotropica* **25**, 478–480. doi:10.2307/2388872

Williams GA, Adam P (1994) A review of rainforest pollination and plant–pollinator interactions, with particular reference to Australian subtropical rainforests. *Australian Zoologist* **29**, 177–212. doi:10.7882/AZ.1994.006

Williams GA, Adam P (1995) Records of aculeate wasps from flowering subtropical rainforest trees. *Australian Entomologist* **22**, 51–58.

Williams GA, Adam P (1997) The composition of the bee (Apoidea: Hymenoptera) fauna visiting flowering trees in New South Wales lowland subtropical rainforest remnants. *Proceedings of the Linnean Society of New South Wales* **118**, 69–85.

Williams GA, Adam P (1998) Pollen loads collected from large insects in Australian subtropical rainforests. *Proceedings of the Linnean Society of New South Wales* **120**, 49–67.

Williams GA, Adam P (1999) Pollen sculpture in subtropical rain forest plants: is wind pollination more common than previously suspected? *Biotropica* **31**, 520–524. doi:10.1111/j.1744-7429.1999.tb00396.x

Williams GA, Adam P (2001) The insect assemblage visiting the flowers of the subtropical rainforest pioneer tree *Alphitonia excelsa* (Fenzl) Reiss. ex Benth. (Rhamnaceae). *Proceedings of the Linnean Society of New South Wales* **123**, 235–259.

Williams GA, Adam P (2010) *The Flowering of Australia's Rainforests*. CSIRO Publishing, Melbourne.

Williams GA, Adam P (2019) A preliminary checklist of flower-visiting insects from *Syzygium floribundum*, *Syzygium smithii* and *Tristaniopsis laurina*: three members of the Myrtle Rust-vulnerable plant family Myrtaceae. *Cunninghamia* **19**, 57–74.

Williams G, Walker K (2003) Pollination of the wet forest herb *Pollia crispata* (Commelinaceae). *Cunninghamia* **8**, 141–146.

Williams GA, Watkins S (1986) A revision of the genus *Xyroscelis* Saunders (Coleoptera: Buprestidae). *Journal of the Australian Entomological Society* **25**, 295–299. doi:10.1111/j.1440-6055.1986.tb01118.x

Williams GA, Adam P, Mound LA (2001) Thrips (Thysanoptera) pollination in Australian subtropical rainforests, with particular reference to pollination of *Wilkiea huegeliana* (Monimiaceae). *Journal of Natural History* **35**, 1–21. doi:10.1080/002229301447853

Williams JH (2009) *Amborella trichopoda* (Amborellaceae) and the evolutionary development origins of the angiosperm progamic phase. *American Journal of Botany* **96**, 144–165. doi:10.3732/ajb.0800070

Williams JH (2012) The evolution of pollen germination timing in flowering plants: *Austrobaileya scandens* (Austrobaileyaceae). *AoB Plants* **010**, doi:10.1093/aobpla/pls.

Williams PR (2000) Fire-stimulated rainforest seedling recruitment and vegetative regeneration in a densely grassed wet sclerophyll forest of north-eastern Australia. *Australian Journal of Botany* **48**, 651–658. doi:10.1071/BT99020

Williams PR, Parsons M, Jensen R, Tran C (2012) Mechanisms of rainforest persistence and recruitment in frequently burnt wet tropical eucalypt forests. *Austral Ecology* **37**, 268–275. doi:10.1111/j.1442-9993.2011.02271.x

Williams SE, Bolitho EE, Fox S (2003) Climate change in Australian tropical rainforests: an impending environmental catastrophe. *Proceedings. Biological Sciences* **270**, 1887–1892. doi:10.1098/rspb.2003.2464

Williams SE, Shoo LP, Henriod R, Pearson RG (2010) Elevational gradients in species abundance, assemblage structure and energy use of rainforest birds in the Australian Wet Tropics bioregion. *Austral Ecology* **35**, 650–664. doi:10.1111/j.1442-9993.2009.02073.x

Willmer PG (2011) *Pollination and Floral Ecology*. Princeton University Press, Princeton.

Willson MF, Sabag C, Figueroa JA, Armesto JJ, Caviedes M (1996) Seed dispersal by lizards in Chilean rainforest. *Revista Chilena de Historia Natural* **69**, 339–342.

Wilson EO, Nowak MA (2014) Natural selection drives the evolution of ant life cycles. *Proceedings of the National Academy of Sciences of the United States of America* **111**, 12585–12590. doi:10.1073/pnas.1405550111

Wilson GW (2002) Insect pollination in the cycad genus *Bowenia* Hook. ex Hook. F. (Stangeriaceae). *Biotropica* **34**, 438–441.

Wilson TC, Conn BJ, Henwood MJ (2017) Great Expectations: correlations between pollinator assemblages and flora characters in Lamiaceae. *International Journal of Plant Sciences* **178**, 170–187. doi:10.1086/690023

Winter JW (1988) Ecological specialization of mammals in Australian tropical and sub-tropical rainforest: refugial or ecological determinism? *Proceedings of the Ecological Society of Australia* **15**, 127–138.

Wolda H (1978) Seasonal fluctuations in rainfall, food and abundance of tropical insects. *Journal of Animal Ecology* **47**, 369–381. doi:10.2307/3789

Woodcock TS, Larson BM, Kevan PG, Inouye DW, Lunau K (2014) Flies and flowers II: floral attractants and rewards. *Journal of Pollination Ecology* **12**, 63–94. doi:10.26786/ 1920-7603(2014)5

Worboys SJ, Jackes BR (2005) Pollination processes in *Idiospermum australiense* (Calycanthaceae), and arborescent basal angiosperm of Australia's tropical rainforests. *Plant Systematics and Evolution* **251**, 107–117. doi:10.1007/s00606-004-0226-z

Wright GA, Schiestl FP (2009) The evolution of floral scent: the influence of olfactory learning by insect pollinators on the honest signalling of floral rewards. *Functional Ecology* **23**, 841–851. doi:10.1111/j.1365-2435.2009.01627.x

Wright SJ, Calderón O (2006) Seasonal, El Nino and longer term changes in flower and seed production in a moist tropical forest. *Ecology Letters* **9**, 35–44.

Wright SJ, Cornejo FH (1990) Seasonal drought and the timing of flowering and leaf fall in a neotropical forest. In *Reproductive Ecology of Tropical Forest Plants*. Vol. 7. (Eds KS Bawa and M Hadley) pp. 49–61. Unesco, Parthenon, Carnforth.

Xi Z, Liu L, Rest J, Davis CC (2014) Coalescent versus concatenation methods and the placement of *Amborella* as sister to water lilies. *Systematic Botany* **63**, 919–932.

Xiang X-G, Wang W, Li R-Q, Lin L, Liu Y, Zhou Z-K, Li Z-Y, Chen Z-D (2014) Large-scale phylogenetic analyses reveal fagalean diversification promoted by the interplay of diaspores and environments in the Palaegene. *Perspectives in Plant Ecology, Evolution and Systematics* **16**, 101–110. doi:10.1016/j.ppees.2014.03.001

Xic P-W, Luo Z-L, Zhang D-X (2012) Syrphid fly pollination of *Guihaiothamnus acaulis* (Rubiaceae), a species with "butterfly" flowers. *Journal of Systematics and Evolution* **51**, 86–93.

Xing S, Bonebrake TC, Tang CC, Picket EJ, Cheng W, Greenspan SE, Williams SE, Scheffers BR (2016) Cool habitats support darker and bigger butterflies in Australian tropical forests. *Ecology and Evolution* **6**, 8062–8074. doi:10.1002/ece3.2464

Xu C, Li Z (2002) Stochastic ecosystem resilience and productivity: seeking a relationship. *Ecological Modelling* **156**, 143–152. doi:10.1016/S0304-3800(02)00145-X

Xu X, Zhou Z, Dudley R, Mackem S, Chuong C-M, Erikson GM, Varricechio DJ (2014) An integrative approach to understanding bird origins. *Science* **346**, 1253293. doi:10.1126/science.1253293

Yamasaki E, Sakai S (2013) Wind and insect pollination (ambophily) of *Mallotus* spp. (Euphorbiaceae) in tropical and temperate forests. *Australian Journal of Botany* **61**, 60–66. doi:10.1071/BT12202

Yap SK, Chan HT (1990) Phenological behaviour of some *Shorea* species in Peninsular Malaysia. In *Reproductive Ecology of Tropical Forest Plants*. Vol. 7. (Eds KS Bawa and M Hadley) pp. 21–35. Unesco, Parthenon, Carnforth.

Young PAR, McDonald WJF (1987) The distribution, composition and status of the rainforests of southern Queensland. In *The Rainforest Legacy. Australian National Rainforests Study*. Vol. 1. (Eds G Werren and AP Kershaw) pp. 119–141. Australian Government Publishing Service, Canberra.

Yuan L-C, Luo Y-B, Thien LB, Fan J-H, Xu H-L, Chen ZD (2007) Pollination of *Schisandra henyri* (Schisandraceae) by female pollen-eating *Megommata* species (Cecidomyiidae, Diptera) in south-central China. *Annals of Botany* **99**, 451–460. doi:10.1093/aob/mcl287

Yumoto T, Momose K, Nagamasu H (2000) A new pollination syndrome: squirrel pollination in a tropical rainforest in Lambir Hills National Park, Sarawak, Malaysia. *Tropics* **9**, 147–151. doi:10.3759/tropics.9.147

Zanis MJ, Soltis DE, Soltis PS, Mathews S, Donoghue MJ (2002) The root of angiosperms revisited. *Proceedings of the National Academy of Sciences of the United States of America* **99**, 6848–6853.

Zapata TR, Arroyo MTK (1978) Plant reproductive ecology of a secondary deciduous tropical forest in Venezuela. *Biotropica* **10**, 221–230. doi:10.2307/2387907

Zavodna M, Compton SG, Biere A, Gilmartin PM, Van Damme JMM (2005) Putting your sons in the right place: the spatial distribution of fig wasp offspring inside figs. *Ecological Entomology* **30**, 210–219. doi:10.1111/j.0307-6946.2005.00674.x

Zeng L, Zhang Q, Sun R, Kong H, Zang N, Ma H (2014) Resolution of deep angiosperm phylogeny using conserved nuclear genes and estimates of early divergence times. *Nature Communications* **5**, 4956. doi:10.1038/ncomms5956

Zhang Q, Wang B (2017) Evolution of Lower Brachyceran flies (Diptera) and their adaptive radiation with angiosperms. *Frontiers of Plant Science* **8**, 631. doi:10.3389/fpls.2017.00631.

Zhang S, Soltis DE, Yang Y, Li D-Z, Ting-shuang Y (2011) Multigene analysis provides a well-supported phylogeny of Rosales. *Molecular Phylogenetics and Evolution* **60**, 21–28. doi:10.1016/j.ympev.2011.04.008

Zimmer HC, Auld TD, Hughes L, Otford CA, Baker PJ (2015) Fuel flammability and fire responses of juvenile canopy species in a temperate rainforest system. *International Journal of Wildland Fire* **24**, 349–360. doi:10.1071/WF14054

Zoeller KC, Steenhuisen S-L, Johnson SD, Midgley JJ (2016) New evidence for mammal pollination of *Protea* species (Proteaceae) based on remote-camera analysis. *Australian Journal of Botany* **64**, 1–7. doi:10.1071/BT15111

Zurbuchen A, Cheeseman S, Klaiber J, Müller A, Hein S, Dorn S (2010) Long foraging distances impose high costs on offspring production in solitary bees. *Journal of Animal Ecology* **79**, 674–681. doi:10.1111/j.1365-2656.2010.01675.x

Index

Page numbers in **bold** refer to colour plates.

www.ingramcontent.com/pod-product-compliance
Ingram Content Group UK Ltd.
Pitfield, Milton Keynes, MK11 3LW, UK
UKHW061542051026
15625UKWH00011B/572